IET ENERGY ENGINEERING 129

Renewable Energy from the Oceans

Other volumes in this series:

Volume 1	**Power Circuit Breaker Theory and Design** C.H. Flurscheim (Editor)
Volume 4	**Industrial Microwave Heating** A.C. Metaxas and R.J. Meredith
Volume 7	**Insulators for High Voltages** J.S.T. Looms
Volume 8	**Variable Frequency AC Motor Drive Systems** D. Finney
Volume 10	**SF_6 Switchgear** H.M. Ryan and G.R. Jones
Volume 11	**Conduction and Induction Heating** E.J. Davies
Volume 13	**Statistical Techniques for High Voltage Engineering** W. Hauschild and W. Mosch
Volume 14	**Uninterruptible Power Supplies** J. Platts and J.D. St Aubyn (Editors)
Volume 15	**Digital Protection for Power Systems** A.T. Johns and S.K. Salman
Volume 16	**Electricity Economics and Planning** T.W. Berrie
Volume 18	**Vacuum Switchgear** A. Greenwood
Volume 19	**Electrical Safety: A guide to causes and prevention of hazards** J. Maxwell Adams
Volume 21	**Electricity Distribution Network Design, 2nd Edition** E. Lakervi and E.J. Holmes
Volume 22	**Artificial Intelligence Techniques in Power Systems** K. Warwick, A.O. Ekwue and R. Aggarwal (Editors)
Volume 24	**Power System Commissioning and Maintenance Practice** K. Harker
Volume 25	**Engineers' Handbook of Industrial Microwave Heating** R.J. Meredith
Volume 26	**Small Electric Motors** H. Moczala *et al.*
Volume 27	**AC-DC Power System Analysis** J. Arrillaga and B.C. Smith
Volume 29	**High Voltage Direct Current Transmission, 2nd Edition** J. Arrillaga
Volume 30	**Flexible AC Transmission Systems (FACTS)** Y.-H. Song (Editor)
Volume 31	**Embedded Generation** N. Jenkins *et al.*
Volume 32	**High Voltage Engineering and Testing, 2nd Edition** H.M. Ryan (Editor)
Volume 33	**Overvoltage Protection of Low-Voltage Systems, Revised Edition** P. Hasse
Volume 36	**Voltage Quality in Electrical Power Systems** J. Schlabbach *et al.*
Volume 37	**Electrical Steels for Rotating Machines** P. Beckley
Volume 38	**The Electric Car: Development and future of battery, hybrid and fuel-cell cars** M. Westbrook
Volume 39	**Power Systems Electromagnetic Transients Simulation** J. Arrillaga and N. Watson
Volume 40	**Advances in High Voltage Engineering** M. Haddad and D. Warne
Volume 41	**Electrical Operation of Electrostatic Precipitators** K. Parker
Volume 43	**Thermal Power Plant Simulation and Control** D. Flynn
Volume 44	**Economic Evaluation of Projects in the Electricity Supply Industry** H. Khatib
Volume 45	**Propulsion Systems for Hybrid Vehicles** J. Miller
Volume 46	**Distribution Switchgear** S. Stewart
Volume 47	**Protection of Electricity Distribution Networks, 2nd Edition** J. Gers and E. Holmes
Volume 48	**Wood Pole Overhead Lines** B. Wareing
Volume 49	**Electric Fuses, 3rd Edition** A. Wright and G. Newbery
Volume 50	**Wind Power Integration: Connection and system operational aspects** B. Fox *et al.*
Volume 51	**Short Circuit Currents** J. Schlabbach
Volume 52	**Nuclear Power** J. Wood
Volume 53	**Condition Assessment of High Voltage Insulation in Power System Equipment** R.E. James and Q. Su
Volume 55	**Local Energy: Distributed generation of heat and power** J. Wood
Volume 56	**Condition Monitoring of Rotating Electrical Machines** P. Tavner, L. Ran, J. Penman and H. Sedding
Volume 57	**The Control Techniques Drives and Controls Handbook, 2nd Edition** B. Drury
Volume 58	**Lightning Protection** V. Cooray (Editor)
Volume 59	**Ultracapacitor Applications** J.M. Miller

Volume 62 **Lightning Electromagnetics** V. Cooray
Volume 63 **Energy Storage for Power Systems, 2nd Edition** A. Ter-Gazarian
Volume 65 **Protection of Electricity Distribution Networks, 3rd Edition** J. Gers
Volume 66 **High Voltage Engineering Testing, 3rd Edition** H. Ryan (Editor)
Volume 67 **Multicore Simulation of Power System Transients** F.M. Uriate
Volume 68 **Distribution System Analysis and Automation** J. Gers
Volume 69 **The Lightening Flash, 2nd Edition** V. Cooray (Editor)
Volume 70 **Economic Evaluation of Projects in the Electricity Supply Industry, 3rd Edition** H. Khatib
Volume 72 **Control Circuits in Power Electronics: Practical issues in design and implementation** M. Castilla (Editor)
Volume 73 **Wide Area Monitoring, Protection and Control Systems: The enabler for smarter grids** A. Vaccaro and A. Zobaa (Editors)
Volume 74 **Power Electronic Converters and Systems: Frontiers and applications** A.M. Trzynadlowski (Editor)
Volume 75 **Power Distribution Automation** B. Das (Editor)
Volume 76 **Power System Stability: Modelling, analysis and control** B. Om P. Malik
Volume 78 **Numerical Analysis of Power System Transients and Dynamics** A. Ametani (Editor)
Volume 79 **Vehicle-to-Grid: Linking electric vehicles to the smart grid** J. Lu and J. Hossain (Editors)
Volume 81 **Cyber-Physical-Social Systems and Constructs in Electric Power Engineering** S. Suryanarayanan, R. Roche and T.M. Hansen (Editors)
Volume 82 **Periodic Control of Power Electronic Converters** F. Blaabjerg, K.Zhou, D. Wang and Y. Yang
Volume 86 **Advances in Power System Modelling, Control and Stability Analysis** F. Milano (Editor)
Volume 87 **Cogeneration: Technologies, optimisation and implementation** C.A. Frangopoulos (Editor)
Volume 88 **Smarter Energy: From smart metering to the smart grid** H. Sun, N. Hatziargyriou, H.V. Poor, L. Carpanini and M.A. Sánchez Fornié (Editors)
Volume 89 **Hydrogen Production, Separation and Purification for Energy** A. Basile, F. Dalena, J. Tong and T.N. Veziroğlu (Editors)
Volume 90 **Clean Energy Microgrids** S. Obara and J. Morel (Editors)
Volume 91 **Fuzzy Logic Control in Energy Systems with Design Applications in MATLAB®/Simulink®** İ.H. Altaș
Volume 92 **Power Quality in Future Electrical Power Systems** A.F. Zobaa and S.H.E.A. Aleem (Editors)
Volume 93 **Cogeneration and District Energy Systems: Modelling, analysis and optimization** M.A. Rosen and S. Koohi-Fayegh
Volume 94 **Introduction to the Smart Grid: Concepts, technologies and evolution** S.K. Salman
Volume 95 **Communication, Control and Security Challenges for the Smart Grid** S.M. Muyeen and S. Rahman (Editors)
Volume 96 **Industrial Power Systems with Distributed and Embedded Generation** R Belu
Volume 97 **Synchronized Phasor Measurements for Smart Grids** M.J.B. Reddy and D.K. Mohanta (Editors)
Volume 98 **Large Scale Grid Integration of Renewable Energy Sources** A. Moreno-Munoz (Editor)
Volume 100 **Modeling and Dynamic Behaviour of Hydropower Plants** N. Kishor and J. Fraile-Ardanuy (Editors)
Volume 101 **Methane and Hydrogen for Energy Storage** R. Carriveau and D.S.-K. Ting
Volume 104 **Power Transformer Condition Monitoring and Diagnosis** A. Abu-Siada (Editor)
Volume 106 **Surface Passivation of Industrial Crystalline Silicon Solar Cells** J. John (Editor)
Volume 107 **Bifacial Photovoltaics: Technology, applications and economics** J. Libal and R. Kopecek (Editors)

Volume 108 **Fault Diagnosis of Induction Motors** J. Faiz, V. Ghorbanian and G. Joksimović
Volume 110 **High Voltage Power Network Construction** K. Harker
Volume 111 **Energy Storage at Different Voltage Levels: Technology, integration, and market aspects** A.F. Zobaa, P.F. Ribeiro, S.H.A. Aleem and S.N. Afifi (Editors)
Volume 112 **Wireless Power Transfer: Theory, technology and application** N. Shinohara
Volume 115 **DC Distribution Systems and Microgrids** T. Dragičević, F. Blaabjerg and P. Wheeler
Volume 117 **Structural Control and Fault Detection of Wind Turbine Systems** H.R. Karimi
Volume 119 **Thermal Power Plant Control and Instrumentation: The control of boilers and HRSGs, 2nd Edition** D. Lindsley, J. Grist and D. Parker
Volume 120 **Fault Diagnosis for Robust Inverter Power Drives** A. Ginart (Editor)
Volume 123 **Power Systems Electromagnetic Transients Simulation, 2nd Edition** N. Watson and J. Arrillaga
Volume 124 **Power Market Transformation** B. Murray
Volume 126 **Diagnosis and Fault Tolerance of Electrical Machines, Power Electronics and Drives** A.J.M. Cardoso
Volume 128 **Characterization of Wide Bandgap Power Semiconductor Devices** F. Wang, Z. Zhang and E.A. Jones
Volume 130 **Wind and Solar Based Energy Systems for Communities** R. Carriveau and D.S.-K. Ting (Editors)
Volume 131 **Metaheuristic Optimization in Power Engineering** J. Radosavljević
Volume 132 **Power Line Communication Systems for Smart Grids** I.R.S. Casella and A. Anpalagan
Volume 139 **Variability, Scalability and Stability of Microgrids** S.M. Muyeen, S.M. Islam and F. Blaabjerg (Editors)
Volume 155 **Energy Generation and Efficiency Technologies for Green Residential Buildings** D. Ting and R. Carriveau (Editors)
Volume 157 **Electrical Steels, 2 Volumes** A. Moses, K. Jenkins, Philip Anderson and H. Stanbury
Volume 905 **Power System Protection, 4 volumes**

Renewable Energy from the Oceans

From wave, tidal and gradient systems to offshore wind and solar

Edited by
Domenico Coiro and Tonio Sant

The Institution of Engineering and Technology

Published by The Institution of Engineering and Technology, London, United Kingdom

The Institution of Engineering and Technology is registered as a Charity in England & Wales (no. 211014) and Scotland (no. SC038698).

First published 2019

The Institution of Engineering and Technology
Michael Faraday House
Six Hills Way, Stevenage
Herts, SG1 2AY, United Kingdom

www.theiet.org

British Library Cataloguing in Publication Data
A catalogue record for this product is available from the British Library

ISBN 978-1-78561-766-9 (hardback)
ISBN 978-1-78561-767-6 (PDF)

Typeset in India by MPS Limited
Printed in the UK by CPI Group (UK) Ltd, Croydon

Contents

Preface

Energy is essential to humanity, and our society cannot operate without it. Energy has been the prime mover of the industrial revolution to develop the world economy as we know it today. Yet, our energy supply system, which is still heavily based on fossil fuels for both electricity generation and transport, has now become a threat to mankind and the global ecosystem we live in. Massive amounts of carbon dioxide emitted into the atmosphere every day are changing our climate. Scientific evidence has confirmed that the induced greenhouse effect is driving a rise in air and water temperatures at a rate that is much larger than that observed in the pre-industrial era, leading to the destruction of natural habitats, loss of sea ice and ice sheets, higher sea levels, drought and extreme weather events.

While decarbonising our present energy supply system has emerged to be one of the largest challenges ever faced by mankind, the availability of natural and clean energy resources that are large enough to meet the demand of the world populations is definitely not an issue. The amount of solar energy alone that reaches Earth every day, in the form of light as well as in the form heat that in turn creates wind, sea waves and marine thermoclines, is already orders of magnitude larger than mankind's energy needs.

With over 71 percent of the Earth's surface covered by water, our oceans offer by far the major contribution of clean energy through various renewable forms: offshore wind, wave, tidal and current, and thermal and salinity gradient energy. It is worth noting that the majority of the people in the world live in close proximity to the sea, in coastal cities and islands. Limited space on land in these often densely populated areas and accessibility to onshore space that is large enough to meet the requested energy demand through renewable energy exploitation are primary factors driving coastal populations to exploiting energy resources at sea.

The exploitation of renewable energy at sea brings about a number of engineering challenges making the design of marine systems more difficult as compared to those on land. The structurally integrity requirements offshore are significantly more demanding as a result of more extreme winds and the rough waves experienced over the typical design lifetime of 20 to 30 years. The corrosive marine environment also dictates more drastic measures to protect the marine systems against material degradation. Other practical considerations, such as the presence of marine growth, moving ice, accidental impacts by sea vessels and scour induced by sea currents, imply the need for more rigours engineering analysis during the design process. Installation, operation and maintenance procedures at sea are logistically more complex than on land not only because of site accessibility that is often impaired by stormy weather conditions but also by the need of good port facilities and adequate sea vessels.

This book presents an overview of the development of ocean energy technologies, renewable energy resources and latest emerging trends. It presents a broad perspective, covering important aspects from energy conversion, installation, operation and planning to grid connection and storage. Latest literature sources in various relevant fields are presented together with theoretical fundamentals for system engineering analysis.

The book is organised in 12 chapters: Chapter 1 reviews the progress achieved in ocean energies and the levels of technology maturity reached by the different technologies. The following three chapters, Chapter 2 to 4, are dedicated to ocean energy technologies designed to extract energy directly from the sea water. Chapters 2 and 3 deal with the extraction of potential and kinetic energy available in sea waves and current flows. Chapter 4 is dedicated to energy extraction from temperature and salinity differences present in the oceans. The next two chapters discuss two technologies that utilise marine space, yet they only exploit the renewable above the sea surface: offshore wind and marine solar energy. Chapter 7 covers offshore support structure design, including both bottom-mounted and floating platform concepts, that is necessary for safely supporting different renewable energy systems in the harsh marine environment.

Chapters 8 and 9 are dedicated to two equally important aspects for renewable energy systems at sea: electrical power transmission and offshore energy storage. Ocean energy conversion systems may be located well offshore, at large distances from the nearest grid connection point onshore. In such cases, power transmission to shore and connectivity to the electricity networks on land become critical aspects affecting both the overall efficiency of energy conversion and the cost of generation. Furthermore, although renewable energy technologies have made significant progress, many natural energy sources remain intermittent and will pose important challenges when integrated into electricity grids at high penetration levels. Countries nowadays relying on large amounts offshore wind capacity are already known to experience low and often unsustainable electricity prices due to congestion problems encountered in the electricity grids during periods of high wind availability and low energy demand. Energy storage is now regarded as the missing link for solving such problems and to enable a fully decarbonised energy supply system. Given its strategic importance nowadays, we believed that a whole chapter in this book should be solely dedicated to offshore energy storage concepts.

Chapter 10 discusses multi-purpose platforms integrating different forms of renewable energy generation and other activities, such as aquaculture, to improve the overall economic viability of offshore-based environmentally sustainable activities. Chapter 11 addresses installation, operational and maintenance aspects of offshore renewables. Finally, Chapter 12 discusses challenges in the offshore renewable energy sector and future expectations.

We would like to thank Ing. Gaetano Gaudiosi who served as the president of the OWEMES[1] society for many years and who had encouraged us to take-up the task of compiling this book. This work would not have been possible without the

[1]Offshore wind and other marine renewable energies in Mediterranean and European Seas.

contribution of many authors, experts in various fields of offshore renewables and who have brought together a wealth of knowledge and experience for the benefit of our readers. We worked closely with them to edit the book in a coherent manner. We are indebted for their hard work and dedication in preparing their chapters in a timely manner. Finally, we would like to express our sincere gratitude to the Institution of Engineering and Technology (IET) of the UK, especially the editorial and production teams, for their continuous guidance, support and meticulous work to produce the book on time.

Domenico P. Coiro and Tonio Sant

List of contributing authors

Chapter 1 – A review of progress in ocean energies

Neil Kermode, The European Marine Energy Center, UK

Chapter 2 – Wave energy

Gianmaria Sannino, ENEA, Italian National Agency for New Technologies, Energy and Sustainable Economic Development, Italy
Adriana Carillo, ENEA, Italian National Agency for New Technologies, Energy and Sustainable Economic Development, Italy
Arne Vogler, University of the Highlands and Islands – Lews Castle College, Stornoway, Isle of Lewis, UK
Giuliana Mattiazzo, Politecnico di Torino, Torino, Italy
Giovanni Bracco, Politecnico di Torino, Torino, Italy
Diego Vicinanza, Inter-University National Consortium for Marine Sciences (CoNISMa), Italy
Pasquale Contestabile, University of Campania Luigi Vanvitelli, Aversa (Caserta), Italy
Domenico P. Coiro, University of Naples Federico II, Naples, Italy
Giancarlo Troise, SeaPower Scarl – Research Consortium, Naples, Italy
Luca Castellini, UMBRAGROUP S.p.A., Foligno, Italy
John Ringwood, National University of Ireland, Maynooth, UK

Chapter 3 – Tidal and current energy

Brian Kirke, University of Australia, Barbara Hardy Institute, Australia
Domenico P. Coiro, University of Naples Federico II, Italy
Giancarlo Troise, Seapower Scrl – Research Consortium, Italy

Chapter 4 – Thermal and salinity gradient systems

Gianfranco Rizzo, University of Salerno, Fisciano, Italy
Francesco A. Tiano, University of Salerno, Fisciano, Italy

Chapter 5 – Offshore wind energy

Graeme McCann, Goldwind Denmark A/S, Bristol, UK

Chapter 6 – Marine solar energy

Luciano Mule' Stagno, University of Malta, Malta

Chapter 7 – Offshore support structure design

Erin Bachynski, Norwegian University of Science and Technology, Trondheim, Norway
Maurizio Collu, University of Strathclyde, Glasgow, UK

Chapter 8 – Electrical power transmission and grid integration

Elisabetta Tedeschi, Norwegian University of Science and Technology, NTNU, Norway
Abel A. Taffese, Norwegian University of Science and Technology, NTNU, Norway

Chapter 9 – Offshore energy storage

Seamus D. Garvey, University of Nottingham, UK
Rupp Carriveau, University of Windsor, Canada

Chapter 10 – Multipurpose platforms

Maurizio Collu, University of Strathclyde, Glasgow, UK
Erin Bachynski, Norwegian University of Science and Technology, Trondheim, Norway

Chapter 11 – Installation, operation and maintenance of offshore renewables

Vincenzo Nava, Tecnalia Research and Innovation, Energy and Environment Division, Derio, Spain / Basque Centre for Applied Mathematics, CFD-MS Group, Bilbao, Spain
Pablo Ruiz-Minguela, Tecnalia Research and Innovation, Energy and Environment Division, Derio, Spain
Germán Pérez-Morán, Tecnalia Research and Innovation, Energy and Environment Division, Derio, Spain
Raúl Rodríguez-Arias, Tecnalia Research and Innovation, Energy and Environment Division, Derio, Spain
Joseba Lopez-Mendia, Tecnalia Research and Innovation, Energy and Environment Division, Derio, Spain
José-Luis Villate-Martínez, Tecnalia Research and Innovation, Energy and Environment Division, Derio, Spain

Chapter 12 – Challenges and future research

Neil Kermode, The European Marine Energy Center, UK

Chapter 1

A review of progress on ocean energies

Neil Kermode[a]

1.1 A risky business

Writing the history of marine energy is a risky process. It is a task that is fraught with the dangers of personal bias and limited experience, and because there is no presently accepted historical narrative, it is a space that is open to debate and discovery.

As such, this is a subjective analysis, and, of course, I am viewing the world through a particular geographical lens: I have lived most of my life in the UK and have spent the last 10 years in Orkney, witnessing the birth and growth of the European Marine Energy Centre (EMEC).

So, much of my most recent experience has been in the commercialisation of wave and tidal stream technologies – and therefore I will have a natural bias to talk about that which I know best.

In reviewing the progress of ocean energies, wave energy is a good place to start. The idea of harnessing energy from waves has been around for centuries, yet today it remains a major engineering challenge to do this successfully at scale. As such, wave energy 'brackets' our story nicely.

Here in the UK, it is popularly accepted that Professor Stephen Salter is the 'father of wave energy', and that his 1970s 'Nodding Ducks' were the world's first wave energy machine. However, this ignores the facts that Stephen was not the first to think of this energy source, nor was he alone at the time when he did his sterling work.

And as soon as you say 'this is where it started', you find precedents that prove you wrong. At best you can be very specific about the claim such as 'Pelamis was the world's first floating deep-water wave converter to generate into the National Grid', but it is less elegant than 'it started here'.

So, as we journey through the history of marine energy, there will need to be occasional digressions and eddies to the narrative. Often these may be due to partially told stories and may provide clues for further research. However, the intent is to show as complete a record of the development journey as it appears to me in late

[a]The European Marine Energy Centre, Orkney, UK

2018, secure in the knowledge that the sector will continue to progress across many fronts even before the ink in this chapter has had time to dry.

1.2 In the beginning

So rather than ask 'where did it all start', it is more illuminating to ask '*why* did it all start'?

Energy. Society needs energy to produce or process material into more valuable goods. This has always been the case and over history energy has been produced by the manual labours of humans, or their domesticated animals.

For most of human history, the most valuable commodity was food and nearly all energy harvesting systems went into this. In most cases, some natural energy flow was harvested, initially the wind and in due course the flow of rivers. All of these techniques require technology to turn flowing fluids into useful work and most relied upon grinding or processing plant material into food stuffs.

This gives the first example of a tidal site I have encountered: Eling Mill in Hampshire. The mill is believed to have Roman foundations. It has been built across the river at Eling where it enters Southampton Water and the sea. The gates of the mill are left open as the tide rises so allowing the mill runs releasing pond to be filled with the incoming tide. At high tide the gates are closed, so impounding the tidal water in the mill pond. Once the tide has fallen outside the pond, the mill is free to run by releasing the impounded sea through its mill wheels. It has been doing this for nearly 2,000 years.

There must be other sites where this has been done but forgotten. Indeed, there are multiple tidal mills shown on maps in Wales and Cornwall, but the dates of them are often unclear. However, they never achieved the ubiquity that wind mills achieved, so they have often been 'the first of a kind' when the idea was re-discovered again and again during history. However, as an idea they were eclipsed by the bigger and more powerful forces of fossil fuels. There are practically no examples of new tidal schemes being built, commissioned and successfully run for any period of time once the technology to burn fossils was developed. Where there are examples then they tended to be commissioned at times of crisis when oil supplies were looking fragile, or when the costs looked to be rising uncontrollably.

Another driver for the development of renewables seems to have been the development of the means to generate and use electricity. This occurred around 1880 when shaft power could be turned into kilowatts and gave rise to a flurry of schemes.

This leads to my earliest wave scheme: Herring Cove, Nova Scotia. The Parsons Ocean Power Company was incorporated in 1922 [1] and by 1925 had built a machine which successfully generated power, until its later destruction in a storm. This scheme relied upon capturing the movement of a tethered float. The movement of the float caused machinery to rotate by way of a lever arm and gear mechanism and generated sufficient electricity to light 'several 60 watt bulbs'.

The Pacific coast also seemed to attract those who saw the opportunity and there are several examples of around a dozen schemes such as the oscillating water column at Parallel Point, San Francisco.

Schemes like Herring Cove aimed to generate electricity for the local area, but the advent of the electricity supply grid generally marked the end of their usefulness, or certainly their competitiveness.

In the space of a few short decades, a new era of coal, steam and electric power, together with the creation of an electricity grid, disconnected the need to co-locate generation and the users of energy and heralded a long hiatus in the need for, or an interest in, renewable power.

Industrial economies became used to plentiful and relatively cheap electricity provided through increasingly large fossil fuel power plants streaming electrical energy to people's homes. Motor transport grew on the back of plentiful, cheap oil. This meant that, with very few exceptions, there was little need to consider renewables.

Yet in some places renewables found a role. In France, La Rance Tidal Power Station was opened in 1966 as the world's first tidal power station, feeding power into the country's grid. It was for decades the largest tidal power station in the world.

Whilst in the Highlands of Scotland – geographically remote and poorly connected – massive hydro schemes were built throughout the middle of the twentieth century to 'power the glens' and provide economic stimulus to what was until then a socio-economic backwater. Likewise in the USA, where, by 1920, 40 per cent of all electricity came from hydro [2]. Similarly, other one-off renewable projects could be found around the world. But by and large, there was limited interest or need to consider new forms of power until, like the industrial revolution, another tectonic shift occurred.

1.3 Shocks to the system

The first OPEC oil price shock in 1973–74 [3], which saw the world oil price quadrupled in less than a year (from $3 to $12 per barrel), was a major blow to the world's economy and meant – for the first time, that politicians, policymakers and politicians would need to consider new forms of power. This shock sparked a new interest in alternative sources of energy, with wave power leading the charge for ocean technologies.

A further shock in 1979 saw the price double again (to nearly $40 a barrel), whilst by the 1980s, a growing recognition of climate change demanded action on the world's stage.

Interest in onshore wind began to grow.

By 1987 the Montreal Protocol was agreed, restricting chemicals that damage the ozone layer, then in 1989 the UK Prime Minister Margaret Thatcher – possessor of a chemistry degree – warned in a speech to the UN that

> We are seeing a vast increase in the amount of carbon dioxide reaching the atmosphere... The result is that change in future is likely to be more fundamental and more widespread than anything we have known hitherto.

It has been this recognition – and the resulting global treaties on climate change – that has spurred action around the world and has sparked the massive growth in renewables we see today.

The industrial revolution, the growth of grids and the recognition of climate change have all resulted in radical shifts in how we make and consume energy. At a political level, nations states must decide – for political and social ends – how energy is created and used, and the plain numbers in $ per kWh must somehow or other stack up.

And these decisions are apt to change. In the last decade alone two further shocks – the Fukushima nuclear disaster and the U.S. shale gas revolution – have marked ground-breaking shifts in the ways in which different countries view energy and energy security.

Mature economies including Japan and Germany have decided to move away from nuclear energy, whilst China – now living with the very real health and climate risks caused by an over-reliance on coal – has become the world's largest developer of onshore wind.

Looking ahead it is clear we cannot know what future changes we might see. Climate change continues to accelerate, the move towards electric vehicles can only grow and we must all find new ways to lighten our load on our crowded planet. Other 'unknown unknowns' – to paraphrase Donald Rumsfeld – inevitably lurk around the bend.

Renewable energy is making an incredible contribution having gone from practically zero in the 1940s to now being the dominant electricity source in several countries such as Norway and Scotland. In 2016, cumulative grid-connected onshore wind capacity reached 451 gigawatts (GW) and wind power accounted for almost 4 per cent of global electricity generation [4].

Onshore wind capacity is now expected to reach almost 750 GW by 2022 – and although ocean energy technologies lag significantly behind, once each sector develops a commercially available product, we can with some confidence expect similar long-term exponential growth.

1.4 The challenge of ocean energy

More than 70 per cent of our planet is covered in ocean, and whilst in many areas our seas are already heavily exploited, for food, oil and gas – they still represent a massive bounty of renewable energy waiting to be harvested.

Waves, tides, wind, salinity, temperature gradients and the sun all offer tempting new sources of power – power which can often be generated offshore, with little or no conflict with other sea users.

In some cases, these technologies are already in the market – indeed some have been around for more than a century, and in others they are not quite there yet.

But what is clear is that – in an ever-changing world – we cannot afford to shut out alternative new sources of energy, whatever they may be.

The challenge – as will become clear in this book – is not only to develop new technologies that work, but also to develop new technologies that work *commercially*. And to understand this, we need a brief exploration of technology readiness.

1.5 Technology Readiness Levels

Trying to describe how ready a technology is for deployment has always been difficult. Every inventor is enthusiastic about their project, often overly so, and this tends to lead to wild claims. Fortunately, NASA came up with a Technology Readiness Levels (TRL) scale to measure the readiness of any technology and this has now been widely adopted as a means of comparison and categorisation.

Technological and commercial maturation occurs over several phases, from concept design to commercial deployment at sea. Each technology project is evaluated against the parameters for each technology level and is then assigned a TRL rating based on the project's progress. Moving from one phase to the next requires increased deployment, leading to technological and hence economic improvements.

There are nine technology readiness levels. TRL 1 is the lowest and TRL 9 is the highest.

NASA defines TRLs as follows [5]:

When a technology is at TRL 1, scientific research is beginning, and those results are being translated into future research and development.

TRL 2 occurs once the basic principles have been studied and practical applications can be applied to those initial findings. TRL 2 technology is very speculative, as there is little to no experimental proof of concept for the technology.

When active research and design begin, a technology is elevated to TRL 3. Generally, both analytical and laboratory studies are required at this level to see if a technology is viable and ready to proceed further through the development process. Often during TRL 3, a proof-of-concept model is constructed.

Once the proof-of-concept technology is ready, the technology advances to TRL 4. During TRL 4, multiple component pieces are tested with one another. TRL 5 is a continuation of TRL 4; however, a technology that is at 5 is identified as a breadboard technology and must undergo more rigorous testing than technology that is only at TRL 4. Simulations should be run in environments that are as close to realistic as possible.

A TRL 6 technology has a fully functional prototype or representational model.

TRL 7 technology requires that the working model or prototype be demonstrated in a space environment. TRL 8 technology has been tested and 'flight qualified' and it's ready for implementation into an already existing technology or technology system. Once a technology has been 'flight proven' during a successful mission, it can be called TRL 9.

To look specifically at ocean energy, the EU has described a number of stages for the development of ocean energy technologies as they pass through these levels (Figure 1.1).

A crucial question is how *quickly* any given technology will progress through these stages. It is essential to understand that any timeline is highly dependent on overcoming the barriers faced by ocean energy developers and the level of public support offered in the short and medium terms by national and regional governments and international institutions such as the EU.

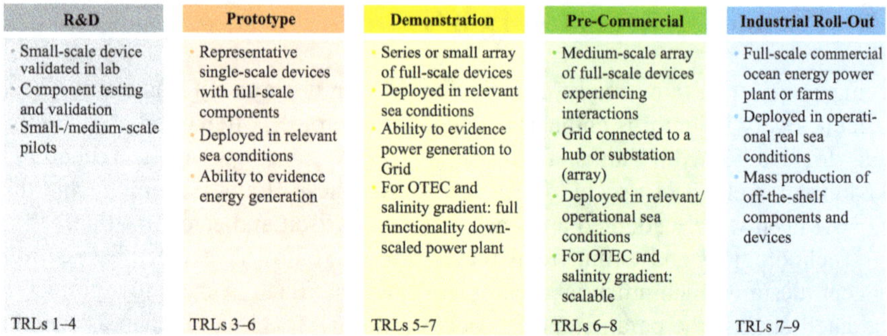

Figure 1.1 From EU Ocean Energy Forum Strategic Roadmap

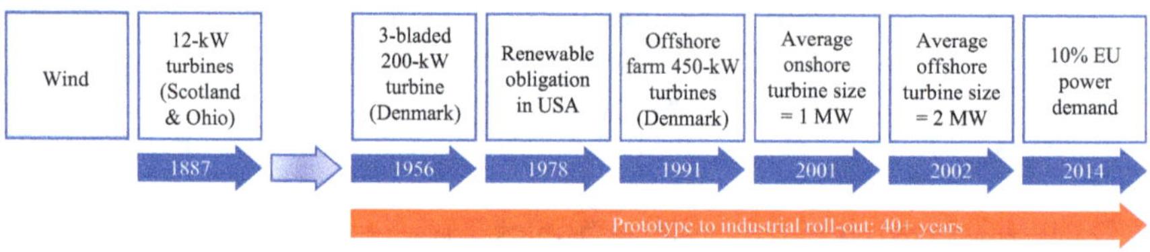

Figure 1.2 Development of wind turbines, from early experiments to industrial rollout, Ocean Energy Europe

To put above in context, if we take the example of onshore wind, it took 40 years from the first experiments until early rollout (Figure 1.2) – and the speed of rollout was highly dependent on the public policy framework.

One should also note that it was not the first inventors (Scotland and the USA) which capitalised on the industrial rollout and economic benefit which ensued that prize went to Denmark. They put in place the specific tariffs, investment vehicles and consents processes that gave the fledgling technology somewhere to take flight. The lesson here is the benefit went to the country which *commercialised* the technology, not the one that invented it, and there are parallels here for ocean energy policymakers [6].

Wave Energy Scotland, an organisation established to bring wave energy to the point of commercialisation, has taken the TRL model and matched it to its own four-stage progression. EMEC provides an overview of the technology development pathway described here (Figure 1.3).

Applying this discipline to marine energy is useful. It provides a clear pathway from the lab to the sea, and all of the steps that need to be taken in between.

It is not, however, the end of the story. As in the example of the onshore wind industry, it is not just about inventing a technology that works, it is about finding a path to make a technology work *commercially* and this can take many years. This is a theme I will return to in the closing chapter.

In the last decade, a number of ocean energy technologies have made tremendous strides towards TRL 9 and here we will look at them all in turn.

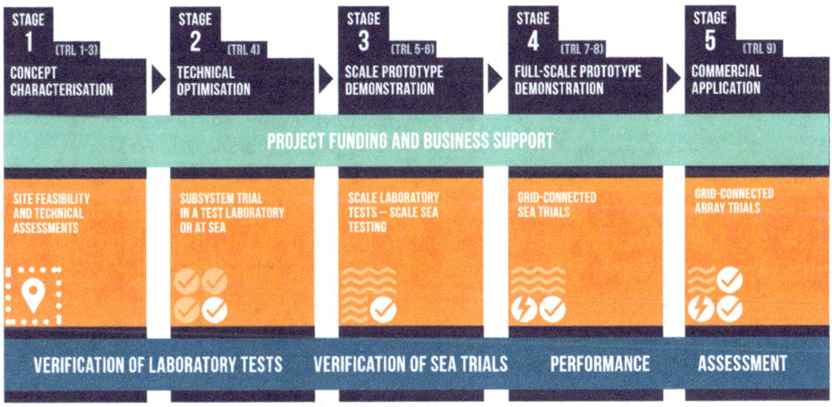

Figure 1.3 EMEC pathway to commercialisation

1.6 Wave energy – TRL 7

Wave energy remains one of the world's last great untapped sources of renewable energy – with a total theoretical wave energy potential of 32 PWh/year, roughly twice the global electricity supply [7]. But thus far, it has proved the most challenging to develop.

Today there are an estimated 218 active wave energy developers around the globe [8], ranging from start-ups and small-to-medium enterprises, to utilities, original equipment manufacturers and industrial engineering firms, each seeking to develop one of at least nine different types of wave energy converter. Yet we still have not seen a dominant technology emerge at full-scale TRL 8, with various wave energy concepts in development.

A full explanation of the different types of wave energy converters can be found on the EMEC website at the following link: *www.emec.org.uk/marine-energy/wave-devices/*.

In Europe, it is estimated ocean energy could meet 10 per cent of EU electrical demand by 2050 [9], whilst worldwide it is recognised that 'blue growth' has the potential to regenerate peripheral areas affected by declining traditional industries [10]. This is particularly true for wave energy, where areas of high resource are closely aligned with remote and economically fragile areas – places where new, high value jobs have a disproportionate impact.

I can vouch personally for the massive impact the industry has had in Orkney, where it is estimated that EMEC has spent over £16 million in Orkney with around 200 people currently employed in the marine renewables sector [11]. An economic impact assessment commissioned by government agency Highlands and Islands Enterprise estimates that EMEC has generated a gross value added to the wider UK economy of £284.7 million, with 4,224 full-time equivalent (FTE) job years so far [12].

In the modern era, I consider the wave energy story began in earnest in the 1970s in the UK at the Wave Power Department at the University of Edinburgh, led by Prof Stephen Salter whose invention – known as the Edinburgh or Salter Duck – excited considerable international attention and showed that the words 'wave' and 'energy' do actually fit together. However, his work was not done in isolation as he gathered a team of researchers and there were others such as hovercraft inventor Sir Christopher Cockrell also working on schemes in parallel.

By the late seventies however, the UK had struck oil in the North Sea and public support for wave energy waned. It is regrettable that Salter's work was undermined by inaccurate reporting by UK government's advisors of the time which led to the programme's cancellation. And although wave energy remained a significant topic at an R&D level, it was not until the 2000s that we saw a significant amount of progress – again led by a Scottish concern.

The rise of Pelamis in the early years of this century was a tremendous story of endeavour, and one of the firsts in the world. It dovetailed seamlessly with the establishment of EMEC in 2003 and enabled the Edinburgh pioneers to successfully pilot the world's floating first grid-connected device.

For the next decade, Scotland and the UK gave tremendous support to the sector until – by 2014, the technological and financial challenges in creating energy from waves overwhelmed the leaders in the sector – with the high-profile collapses of Pelamis and, shortly thereafter, Aquamarine Power.

What these financial collapses, or as I would prefer to term them *lessons*, have shown us is that some technological challenges are too big or too complex for single private firms to tackle alone, even with public support.

1.7 Tidal and current energy

1.7.1 Tidal range – TRL 9

Whilst wave energy continues to demonstrate its ability to generate electricity, tidal range power is already well proven.

As I mentioned earlier, Eling Mill in Hampshire has been utilising tidal range power for over 2,000 years, whilst the 240-MW la Rance Tidal Power Station in France has been operating since 1966. In Canada, the 20-MW Annapolis Royal Generating Station Opened in 1984, whilst the 254-MW Sihwa Lake Tidal Power Station in South Korea has been operational since 2011.

Tidal range technology shares characteristics similar to hydropower, essentially utilising the height difference of two bodies of water created by a dam or barrier in order to produce electricity. Its main advantage is that is it very predictable. However, its capacity factor of 25 per cent is not as high as offshore wind due to the nature of tidal cycles and turbine efficiency [13].

In recent years, tidal range has evolved with the concept of tidal lagoons: these are artificial basins built in bays and estuaries and are considered less intrusive than tidal barrages (which tend to span entire estuaries and alter the salinity and ecology). In the UK, the 320-MW Swansea Bay Lagoon project in Wales has (at the time of writing) failed to secure the government support it requires to proceed.

This was intended to be a pathfinder project, leading to five larger proposed tidal lagoon projects in the United Kingdom: Cardiff (3,000 MW), Newport (1,800 MW), Colwyn Bay, West Cumbria and Bridgwater Bay.

One recent review [14] estimates that 5,792 terawatt-hours (TWh, around a third of the total global electricity supply) could be produced by tidal range power plants. However, 90 per cent of the resource is distributed across just five countries, with both the UK and France having a significant share of that resource.

In common with other forms of energy, it is not the technology *per se* that is an obstacle to progress – tidal range clearly works. The issue appears to be its uncompetitiveness versus unsustainable and incumbent fossil-based fuels, or heavily subsidised nuclear generation. However, the technology does require a very specific set of physical and geographical circumstances to make a scheme viable – and the necessary revenue stream or public support to offset the high initial capital costs.

1.7.2 Tidal stream – TRL 8

In contrast to wave energy, the tidal stream sector has made significant advances towards commercialisation in recent years. Orkney technology firm Orbital Marine Power has developed a floating turbine already providing significant power to Orkney Mainland, whist in the Pentland Firth the MeyGen scheme, owned by SIMEC Atlantis, has the world's largest operational array, powered by Andritz Hydro Hammerfest and Atlantis Resources turbines.

Tidal stream generators draw energy from water currents in much the same way as wind turbines draw energy from air currents. However, the higher density of water relative to air (water is 800 denser) means that tidal generators have smaller blades than wind turbines and operate at lower velocities.

Given that power varies with the density of medium and the cube of velocity, water speeds of nearly one-tenth the speed of wind provide the same power for the same size of turbine system; however, this limits the application in practice to places where tide speed is at least 2 knots (1 m/s) even close to neap tides.

In Europe, the majority of high resource tidal stream sites which meet this criteria lie around the coasts of the UK, Ireland and France with some hotspots in the Mediterranean. Worldwide other tidal hotspots include Nova Scotia, Chile, the Philippines and Indonesia.

In the modern era, the development of tidal energy was effectively led by Dr Peter Frankel when he installed his 15-kW machine in Loch Linnhe, Scotland, in 1994–95 [15]. This two-bladed rotor project by IT Power had been sponsored by Scottish Nuclear Electric and was the forerunner of the 300-kW SeaFlow project later installed off Lynmouth in Devon in 2003 by Peter's company Marine Current Turbines (MCT).

That device was not grid connected but provided invaluable information on the practicalities of a stand-alone generating structure at sea. It led directly to the iconic twin rotor 1.2-MW SeaGen device at Strangford Loch in Northern Ireland installed in 2008. This turbine operated successfully for several years and generated over 350 MWh into the grid until it was shut down in 2017.

In parallel with MCT's work, a number of other companies also began development on their own tidal turbines, with several notable examples.

In the Straits of Messina, Ponte de Archimede developed a 150-kW vertical axis turbine (KOBOLD). This largely forgotten device was the first to generate into the grid in Europe in 2001 [16]. The company was owned by ship owner Dr Elio Matacena and used his experience of the Voith Schneider vertical propellers that powered his car ferries to come up with the design.

In Hammerfest, Norway, a fully submerged horizontal axis turbine was installed in 2003 [17]. This 300-kW machine generated into the local grid and led to the development of the 1-MW horizontal axis turbine installed at EMEC in 2011.

Herbert Williams in Florida developed a rim generator in the 1990s to work in the Gulf Stream [18] which eventually became Irish-owned OpenHydro. They installed a test rig at EMEC in 2006 and gird-connected their first 6-m diameter open-centred turbine in 2008. They narrowly beat MCT to the accolade of the first to generate to the GB National Grid and continued to test at EMEC until the company's demise in 2018. OpenHydro had gone on to install a 10-m diameter machine in the Bay of Fundy in Canada in 2009 before installing a 16-m diameter machine at a nearby site and also two machines at Roche Blanchard in France [19].

In the middle of all this Tidal Generation Limited, a company made up in part of ex-MCT employees, developed a buoyant sub-sea turbine. This initially 500-kW machine led to a 1-MW machine through Rolls Royce's support and eventual ownership of the company before its sale to Alstom.

In more recent times, Scotland has seen three most notable successes.

In 2016, Nova Innovation deployed the world's first fully operational, grid-connected offshore tidal energy array, at Bluemull Sound in Shetland. The first Nova M100 turbine was deployed at the site in March 2016, and the second was deployed in August 2016, making this the first offshore tidal array in the world to deliver electricity to the grid. A third turbine was added to the array in early 2017.

The first phase of the MeyGen project passed the 8-GWh generation milestone in July 2018 [20] through its use of two Hammerfest turbines and an Atlantis machine – both companies having undertaken testing work at EMEC in the previous years. This scheme is set to expand its site with the next phase being planned at the time of writing.

And finally, the device of which Orkney is most proud is the locally envisaged, designed, managed and installed SR1-2000 made by Scotrenewables (now known as Orbital Marine Power), which repeatedly generated around 7 per cent of Orkney's electricity through much of 2018. This machine is the first pre-commercial prototype of the company and follows years of development by the local team. The first open water tests of a large-scale model were undertaken within sight of EMEC's offices and led to the 250-kW machine being developed and connected in 2012 [21]. The success of that device led to the 2-MW machine that effectively powered Orkney for a day a fortnight in 2018 [22].

And there have been many other successes and firsts achieved in this period that are too numerous to mention here with both devices built and tested around the world, and also the experience of some of the work above being taken to Asia and South America to join with their local initiatives.

Now after decades of dreaming and sporadic one-off projects, we have reached the point where several developers have built more than one device. This enables the essential 'learning by doing' to begin to take place. Up to this point, most schemes were the first (and often last) of a kind. Now we have some serious learning being undertaken. We will return to this point in the final chapter.

So in the last 17 years, we have gone from the first grid-connected machine, to the delivery of industrially useful and dependable electricity day in and day out. This is remarkable progress. However, we have to recognise that it is just the start.

1.8 Thermal and salinity gradient systems

1.8.1 Ocean thermal energy conversion – TRL 8

Ocean thermal energy conversion (OTEC) is unusual in that its first recorded presence is in fiction. In 1870, Jules Verne introduced the concept in Twenty Thousand Leagues Under the Sea: 'I owe it all to the ocean; it produces electricity, and electricity gives heat, light, motion, and, in a word, life to the Nautilus.'

His idea – which relied upon utilising the different temperatures of water in the sea – was spot on. The potential for ocean thermal energy arises from the temperature difference between near-tropical surface seawater, which may be more than 20°C hotter than the temperature of bathyal (>1,000 m) ocean water, which is relatively constant at about ~4°C. Bringing large quantities of this cold seawater to the surface enables a heat exchange process with the warmer surface waters, from which energy can be extracted.

By the 1880s American, French and Italian scientists are all believed to have been working on the concept. But it is the Frenchman, physicist Jacques-Arsene d'Arsonval, who is generally credited as the father of OTEC to create power.

One of the students of D'Arsonval, Georges Claude, built the first working OTEC power plant in 1930 in Cuba, which produced 22 kW of electricity. This led to an onshore open cycle plant, with a pipe extending out to sea and he went on in 1935 to construct another plant, this time aboard a 10,000-ton cargo vessel moored off the coast of Brazil. Unfortunately, both plants were destroyed by weather before they could become operational and French research continued in earnest through the 1940s and into the 1950s. Research also began in California in the 1940s. In all cases, work was slowed or halted by cheaper alternatives to power generation.

In the 1960s, J. Hilbert Anderson and his son James Anderson designed a closed-cycle OTEC power plant, aimed to be more practical, compact and economic. This cycle pumps warm surface water through heat exchangers to boil a working fluid into a vapour. The vapour expands to power turbines and drive generators. Cold water pumped from the deep ocean condenses the vapour back into its liquid state. As in the story of wave energy, the Arab oil embargo and skyrocketing oil prices in the mid-1970s drove away interest in the Andersons' and other OTEC models.

Today, with a much more visible threat posed by climate change, and with island communities threatened by rising sea levels, we see a renewed interest in OTEC with a handful of state-sponsored projects now taking place around the world. These include the following.

Hawaii, USA

In 2015 Hawaii became the first state in the USA to generate electricity using their 105-kW OTEC plant on Big Island. The plant cost was $5 million and was funded and developed through a collaboration between the U.S. Office of Naval Research and Hawaii's Makai Ocean Engineering.

Okinawa, Japan

This 2013 scheme aims to combine power production from OTEC with the use of deep seawater for other uses to improve the economics of the system as a whole.

The project has been built in the Okinawa Prefecture Deep Sea Water Research Institute and utilises an existing OTEC plant which has been expanded with the construction of post-OTEC seawater pipelines in 2017.

The worldwide potential of ocean thermal power conversion has been conservatively estimated at 44,000 TWh/year across the Tropics – however, the challenge lies in realising this potential.

This hurdle has been summarised neatly by Gérard C. Nihous, an OTEC expert at the University of Hawaii who said: 'The technology is simple to understand but very difficult to implement in the field. There are engineering challenges, but most of the reasons for its incomplete development are economic.'

1.8.2 Salinity gradient – TRL 4

The idea of exploiting the osmotic pressure which arises when freshwater and saltwater meet was first conceived in the 1970s. However, a quarter of a century had to pass before market conditions made several independent public and private enterprises take up the idea and start developing the technology further.

Seawater is approximately 200 times more saline than freshwater, and this relatively high salinity establishes a chemical pressure potential with fresh river water, which can be used to generate electricity.

The two leading methods for generating power from salinity gradients are pressure-retarded osmosis (PRO) and reverse electrodialysis (RED). Other technologies such as hydrocratic generators, solar ponds (vapour compression) and osmotic heat engines have also been proposed.

PRO uses a membrane to separate a concentrated salt solution (like seawater) from freshwater. The freshwater flows through a semipermeable membrane towards the seawater, which increases the pressure within the seawater chamber. A turbine is spun as the pressure is compensated and electricity is generated.

RED uses the transport of (salt) ions through membranes. RED consists of a stack of alternating cathode and anode exchanging permselective membranes. The compartments between the membranes are alternately filled with seawater and freshwater. The salinity gradient difference is the driving force in transporting ions that results in an electric potential, which is then converted to electricity.

Two main applications exist: as stand-alone plants in estuaries where freshwater rivers run into the sea and as hybrid energy generation processes recovering energy from high-salinity waste streams, which could be, for example, brine from desalination or salt mining. A possible third application is salinity gradient technologies applied to land-based saltwater lakes or other types of saltwater reserves [23].

The total technical potential for salinity gradient power is estimated to be around 647 GW globally; however, despite this high theoretical potential, challenges both in proving the technology and in finding a route to commercialisation means current salinity gradient schemes are currently on hold.

Norwegian hydro utility Statkraft was an early champion of the sector and developed its first PRO power plant at Tofte on Hurum, near Oslo in 2009. It used 2,000 m^2 of flat sheet membranes and could theoretically produce 10 kW, although the actual production was around 5 kW and powered an electric kettle for demonstration purposes.

Yet by 2013, Statkraft had decided to withdraw from the sector, suggesting that the technology would not be sufficiently developed to become competitive within other technologies.

Making his assessment of the technology, Statkraft department manager Stein Erik Skilhagen said:

> We have proven that the technology works and have achieved substantial improvements through our efforts. Our main challenge has been to make the technology efficient enough to achieve energy production costs on par with competing technologies. With the current market conditions, we see that we cannot achieve this in the foreseeable future. There are other technologies which have developed enormously in recent years. These are more competitive and relevant investments for us in the future [24].

In 2013, Dutch firm REDstack officially opened 50-kW RED pilot, situated on the sea defence causeway, which separates relatively clean freshwater on one side from relatively clean seawater present in the Wadden Sea/North Sea [25]. This project is still operational, although has not published plans for any further development.

1.9 Offshore wind – TRL 8

Unique amongst the other forms of ocean energy explored in this book, offshore wind is already a highly successful competitive product, fully proven at TRL 8 and close to TRL 9.

Compared to other forms of power, offshore wind remains a relative newcomer – but already this decade deployment of offshore wind has more than quintupled from

3.2 GW in 2010 to 18.7 GW in 2017 by which time it contributed some 56 TWh or 0.3 per cent of global electricity generation [26].

The key factor behind the rise of the offshore wind market is a concerted series of public/private initiatives undertaken by countries bordering the North Sea in Europe. More than 80 per cent of global offshore wind capacity is located in Europe, of which the United Kingdom with installed capacity of 6.8 GW and Germany with 5.4 GW are the two largest countries.

Beyond Europe, only the People's Republic of China has large-scale offshore wind capacity installed, at 2.7 GW, while smaller offshore wind facilities are located in the United States, Korea and Japan.

Although it uses a fundamentally similar technology to onshore wind, offshore wind enjoys some distinctive advantages: the main ones are that offshore installations are able to tap more consistent and higher winds speeds, and there are fewer restrictions on ground area and height. As a result, project sizes and turbines are typically larger and performance is higher.

Costs relative to other forms of renewables are higher. However, recent energy auctions across Europe have seen offshore wind energy prices fall to below £58 per MWh [27].

And despite the vast resource potential that could be tapped with fixed-bottom configurations (i.e. turbines with foundations on the seabed), studies have found that 80 per cent of estimated offshore wind resources in Europe are located in waters at depths greater than 60 m, with the corresponding figure being 80 per cent in Japan and 60 per cent in the United States [28].

This has sparked a new wave of innovation – to site offshore wind farms in ever-deeper waters – with a growing appetite to surmount any technological challenges. Already there is a vast supply chain engaged in the development of huge offshore wind parks, and the ever-present pressure to reduce cost and exploit new resources will continue to spur new advances.

In Japan, they are successfully testing a number of floating turbine concepts at scale at the Fukushima Floating Offshore Wind Farm [29], whilst in late 2017 Statoil switched on the world's first commercial floating wind farm – Hywind – off Peterhead in Scotland. This comprises five 6-MW floating turbines anchored by means of three suction anchors in 95–120 m of water.

Industry voices expect we will see ever larger turbines – potentially 20 MW or more by the middle years of the next decade – and it is clear offshore wind is here to stay.

1.10 Marine solar

Marine solar on the other hand is a new concept. It relies upon the open and unfettered expanse of water providing a flat and unshaded area to place floating solar panels. This technology may find a better home on lakes and reservoirs where the added benefit of reducing evaporation may strengthen water resources in arid areas.

It is a technology to watch and will clearly owe much to the technical developments of terrestrial solar and its application is expected to centre in more equatorial regions.

1.11 Enabling technologies and actions

For the purposes of this opening chapter, I do not propose to go into detail in what I have termed 'enabling technologies and actions' each of which merits their own section within this book, including the following:

- offshore support structure design
- electrical power transmission and grid integration
- offshore energy storage
- multi-purpose platforms
- installation, operation and maintenance of offshore renewable.

What I do observe, however, is that all of these topics – and they are major ones – involve what we do with ocean energy technologies *at sea*.

How do we build structures that can withstand the extraordinary forces they will experience in winter storms?

How do we harness new technologies such as High-Voltage Direct Current (HVDC) to bring power ashore? (Or can we, indeed store energy at sea?) Can we reduce cost through multi-purpose platforms, and how can we install, maintain and operate these new technologies in real ocean conditions?

For someone starting off on their journey of creating energy from the oceans with a radical new idea, these enabling technologies and actions can appear far away from the early days spent in the laboratory.

But as soon as a technology crosses the shoreline in earnest, and aims to go to sea, then these are the areas which present some of the biggest challenges we face. In order to operate at sea, each technology must be able to operate – and operate reliably – whilst experiencing forces far beyond the average sea state.

Robustness and reliability are key, and they do not come cheap. So, whilst in the ocean the first challenge is to prove 'it works' the next challenge is to show it works *reliably*. Then we must show there is at least a route to making it work *commercially* – given whatever support mechanisms may be in place.

This is a topic I will return to in the closing chapter – where themes such as 'learning by doing' and serial production will come to the fore.

Meanwhile, dip in, and enjoy exploring the fascinating journey to realise renewable energy from our oceans.

References

[1] https://otcns.ca/news/2012-06-01/ocean-wave-energy-a-nova-scotia-first-, accessed on 5 March 2019.

[2] 'History of Hydropower'. U.S. Department of Energy, https://www.energy.gov/eere/water/history-hydropower accessed on 5 March 2019.
[3] https://en.wikipedia.org/wiki/1973_oil_crisis, accessed on 5 March 2019.
[4] https://www.iea.org/topics/renewables/wind/, accessed on 5 March 2019.
[5] https://www.nasa.gov/directorates/heo/scan/engineering/technology/txt_accordion1.html, accessed on 5 March 2019.
[6] The Danish wind industry 1980–2010: Lessons for the British marine energy industry, Kyle Smith, *International Journal of the Society for Underwater Technology*, Vol. 30, No. 1, pp. 27–33, 2011.
[7] https://www.worldenergy.org/wp-content/uploads/2017/03/WEResources_Marine_2016.pdf, accessed on 5 March 2019.
[8] EMEC. http://www.emec.org.uk/services/
[9] https://webgate.ec.europa.eu/maritimeforum/sites/maritimeforum/files/OceanEnergyForum_Roadmap_Online_Version_08Nov2016.pdf, accessed on 5 March 2019.
[10] https://ec.europa.eu/maritimeaffairs/policy/ocean_energy_en, accessed on 5 March 2019.
[11] http://www.hie.co.uk/growth-sectors/energy/marine-energy.html, accessed on 5 March 2019.
[12] Atlantic Marine Energy Test Site (AMETS) Economic Impact Case Study on EMEC in Orkney, December 2017.
[13] WEC, 2016. https://www.worldenergy.org/wp-content/uploads/2016/10/World-Energy-Resources-Full-report-2016.10.03.pdf
[14] https://phys.org/news/2018-05-tidal-range-power-potential-electricity.html#jCp, accessed on 5 March 2019.
[15] https://www.supergen-marine.org.uk/sites/supergen-marine.org.uk/files/publications/taiwan_2009_mct.pdf, accessed on 5 March 2019.
[16] https://tethys.pnnl.gov/annex-iv-sites/enermar-project, accessed on 5 March 2019.
[17] https://www.powerengineeringint.com/articles/print/volume-19/issue-1/features/orkney-tidal-power-a-hotbed-on-the-seabed.html, accessed on 5 March 2019.
[18] https://www.bloomberg.com/news/articles/2016-06-09/the-ex-con-inventor-disrupting-underwater-energy, accessed on 5 March 2019.
[19] https://www.rpsgroup.com/images/UK/PDFs/MARITIME_TIDAL_161012.aspx, accessed on 3 February 2019.
[20] https://simecatlantis.com/2018/07/11/meygen-phase-1a-operational-update/, accessed on 5 March 2019.
[21] http://www.scotrenewables.com/technology-development/sr250, accessed on 5 March 2019.
[22] http://www.scotrenewables.com/technology-development/sr2000, accessed on 5 March 2019.
[23] http://www.irena.org/documentdownloads/publications/salinity_energy_v4_web.pdf, accessed on 5 March 2019.
[24] https://www.statkraft.com/media/news/News-archive/2013/Statkraft-halts-osmotic-power-investments/, accessed on 5 March 2019.

[25] http://www.redstack.nl/en/home, accessed on 5 March 2019.
[26] https://www.iea.org/publications/freepublications/publication/WEO2017 Special_Report_OffshoreEnergyOutlook.pdf, accessed on 5 March 2019.
[27] https://www.gov.uk/government/publications/contracts-for-difference-cfd-second-allocation-round-results, accessed on 5 March 2019.
[28] Carbon Trust, 2015. https://www.carbontrust.com/resources/reports/technology/floating-offshore-wind-market-technology-review/
[29] https://www.4coffshore.com/windfarms/project-dates-for-fukushima-floating-offshore-wind-farm-demonstration-project-(forward)-phase-2-jp13.html, accessed on 5 March 2019.

Chapter 2

Wave energy

Gianmaria Sannino[a], Adriana Carillo[a], Arne Vogler[b], Giovanni Bracco[c], Giuliana Mattiazzo[c], Diego Vicinanza[d], Pasquale Contestabile[e], Domenico P. Coiro[f], Giancarlo Troise[g], Luca Castellini[h], and John V. Ringwood[i]

Wave energy has an estimated global potential of 3.7 TW, almost double that of current world electrical energy consumption, and has, to date, remained virtually untapped [1]. Wave also has a relatively low correlation with, for example, wind and solar [2], meaning that a balance of wave with other renewable technologies can offer greater consistency of supply. Despite these attractive features, the drive to make wave energy economic has made relatively poor progress, due to a combination of the following factors: the ocean environment is hostile, wave energy is reciprocating (rather than unidirectional, therefore requiring rectification) and variable, and the technology is still in its relative infancy. These difficulties are manifest in the current unattractively high levelized cost of energy (LCoE) for wave energy, currently estimated at €225/MW h compared to the LCoE for offshore wind, for example, at €165/MW h [3]. This chapter examines the various facets of wave energy, including resource quantification and wave measurement (Section 2.1), the wide variety of onshore and offshore wave energy devices (Section 2.2), the variety of power take-off (PTO) mechanisms which convert wave power into other useful forms (Section 2.3) and concluding with some insight to how wave energy devices are modelled and controlled (Section 2.4).

[a]ENEA, Italian National Agency for New Technologies, Energy and Sustainable Economic Development, Rome, Italy
[b]Department of Marine Energy Research, University of the Highlands and Islands – Lews Castle College, Stornoway, Isle of Lewis, UK
[c]Dipartimento di Ingegneria Meccanica e Aerospaziale, Politecnico di Torino, Torino, Italy
[d]Inter-University National Consortium for Marine Sciences (CoNISMa), Italy
[e]Department of Engineering, University of Campania Luigi Vanvitelli, Aversa (Caserta), Italy
[f]Department of Industrial Engineering – Aerospace Section, University of Naples Federico II, Naples, Italy
[g]SeaPower Scarl – Research Consortium, Naples, Italy
[h]UMBRAGROUP S.p.A., Foligno, Italy
[i]Centre for Ocean Energy Research, Maynooth University, Ireland

While this chapter can only give a limited overview of the challenges and some potential solutions to successfully harness wave energy, it belies the truly vast and diverse array of prototype wave energy converter (WEC) devices [4]. This diversity demonstrates the lack of technology convergence in WEC design and, combined with the multitude of PTO modalities described in Section 2.3, the potential variety of wave energy systems is truly broad. Nevertheless, from the overview and examples presented in this chapter, it is hoped that the reader will gain a better understanding of the challenges, and potential benefits, of adding wave energy to the current mix of commercial renewable energy technologies.

2.1 The wave resource

2.1.1 *Wave resource assessment*[1]

Waves are generated from the transfer of energy from wind blowing over the oceans and can travel thousands of kilometres from the generation area. During their propagation, the wave characteristics change and waves become progressively more regular and increase their wavelength, becoming the so-called swell waves. Due to the superposition of wind and swell waves, the distribution of wave heights does not coincide with that of winds. The highest latitudes both in the Northern and in the Southern Hemisphere are characterized by the largest values of wave height. Monthly mean values over 5 m are reached in both hemispheres during winter but, while there is a strong seasonality in the Northern hemisphere, values are more constant over the year in the Southern hemisphere [5]. Wave energy is proportional to the square of height and to the mean period. Due to the difference in frequency characteristics of wind waves and swell waves also the relative occurrence of the two types of waves has been investigated at a global scale [6].

Wave energy atlases have been created both on global scale and on particular regions (e.g. [1]). They are based on wave measurements obtained from buoys, satellite data and output from model simulations. In the last decades, satellite altimeters have been launched and are providing significant wave height data covering the entire ocean surface at constantly increasing resolution. This data is complemented by a variety of wave models that are actually running over different domains and at different resolutions. These models simulate numerically the growth, propagation and decay of waves forced by surface winds. They produce homogeneous data over long periods of time and provide a series of wave variables permitting a detailed characterization of the waves. In situ wave measurements from wave buoys represent the reference for the validation of models, recording very high temporal resolution and good quality data. Even if a few centres exist worldwide that collect and maintain buoys networks, observations are sparse and the maintenance of instruments expensive. Time series from wave buoys describe wave climate only locally and often present large data gaps, caused by temporary failure or by routine maintenance operations.

[1]Written by Gianmaria Sannino and Adriana Carillo.

2.1.1.1 Methods

Prior to the installation of a device in a specific site, the assessment of the wave energy resource has to be computed using available data. Due to the variability of waves, local climatology has to be computed over a period of at least a decade, considering seasonal and inter-annual mean values of the principal wave characteristics such as significant wave height, mean direction and period.

Preliminary analysis can be based on large-scale models using data at intermediate and deep water. Then local wave conditions should be evaluated by means of high-resolution models directly validated at the site by ad hoc installed instruments.

Surface elevation of waves can be represented as the sum of a large number of harmonic waves, statistically independent. Actually, most of the sea wave models are spectral and are based on the observation that sea surface is composed of a high number of random waves of different frequency and length, forced by the irregular wind. The models compute the propagation of the variance density spectrum $E(f)$ that is related to the variance of the sea surface elevation by the equation:

$$\langle \eta^2 \rangle = C(0) = \int_0^{\infty} E(f)df \tag{2.1}$$

where $E(f)$ is the Fourier transform of the autocovariance ($C(\tau)$) of the surface elevation, and τ represents the time lag. The energy density spectrum can also be expressed as a function of the wave directions θ, with

$$E(f) = \int_0^{2\pi} E(f,\theta)d\theta \tag{2.2}$$

The integral wave parameters can be expressed as the m-order moment of the energy density spectrum:

$$m_n = \int_0^{\infty} f^n E(f)df \tag{2.3}$$

In particular, the significant wave height (in m) is defined

$$H_S = 4\sqrt{m_0} \tag{2.4}$$

The peak period is the period corresponding to the maximum in the energy spectrum and the mean wave energy period (in s) as

$$T_e = \frac{m_{-1}}{m_0} \tag{2.5}$$

Wave energy flux is the rate of propagation of the energy density across a plane perpendicular to the propagation direction and extending from the surface down to the bottom. In the linear theory, wave energy flux is the product of the energy density times the group velocity.

In deep water, the available wave energy flux per meter of wave crest in Watt can be expressed as

$$J = \frac{\rho g^2}{64\pi} T_e H_s^2 \tag{2.6}$$

where g is the gravity acceleration, ρ the sea water density assumed to be $\rho = 1{,}025\ \mathrm{kg/m^3}$.

Energy matrices are one of the most useful means to synthetize the wave characteristics over a long time period in order to assess the productivity of a specific wave converter. All the data collected are divided into bins of equal period and height, and then the energy in each bin is computed using the previous formula.

The productivity of a device in a specific site can be determined multiplying the distribution of occurrences of the different sea states as a function of H_s and T_e and the power matrix of the device.

2.1.1.2 Mediterranean wave assessment

Global wave energy atlases lack the spatial resolution required to correctly describe the wave energy distribution in small and semi-enclosed basins as the Mediterranean Sea. In such regions, wind waves dominate as the fetch is limited. Wave height and period show substantial spatial variations due to wind variability and to the presence of complex topography. In these regions, specific high-resolution wave models represent the most important tool to assess wave-energy distribution.

Different wave energy climatologies have been realized for the Mediterranean Sea in the last years. They are based on data collected from the Italian Wave Buoys Network, operating since 1989 (e.g. [7]), or from model simulations (e.g. [8]).

The wave energy climatology for the entire Mediterranean Sea presented in [9] covers the period 2001–10. The wave model used is a parallel version of WAM (WAve Model) Cycle 4.5.3 at a horizontal resolution of 1/16° [10]. The model has been forced with six-hourly wind fields obtained from ECMWF operational analysis at 1/4° spatial resolution. The main integral wave parameters, wave height (H_s), mean wave period (T_m), significant wave period (T_e) and mean direction (θ_m) have been collected over the entire model domain, every 3 h.

An accurate validation of the wave parameters obtained from the model simulation has been performed against available buoys data, from the Italian Wave Buoy Network, managed by ISPRA; wave heights have also been compared against satellite radar altimeters data. Both the comparisons have shown very good statistical agreement. Maps of the available seasonal wave power flux per unit crest averaged over the entire 10 years of the Mediterranean simulation are shown in Figure 2.1.

In the Mediterranean basin, the winter season is the most energetic, followed by autumn, while summer is characterized by very low values almost everywhere. The most productive area is located in the Western Mediterranean, between the Balearic Islands and the western coast of Sardinia where an average energy flux of around 24 kW/m is reached in a large area, during winter.

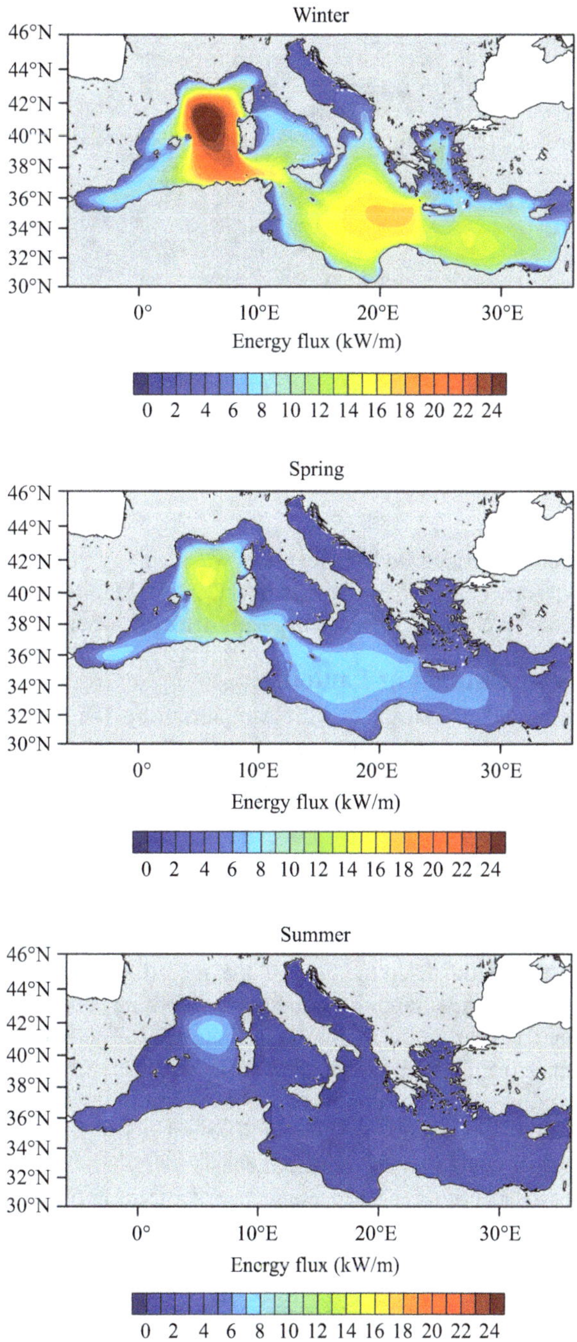

Figure 2.1 *Seasonal distribution of average power per unit crest in the Mediterranean Sea. Averages are calculated seasonally for the entire 10-year simulation*

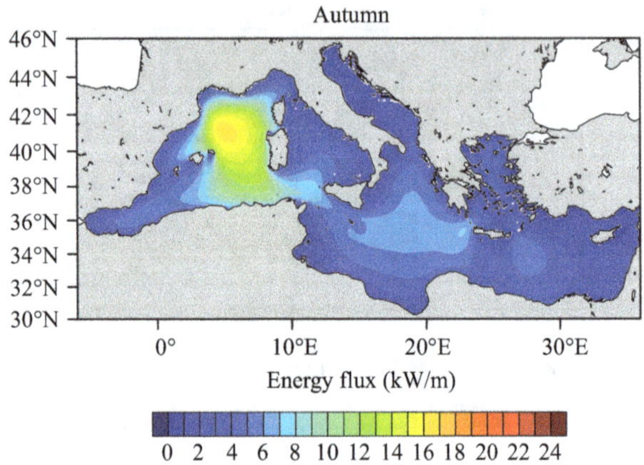

Figure 2.1 (Continued)

Other areas characterized by high levels of wave energy in most of the seasons are the north-western coast of Sicily and the Central Mediterranean. Values of energy lower than 9 kW/m are observed in summer everywhere.

2.1.1.3 Wave assessment at Pantelleria

Results from the model covering the entire Mediterranean Sea have shown that the Sicily Channel represents a promising area for wave-energy production. The area around the island of Pantelleria (located in the Sicily Channel) has been selected to perform a higher resolution study, which can be used in support for the installation of a specific device [11].

Data from the Mediterranean simulation performed using WAM have been used to laterally force a higher resolution version (1/120° corresponding to about 1 km) of the same model. The nested model grid is centred on the Island of Pantelleria and has the same discretization of the directional wave energy density function of the coarse model, and the same surface forcing. The map of the energy averaged over the 10-year model is shown in Figure 2.2. The main contribution to wave energy comes from waves propagating from North-Northwest. The sheltering effect of the island is evident in the decrease of energy in the Southern coast.

The distribution of wave energy among different sea states has been analysed in detail at a near-shore site on the northern coast.

The scatter plot in Figure 2.3 represents the distribution of the yearly average energy in terms of T_e and H_s, evaluated over the 10-year period. The contribution to the total energy given by individual sea states are lumped in 0.25 s wide intervals of T_e and 0.25 m wide intervals of H_s. Rectangles are coloured according to the total energy computed over the 3-h sea states extracted from the model. Lines of constant power are drawn on the scatter plot. The maximum amount of energy is available for significant wave heights between 2 and 3 m and for mean periods between 6 and 7 s.

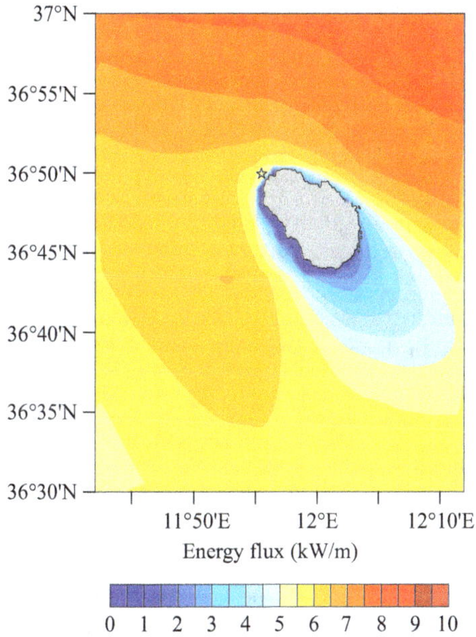

Figure 2.2 Wave energy flux per unit crest around Pantelleria, averaged over the entire high-resolution 10-year simulation

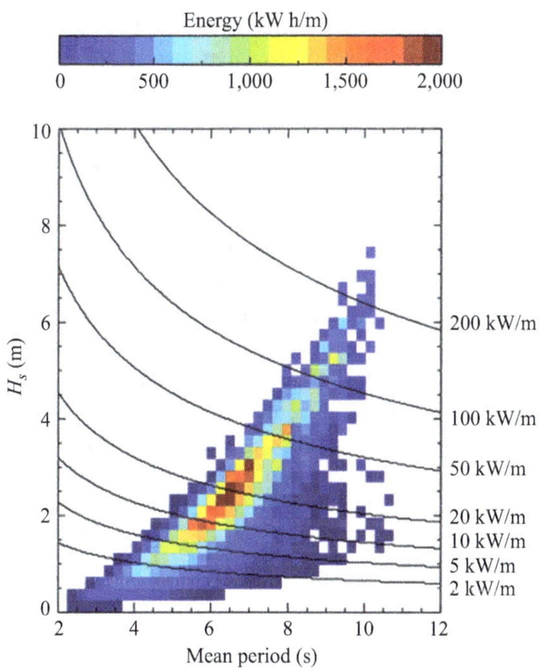

Figure 2.3 Distribution of wave energy among sea states computed over the entire simulation period at the site indicated by a star in the previous figure

2.1.1.4 Conclusions

The Mediterranean is a semi-enclosed sea and it is characterized by lower values of wave energy with respect to the major oceans. Nevertheless, even under these conditions, the conversion of wave energy can represent an economically profitable resource if ad hoc designed energy converters are developed. In these lower energy areas, in fact, devices of reduced size can be more suitable to extract energy. To implement efficient wave energy conversion solutions, accurate high-resolution projections of the resource are essential, as well as devices specifically designed for the harvesting of energy in the range typical of the area.

2.1.2 Wave measurements[2]

To ensure both safe shipping and the integrity of coastal structures such as breakwaters, ports and navigational aids, the observation and measurement of waves has always been an important consideration for seafarers, naval architects and coastal engineers. Initially based on visual observation only, which are limited in the number of distinguishable parameters and accuracy, advances in technology in the second half of the twentieth century have resulted in a range of sensors capable of recording and transmitting wave parameters with a considerably higher level of detail and within specified error boundaries.

Early visual observations on waves and sea states were collated from the eighteenth century onwards, based on passage reports from ships' logbooks, and these resulted in the production of sailing directions for coastal and ocean voyages, together with the so-called pilot charts showing seasonally and spatially discretized probability of occurrence of relevant wind and wave parameters. However, such reports were generally confined to standard shipping routes only, as often insufficient data was available for areas less frequented by ships to allow establishing generically applicable sea state information.

The development of improved wave-sensing capabilities for the shipping, engineering and meteorological science sectors experienced various step changes in recent decades. Visual observations were initially supplemented with rigidly mounted staff poles or lines marked at breakwaters to allow readings of the water level, but also of height and period from passing waves near the coastline. Although this method was later enhanced with the additional installation of electronic sensors such as capacitive or resistance probes to allow continuous logging of the waves without maintaining visual contact, the requirement to rigidly mount wave staff gauges on the seafloor or manmade structures with a pole extending to above the water surface generally allowed for shallow water installations only. This limitation was to some extend overcome with the introduction of submerged pressure sensors for wave measurements, where a clear relation of excursive pressure readings and the wave variable sea surface elevation could be observed. With an ability of pressure sensors to accurately register passing waves at considerably deeper water

[2]Written by Arne Vogler.

than what was previously possible with staff gauges, a reduced ability to capture shorter period waves with an increased deployment depth was also observed.

Significant advances in wave measurement technology were made in the 1960s with the use of accelerometers mounted in ships or floating buoys. Through the process of double integration, the accelerometer output can be converted to distance, thus providing continuous readings of vertical wave displacement. Limitations to this system were found in the inertia and size of the carrier platform, which prevented accurate readings to be taken for shorter wavelengths and periods as in wind waves. This limitation was to some extent overcome by installing accelerometers in smaller size buoys with a reduced mass, and thus providing an increased ability to follow the surface motion related to smaller short period waves. Another constraint in the case of floating buoys was due to the mooring connection, which limits the ability of a floating body to follow the sea surface freely. By deploying buoys free floating, which provides accurate wave data in a non-stationary way as the buoy moves around with the wind, wave and current drift, or more often by using a flexible and damped mooring configuration, this constraint was largely eliminated.

A shared shortcoming of the methods described in the above section was the non-ability to capture wave directionality or multimodality for situations with waves transiting the sensors from more than one direction. An initial solution to resolving directional wave data was found by installing sensors in an array configuration, often triangular, which allowed the tracing of individual waves through the array, and subsequent calculation of direction and wave speed.

By fitting out a buoy with multiple accelerometers along three perpendicularly aligned axes, two in the horizontal pane and the third one vertically aligned, a breakthrough was achieved in the 1980s, as it was now possible for the first time to gather wave directional data from a single instrument by combining horizontal and vertical motions [12]. Around a similar time, successful results on measuring wave spectra were also obtained from an acoustic sensor system utilizing the Doppler effect [13], and this has now become a standard feature of many acoustic Doppler current profilers (ADCPs) available on the market. Generally, the use of buoys and acoustic profilers to measure waves is confined to providing time series or time-averaged spectral data for single-spot locations. Additional modern wave-sensing technologies and approaches include the use of radar and satellite imagery to also allow continuous wave monitoring across wider spatial areas, and a more detailed introduction into the different sensor types is given in Section 2.1.2.2.

2.1.2.1 Wave characterization for energy applications

To provide annual energy production estimates of WEC farms, a sound understanding of the wave resource is imperative. Such understanding and estimates are an essential part of project design and development and are not only required to match and refine WEC technology against a local wave resource, but also to increase confidence in cash flow projections, thus supporting the investment case.

In addition to the requirement for detailed long-term wave resource assessment during the project development stages, ongoing wave forecasting and monitoring is also important throughout the construction and operational phases of WEC farms. Offshore installation works such as lifting of heavy loads during anchor handling or tasks that involve crew transfer from support vessels onto individual WECs, e.g. to undertake electrical connections between WECs and offshore hubs, can only be safely undertaken during moderate or calm sea states and thus careful weather windowing is required. During operations, the constant tuning of energy converters to the impacting sea states allows an increased energy production by operating the WECs close to their resonance frequencies. A further improvement to the sea state dependent tuning of energy converters is suggested through the implementation of a fast adaptive control approach, based on an individual wave-by-wave input that combines instantaneous measured wave data a short distance up-wave of a WEC with a short-term wave forecasting algorithm [14,15]. Wave monitoring during the operational phase at an energy site is further required to implement mitigation and survival strategies during wave conditions exceeding the operational design window of the deployed technology.

The requirement to use measured wave data for the development of wave resource assessments is defined in standard IEC 62600-101:215 published by the International Electrotechnical Commission [16]. Although the primary standard tools for wave resource assessments are numerical spectral wave models such as SWAN or DHI Mike 21 SW, typically applied for a 10-year hindcasting period, access to measured data for a number of locations within the model domain is considered essential for the calibration and validation of model outputs. The relevant IEC standard prescribes that '*all numerical modelling shall be validated using measured wave data. Whenever possible the numerical model output should be validated using data from one or more locations close to where wave energy converters might realistically be deployed*' [16, p. 16]. The same standard furthermore recommends measured field data be used covering a minimum of a consecutive year to avoid any seasonal bias in the model outputs.

In the context of wave resource assessments, the importance of using measured wave data for model calibration is visualized in Figure 2.4 for an area off the Isle of Lewis, north-west Scotland. Shown is the distribution of significant wave height of a prospective WEC development site, based on a Mike21 spectral wave model, together with the locations of two acoustic wave and current profilers (AWAC 1 and AWAC 2) deployed at a spacing of 600 m between sensors. A considerable difference in wave heights between both sensor locations was found with an energy hotspot at location AWAC 1. By calibrating against the sensor data, it was possible to replicate the hotspot in the numerical model, thus increasing model confidence and applicability for the wider area. This example also highlights the importance to deploy multiple sensors at a development site during the resource assessment process to capture a wider range of localized phenomena. To this end, it is helpful to set up an initial numerical model to identify potential anomalies prior to the deployment of sensors. Once initial results are obtained from such a model, appropriate data acquisition locations can be selected to confirm and help improve initial model outputs. In addition to the two AWACs shown in Figure 2.4, additional three-wave measurement buoys were deployed at the same time slightly

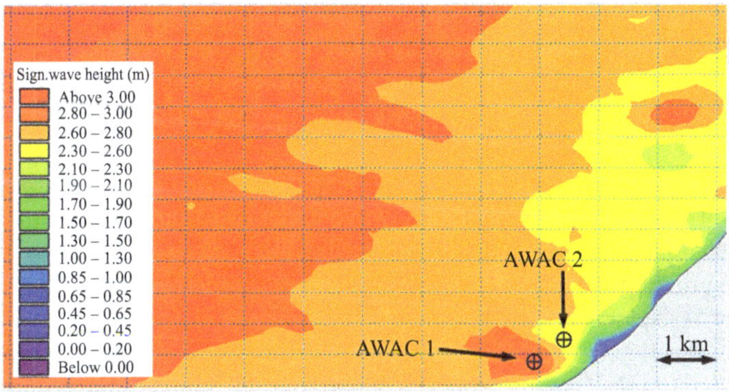

Figure 2.4 DHI Mike 21 SW model at 200 m mesh size. Close up of the AWAC location clearly confirms the energy hot spot 600 m SW of AWAC2 at AWAC1. An additional hotspot is visible 3.5 km to the NNE of the AWACs (adapted from [17])

further offshore to help inform boundary conditions for the localized model, and also as additional calibration points for a larger model domain.

Where the calibration of phase averaged wave models requires spectral wave data, i.e. data related to sea state parameters derived from the wave spectrum and generally with directional energy flux discretized across a range of frequencies, the analysis of statistical parameters from wave time series data is relevant to some other applications. Although standard wave models allow for extreme value analysis including assessment of a 50-year return periods to assess probabilities of occurrence of severe sea states, model outputs are limited to the assessment of wave geometry, such as differences between back and front wave steepness, maximum individual wave heights or detection of rogue waves across the model domain. If the interest is on assessment of individual wave-by-wave parameters, this is best facilitated through directional time series data from in situ sensors followed by statistical processing. However, the decomposition of spectral data following theoretical models such as Jonswap [18] or [19] wave spectra offers an alternative route for such analysis if no time series data is available.

An overview of different wave sensor types suitable for measurement of time series or spectral data at single spot locations or across wider domains is given in the following section.

2.1.2.2 Instrumentation for wave measurement

Where the earliest method to parameterize sea states relied on visual observations by trained, often ship based, observers, a range of technological solutions is now available to provide vastly improved detail and accuracy. Although it is widely accepted that a visual observer is well able to estimate the significant wave height (H_s), defined as the average of the highest third of waves during an observation period, this method provides poor results with view to detection of wave period, wave length, and to some extent direction and directional scatter. More advanced systems are discussed below, including solutions for time series and spectral analysis based on single spot in situ sensors or remote sensing tools providing cover across a wider domain.

2.1.2.3 Staff gauges

Although primarily used in laboratory wave tanks and flumes, wave staff gauges can also be found installed at harbour breakwaters or fixed offshore structures. Primarily utilizing conductive probes with an immersion-dependent variable capacity or resistance between probes, the sea-surface elevation can be monitored by logging the fluctuating electrical parameters as waves pass by. Some systems also utilize multiple optical sensors (e.g. photodiodes) aligned along the vertical axis, where the electrical output changes as the sensors are immersed by water. However, photodiodes suffer from built up of salt spray or biofouling, and are thus not widely used in this context. Some staff gauges also feature pressure sensors close to the surface, where the wave height is directly linked to the changing head of water above the sensor under the waves. When installed as a single sensor only, no directional data is obtained, and this can be achieved by installing multiple staff gauges at relatively close (wave length dependent) proximity in an array configuration, often in a triangular layout. The initial output of wave staff gauges is in time series format, and this can be converted to wave spectra by applying fast Fourier transform (FFT) processing. In addition to measuring waves, staff gauges are also suitable to monitor the tidal elevation at an installation site.

2.1.2.4 Laser range finder

Suitable for similar applications as staff poles, laser range finders or downwards facing radar systems are installed above the water surface with the sensor head pointing vertically towards the sea surface. Laser range finders record time series data by monitoring the distance between the transceiver and the sea surface, based on the time lapse between the emission of an electromagnetic pulse and return of the reflected signal. These systems are widely used in the offshore sector, e.g. installed on oil rigs, where the installation of multiple units on the platform perimeter also allows data processing for wave directionality. Although laser range finders are not in direct contact with the sea surface, they are still considered in situ sensors, due to the stationary nature of the installation and capability of providing data for single spots directly underneath the unit only. An example for a famous observation by a laser range finder is the New Year's wave, made on 1 January 1995 in the northern North Sea on the Draupner platform. During storm conditions with a significant wave height of approximately 12 m, an extraordinary wave measuring 25.6 m was recorded, and this constitutes the first time ever that a rogue wave was captured with an instrument [20].

2.1.2.5 Pressure sensors

Pressure sensors offer a relatively low-cost solution for in situ spot sensor wave measurements. Non-directional sea surface elevation data is recorded in displacement time series format from a fixed mounted sensor at a distance away from the sea surface. As with staff gauges or laser range finders, the combination of multiple pressure sensors allows for directional processing, and spectral information can be obtained by FFT processing. Although pressure sensors can be installed as standalone systems, they are also often found in combination with ADCPs to supplement or replace vertical beam information. Following [21] the pressure readings can be translated to sea surface elevation by

$$\eta(t) = \frac{p(t)}{\rho g(\cosh(k(d+z))/\cosh(kd))} \tag{2.7}$$

where p is the pressure, ρ is the water density, d is the water depth and z is the height of the sensor above the seabed. The wave number k is calculated from the wavelength λ as

$$k = \frac{2\pi}{\lambda} \tag{2.8}$$

2.1.2.6 Wave measurement buoys

Because of their ease of deployment and reliability of data returns, wave measurement buoys provide a popular means to obtain directional wave data in coastal and offshore environments. Applications range from the use as standalone wave measurement tools to larger systems integrated in multi-use sensor platforms. Wave measurement buoys range in size from light weight units of 0.4 m diameter only for short-term deployments up to self-contained larger offshore systems with a size of several meters. Where the deployment period of smaller units is limited by the on-board battery storage, larger systems are typically fitted out with integrated power-generation systems such as solar power or wind generators, thus allowing extended deployment periods. Some intermediate size units also combine a small number of solar panels with internal battery storage to extend durations of measurement campaigns.

Buoys are generally deployed quasi-stationary tethered to a mooring weight, where the movement of the buoy is confined to a circle around the ground weight, with the radius of movement limited by the length of the mooring cable. For smaller buoys in shallow or intermediate water depth, a flexible bungee cord is often integrated into the mooring to absorb shock loads, thus reducing the impact of the mooring constraint. An alternative option is the deployment in an unconstraint free-floating configuration, which allows measurement of longer period waves compared to a tethered arrangement.

Buoys are often fitted out with a radio link or satellite data transfer system to broadcast live data, and also write data on an internal memory card as backup.

The measurement principle of most wave buoys relies on the use of perpendicularly aligned accelerometers in x, y and z axis where the acceleration data is converted to vertical and horizontal displacement values by double integration of the signal. Data is cleaned up through on-board processing based on additional inputs from pitch and roll sensors, together with heading information from a compass. Depending on wave buoy type additional low and high pass filters are used to limit the data returns to realistic values, e.g. by considering the shortest possible waves that can be measured based on ratio between wave buoy diameter and wave length. In addition to the use of accelerometers in wave buoys, some modern types now entirely utilize GPS data to monitor buoy motion, with results of a comparable quality to previous systems.

Although data is initially processed as time series data, most systems also feature on-board processing to produce and broadcast fully directionally resolved spectral information. Where some systems broadcast and store the entire raw and processed data throughout the deployment period, other models delete the initial raw data and only transmit half hourly statistical and/or spectral sea state parameters.

Wave buoys have been used reliably in a wide range of conditions throughout the oceans in a wide range of conditions. An example of an extreme wave captured by a Datawell Waverider Mk3 buoy with a 0.7 m hull diameter and operated during storm conditions in 60 m depth near the north-west coast of Scotland is shown in Figure 2.5. The significant wave height at the time was 13.5 m with a maximum wave height of 29.41 m shown, thus (depending on the definition used) satisfying the criteria for a rogue wave. This data was carefully quality checked and was found clean. However, there is a possibility of wave buoys to overestimate wave heights in storm conditions, as breaking waves can result in fast accelerations that may falsely be misread as excessive wave heights. Thus a careful screening of the data is recommended in the case of any unusual observations.

The use of wave measurement buoys in the nearshore zone is generally not recommended, as the wave surge motion during higher waves can result in excessive mooring constraints on the buoy, thus resulting in spurious data. The same constraints also increase the risk of mooring failure with the resulting possible loss of a valuable instrument. Wave sensors better suited for nearshore deployment are discussed in the following two subsections.

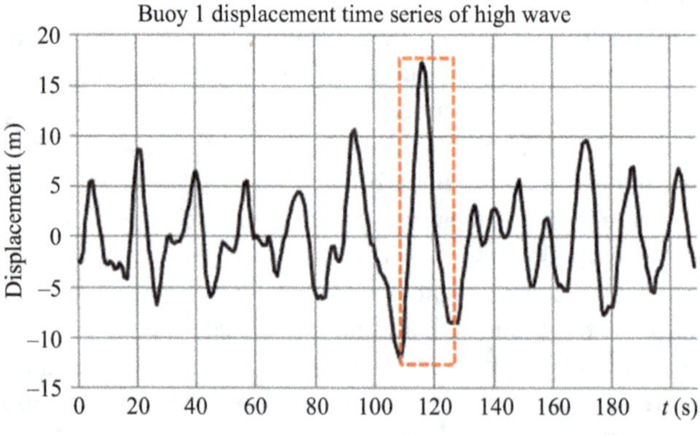

Figure 2.5 Vertical displacement of 29.41 m measured by a wave buoy off the Isle of Lewis on 04/02/2013, 15:52UTC (adapted from [17])

2.1.2.7 Acoustic Doppler current profilers

Initially developed for current measurements in flow environments, ADCPs are now also widely used in applications that require monitoring of wave data in time series and spectral format. The measurement principle is an advancement of standard sonar technology used by single beam echo sounders. These basic sonar systems establish the distance between a transceiver, typically mounted at the bottom of a ship, and seabed from the time lapse between emitting an acoustic pulse and reception of the reflected return signal. The intensity and pulse width of the reflected signal depends on the frequency of the emitted signal, and the composition of the seabed, where high frequencies and a flat, hard seafloor result in very sharp returns. Lower frequencies have better ground penetrating properties and thus allow for improved assessment of the bottom or sub-bottom geology. In ADCP applications, lower frequencies relate to longer range, and units suitable for wave measurements typically operate on frequencies of between 400 kHz and 1 MHz, with measurement ranges of up to 100–30 m, respectively.

To be able to measure currents in addition to the distance between transceiver and seabed ADCPs monitor, the frequency shift between outgoing and return signal of a transceiver slanted away from the vertical axis at user-defined time lapses. Return signals at these defined time lapses can be related to the distance between the data point and sensor, and this allows discretization of the water column in multiple bins to allow capture of a full depth profile.

The frequency shift between outgoing and return signal is related to the beam wise velocity by

$$v = \frac{F_{DOPPLER}}{F_{SOURCE}} \times \frac{c}{2} \tag{2.9}$$

where v is the velocity, F_{SOURCE} the transceiver frequency, $F_{DOPPLER}$ the change of frequency between emitted and returned signal and c the speed of sound in the working environment. ADCPs for current measurements generally combine three or four slanted beams at equal angles around the sensor head, and in the case of wave measurements this is also combined with an additional beam into the vertical direction. An example of the beam geometry is shown in Figure 2.6 for a Nortek AWAC, which features three beams slanted at 25° from the vertical axis and a vertically aligned beam to measure the distance between the sea surface and sensor head. Considering V_H and V_B as horizontal and beam-wise velocities, respectively, and a transceiver angle of 25° for this example, the beam velocities can be transformed to horizontal by applying simple geometry as in

$$V_H = \frac{V_B}{\cos(90 - 25)^\circ} \tag{2.10}$$

The beam aligned horizontal velocities can then be transformed to earth coordinates or x, y, z direction by applying sensor specific transformation matrices. Alternatively, proprietary software is available from the suppliers, where the output formats and directional conventions can be preset by the user.

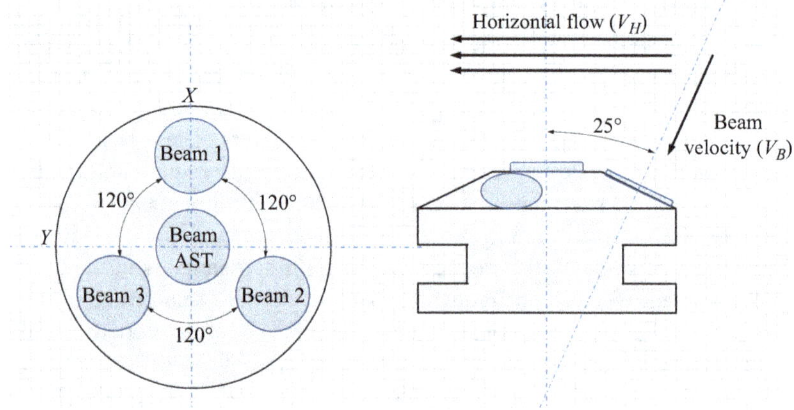

Figure 2.6 ADCP Beam angle setup example: Nortek AWAC

Early ADCPs that combined wave and current measurements were based on the PUV principle, where an in-built pressure sensor (*P*) was combined with orbital wave velocities under the surface in *U* and *V* direction. This method is very reliable and is still being used by some devices in cases when the acoustic data returns are compromised, e.g. by aeration in the water column during storm conditions. However, by combining the slanted beams with a dedicated vertical beam operating in acoustic surface tracking (AST) mode, a better resolution can be achieved, thus it is possible to capture shorter wavelengths compared with the PUV method. For wave energy applications, the interest is generally on longer period swell waves, and thus the PUV method offers a valid alternative to the AST mode.

Where wave measurement buoys are typically the preferred option for deeper water, ADCPs have been deployed successfully in the nearshore shallow water environment at high energy sites. To achieve good data returns, it is important to devise a suitable ADCP carrier platform, and this includes consideration of station keeping and movement of the sensor head for gravity-based sensor frames that utilize a gimbal mount. For sites that are exposed to breaking waves and high surge action, a fixed mounted solution might offer advantages, and a frame solution that was deployed at 13 m water depth at one of the highest energy wave sites in Europe is shown in Figure 2.7. This frame was deployed for a 2-year period at what was then the world's largest fully consented wave power site off the Isle of Lewis in Scotland with close to 100% successful data capture. The frame was rigidly mounted to the bedrock and was adequately aligned to the horizontal pane using divers. Data of this standalone system was recovered and batteries replaced at quarterly intervals with help from a dive team during short recovery missions. Figure 2.8 shows some examples of ADCPs widely used for wave energy measurements at prospective development sites.

2.1.2.8 X-band radar

By combining a standard navigational X-band radar antenna with an AD-converter and by processing the backscatter data for sea state analysis, it is possible to extract

Figure 2.7 ADCP rigidly mounted to sea bed at high energy site

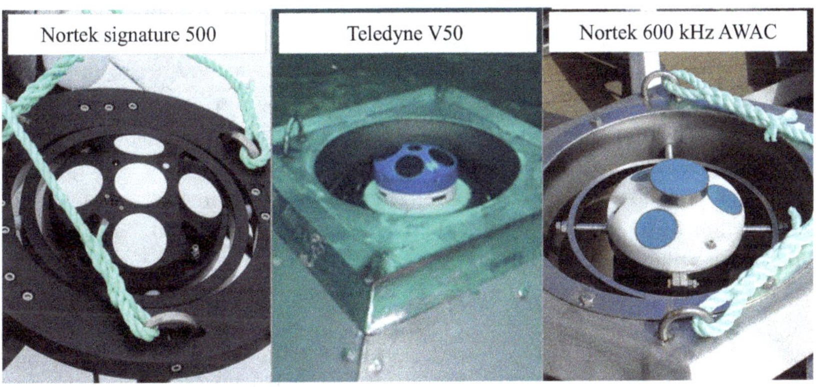

Figure 2.8 Examples for ADCPs suitable for wave measurements

directional spectral wave information. Although parameters such as wave length (λ), wave speed (c) and wave period (T) can be extracted relatively easily based on time variable geometry of the returned backscatter, extraction of wave height requires further processing. This is normally achieved by modifying the signal to noise (SNR) ratio with two calibration parameters ($a+b$), which can be identified by calibrating the data returns against in situ sensors during system commissioning. Once data is received from both in situ sensor and radar backscatter, the wave height is related to the sea state specific SNR by

$$H_s = a + b\sqrt{SNR} \tag{2.11}$$

Waves that are perpendicularly approaching the radar antenna result in a stronger backscatter signal, and thus the wave direction is established by identifying the direction with the strongest return signal. The depth in intermediate and shallow water across the scanned domain can be calculated from the wave dispersion relationship:

$$\lambda = \frac{gT^2}{2\pi} \tanh\left(\frac{2\pi d}{\lambda}\right) \tag{2.12}$$

Wave speed and current velocity from the radar backscatter signal can be extracted by applying the wave celerity equation:

$$c = \frac{gT}{2\pi} \tanh\left(\frac{2\pi d}{\lambda}\right) \tag{2.13}$$

A number of commercial solutions applying the above approach are readily available, and research efforts are well underway to also extract time series wave-by-wave information from the radar signal by continuously relating the backscatter data to individual wave heights. A particular challenge in this respect is related to wave troughs that might not be captured by the radar signal due to shadowing from the neighbouring waves. A solution to this problem of wave shadowing has been established with good comparison against in situ sensors by Greenwood *et al.* [22] who have supplemented the backscatter with an artificial second-order Stokes wave to fill in the missing troughs.

X-band radars are powerful tools for fixed installations where spatial coverage across a spatial domain of up to 12 km radius is required, with larger domains resulting in reduced resolution. Extraction of near real-time time series data appears feasible, but at this stage, it is still a field of progressing research. An example of X-band radar imagery for a system operating from a lighthouse on the Isle of Lewis in parallel with two wave buoys at 3 km radius is shown in Figure 2.9, where the approaching waves from the left side (i.e. west) can clearly be identified. Figure 2.10 shows the system from the seaward side, with one of the wave buoys visible in the foreground and the radar installation at the lighthouse in the back.

2.1.2.9 HF-radar

Spectral directional wave data and sea surface current information on larger scales can be obtained from HF-radar systems, based on the sea surface roughness. These systems require typically two antennas positioned at different angles to the incoming wave direction. HF-radar can be operated across domains of more than 100 km range and is an interesting tool for oceanography applications. The relatively coarse resolution makes these systems less interesting for wave power applications, although useful information for coastal or global wave models is provided.

2.1.2.10 Satellites

There are two different types of satellite-based wave measurement systems available. Basic systems feature an altimetry system and are able to provide the non-directional significant wave height along the satellite transect. This is based on the time lapse between an emitted and returned signal, as reflected from the ocean surface. This method is improved by applying synthetic aperture radar, which

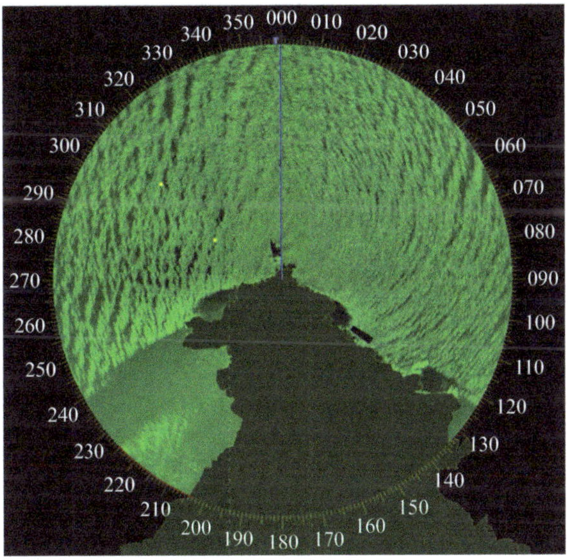

Figure 2.9 Example X-band radar image from the Butt of Lewis, Scotland

Figure 2.10 Combined deployment of two wave measurement buoys (one visible in the foreground) with an X-band radar system (on the lighthouse) in 2016

returns fully directional spectral information through the combination of multiple sidewards looking beams. Satellite technology for wave measurements offers a good solution for monitoring of very remote areas, such as the southern ocean, but with a resolution that is inferior to spot sensors or coastal X-band radar systems. As satellites are non-stationary, no continuous data is returned for individual spots, but wave data is returned at regularly occurring intervals for areas transected by the orbiting system. Due to the intermittency and reduced resolution of wave measurement satellites, applications in the wave energy context are limited.

2.2 Wave energy devices

2.2.1 The ISWEC plant at Pantelleria[3]

2.2.1.1 Introduction

As presented in Section 2.4, most of WECs have been developed with reference to Oceanic wave climate. Section 2.1.1 shows that the Mediterranean Sea has different characteristics with respect to the Ocean, in particular the Mediterranean Sea presents a lower energetic potential with a wave climate composed of faster (higher frequency) waves with shorter wavelengths. Given this resource assessment, in 2006 at the Politecnico di Torino, a research on inertial WECs started. As a matter of fact, it can be demonstrated that WECs exploiting inertial working principles such gyroscopes or pendulums have productivity proportional to the wave frequency [23,24] thus providing the basis for efficient wave exploitation in the Mediterranean Sea.

The use of gyroscopes, as means of obtaining inertia to provide the torque reactions to the power conversion system, has been implemented in other devices: the most prominent examples are the Duck developed at the University of Edinburgh in Scotland, the floating device of the Tottori University in Japan and the Oceantec WEC, developed by Oceantec in the Basque Country, Spain. The Duck was proposed by Salter in 1974 and it is one of the first WECs studied in Europe [25]. In one of his versions, the duck uses a 'gyro canister' technology to transform the floater rocking motion into electrical power, without any mean of reaction on the external connecting body 'spine'. The device conceived by Tottori University uses gyroscopic moment produced by the rotation of large flywheels and the swing of a float excited by wave motion [26]). The Oceantec WEC is similar to the Inertial Sea Wave Energy Converter (ISWEC), being an offshore floating structure, whose capture principle is based on a relative inertial movement produced by a gyroscopic system [27]. All the devices previously illustrated possess similar conversion technology when compared to the ISWEC, but were developed for use in the Oceans, thus with less favourable conditions with respect to their inertial nature.

2.2.1.2 The ISWEC

ISWEC uses gyroscopes to create an internal inertial reaction able to harvest wave power without exposing mechanical parts to the harsh oceanic environment. As shown in Figures 2.11 and 2.12, the hull must be aligned to the incoming wave on the axes X, thus provoking the pitch motion δ around axes Y. This motion combines with the flywheel speed $\dot{\varphi}$ and produces a torque on the precessing axis ε that is finally used to drive the PTO and produce electric energy. The gyroscope also acts as a gearbox, increasing the small pitching angle (~10°) to a bigger precession angle (~60°).

The ISWEC externally presents as a monolithic float, and it is intrinsically more reliable with respect to other WECs, because the gyroscopic system and all the moving parts are sealed inside the hull and protected from the harsh marine environment.

[3]Written by Giovanni Bracco and Giuliana Mattiazzo.

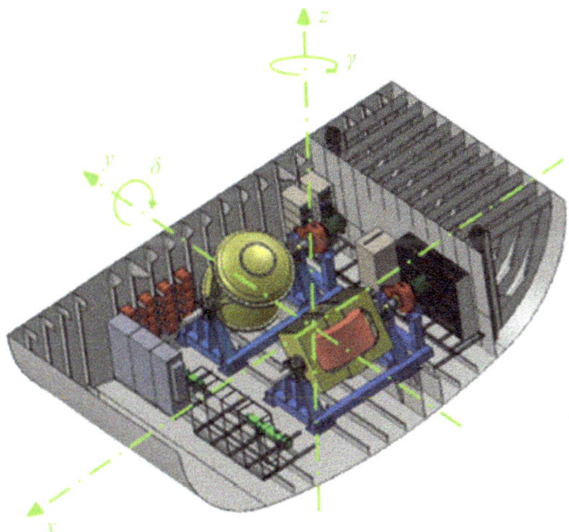

Figure 2.11 ISWEC Pantelleria prototype drawing

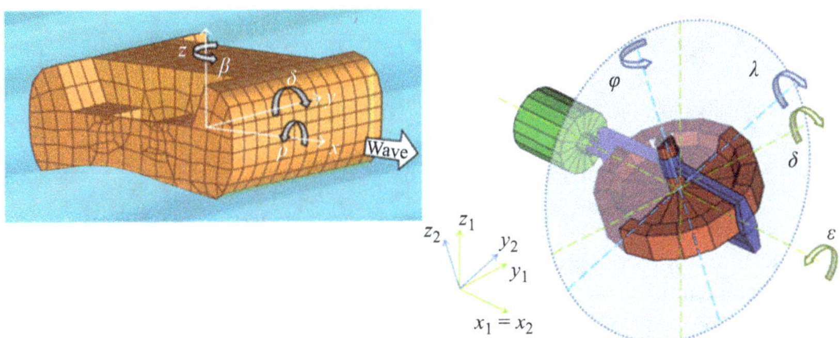

Figure 2.12 The hull and gyroscopic system reference frames

Furthermore, by switching off the gyroscopic system, ISWEC behaves as a bulky dead body, thus increasing its chances of survival in the case of storm.

- *The gyroscope mathematical model*
  By following a Newtonian approach, the equations of the angular momentum of the gyroscope with respect to the Global System of Axes described in Figure 2.12 are derived:

$$M_{E,x1} = \overline{M}_E \cdot \vec{i}_1 = \overline{M}_E \cdot \vec{i}_2 = I_g\ddot{\varepsilon} - (J - I)\dot{\delta}^2 \sin\varepsilon\cos\varepsilon + J\dot{\varphi}\dot{\delta}\cos\varepsilon \tag{2.14}$$

$$\begin{aligned} M_{E,y1} &= \overline{M}_E \cdot \vec{j}_1 = \overline{M}_E \cdot \left(\vec{j}_2 \cos\varepsilon - \vec{k}_2 \sin\varepsilon\right) \\ &= \left(I\cos^2\varepsilon + J\sin^2\varepsilon\right)\ddot{\delta} + 2(J - I)\dot{\delta}\dot{\varepsilon}\sin\varepsilon\cos\varepsilon \\ &\quad - J\dot{\varphi}\dot{\varepsilon}\cos\varepsilon - J\ddot{\varphi}\sin\varepsilon \end{aligned} \tag{2.15}$$

$$\begin{aligned} M_{E,z1} &= \overline{M}_E \cdot \vec{k}_1 = \overline{M}_E \cdot \left(\vec{j}_2 \sin\varepsilon + \vec{k}_2 \cos\varepsilon\right) \\ &= -(J-I)\ddot{\delta}\sin\varepsilon\cos\varepsilon + \left[(J-2I)\sin^2\varepsilon - J\cos^2\varepsilon\right]\dot{\delta}\dot{\varepsilon} \\ &\quad - J\dot{\varphi}\dot{\varepsilon}\sin\varepsilon + J\ddot{\varphi}\cos\varepsilon \end{aligned} \tag{2.16}$$

where $\overline{M}_E$ is the angular momentum of the flywheel; $\vec{i}_1, \vec{j}_1, \vec{k}_1$ are the versors for x_1, y_1, z_1; $\vec{i}_2, \vec{j}_2, \vec{k}_2$ are the versors for x_2, y_2, z_2; J is the gyro moment of inertia around its axis; I is the gyro moment of inertia around the precession axis of the PTO; $\dot{\varphi}$ is the flywheel angular velocity; ε is the gyroscope angle of precession; δ is the hull pitching angle.

- *Equations simplification*
  The following expressions can be simplified in order to get more handy and meaningful equations. In particular, the flywheel moments of inertia I and J have the same order of magnitude and in most cases are quite similar; the gyroscope is run at constant speed and the system oscillates around the position ε=0. Equations (2.14), (2.15) and (2.16) can be then simplified as follows:

$$M_{E,x1} \cong I\ddot{\varepsilon} + J\dot{\varphi}\dot{\delta}\cos\varepsilon \xrightarrow{linearization} I\ddot{\varepsilon} + J\dot{\varphi}\dot{\delta} \tag{2.17}$$

$$M_{E,y1} \cong I\ddot{\delta} - J\dot{\varphi}\dot{\varepsilon}\cos\varepsilon \xrightarrow{linearization} I\ddot{\delta} - J\dot{\varphi}\dot{\varepsilon} \tag{2.18}$$

$$M_{E,z1} \cong -J\dot{\varphi}\dot{\varepsilon}\sin\varepsilon \xrightarrow{linearization} -J\dot{\varphi}\dot{\varepsilon}\varepsilon \tag{2.19}$$

The first equation represents the torque unloaded from the gyro to the hull on the floater pitch degrees of freedom (DOF), the second equation expresses the torque equilibrium around the axis of the PTO while the third represents the yaw gyro reaction on the floater.

If the PTO is controlled to act as a spring damper system, (2.18) can be rewritten as follows:

$$I\ddot{\varepsilon} + c\dot{\varepsilon} + k\varepsilon = J\dot{\varphi}\dot{\delta} \tag{2.20}$$

Such control is called proportional–derivative (PD), but more commonly in literature, it is known as complex conjugate control [28], and it provides an action on the system proportional to the position (k) and the velocity (c). In the case of pure sinusoidal wave, it is possible to evaluate the harvested power in a closed formulation ($\omega_n = k/I$):

$$P_d = \frac{c}{2}\omega^2\varepsilon_0^2 = \frac{c}{2}\frac{(I\dot{\varphi}j\omega^2\delta_0)}{J^2\left(\omega_n^2 - \omega^2\right)^2 + c^2\omega^2} \xrightarrow{resonance} P_d = \frac{1}{2}J\dot{\varphi}\omega^2\delta_0\varepsilon_0 \tag{2.21}$$

The typical wave period in the oceans is comprised between 6 and 10 s, meaning the order of magnitude of ω is 1 rad/s. The order of magnitude of d is of some degrees and therefore 10^{-1} rad, whereas ε if the device is controlled as described, has order of 1 rad. That means that if there is the need to build a device with 100 kW of rated power, the needed angular momentum will be of the order of 10^6 kg m^2 rad/s. Being P_d proportional to ω^2, the extractable power from a given gyroscope is increased if the device is deployed in seas with shorter wavelength and therefore shorter periods, like the Mediterranean Sea.

- *Hydrodynamics*
  As described in Section 2.4, the hydrodynamic behaviour of the floater can described by the equation of Cummins:

$$(M + A_\infty)\ddot{X} + \int_0^t h_r(t - t')\dot{X}\,dt' + \beta\dot{\delta}|\dot{\delta}| + KX = F_w + F_m + F_c \qquad (2.22)$$

  where M is the inertia matrix, K is the constant hydrostatic restoring matrix due to buoyancy and gravity, A_∞ is the constant positive definite added inertia matrix for infinite frequency, h_r is the impulse response functions of the radiation forces, F_m is the force related to the moorings, F_w is the force related to the wave, F_c is the force related to the control torque on the floater, in this case, the gyroscopic reaction corresponding to $M_{E,x1}$. β is the Morison viscous correction (experimentally identified).

- *Control*
  The PD control previously described is an oscillating control strategy where a stiffness term keeps the flywheel axis oscillating around the vertical position, while a damping component enables the power extraction. The stiffness also provides a mean to resonate the gyroscopic system with respect to the incoming waves.

  Other types of control are δREF (DREF in figure), linear quadratic regulator (LQR), linear quadratic integral (LQI), SpeedREF (SREF in Figure 2.13) [29,30].

  δREF is a position feedback control that opportunely scales the measured pitch values and gives the value as a set point for the internal precession axis. In this way, a synchronized oscillating control is realized, mimicking the complex-conjugate control working in resonance. A problem of this strategy relies in choosing the optimal gains for the same set point tracking, like in the complex-conjugate, where the gains are sea states dependent. LQR is another oscillating

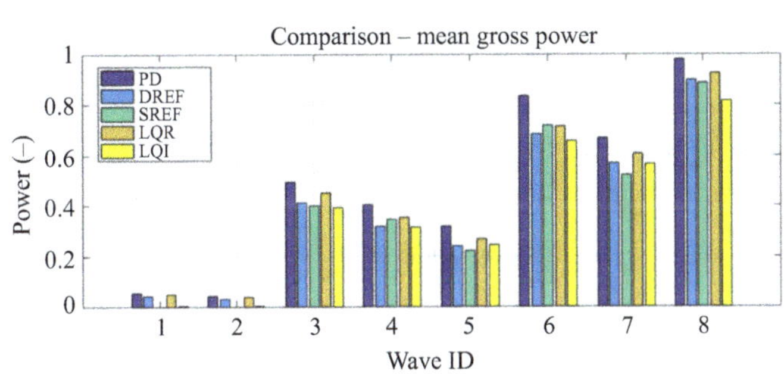

Figure 2.13 A dimensional mean gross absorbed power. Sea states describing Pantelleria site

control strategy that developed and adopted the optimal control theoretical background. The cost function gives the opportunity to directly weight the product of the PTO speed and the PTO torque, thus directly optimizing the power adsorption.

For what concerns the continuous rotation control, a model-less (SREF) and a model-based control (LQI) is presented. The model-less control takes as reference a speed-computed starting from the single identified wave period or even a statistical number given by forecasts as the energy period. The model-based continuous rotation control uses instead a formulation of the LQI control, giving always as reference the same as above but also enabling the cross weight of the PTO speed and torque, as for the LQR case.

With the perspective of optimizing the system via maximizing the gross power absorption, the most productive architecture is the PD control. This technologically robust architecture reaches the best performances within the numerical environment. There are three control parameters: the damping coefficient, c, the stiffness coefficient, k, and the flywheel speed, $\dot{\varphi}$. The main limitation of the application of the PD control lays in the need to know the current sea state to determine the coefficients. The control parameters are determined on a model-based approach with an a priori offline optimization as a function of the wave climate, generally described by the significant wave height, H_s, and the energy period, T_e. To choose the correct control parameters, the incoming sea state must be known, by measurements or forecasts. In the full-scale operation at Pantelleria, the wave gauge data was not instantly available, and therefore the parameters were set according to the wave climate forecast. A 5-day, hour-based, weather forecast is transmitted every morning by the ENEA-UTMEA laboratory to the ISWEC. A recent development allows to determine the sea state from the heave motion of the hull, thus freeing the device from external information [31].

2.2.1.3 ISWEC evolution: from model to full-scale device

According to European Marine Energy Centre (EMEC) guidelines based on the Equimar Project [32], a WEC development from concept to commercial viability should follow the five stages described in Table 2.1, each stage corresponding to a technology readiness level (TRL). ISWEC is currently at Stage 4 and its TRL is 7.

Table 2.1 Development phase overview [32]

Stage	TRL	Scale	Objective	Description
1	1–3	1:25–1:100	Concept validation	Prove the basic concept from wave flume test in small scale
2	4	1:10–1:25	Design validation	Subsystem testing at intermediate scale
3	5–6	1:3–1:10	Testing operational scaled models at sea	Extended test period to ensure all seaways included
4	7–8	1:2–1:1	Full-scale prototype tested at sea	Full scatter diagram for initial evaluation, continuous thereafter
5	9	1:1	Economic validation	

During its development, knowledge on the device was gained and the main result was the increase of the ISWEC efficiency, represented in Figure 2.14 with the relative capture width parameter, the ratio between the harvested power and the net incident power. The first trials with a preliminary hydrodynamic shape granted around 10% of relative capture width, whereas the final shape allowed more than 50% of energy harvesting in regular waves.

2.2.1.4 The ISWEC prototype plant at Pantelleria

After almost a decade of long development track, the launch of the full-scale ISWEC prototype at Pantelleria took place in August 2015. In order to reduce the time to sea, the prototype presented some limitations with respect to the final version: there was no connection to the network (therefore, the machine was in stand-alone operation with breaking resistors to dissipate power), and the system had only one of the two gyroscopic units. The decision to proceed with subsequent steps of experimentation was born also with the aim of proceeding to intermediate checks of the design choices been made.

- *Positioning and mooring system*

  The mooring system final design is related to the installation point 800 m from the shore. Being Pantelleria in the Channel of Sicily, between Italy and Tunisia and oriented along the Mistral direction, the wave resource is directional and the energy is mostly concentrated from north-west direction (Figure 2.15).

  Even if the resource is directional, since the device is directional, the selected mooring configuration possesses weathervaning capabilities and self-aligning

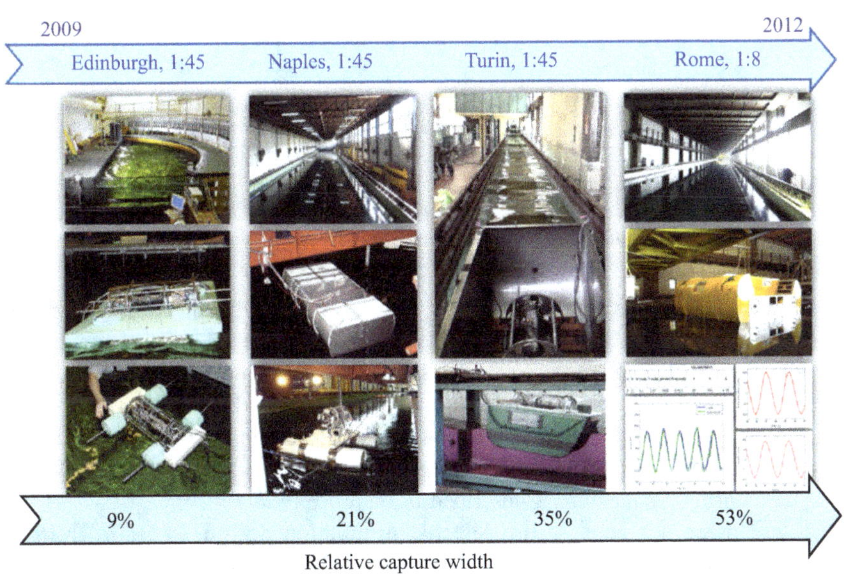

Figure 2.14 Evolution of ISWEC concept from 2009 to 2012

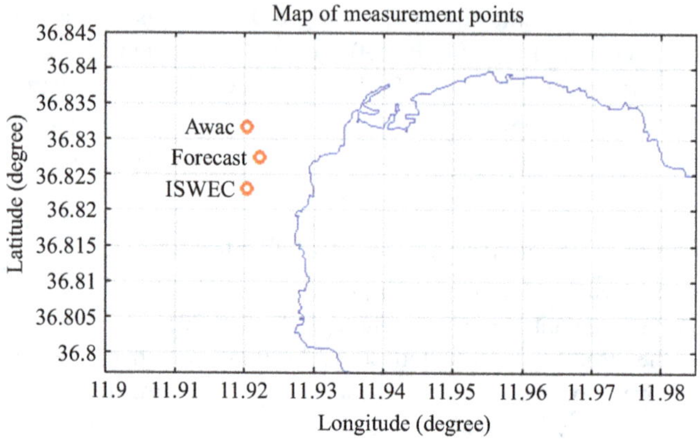

Figure 2.15 Positioning of the ISWEC and Wave gauge AWAC and ENEA grid point with respect to the Pantelleria north-west coast

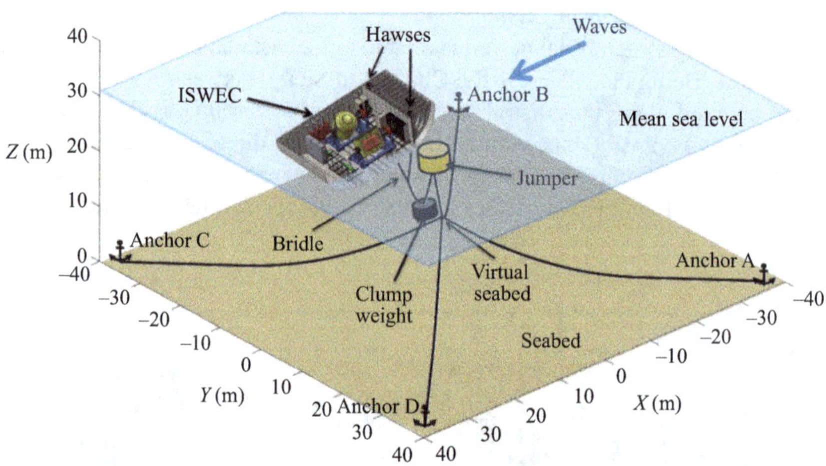

Figure 2.16 The mooring system: Pantelleria architecture scheme

properties. The mooring system must also have little influence on the dynamic behaviour of the device. The most important degree of freedom of the ISWEC not to be influenced by the mooring is the pitch, the main responsible for power absorption. A mass-jumper configuration is chosen for this purpose, able to give little load on moderate mooring extension (corresponding to middle power wave climate) and high restraint capability on extreme mooring extension/wave climate. The mooring architecture is described in Figure 2.16.

The mean sea level in the selected installation point is at 31 m from the seabed. Four 4 t Hall type anchors are installed in an axisymmetric configuration. From these, four catenary lines are linked to a single point, the virtual seabed, 10 m above the seabed. From this point, a chain line is connected to the jumper, a body with a positive buoyancy of 12.9 t. After this, another chain line reaches the clump weight, a negative buoyancy of 8 t. From this body, a line starts towards the hull and, to enhance weathervaning, it splits in a bridle shape and the two lines reach the hull in proximity of the two hawseholes. From a dynamical point of view, the mooring force is thus a non-linear term, mainly function of the floater surge position, as shown in Figure 2.17.

- *Experimental results*

  The test campaign lasted from September to December 2015. The ISWEC was towed and anchored to the mooring line, as shown in Figure 2.18. During the

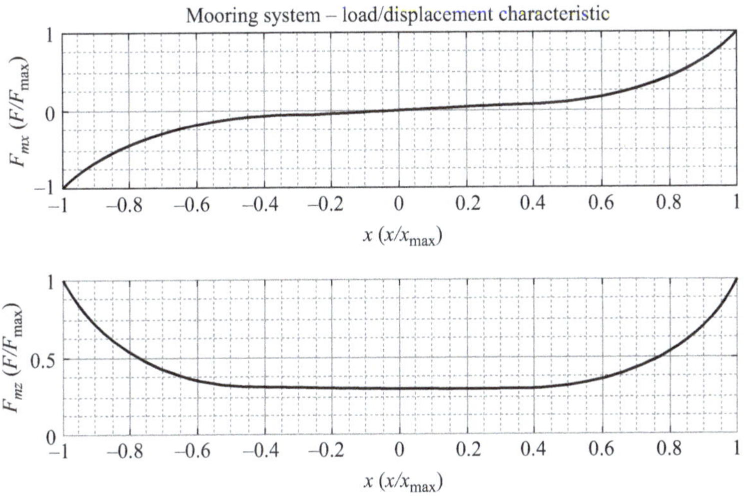

Figure 2.17 ISWEC prototype, bow mooring system load vs displacement characteristic

Figure 2.18 Transportation of ISWEC to the installation site in Pantelleria. August 2015 (up) and ISWEC deployed at sea (down)

campaign, the ISWEC overcame most of the wave climates typical of the Pantelleria site and also extreme events, as shown in Figure 2.19.

During the test campaign several control logics were tested and the device proved to be able to produce energy according to the mathematical model developed in the experimental track. Hereafter a model validation is shown, demonstrating that the mathematical model of the hydrodynamics is able to replicate the experimental behaviour with acceptable error (Table 2.2).

2.2.1.5 Conclusions

At the time of writing the prototype is in the process of completing the upgrade procedure and received the addition of the second gyro group. At the Pantelleria site, the installation of the electrical infrastructure (submarine cable and connection to the grid) was performed. In 2019, it is expected a new launch, and the second test

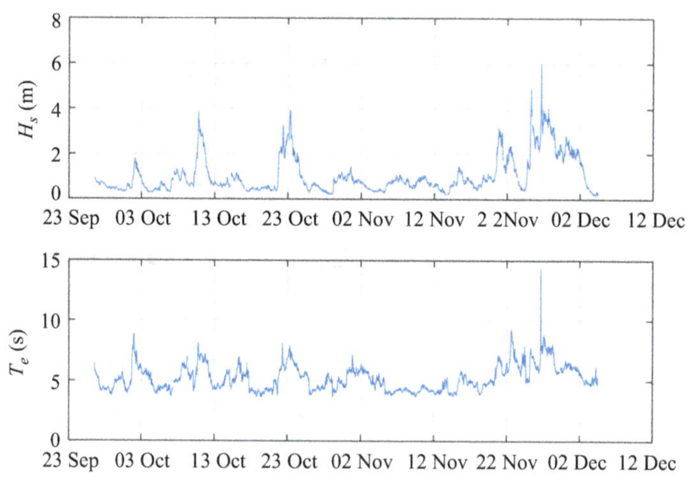

Figure 2.19 Wave climate during autumn 2015 in Pantelleria

Table 2.2 ISWEC hydrodynamics model validation

Test number	**Roll RMS (degree)**	**Angle pitch RMS (degree)**			**Heave RMS (m)**		
–	–	EXP	SIM	Relative Errors (%)	EXP	SIM	Relative Errors (%)
–	–	–	3DOF	3DOF	–	–	–
1	2.64	5.24	5.50	5	0.57	0.54	−5
2	2.51	4.73	4.97	5	0.5	0.54	8
3	2.29	5.15	5.43	5	0.47	0.49	4
4	1.87	5	5.14	3	0.49	0.51	4
5	2.16	5.42	5.5	1	0.46	0.50	8
6	2.18	5.49	5.28	−4	0.45	0.46	2

phase will begin. Novel control logics are under development and in particular model predictive control seems a good alternative to the PD granting superior power absorption and independence from the knowledge of the sea state.

In parallel, an omnidirectional version of the device is under study to simplify the mooring layout and ease power absorption in multidirectional wave climate.

Finally, is the gyroscope suitable only for closed seas? Maybe not if some help can be given to the frequency response of the floater. As a matter of fact, by inserting a pitching resonating system, such as a U-tank, the frequency response of the system can change and adapt to longer waves [33]. A U-tank is commonly used as a ship stabilizer, being able to kill the roll resonance. However the U-tank creates a new peak in the response at longer periods, thus enhancing the gyro response and increasing the power absorption for longer waves. The novel device was already tank tested at small scale, and it is under optimization to tackle high power sea states, such as Chile and Australia.

2.2.2 Harbour breakwaters for wave energy conversion[4]

2.2.2.1 Breakwater-integrated WECs

One of the main tasks and challenges for coastal engineers involved in the wave energy field is the design of hybrid structures, i.e. WECs integrated into traditional harbour protections. This idea is not very recent and concerns the development of innovative coastal structures with the benefit of the energy production. Obviously, hydraulic performances and stability should be similar to the one provided by traditional breakwaters.

In the general classification of WEC technologies, i.e. oscillating water column (OWC), OverTopping Devices (OTDs) and wave-activated bodies (WABs), only the first two typologies can be considered ‘embedded’. In fact, for WAB systems, the breakwater represents a support structure, without a real integration: in this sense, such a device can be considered just onshore WECs.

If compared to offshore standalone WECs, the integration of WECs into coastal structures can provide some advantages:

- Easier access for the installation and maintenance.
- Lower construction cost, thanks to the sharing of the infrastructures and their construction costs (for the function of coastal defence).
- The construction techniques to be adopted are similar to those used in standard maritime and coastal engineering structures.

Moreover, a distinguishing feature of WECs embedded in coastal defence structures is their low marginal environmental impact. Finally, their characteristics of adaptability and multifunctionality (e.g. the possibility to build smart coastal defence equipped with a wave energy-based desalination plant, as shown by Contestabile and Vicinanza [34] make them even more interesting).

[4]Written by Diego Vicinanza and Pasquale Contestabile.

2.2.2.2 Site-specific design challenges

The identification of a site-specific breakwater-integrated WEC design is a very complex process, involving:

- geometric optimization according to hydrodynamic performance assessment;
- geometric optimization according to wave loading reduction, structural behaviours and fatigue analysis;
- preliminary environmental constrains;
- inclusion of power absorption, electrical cabling and secondary hydro-electrical equipment;
- design of PTO system, including power smoothing and conditioning equipment, control system and strategy;
- transportation operation and maintenance feasibility analysis and optimization;
- environmental impact assessment, also through a life cycle assessment (i.e. from raw material extraction to decommissioning).

It is well known that the executive WEC design is an outcome of improving a number of intermediate project ideas, satisfying a set of technical and non-technical requirements and subject to constraints [35]. The technical design of a breakwater-integrated WEC could be synthesized in Figure 2.20.

2.2.2.3 Breakwater-integrated oscillating water column

The concept of an OWC integrated into a breakwater was conceived in the 1940s, and over the last 30 years, several new pilot devices were constructed [36]. The

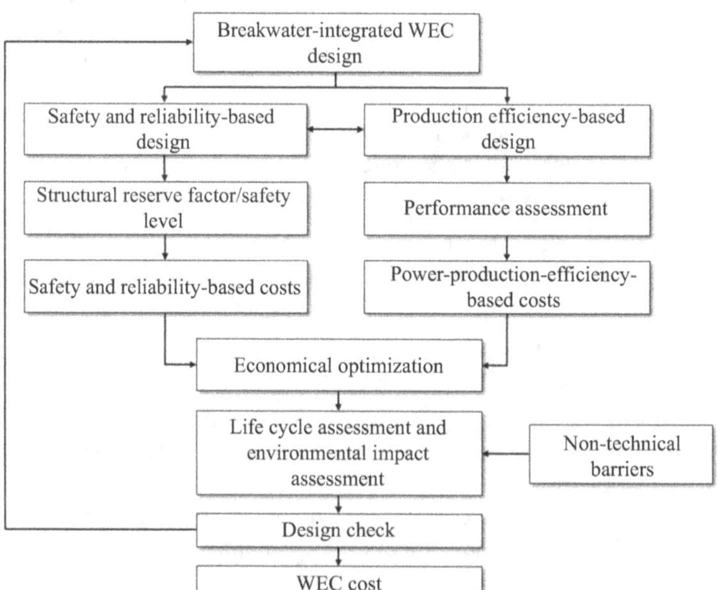

Figure 2.20 Site-specific breakwater-integrated WEC design flowchart

technology consists of a chamber, submerged in the seawater and opened below the water surface, in which the wave motion alternately compresses and decompresses the trapped air (Figure 2.21). This bidirectional flow is driven towards a turbine (or a set of air turbines) coupled to a generator. The most recent turbine technology used for OWC devices is able to self-rectify the turbine spin, allowing for energy to be continuously generated regardless of the direction of airflow.

Since the inertia of the air turbine is very low, easy start/stop of the PTO system is ensured. Moreover, the seawater does not impinge directly on the turbines, allowing several corrosion or fatigue problems to be avoided.

Although many studies on the hydraulic performances of the OWC-Caisson Breakwaters are available in literature, there is no an accepted and well-defined empirical method to design this innovative structure. The earliest studies on the wave-OWC interaction can be attributed to [37–40], and [41,42] conducted analysis on an OWC with an innovative shape that includes an additional vertical duct. The device is termed U-OWC type and shows the advantages to obtain a natural resonance without any device for phase control, amplifying the overall performance (Boccotti *et al.*, 2007a) and better performances with swells and large wind waves [43]. Other novel theoretical models describing water column oscillations were proposed by Strati *et al.* [44], Malara *et al.* [45] and Malara *et al.* [46]. The design method in terms of structure sizing were studied by several authors (e.g. [47–52]. Allsop *et al.* [53] and Viviano *et al.* [54]) presented the results of a unique large-scale experiments carried out on the OWC device integrated into vertical breakwater. Ashlin *et al.* [55] conducted an experimental campaign to study the nature and magnitude of the dynamic pressures and forces exerted on a circular bottom profile OWC-integrated

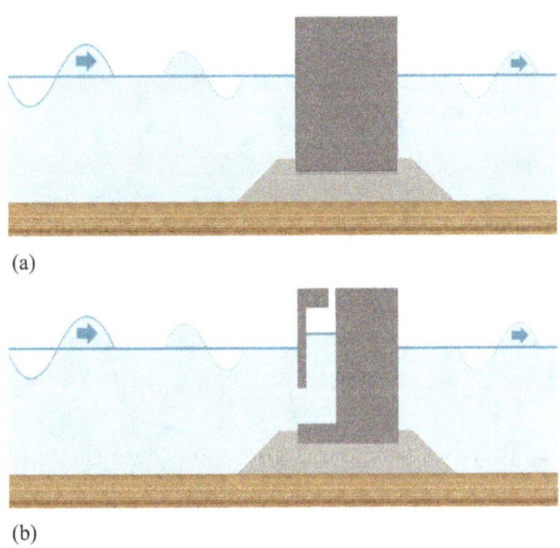

Figure 2.21 (a) Scheme of a traditional caisson breakwater; (b) working principle of a breakwater-integrated oscillating water column

caisson breakwater exposed to the action of regular waves. The authors indicated that the horizontal wave force acting on the OWC structure is more than 2.5 times the vertical wave force acting on the structure. Although no direct comparison has been presented, a curved front wall may also provide an auspicious alternative to the vertical structure, due to the increase of the energy extraction rate for small-height waves [56]. It is worth noting that the structural design of the device may not necessarily agree with the optimum design for efficiency or constructability [57].

2.2.2.4 Breakwater-integrated overtopping device

A breakwater-integrated OTD utilizes a frontal sloping plate designed with the aim of capturing the overtopping waves (Figure 2.22). This impermeable front ramp leads the overtopping waves into one or more reservoirs located immediately behind it. The reservoirs, placed at a level higher than the sea water, act as a storage basin. Hence, using the difference in water levels between the reservoir and the mean sea water level, energy is extracted via low head turbines. An OTD-integrated breakwater shall retain construction operations similar to those of a traditional rubble mound structure, except for the installation of the sloping ramp (in particular the ramp below the mean water level). In fact, the idea of a ramp-reservoir system (also defined 'stilling wave basin') was first introduced by Phoenicians and ancient Romans for their breakwaters. Several authors [58–62] investigate this non-conventional cross section, demonstrating the better hydraulic (in terms of overtopping flow rate at the rear of the structure or wave reflection) and structural (in terms of stability of the crown wall) efficiency than those of conventional structures. One of the most important concepts of OTD-integrated breakwater is the SSG – seawave slot-cone generator [63–65]. The device consists of an OTD-device with three (or more) frontal reservoirs located one on the top of each other [63] identified two different behaviours of wave loading acting on the SSG: surging waves on the front sloping plate and impact waves on the vertical recessed wall

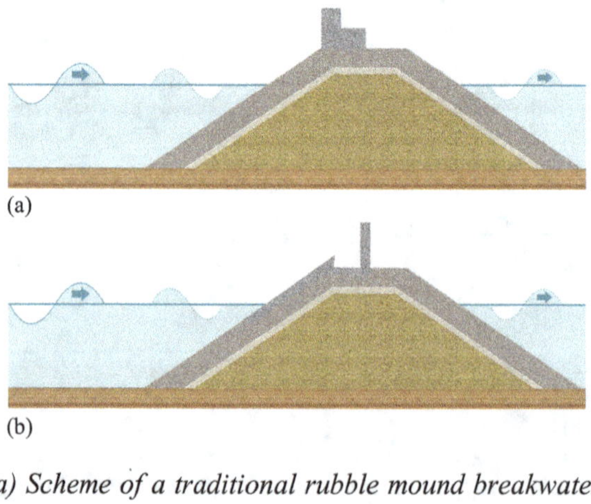

Figure 2.22 (a) Scheme of a traditional rubble mound breakwater; (b) scheme of a breakwater-integrated overtopping device

with a typically '*church-roof*' shape [66] of the pressure time series. Margheritini *et al.* [64] showed that the measured wave pressure on the three ramps does not vary substantially from one plate to the other, confirming a quasi-static loading time history of the pressure signals. Vicinanza and Frigaard [65] showed that maximum pressures on front plates are quasi-static or pulsating loads generated by non-breaking waves ($p \sim \rho_w g H_s$), where ρ_w is the water density, g is the acceleration due to the gravity and H_s is the significant wave height. Loading on the rear vertical wall exhibits a small impact pressure ($p \sim 2-3\rho_w g H_s$), which is damped by the preceding foamy mass. The authors compared pressure on the three ramps with the prediction design formulas used for caisson breakwaters with the sloping top [67], showing that the formula underestimates the forces between 20% and 50%. A pilot project to be built in Svåheia in Norway [7] was designed, but the prototype was never built. However, the results obtained for the SSG was the starting point for a device with only one ramp and frontal reservoir, denominated Overtopping BReakwater for Energy Conversion (OBREC). The results of the structural response of the OBREC device in terms of wave pressure distribution and wave forces are provided in [68,69] and [35]. Hydraulic performance, in terms of wave reflection, and overtopping in the reservoir and at the rear of the structure are shown by Vicinanza *et al.* [70], Vicinanza *et al.* [69] and Iuppa *et al.* [71]. The latter found also that the two configurations investigated in the laboratory tests, i.e. with a curved and a flat ramp, shown a mean difference of around 20% in the overtopping discharge in favour of the linear (constant) slope configuration.

2.2.2.5 Full-scale devices

The first OWC-caisson breakwater capable of converting wave energy into electricity was installed in Sakata Port, Japan in 1992 [72–76]. This caisson had a machine room with turbines coupled to a 60 kW-generator. Due to the limitation of research funds, the turbine was set to be smaller than the ideal turbine for that wave climate (of about 20 kW/m). Tests conducted from October 1990 to March 1991 show an average electric power of the system was 13.25 kW [74], which corresponds to 0.663 kW/m.

The LIMPET 500 (*Land Installed Marine Powered Energy Transformer – 500 kW*) was installed between 1998 and 2000 on the rugged coastline of the Scotland isle of Islay, in Scotland [77]. It can be considered a full-scale OWC integrated in a coastal structure at the embryonic stage.

The pilot consists of a collector that contains three chambers inclined at an angle of 40° to the horizontal. Contrary to the OWC with vertical walls, the inclined chamber offers an easier path for water ingress and egress, resulting in a less turbulence and hence a lower energy loss [77,78]. Tests carried out by McStay [79] confirmed the better hydrodynamic performance on the use of sloping wall compared with traditional vertical OWC device.

With the same aim to prove the technical viability of an OWC device for remote locations (i.e. isolated small islands) the device 'Pico' was installed, from the name of the hosting island, in the Central Group of the Portuguese Azores [80]. The pilot consists in a fixed concrete chamber, with a frontal sloping wall having a width of 12 m and a height of 15 m above sea water level. A horizontal-axis Wells turbine coupled with a generator with a rated power of 400 kW was installed. The pilot was estimated to deliver around 0.5 GW h/year, having an overall (estimated) efficiency of 35% [81]. Although commissioned in 1999, the Pico plant was reactivated and connected to the local power grid only in 2005 [80] due to several technical problems and damages of the electrical equipment. Unfortunately, plant operation remained problematic and infrequent mainly due to the persistence of serious technical limitations of the turbo-generation apparatus, limiting it to power production in the range of 20–70 kW. However, the 16 years of power production demonstrate the survivability of the OWC technology under very extreme natural condition [82,83].

The first European OWC embedded into a breakwater was installed in 2008 at the port of Mutriku in Spain [84]. It consists of 16 chambers, each connected to a turbogenerator, with a total capacity of 296 kW. Despite the expectation, the average yearly output for the period that runs from 1 January 2014 to 7 October 2016 was only 41% of initial estimate [85]. A new kind of OWC embedded into a caisson breakwater, named Resonant WEC 3 (REWEC3), is installed at the harbour of Civitavecchia in Italy [86]. The entire plant in Italy includes 136 chambers for a total length of 578 m [87]. A preliminary study [88] indicates that the expected annual average electrical power delivered by the REWEC3 in Civitavecchia Port (assuming each chamber equipped with Wells turbine of 20 kW) can be 2,800 MW h/year, with an average power per unit length of 0.55 kW/m.

Relatively to OTD-integrated breakwater, the first and unique demonstration pilot is the full-scale of the OBREC, installed at the end of 2015 in the San Vincenzo breakwater, Naples port (Italy). The pilot was built substituting a part of the seaward armour layer for a total area of ~75 m^2 [89]. A specially designed concrete structure, consisting of a sloping impermeable front ramp, leads the overtopping waves into a reservoir located immediately behind it. In the machine room, a set of low-head turbines convert potential energy (water stored in the reservoirs) into kinetic energy and then into electrical energy by mean generators. The device is composed by two adjacent structures, which are similar to each other, except for the freeboard of the ramp.

2.2.2.6 Conclusions

WECs embedded in breakwaters could represent the first commercial wave energy technologies in a next future, providing a new form of a clean energy source in the core of coastal cities. In fact, thanks to the low marginal costs and low marginal environmental impact than building traditionally conceived breakwaters, conversion technologies integrated in harbour breakwaters are revealing an unexpected value adding respect on the whole wave power sector. Especially for harbours serving small islands or remote areas, these devices are a promising solution to replace diesel generated electricity. Obviously, high-conversion performance and competitive costs are a fundamental prerequisite. Therefore, such technologies need a design-optimization procedure to work under specific wave climates. Actually, it is difficult to correctly estimate their costs and performance as well as the rest of the installation. Most of current economic studies are oversimplified, which consequently creates uncertainty and difference on investors. Due to these reasons, a large part of the entire sector of wave energy relies heavily on research on full-scale prototypes. Hence, the challenge for the next round of pre-commercial phase has been established, and a great deal of work ahead of us but the way forward has been mapped out.

2.2.3 *Overview of the development of a pivoting buoy system*[5]

As many research results have pointed out, wave energy shows several attractive features with respect to alternative renewable sources. However, several issues still need investigations and the development of WECs is an active field of research. A review of some general features of wave energy conversion may be found in [90].

Many different concepts and solutions have been developed. The present chapter reports some results related to a specific type of WEC, based on a wave actuated pivoting body, mainly intended for onshore installation.

Reported results refer to the research activities performed at the University of Naples during the development of a WEC system, in cooperation with Umbra Group Spa, under several research programs. The development of the system required both numerical modelling and experimental verification of the system working principle and of the energy-conversion performance. The related research activities are reviewed in the following sections.

2.2.3.1 System configuration and operation

In the field of onshore installed WEC systems, several concepts may be applied for wave energy exploitation. Among the main system typologies, we may recall:

OWC systems, which use the forced motion of air generated by the water oscillation, alternatively raised and lowered by wave action in a confined chamber, to actuate a suitably designed turbine (generally a Wells type turbine); a review of such kind of devices may be found in [83];

OTDs, which use the water, captured in a reservoir above sea level due to wave movements, to feed low-head turbines; examples are reported in [91,92];

Wave actuated body systems, which use the wave induced motion of bodies to feed several possible kinds of mechanically actuated PTO systems.

Following the experience attained at the University of Naples, the study reported here refers essentially to wave actuated bodies. Different implementations of such concept are possible and several systems have already been developed. Several other studies may also be found in literature. For example, an implementation of such working principle is represented by a heaving point absorber; together with other general aspects of WEC devices, examples of such kind of devices are

[5]Written by Domenico P. Coiro and G. Troise.

described in [93] or [94]. Numerical and experimental studies about the same topic may also be found in [95] or [96]. Based on the same general-working principle, but also using different motion DOFs, pivoting and hinged devices have been developed (see, for example [97–99]).

In the case of on-shore installations, the relative motion of a floating body with respect to fixed coastal structures may be exploited for energy conversion.

The system considered in this study comprises a floating body of suitable shape (the pivoting buoy), free to rotate around a hinge under the actions exerted by the incoming waves. The floating body, which converts the energy transported by waves in mechanical energy available for further transformations, acts as the prime mover of the system and is connected to its rotation axis by means of oscillating support arms. The hinge, on turn, is linked to an external support structure, generally fixed on-shore, for example, in installations on a breakwater or on other suitable port structures. Near-shore or off-shore deployment may also be considered by using, for example, a floating external support structure. Previous results and recent development about the considered system are also reported by Coiro *et al.* in [100,101].

Although, in principle, several PTO devices may be integrated within the system, in the actual development, a configuration has been defined and optimized for the coupling with a specific linear electromechanical conversion device, provided by Umbra Group, based on the recirculating ballscrew technology. In such device, an oscillating piston transfers mechanical power to a rotating electrical generator through a suitable conversion mechanism. Two other hinges, placed on the oscillating arms and on the fixed structure, provide the connection to the ballscrew-based electromechanical generator, in such a way to allow PTO rotation according to the buoy oscillations.

A schematic representation of the system is shown in Figure 2.23.

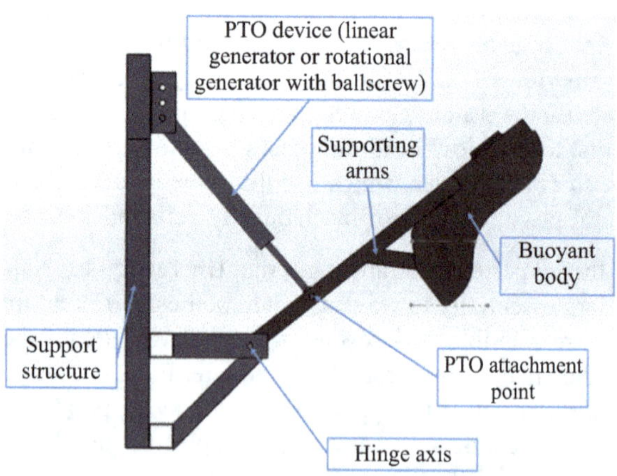

Figure 2.23 Schematic of the system

2.2.3.2 System modelling

Several approaches may be used in modelling wave body interaction, with different levels of accuracy and computational cost. Potential flow theory analyses have been primarily used in this study, although generally less accurate with respect to other methods, due to their reduced resource requirements. Anyway, for some specific investigations, the more accurate and resource demanding computational fluid dynamics (CFD) approach may be used.

The dynamic behaviour of the system may be described by the use of a simple equivalent 1 DOF equation that represents the equilibrium of moments around the hinge axis:

$$I\ddot{\theta} = M_{ext} + M_{rad} + M_{hs} + M_{PTO} \tag{2.23}$$

where I is the rotational inertia around the hinge axis, accounting for the hydrodynamic added mass also; M_{ext} is the external moment due to waves excitation forces (diffractive and Froude–Krylov actions); M_{rad} is a term accounting for the radiation action, which should be corrected for viscous contribution; M_{hs} is a term related to hydrostatic stiffness; M_{PTO} is the moment due to the point pivoted PTO device; θ is the rotation angle of the body around its pivoting hinge.

Equation (2.23) represents a general expression of the dynamic equilibrium of the system, accounting for the different terms contributing to the definition of system response, due to wave–body interaction and to PTO device actions.

As stated before, the problem of modelling wave-body interaction may be studied using different approaches of varying complexity. A well-established technique is based on the use of potential flow theory to describe the wave motion field and its action on floating bodies. In such an approach, generally, a superposition technique is used, subdividing the overall problem in a set of subproblems related to different physical effects (some details may be found, for example, in [102]). The potential of the unperturbed wave field is summed with the perturbation field due to the presence of the body and to its motion. The wave field is described as the superposition of the solution to the following subproblems: unperturbed flow (which determines the hydrostatic and Froude–Krylov forces), wave diffraction (in which the body is considered fixed under the incoming wave), radiation (in which the body is oscillating in still water). The proper boundary conditions are associated to each subproblems. The boundary element method, based on a discrete representation of the domain boundaries, is generally used to solve the differential problems for each contribution to wave–body interaction, calculating the hydrodynamic coefficients as a function of the incident wave frequency. The hydrodynamic actions are represented in terms of the Froude–Krylov and diffraction coefficients together with the matrices of added mass and radiation damping. Another contribution to the overall action is related to the hydrostatic stiffness, which accounts for the reaction of the system to perturbing actions moving the body away from its floating equilibrium position and is directly related to weight and buoyancy.

Finally, in the definition of system response, a fundamental role is played by the action of the PTO system, which exerts an action on the floating body during wave-induced oscillation. Generally, a control law is assigned in order to define the action required to the PTO system as a function of chosen operating parameters. For the specific system under consideration, the force acting on the PTO piston is related by the control law to the instantaneous oscillation velocity of the device. In this case, moreover, a simple linear relation between piston force, F_{pist}, and speed, V_{pist}, is assumed, with an assigned force–speed gain, K, as in the following equation:

$$F_{pist} = KV_{pist} \tag{2.24}$$

It is remarkable that the choice of the control strategy has a significant effect on performance and choosing other types of control force law may improve conversion efficiency. The assumed linear control law has the advantage of simplicity, both for simulation purposes and for control implementation. More detailed models of the PTO may consider also inertial and frictional properties of the power-conversion device.

It has to be noted that, due to the kinematic of the system, a non-linearity in system response arises, particularly in the case of a high amplitude oscillation response.

After the definition of the required hydrodynamic coefficients, in the potential flow approach, and the definition of the control law, time domain simulations may be performed for the cases of interest in order to determine the applied forces and output power.

2.2.3.3 Small-scale model test of the pivoting system

The working principle and possible configuration issues of the considered system have been experimentally studied in a controlled environment on a small-scale model (with a width of 1 m, in a 1:5-scale ratio with respect to a future planned installation). Figure 2.24 reports a picture of the model installed in the towing/wave tank of the University of Naples 'Federico II'.

Using the wave maker installed in the test plant, different sea states may be generated over a range of frequencies (~0.35 to ~1.2 Hz) and amplitudes (with a maximum of about 0.6 m, variable with frequency).

Figure 2.24 PIVOT WEC small-scale test set-up in the test basin of the University of Naples

The scaled model reproduces the main kinematic arrangement of the full-scale system, with a pivoting floating body oscillating around a hinge and a linear-power absorbing device. The action of the PTO device, which has a major role in the definition of the behaviour of the system, was replaced with the force exerted by a controlled pneumatic piston, for most of the performed tests. A PID control system, with adjustable gain, was used to apply the desired force control law. The value of the force–speed gain K in (2.24) was varied throughout the tests, allowing the observation of the effects of such parameter on system response. A feedback control was used in order to ensure the requested linear relation between piston force and oscillation speed. The control feedback, aimed to simulate a purely passive damping device, is based on the measurement of piston oscillation velocity (obtained by derivation of an LVDT potentiometer displacement) and of piston force (obtained by means of a load cell).

The power output of the system was estimated indirectly from the measurements of force and velocity, while water surface elevation was measured using ultrasonic probes, thus allowing an assessment of the mechanical energy conversion performance of the system. Observations have been collected and compared with numerical predictions.

The geometry of the kinematic system may have also an important effect on overall behaviour. The use of different geometrical configurations has been investigated by developing two different models (as indicated in in Figure 2.25), with similar components but different arrangements. In particular, the position of the main oscillation hinge has been taken into account. In one of the models, the main hinge was placed underwater with the oscillating support arms inclined with respect to the horizontal, whereas in another tested model, the position of the buoy oscillation hinge was set above the still water level, with the supporting arms almost horizontal. The PTO piston is mounted differently in the two configurations, preserving the relative angle between arms and piston (almost 90° at rest).

Although the horizontal arms solution may have the advantage of taking the hinge out of water, possibly reducing some structural difficulties, the performed tests have shown a better behaviour of the inclined arms configuration, with sensibly higher

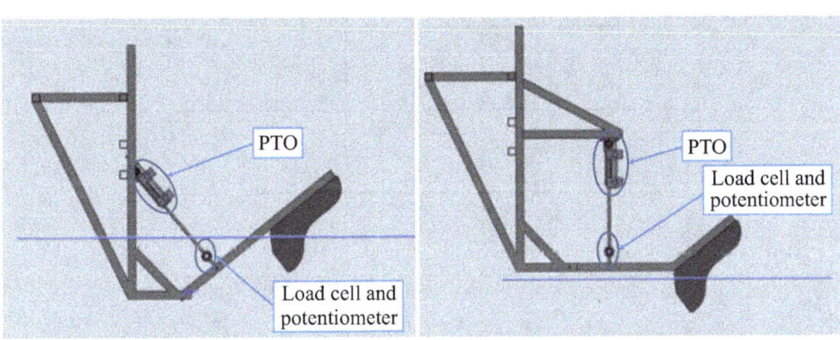

Figure 2.25 PIVOT test model, two tested configurations: inclined arms (left), horizontal arms (right)

power output. A possible explanation of these results may be related to the possibility of exploiting both the horizontal and vertical components of the wave force in the case of inclined arms, whereas for the horizontal arms case, only the vertical component may generate a significant moment around the oscillation hinge.

Such results may be, however, dependent on the specific size, configuration and operating conditions and nevertheless the inclined arms configuration has been adopted for further system development.

The main results obtained from the test campaign are represented in terms of power output and system oscillation response amplitude. An acquisition system recorded the main observed parameters (oscillation velocity, piston force, wave elevation) and the derived instantaneous output power estimation. Time histories of the instantaneous power were logged under different wave conditions (amplitude and frequency) and, for each test, different values of the proportionality coefficient K were used.

To illustrate the effect of force control law, Figure 2.26 shows the measured average power output, for the inclined arms configuration, as a function of force–speed gain and different wave frequencies. A peak in power output is observed for frequency in the proximity of the system natural frequency (about 0.8 Hz for the scaled model). An optimal piston force–speed gain of approximately 600 N/(m/s) was found, with a power peak value of about 8 W, for a 5 cm amplitude incident wave. It has to be noted that the peak frequency is generally dependent on the force–speed gain value.

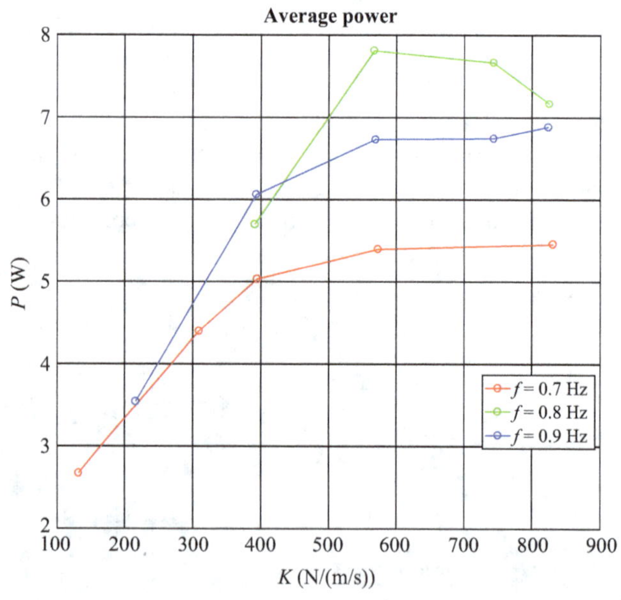

Figure 2.26 Experimental mean power as a function of piston force–speed gain for different incident wave frequency (5 cm wave amplitude)

Numerical–experimental comparison

Using a potential flow model, a prediction of the behaviour of the system may be obtained. A time domain simulation of the oscillation response and power production can be set up, after the hydrodynamic coefficients have been obtained by means of BEM analysis.

Comparisons of the simulated and measured time histories are reported in Figures 2.27 and 2.28, under the peak response test conditions, for piston velocity and power output.

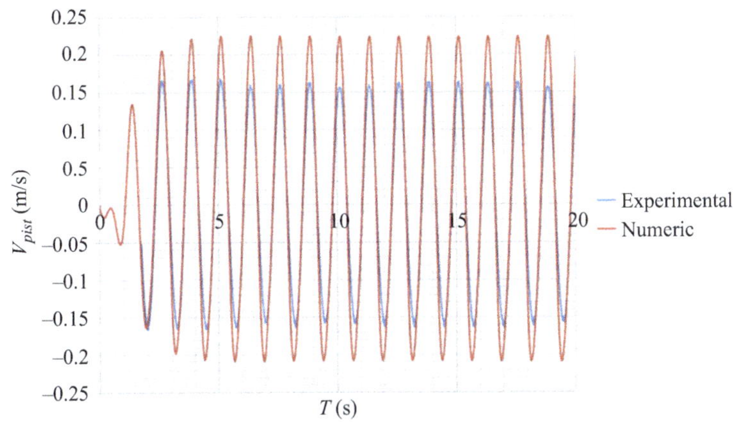

Figure 2.27 Numerical (potential theory) – experimental comparison. Piston velocity time histories (wave amplitude $A = 5$ cm, $f = 0.8$ Hz, $K = 570$ N/(m/s))

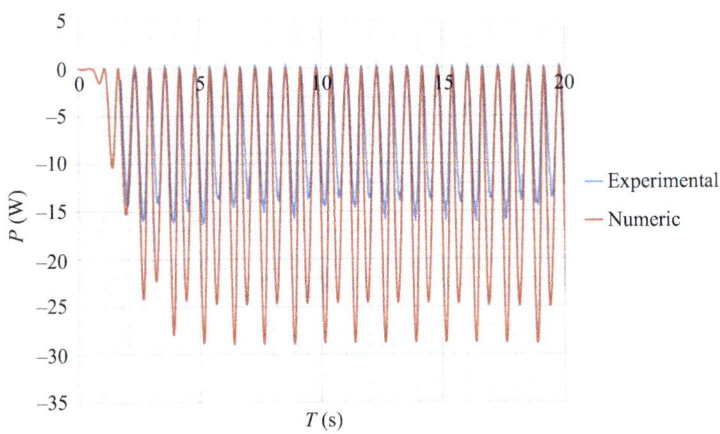

Figure 2.28 Numerical (potential theory) – experimental comparison. Instantaneous power time histories ($A = 0.05$ m, $f = 0.8$ Hz, $K = 570$ N/(m/s))

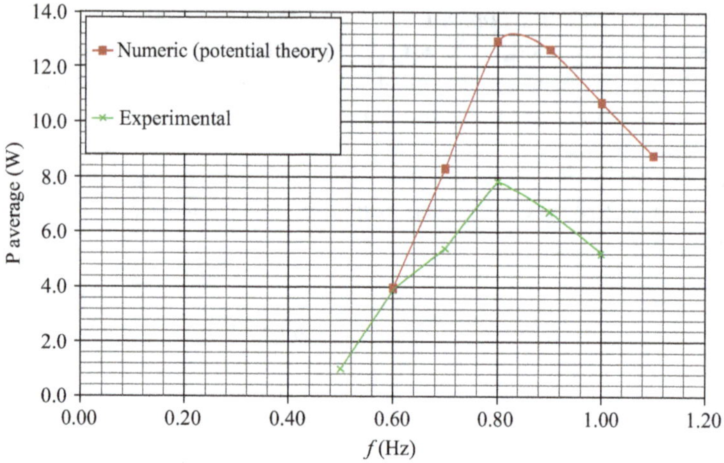

Figure 2.29 Numerical (potential theory) – experimental comparison, average power ($A = 5$ cm, $K \cong 500 \div 570$ N s/m in experimental tests, $K = 570$ N s/m in simulations)

Figure 2.29 reports a numerical–experimental comparison in terms of average power output, for fixed wave height, as a function of wave frequency.

It has to be noted that, for the experimental set-up using a pneumatic device to simulate PTO action, some difficulties arise in keeping a constant value for the force–speed gain.

A large overestimation of the output by potential flow theory may be seen, even if a relatively good qualitative agreement between numerical and experimental results exists. Predicted maximum power is approximately 1.7 times larger than observed data. It may be noted that the potential flow approach, used to obtain the reported results, neglects the effects of viscosity, leading to an overestimation of the predicted response. This is likely a possible reason of the observed discrepancy. Such difference seems to be more pronounced around power output maximum.

2.2.3.4 Optimization based on potential flow simulation

Using optimization software and the numerical model based on the potential flow approach, an optimization procedure has been developed in order to search for a suitable configuration of a larger scale prototype to be used in further project development. In particular, a set of tests in a naval tank was planned within a research program funded by the WES (Wave Energy Scotland) consortium.

A simulation-based approach is adopted for the optimization procedure. Some further details on the defined procedure are given in [103].

The optimization search was performed with the objective to increase the output power in a defined wave condition. A simplified regular sea state with a given wave amplitude and frequency has been considered, with the purpose of reducing simulation time, in view of the large number of runs required during the optimization process. It is, however, worth to note that the choice of the considered wave condition may have a significant effect on optimization results. Moreover, irregular sea states may influence the effective output power and energy conversion efficiency.

At each step of the optimization search, a new configuration is generated, as defined by the following parameters: (a) mass (and immersed volume at equilibrium); (b) shape of the body; and (c) proportionality constant between PTO force and oscillation speed. The optimization goal is to increase the average or peak power output, under a constraint to limit the maximum immersed volume. For each generated configuration, a BEM analysis is performed to derive the hydrodynamic coefficients at the frequency of interest. A geometrical preprocessing is needed, in order to generate the body geometry at the beginning of a new optimization step, and, for the new configuration, the initial equilibrium condition is searched for, distinguishing diffracting and non-diffracting elements (below and above water level, respectively). Around such condition, the BEM linearized hydrodynamic analyses are executed.

The objective function of the optimization process is defined by the peak output power obtained from a time-domain simulation of the system. During time simulations, the actual hydrodynamic forces and moments on the bodies are calculated using the hydrodynamic coefficients (some corrections to linearized BEM scheme may be introduced by considering the actual wetted surface in the calculation of hydrostatic and Froude–Krylov actions) and non-linear kinematic effects are also accounted for. A post-processing phase retrieves the power output data to evaluate the objective function.

Assuming a regular monochromatic wave, the convolution method, more appropriate for irregular seas, has not been used for the analysis of the radiation forces which have been estimated using the response amplitude operators and the hydrodynamic coefficients related to the prescribed incoming wave frequency.

Other important factors are represented by the inertial characteristics of the body and by the geometry of the kinematic system. In the approach here adopted, the mass of the body is assumed as a parameter, whereas initial guess values are assumed for the moments of inertia and the position of the centre of gravity. Moreover, the geometry of the kinematic arrangement (position of piston attachment points, hinge locations, length of the oscillating support arms) is constant along the optimization process. The width of the body, which has a prismatic shape, is also constant (equal to 5 m in the present studies).

The cross section of the considered prismatic body is represented, for ease of manipulation, by a B-spline curve, and can be varied along the process by changing the spline coefficients. The B-spline is defined using 15 coefficients, useful to parameterize the cross section shape, but only three coefficients are actually used as optimization parameters, the ones placed on the front face of the GEMStar large-scale model. Unfeasible geometries, eventually generated, are rejected. It may be noted that some problems may arise using the proposed approach for geometry generation, and sometimes unfeasible configurations may be generated. Reducing the number of parameters and the range of variation may have a positive effect on such issue.

A schematic representation of the optimization process is reported in Figure 2.30. The algorithm will stop when it cannot find solutions with better improvements than an assigned convergence value.

2.2.3.5 Large-scale prototype

In a successive phase of project development, a larger test model, with 5 m width, was designed and built for a further test campaign, in cooperation with Umbra Group Spa and under the funding of a European project financed by the Scottish WES consortium. This second campaign was aimed to investigate the effective production capability and highlight the main issues of the studied technology, at least in a controlled environment, on a dimensional scale closer to a possible real installation.

Figure 2.31 shows the large-scale model mounted on the dynamometric cart in the naval towing tank of the University of Naples.

The test model was provided with a PTO comprising a recirculating ballscrew device coupled with an electrical generator [103].

The developed optimization procedure was applied to the search of a suitable section shape associated with optimal values of the floating body mass and of the PTO force–speed gain. Optimization was performed for a 0.24 m wave amplitude regular wave condition with 0.35 Hz frequency, with the objective to

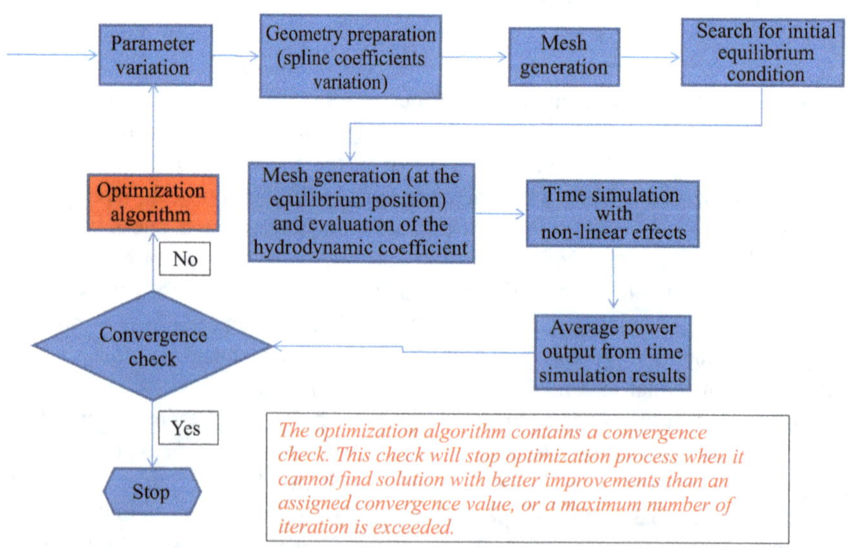

Figure 2.30 Optimization process schematic overview

Figure 2.31 Naval basin experimental test set-up on large-scale model

increase maximum power output. A fixed width of 5 m has been considered and a constraint on maximum immersed volume (5 m^3) was set, in order to limit the weight of the system.

Some enhancements to the numerical model were also implemented. In particular, in the estimation of power output, for the PTO based on the recirculating ballscrew, the model of the PTO force was improved accounting for its inertial and frictional characteristics (estimating the so-called back-drive force).

The previous mentioned optimization procedure allowed to design the optimal shape of the section for the floating body and to determine a value of the required mass and submerged volume, associated with the value of the PTO force–speed gain. The effectively tested configuration was further slightly modified with respect to the direct optimization output to simplify the manufacturing process and to comply with the effective PTO characteristics (particularly, the length of the piston).

The linear PTO control law was implemented by using an adjustable bench of electrical resistors. By changing the resistance, different values of the force–speed gain could be tested. Such a solution allows realizing, at least approximately, the desired linear relation between PTO force and oscillation velocity (it has to be noted that a certain deviation from perfect linearity has been observed, due to variations of the gain K with piston speed for a fixed electrical load).

A potentiometer was used to measure the displacement of the piston (and its velocity) and a load cell measured the action on the PTO piston. The mechanical input power to the PTO could therefore be estimated. Using capacitive wave probes, it was also possible to measure the water surface elevation. By comparing the output power to the wave available power, it was possible to estimate the conversion efficiency of the system.

For regular waves, available wave power per unit crest width may be expressed using the following relation:

$$P_w = \frac{1}{2}\rho g A_w^2 c_g \tag{2.25}$$

where ρ is the density of water, A_w is the wave amplitude and c_g is the group velocity [94]. The overall conversion efficiency (wave-to-wire) is defined as the ratio of the electrical output power, P_{elec}, to the available wave power:

$$\eta_g = \frac{P_{elec}}{P_w B} \tag{2.26}$$

where B, the capture width, is the assumed equal to the width of the body. The global efficiency η_g accounts for the conversion efficiency of the buoy and for the electrical conversion efficiency of the generator.

The peak mechanical power, observed during the regular wave test with 0.24 m wave amplitude (the highest tested amplitude) and 0.35 Hz, was equal to about 6.5 kW. The corresponding average mechanical power was approximately equal to 2.6 kW. A typical time history of the output power is reported in Figure 2.32, in the condition of the maximum tested wave amplitude.

A strong dependence of power-production performance on wave frequency has been observed, as possible for resonating systems. Power output decreases away from the optimal frequency (about 0.35 Hz, but variable with the PTO damping).

The global conversion efficiency is reported in Figure 2.33, for regular waves, in percentage and as a function of wave frequency, for several electrical loads and for the case of a 0.15 m wave amplitude. A noticeable effect of both frequency and electrical load can be observed, with a maximum value of about 60%–64%, for the reported wave amplitude.

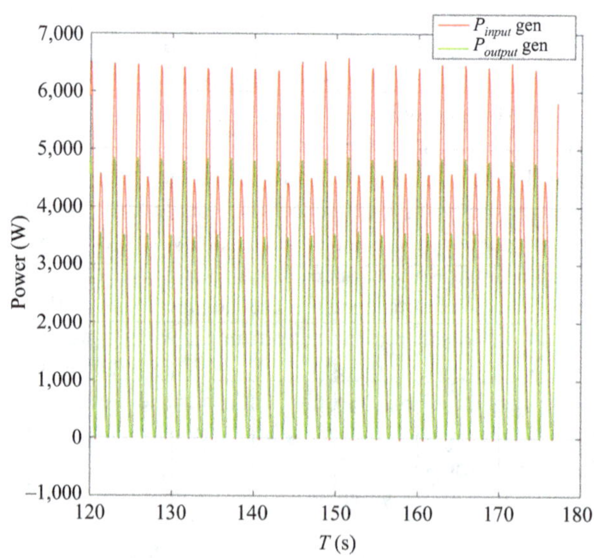

Figure 2.32 Typical time histories of measured power at the generator input shaft (red) and at the electrical output (green). Wave amplitude 0.24 m, frequency 0.35 Hz, electrical load 3.5 Ω (K=90 kN s/m)

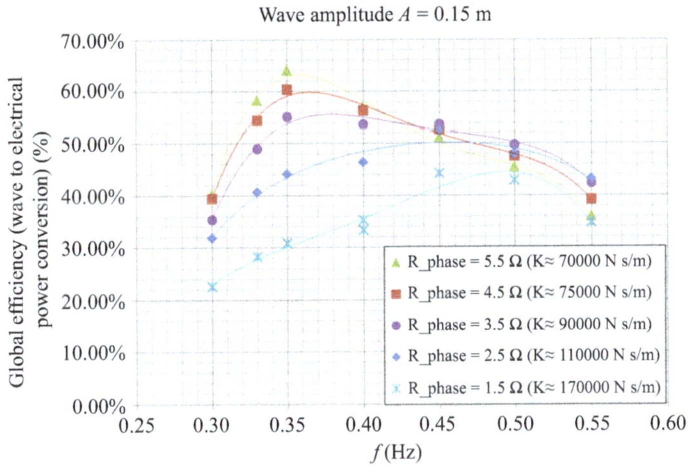

Figure 2.33 Regular waves. Global conversion efficiency (per cent) as a function of wave frequency for several force–speed gain values (electrical load). Wave amplitude 0.15 m

Considering irregular sea states, represented by wave spectra, the actual power output can show large fluctuations due to the irregular variations of wave-surface elevation. Moreover, a large difference between average and peak values of power output and PTO forces may be observed in a given sea state.

Wave spectra are defined by characteristic energy spectral density distributions and the related spectral parameters (in particular significant height and peak frequency).[6] Widely used analytical spectral models have been considered such as JONSWAP and Pierson–Moskowitz spectra [105].

As can be seen in Figure 2.34, which shows the average and peak power in a test under a Pierson–Moskowitz spectrum with $H_s = 0.25$ m significant height for varying peak frequency, the peak power value may be several times larger than the average value (in the reported case, a peak-to-average ratio of about 20 can be observed).

The effect of peak frequency is also shown in Figure 2.35, which reports the global efficiency as a function of peak frequency for two different types of spectrum (JONSWAP and Pierson–Moskowitz). Both trends show a slight reduction of maximum conversion efficiency with respect to the regular sea state case, with a maximum observed global efficiency of about 54%.

[6]Significant height is a spectral characteristic parameter, defined in terms of the standard deviation of wave surface elevation, $\sqrt{m_0}$, as $H_s = 4\sqrt{m_0}$. The peak frequency of the spectrum is the frequency at which the maximum value of power spectral density is reached.

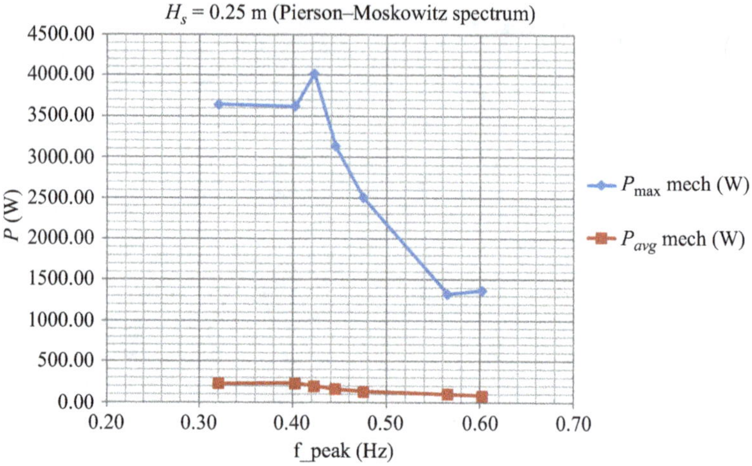

Figure 2.34 Irregular sea states. Average and max mechanical power (constant significant height (H_s = 0.25 m), variable peak frequency, Pierson–Moskowitz spectrum)

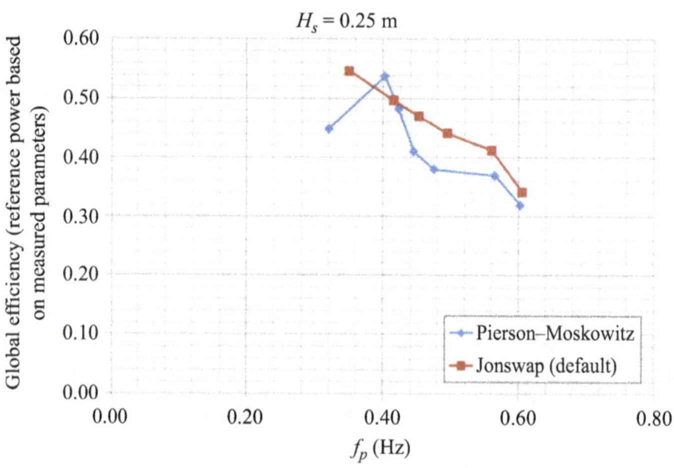

Figure 2.35 Irregular sea states. Global conversion efficiency for two different types of spectra

2.3 PTO development[7]

PTO mechanisms play a crucial role in the ability for WECs to become both technically sound and economically attractive [103,104]. After exploring the use of conventional systems transferring the mechanical wave power into a useful form of

[7]Written by Luca Castellini.

energy, novel or custom-made PTO have emerged over the past 15 years. This new generation of PTO mechanisms for WECs has inspired a growing interest in establishing stage-gate metrics capable of measuring the attractiveness of WEC developers at system and subsystem levels, the appraisal covers cross-cutting engineering, economic and environmental aspects. Conventional PTO systems and innovative PTO solutions for WECs are often derived from development in other sectors of the electromechanical industry. Based on the lessons learnt in previous developments and from the experience gained in the pioneering development of WECs, the new generation of PTOs is expected to be more robust and mature. Evidence on this needs to be shown in terms of test results and correlation with analysis.

2.3.1 Introduction

Among the ocean-energy technologies under development, WECs can be described in four main subsystems. These categories and indicative capital costs (CAPEX) share based on [3,103,105] are depicted in Table 2.3. PTO mechanisms are the main subsystem responsible for the energy transformation process from wave-to-wire (W2W) which not only impacts heavily the CAPEX but also largely affects the energy yield [103]. Indeed, the efficiency, the ability to apply different damping factor, the capability to store and smoothing the output or send back (reactive control) the energy can be governed by the PTO. This is why improvement in PTO systems has been identified as one of the key route for wave energy cost reductions [105]. In addition to representing a large CAPEX share, the PTO mechanism also impacts the OPerational EXpenditures (OPEXs). By using relatively complex electromechanical components, the reliability of a PTO system requires careful attention to achieve economic viability of a wave energy project.

PTO systems are composed of moving parts subject to both mechanical and electrical loads; moreover, control strategies applied to the PTO may impose additional constraints [106]. For these reasons, the trade-off between efficiency, fatigue life, robustness and complexity is pivotal when selecting and designing the PTO of a WEC. Among the wave energy technology developers that have

Table 2.3 WEC subsystems categories and their CAPEX contributions

WEC subsystem	CAPEX share (%)	Description
Structure and prime mover (SPM)	30–40	Direct interface with the water and is responsible for collecting the energy from the waves
Power take-off (PTO)	20–25	Converts the mechanical energy captured by the SPM into a useful form of energy
Moorings and foundations (M&F)	15–20	Reaction system responsible for keeping the WEC in the desired position and orientation
Connection system	10–15	On-board electrical infrastructure, sensors and instrumentation

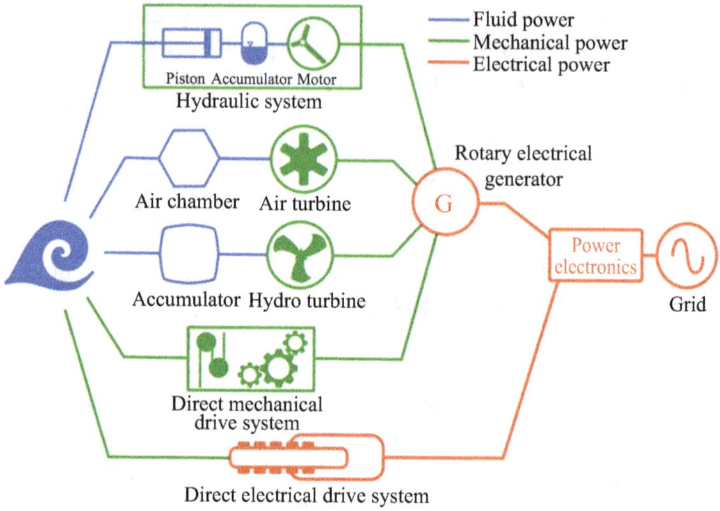

Figure 2.36 Different power-flow paths for wave energy to electricity conversion adapted from [107]

undergone sea trials, four major categories of conventional PTO mechanism should be highlighted [107,108], namely hydraulic systems, air or hydro turbines, direct mechanical or electromechanical systems and linear generators (see Figure 2.36).

Over the last decade, several R&D projects investigating innovations in this field have reshaped the landscape of PTO solutions for WECs demonstrating the need for step-changing PTO mechanisms, an example among the others is WES that opened its first call specifically on this topic [109]. As a result of these R&D projects, novel solutions were investigated and are now considered particularly innovative representing a new generation of PTO mechanisms. These are not mentioned in official classification as expressed in Figure 2.36.

2.3.2 Design and test methods

The conversion of wave energy to useful energy is the main goal of the wave-energy conversion process. Most of the focus in the design of a WEC is to harvest power and make it available as electric power to be transferred into the utility grid. There have also been some studies on small-to-medium-scale systems as stand-alone systems, where the captured energy is consumed in situ or very close to the harvesting area with very little or no need of extensive storage and transmission systems. There are many different existing technologies for wave energy conversion. The choice of PTO is diverse as well. When designing a WEC, the type of harvesting technique should be chosen first. To decide the harvesting technique, a characterization of the ocean wave energy properties and resources, site assessment and other factors that determine the survivability and successful operation of the WEC in the open sea should be considered. Then the harvesting technique defines the PTO technology [107].

Most WEC designers initially design and manufacture laboratory models in wave tanks, taking in consideration the hydrodynamic aspects first. After successful tests, some progressed to testing of prototypes in large wave tanks. While the testing of laboratory models gave a good insight of a WEC's performance, it was still inadequate to fully predict the performance in the ocean environment and the full interaction of the PTO. Most tests at laboratory scale do not include final PTOs and few examples are in the literature of design and test methods considering the PTO development since the early stages.

Often a challenge in the development of PTOs is the slow rate of motion, and a standard approach is to multiply the speed thanks to a mechanical drive system or modification in the WEC's dynamic. The flexibility in the WEC specific parameters and PTO's integration is crucial to optimize the design satisfying conceptual, technical and economical requirements. This can be solved designing the system following an integrated approach. In terms of design, the classic approach is to establish a wave-to-wire (W2W) model identifying all the necessary components or logics, and associated parameters, of the power-conversion chain. A comprehensive W2W numerical model must include a description of the PTO behaviour. To date, there exist a variety of techniques to couple a PTO model to the global dynamic simulation tool of a WEC system [110]. The degree of accuracy of such wave-to-wire numerical models depends on the ability to account for non-linearities and losses. Comparison of numerical models against experimental data is crucial in calibrating and gaining confidence in the validity of simulation results from the W2W model.

To collect data, the use of a mechanical and/or electrical test infrastructure is decisive in prototype testing before sea trials since it allows extensive test campaigns in a least expensive manner via a controllable, repeatable and dry environment. A PTO hardware-in-the-loop (HWiL) test rig is a real-time system that can be used in the development and test. HWiL simulations are performed to provide feedback for a more robust design with the typical configuration. The benefits include increased system reliability, more efficient development and system integration, and lower costs to reduce faults and improve the design. Additional HWiL objectives may focus on hardware and software integration, fault-simulation testing, determining and developing design centring parameters, identifying design weaknesses and limitations, evaluating transient performance including linear steady-state and transient testing over the operating envelope and establishing calibration tables for electronics subsystems.

The HWiL test rig is composed of an emulator representing the prime-mover subsystem and a PTO unit along with the auxiliary equipment, sensors and embedded controller. Some examples of dedicated PTO test rigs for wave energy include a reduced-scale hydraulic system of the Wavebuoy, a full-scale test rig of the Wavestar, a one-eighth-scale gyroscopic PTO of the ISWEC, a scaled test rig of the CorPower, a small-scale test rig of the PolyWEC and a full-scale test rig of ballscrew PTO of UMBRAGROUP.

2.3.3 Innovative PTOs

When only mechanical and electrical power circulate through the conversion path (i.e. no transfer of fluid), the PTO mechanism is considered to be a '*direct-drive train*' systems. Direct-mechanical PTO drive trains employing rotary electrical generators have attracted sharp interest over the past decade. Consequently, different types of mechanical arrangements to spin the shaft of the generator are proposed along the years, among them magnetic gear (MG) systems and screw mechanisms are considered the most innovative solution in the sector. In the absence of a rotary generator, PTOs can be described as '*direct-electrical drive train*' mechanisms. Dielectric Elastomer (DE) Generators (DEGs) respond to this definition.

2.3.4 Magnetic gears

MGs fall in to the direct, or pseudo direct, drive power conversion path and operate by the contactless transfer of torque between rotors through the manipulation of the magnetic field in the rotor air gaps. There are multiple variations on the MG [111] but the current leading systems that are applicable to wave energy are the Field Flux Modulated MG (FFMMG) [112], the MG Lead Screw (MGLS) [113] and the Trans-rotary MG (TROMAG) [114].

The FFMMG operates by employing ferromagnetic pole segments in the air gap between rotors in order to modulate the magnetic flux in such a way that a gearing effect is established. Capable of operating in concentric, axial and linear orientation, this type of MG is not only highly adaptable but is capable of high torque densities in the range of 70–150 kN m/m^3. The MGLS and the TROMAG are the two types of MG that allow for linear to rotary motion conversion with a gearing effect. This has great potential in marine energy where, in a linear heave type device, a traditional rotating generator can be used. The TROMAG operates on the same principles as a mechanical leadscrew while the MGLS uses both leadscrew principles and elements of the FFMMG. The MGLS and the TROMAG have been shown to have force densities of 1.87 kN/length and 2.6 kN/length, respectively.

Currently, MG technology can be described as Stage 2 – TRL 4 according to the wave-energy development protocol (EMEC [115]). The leading designs have been validated over the past 15 years with intermediate-scale prototypes now being developed. Work is ongoing to assess the technologies feasibility and cost. As MGs are capable of rotary, linear and trans-rotary orientation, the technology is highly applicable to a wide variety of wave energy devices which use an intermediary speed/torque converter element [116]. The MG could potentially benefit all of the device types listed in Aquaret with the most suitable being the point absorber, the submerged pressure differential and OWSCs.

The reliability of MGs is a key driver in the technology's development for wave energy as there are no mechanically meshing parts, unlike traditional mechanical gears. These factors are expected to greatly reduce operation and maintenance issues over a device's lifetime. The high-quality magnetic material has a very low demagnetization rate, so the high efficiency should remain for a standard device's deployment life of 20+ years. Additionally, MGs have the advantage of inherent overload protection meaning that, as the torque is established through the magnetic field, in the event of excessive force being applied to a system the rotors will slip and realign rather than break. This gives to the technology an intrinsic self-protection function decoupling the PTO from the WEC.

CAPEX is one of the main obstacles facing MG technology. China produces 76% of the world's rare earth magnets and produces more than 95% of the rare earth material required. This makes the chain of supply limited and the price highly variable ranging from a peak of over $450/kg in 2011 to under $100/kg as of 2018. The OPEX, however, is expected to be greatly reduced when compared to a similar hydraulic or mechanically geared system as a result of the contactless operation as mentioned and based only on the bearings functionality.

MGs have many attributes which make them attractive for marine deployment. The contactless operation has greatly reduced noise over mechanical gears, and the reduction in lubrication requirements lessens the potential for pollutants. Furthermore, the magnetic material which is capable of maintaining its magnetism for decades can be recycled and repurposed at the end of a device's life time. The process of obtaining the permanent magnet material involves intensive mining however which can have environmental impacts. Also as China produces the vast majority of the world's rare earth elements, there are further environmental impacts from the transportation of the materials worldwide and potential issues with security of supply or recycling as well.

2.3.5 Dielectric elastomer PTO

DEs are polymeric materials which can experiment deformations over 500% of their length under the effect of a mechanical load; due to their mechanical compliance and to their non-conductive electric behaviour, they are exploited as transducers, meaning that they operate as devices that convert energy form one form to another [117]. Aspects of interest concerning the application of DEs to wave energy conversion can be summarized in the following points:

- DEs are soft, resilient, lightweight. Thanks to their high deformability, they are suited to be directly coupled with moving oscillating WECs that undergo large displacements without the necessity of any intermediate mechanic transmission.
- Polymeric materials are corrosion resistant and, thus, they appear particularly suitable for applications in aggressive environments (like sea water).
- Many common polymeric materials are already widely spread and largely used; thus, they are available at relatively low price. Natural rubber, for example, is mainly used in the transportation sector (pneumatic tires, automotive belts) and in the clothing sector (gloves, footwear).
- DEGs application would significantly reduce the amount of mechanical moving parts that can potentially meet mechanical failure.
- As already mentioned, a large variety of DEGs can be designed, each exploiting a particular motion condition (angular pitch oscillation, reciprocating motion, etc.); therefore, for a particular design of a WEC device, a proper design for the corresponding polymeric PTO can be investigated and identified.
- DEGs are suitable to work at low frequencies, like those encountered in sea waves (in the neighbourhood of 1 Hz); therefore, their application would allow

to suppress the troublesome problem of using frequency changers to interface a natural low-frequency system (the WEC moving within the waves) and a system that requires high frequency (the traditional electric apparatus, since traditional electric generators usually have high nominal frequency). DEGs convert the energy from the waves into high-voltage direct-current electricity that is immediately ready to be delivered to the coastlines (Figure 2.37).

Of course, as a counterpart, some criticalities must be underlined [118].

- The polymer may be subjected to material aging and performance losses during the life cycle. Fatigue tests on natural rubber samples, subjected to loading cycles with frequency of 5 Hz, at different strain levels, at ambient temperature, showed visible loss of tension or mechanical failure after 106–107 loading cycles. For the specific application of DEGs, anyway, specific fatigue tests should be carried out at loading frequency and stretch levels comparable to those employed in elastomeric generators.
- Elastomeric materials present both electric and mechanic hysteretic behaviour during loading–unloading cycles; therefore, part of the energy absorbed by the generator is dissipated due to viscous phenomena. A proper analysis of the limitations to the amount of convertible energy due to dissipative phenomena has to be carried out and a targeted choice of the materials has to be done.
- As already mentioned, since DEs use voltages in the order of some kV, to now, a lack of proper electronic components may be encountered.
- In order to avoid the rupture of the polymeric material, DEGs must be subjected to limited displacements (in order to keep the strain on the elastomeric membranes below the rupture value). WECs operating in real sea are subjected to storms and gales that tend to provoke very large oscillation amplitudes of the converter moving body. In any case, this particular problem is faced by all existing technologies, since also for WECs equipped with traditional generators, safety devices are provided.

Figure 2.37 The wave flume tests of poly-OWC

DEGs are a relatively new concept, in fact their study and application is limited to experimental laboratory-scale applications or small-scale and limited coastal test. Many of the weak points of this technology have not yet been identified because of leakage of sufficient analysis in a real operating contest.

2.3.6 Electromechanical ballscrew-based PTO

A ballscrew is a mechanical device in which balls, acting as low friction load-carriers, efficiently convert the low-speed linear motion of a translating element (either the nut or the screw) into a high-speed rotary motion of a rotating element (either the nut or the screw). As balls travel at an axial average speed which is different than that of the translating element, they typically require a recirculation system to allow then to continuously re-enter the tracks during motion ('recirculating ballscrews'). Ballscrews can both work in extension and in retraction at the same performance. The ratio between the rotational speed of the rotating element and the axial speed of the translating element is a function of the screw pitch, i.e. the distance between two adjacent threads. The rotating element of the ballscrew can then be connected to a rotary electrical generator:

- directly;
- indirectly, through a flywheel to store mechanical energy and absorb peaks; and
- indirectly, through a lever mechanism to convert reciprocating linear motion into unidirectional rotational motion.

Ballscrews have now been commercialized for decades in a wide range of both safety-critical and high-load applications. In the aerospace sector, ballscrews are used for aircraft stabilizers, spoilers and flaps; in the industrial sector, for laser or water cutting, punching, forging; in the automotive sector, for steering systems; in the oil and gas sector, for gate valves. In aerospace and industrial applications, ballscrew systems are designed for a lifespan of 20 years. For such applications, the ballscrew system has a TRL 9 with dedicated production facilities and large-scale mass production.

Use in the marine energy sector is present at lower TRL values. Solutions based on the integration of a recirculating ballscrew and a permanent magnet generator are currently rated at TRL 6 for medium-scale (100 kW peak electrical power) above-water applications and TRL 7 for small-scale (5 kW peak electrical power) submerged applications. A ballscrew-based PTO solution is applicable to a wide range of WEC. Its integration is straightforward into all devices that require conversion of linear motion into electricity, namely point absorbers, point-pivoted buoys and submerged pressure differential devices. Furthermore, with the introduction of a lever mechanism, integration is also possible on oscillating wave surge converters and attenuators.

As for other linear systems, end-stops reaching is one of the challenges in terms of availability of ballscrews. The ballscrew architecture is typically characterized by a small inertia, short energy conversion chain that includes only rolling-friction elements and a small reaction time to input variations. This allows reaching high values of efficiency even at low speeds. Furthermore, a very rigid direct-drive solution allows an efficient energy conversion regardless of the peak period and spectrum shape of the input force/speed conditions [102].

Poor data is available on ballscrew systems for WECs in terms of affordability. It is surely true that the requirement for long strokes and protection against the marine environment makes ballscrew costs higher than for more 'traditional' applications. Data provided by UMBRAGROUP S.p.A. for their EMG suggests that capital cost (for a series production of industrialized and marinized products) can be in the interval of 450–650 £/kW for a 100 kW (peak electrical power) generator. For the same system, the suggested inspection interval is 48 months: associated OPEX depends on the installation type (offshore, nearshore, onshore) and consequent cost for accessing, retrieving and inspecting the system.

The ballscrew architecture has already been investigated and tested as an actuator in different operational conditions for several years. Its survivability is mainly dictated by the axial load at the ballscrew. During previous wave-tank experiences in wave energy applications, ballscrew-based systems survived a load ratio of 5 and would have survived load ratios between 7 and 15, depending on the wave conditions. With regards to corrosion and, considering industrial-scale series production, proven materials and surface treatments can be used to ensure survivability in the marine environment.

Ballscrew-based systems, compared to state-of-the-art PTO systems for wave energy, sensibly reduce life cycle environmental impact. Indeed, compared to hydraulic system, ballscrew PTOs use no pressurized fluids, thus avoiding the risk of leakage during operation and simplifying disposal. Furthermore, compared to purely electrical systems, the ballscrew PTOs reduce the use of ferromagnetic materials, thus reducing pollution impacts associated with extraction processes [119] (Figure 2.38).

Figure 2.38 A set of ballscrew PTOs ready for shipment

2.3.7 Conclusions

This section introduced some consideration on the design and test of PTO system for a WEC and described novel types of PTO systems with the overall objective of showing how crucial this subsystem is. The efficiency of the PTO system directly affects the annual energy production of the machine, and the choice of components has a direct influence on the cost of the whole converter and the maintenance cost of the system. An efficient, maintenance-free and reliable PTO system is fundamental in order to reach the goal of cost-effectiveness for wave energy conversion. Covering engineering, economic, environmental and cross-cutting technological issues, a presentation framework for the new generation of PTO mechanisms is established taking in consideration the particularly innovative aspect for this technology. Among the others MGs, Ballscrew-based PTO and DEGs were presented.

2.4 Modelling and control[8]

Modelling and control comprise some of the 'soft' technologies associated with wave energy. Since wave energy systems span several technical disciplines, including hydrodynamic, mechanical, hydraulic, electrical and electronic, there is a considerable challenge in determining mathematical models for WECs which are equally accurate across all discipline domains. However, accurate WEC models have considerable utility, not least due to the considerable size and expense of full-scale prototypes, and the need to fully understand the implications of various design decisions. In particular, wave energy project developers need an assessment of the level of power production, for a particular deployment in a specific wave site, before committing to a wave project. The level of fidelity will depend on the degree to which a particular project has progresses, ranging from feasibility studies to detailed design decisions. Such assessments require both a good knowledge of the wave climate in a particular location (the subject of Section 2.1) and a mathematical/simulation model of the device, which is dealt with in Section 2.2. In addition to power assessment, there are several other important uses of WEC mathematical, including

1. assessment of loading forces under extreme sea conditions,
2. simulation of device motion, including evaluating the effectiveness of control strategies and
3. for use as a basis for model-based control design

WEC control presents itself as a possibility to ensure that, for a given WEC structure (i.e. capital cost), maximum power is captured from incident waves. Since most WECs are designed to resonate with the incoming waves and mechanical systems can only be designed to resonate at one primary frequency, control technology can alter the WEC dynamics in an adaptive way, thus modifying the loading on the device to change its resonant frequency (more on this in Section 2.4.2).

[8]Written by John Ringwood.

The two areas of modelling and control are closely related, since most control strategies are model based, needing a mathematical model of the system to be controlled in order to derive the controller algorithm and/or parameters. In addition, the correct operation of a controller is normally verified on a validated simulation model before being implemented on a prototype device, at any scale. One particular difficulty with WEC controller verification relates to the difficulty of building PTO systems at small scale (due to the exacerbation of friction effects, etc.), placing even more emphasis on the need for high fidelity mathematical/simulation models.

2.4.1 WEC models

A complete mathematical model of a WEC includes all aspects from 'wave to wire', as shown in Figure 2.39. A complete mathematical model therefore consists of various sections which articulate the hydrodynamics (the effect of the waves on the hull of the WEC), the PTO system (which converts the mechanical motion of the WEC into useable (e.g. electrical) power), and might also include a model for the electricity grid. The focus in this section will be on the hydrodynamic model, since this is the component which is relatively unique to wave energy devices. The other mechanical, hydraulic, electrical and electronic components find applications in other areas. The models for them are reasonably well documented. The interested reader is referred to [120,121] for a review on full wave-to-wire models. It is also worth mentioning that full wave-to-wire models can be well articulated using the bond graph modelling approach [122,123]. Finally, mooring systems also affect the dynamics of a wave energy device; for the modelling of WEC moorings, the reader is referred to [124].

Within the limited space of this chapter, only an overview of WEC hydrodynamic models is possible; for a more comprehensive and up-to-date treatment, the reader should refer to [125]. In general, there are three paradigms for developing hydrodynamic models of WECs:

1. Model based on CFD,
2. Boundary-element models and
3. Models determined from recorded data.

A fourth possibility is smoothed particle hydrodynamics, but this is normally reserved for WEC response evaluation in extreme seas, with separated flow, etc. [126].

In this chapter, the focus is on the three main methods listed above.

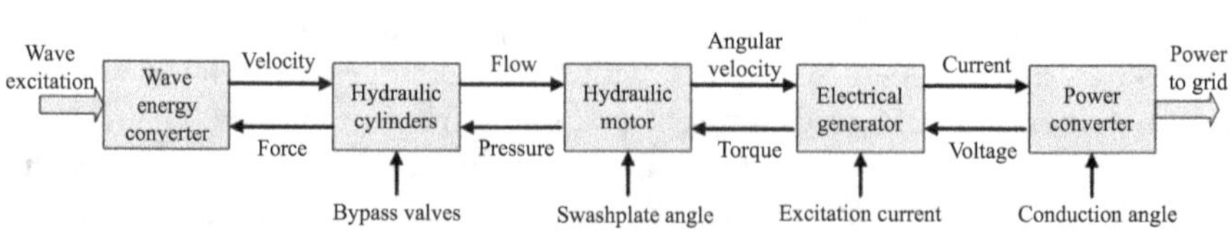

Figure 2.39 WEC system components from wave to wire

2.4.1.1 CFD-based models

In CFD, the full fluid domain is modelled, where pressure variations are modelled within the boundaries of a three-dimensional domain, which includes the WEC as a solid object and air and water fluid phases. A numerical wavemaker is used to generate waves of a user-defined spectrum, and waves are dissipated down-wave of the WEC in order to prevent unwanted reflections. The dynamics of fluids is governed by the transfer of mass, momentum and heat. These three processes are described by the Navier–Stokes equations, a set of partial differential equations derived in the early nineteenth century, which form the basis for the analysis of fluid dynamics [127].

Continuity equation:

$$\frac{\partial \rho}{\partial t} + \nabla \cdot (\rho \boldsymbol{u}) = 0 \tag{2.27}$$

Momentum equation:

$$\rho \frac{\partial \boldsymbol{u}}{\partial t} = \nabla \cdot \tau_{ij} \tag{2.28}$$

Energy equation:

$$\rho \frac{\partial e}{\partial t} = \nabla \cdot (k \nabla T) - p \nabla \cdot \boldsymbol{u} + \tau_{ij}^{v} \frac{\partial u_i}{\partial x_j} \tag{2.29}$$

where ρ is the fluid density, $\boldsymbol{u}$ is the velocity, e is the internal energy, T is the temperature, k is the thermal conductivity and τ_{ij} is the stress tensor comprising the pressure, $-p\delta_{ij}$, and viscous terms, τ_{ij}^{v}:

$$\tau_{ij}^{v} = \mu \left(\frac{\partial u_i}{\partial x_j} + \frac{\partial u_j}{\partial x_i} \right) + \delta_{ij} \lambda \nabla \cdot \boldsymbol{u} \tag{2.30}$$

where μ is the coefficient of viscosity, δ_{ij} is the Kronecker delta function and λ is the bulk viscosity. The coupled continuity, momentum and energy equations, (2.27) to (2.29), are indeterminate and require two more equations to obtain closure which are provided by the ideal gas laws:

$$p = \rho R T \tag{2.31}$$

and

$$e = c_v T \tag{2.32}$$

where c_v is the specific heat at constant volume. In general, these equations have no known analytical solution; however, they may be solved numerically using CFD by discretising the domains of space and time to form a system of linear algebraic equations, which are computer implementable. The level of discretization (the mesh) is usually more refined in the domain region close to the WEC. While CFD provides a high-fidelity non-linear modelling mechanism for WECs, the drawback is the computational intensity of the simulation, typically requiring 1,000s of computation time per second of simulation time. This severely restricts its application in wave energy simulation.

2.4.1.2 Boundary element models

In boundary element methods, the fluid velocity potential is calculated only on the boundary of the WEC. The WEC hull can be modelled as a set of small panels, permitting arbitrary shapes, or specified analytically, restricting the degree of shape variation and number of DOF [128,129]. WEC models based on BEMs can be either linear or non-linear, though linear models are, by far, the most common, due to their simplicity and computational convenience.

Consider a single-body floating system oscillating in heave, schematically depicted in Figure 2.40. Energy is extracted from the relative motion with the sea bottom, through a generic PTO mechanism. The external forces acting on the WEC are the excitation from the waves and the control force produced by the PTO, respectively, $f_{ex}(t)$ and $f_u(t)$. Additional hydrodynamic and hydrostatic forces, which arise due to the motion of the body in the water, are the radiation force $f_r(t)$, the diffraction force $f_d(t)$, the viscous force $f_v(t)$ and the buoyancy force $f_b(t)$ [93].

The radiation force $f_r(t)$ is a damping/inertial force associated due to the fact that device motion, resulting in the production of radiated waves, is affected by the surrounding fluid. Such radiation forces are present even in the absence of incident waves and can be estimated using free response tests. The diffraction (or scattering) force $f_d(t)$ describes the force experienced by the device when scattering incident waves and is independent of the device motion. The viscous damping force $f_v(t)$ is a non-linear force and becomes significant with increased device velocity.

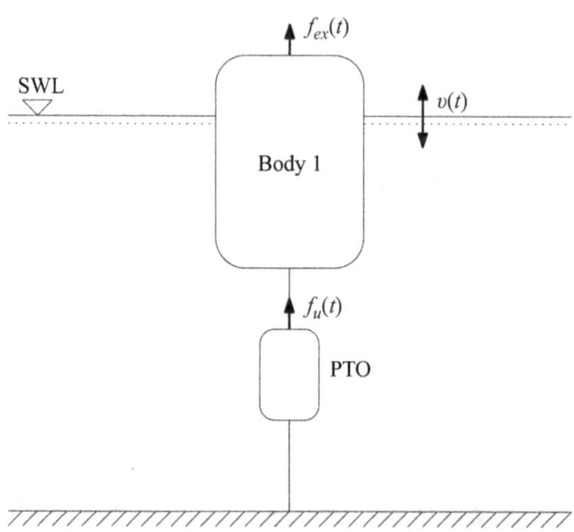

Figure 2.40 One-degree-of-freedom floating system for wave-energy conversion. The lower side of the PTO is anchored to the sea bed, which provides an absolute reference for device motion

It is particularly relevant where the body surface contains discontinuities (such as flanges), which results in the creation of vortices. Finally, the buoyancy force is related to the deflection of the device from its equilibrium (still water) position and is a balance between the Archimedes buoyancy force and the gravity force.

The equation of motion, following Newton's second law, and where a superposition of forces is assumed, in one degree of freedom, is

$$M\dot{v}(t) = f_m(t) + f_r(t) + f_d(t) + f_v(t) + f_b(t) + f_{ex}(t) + f_u(t) \tag{2.33}$$

where $v(t)$ is the heave velocity and M is the WEC mass.

With the assumptions associated with linear potential theory [93], namely that the fluid is irrotational, incompressible and inviscid; the WEC body has a small cross-sectional area (or equivalently, the wave elevation is constant across the whole body); and the body experiences small oscillations (so that the wetted surface area is nearly constant); the equation of motion simplifies to

$$f_{ex}(t) + f_d(t) = \int_{-\infty}^{+\infty} h_{ex}(\tau)\eta(t-\tau)d\tau \tag{2.34}$$

$$f_r(t) = \int_0^t h_r(\tau)v(t-\tau)d\tau - m_\infty\dot{v}(t) \tag{2.35}$$

$$f_b(t) = \rho g S_w \int_0^t v(\tau)d\tau = -K_b x(t) \tag{2.36}$$

$$f_v(t) = 0 \tag{2.37}$$

In (2.34), the excitation (and diffraction) force is related to the incident wave-free surface elevation $\eta(t)$ through the excitation kernel function $h_{ex}(t)$. Equation (2.35) expresses the radiation force as a linear convolution of the radiation kernel $h_r(t)$ with the oscillation velocity $v(t)$. Note that $h_{ex}(t)$ and $h_r(t)$ effectively describe the impulse responses in excitation force and radiation force to impulses in free surface elevation and device motion, respectively. Added mass reflects an effective increase in the device inertia since an accelerating floating body moves some volume of the surrounding fluid. In general, added mass is a frequency-dependent quantity but is often approximated by its infinite frequency asymptote m_∞.

The buoyancy force $f_b(t)$ models the hydrostatic equilibrium, related to the heaving position through a linear coefficient that depends on the gravity acceleration g, the water density ρ, and the surface area of the body cut by the mean water level S_w. Note the noncausality of the expression for the excitation force, where $h_{ex}(t) \neq 0$ for $\tau \leq 0$ [93]. Equation (2.33), excluding the mooring force $f_m(t)$ and the viscous damping force $f_v(t)$, results in the widely used Cummins' equation [130]:

$$(M + m_\infty)\dot{v}(t) + \int_0^t h_r(\tau)v(t-\tau)d\tau - K_b x(t) = \int_{-\infty}^t h_{ex}(\tau)\eta(t-\tau)d\tau \tag{2.38}$$

Typically, $h_{ex}(\tau)$ and $h_r(\tau)$ are calculated numerically using boundary-element potential methods such as WAMIT [131], which performs the calculations in the frequency domain, or ACHIL3D [132], where time-domain calculations are used. Equation (2.38) can also be used to model multibody systems [133] or arrays of devices [125], with the modification that the dimensions of M, m_∞, K and the hydrodynamic parameters represented by $h_{ex}(\tau)$ and $h_r(\tau)$ all increase in dimension accordingly. Typically, for both simulation and control applications, the radiation damping convolution term in (2.35) is replaced by a closed form (finite order) equivalent. This replacement has several advantages. The integro-differential equation in (2.38) is replaced by a higher order differential equation, making analysis more straightforward; the resulting finite-order dynamical system is faster to simulate and the closed-form dynamical equation can be used as a basis for model-based control design.

In general, $h_r(\tau)$ (and its Fourier transform, $H_r(\omega)$) is nonparametric in form, being the result of a numerical calculation on a distributed system. Approximations can be determined in either the time or frequency domain, depending on the manner in which $h_r(t) \leftrightarrow H_r(\omega)$ was determined, and the intended (time/frequency domain) use of the finite-order approximation. For example, WAMIT [131] uses a frequency-domain analysis to determine $H_r(\omega)$ directly, and approximations based on WAMIT data are usually based on frequency-domain error criteria. In such a case, a state-space forms may be determined using frequency-domain identification [134] or via moment matching [135]. Alternatively, if $h_r(\tau)$ is directly produced, for example, from the time-domain code ACHIL3D [132], time-domain impulse-response fitting can be employed, typically using the method in [136].

Cummins' equation has become somewhat ubiquitous in wave energy modelling, but care must be taken in its use. Since the objective is to extract energy from the waves, generally WECs are designed to oscillate (resonate), so the assumption of small (displacement) oscillations may be challenged. In addition, the assumption of an inviscid fluid may also be unreasonable, when the relative WEC/fluid motion becomes significant. In particular under controlled conditions, greater excursions in both displacement and velocity may occur under controlled conditions, as shown in Figure 2.41.

For point absorber WECs, radiation forces can be generally assumed to be linear, while non-linear Froude–Krylov forces (relating to variations in the wetted surface) and non-linear viscous forces can be added to the basic model in (2.38). Non-linear viscous forces can be added (e.g. to (2.38)) using a term experimentally derived by [137], such as

$$f_v(t) = \rho R C_d |v(t)| v(t) \tag{2.39}$$

for a cylindrical shape, where r is the water density, R is the cylinder radius and C_d is the drag coefficient. Empirical validations of (2.39) have proven its validity, and its methods have been proposed to evaluate the coefficient C_d for certain specific shapes [138,142]. In addition, a linear approximation to (2.39) may be derived, using an energy-matching technique [143], if desired. Particularly for heaving point absorbers with a non-uniform cross-sectional area, non-linear Froude–Krylov forces may be relevant, in which case some analytical models

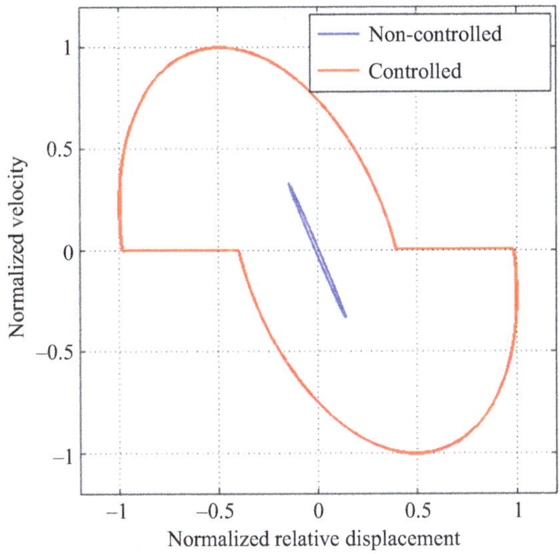

Figure 2.41 Displacement and velocity excursions under uncontrolled conditions and latching control

may be used, where the device is axisymmetric [144], with a mesh-based approach required otherwise [145].

2.4.2 WEC control

In general, WECs are designed to resonate with the waves. However, devices can only usually be designed to have one mechanical resonance period, which is normally placed at the predominant wave frequency. To permit the device to resonate at other wave periods, maximizing captured power, the load on the WEC is adjusted electronically, via a control algorithm [106].

2.4.2.1 WEC control basics

In general, the control objective for a WEC is to maximize captured energy, while observing the physical constraints of the device. Ignoring system constraints for the moment, a start can be made on the energy maximization problem by considering the force-to-velocity model of a WEC, which is obtained from (2.38) in the frequency domain [93] as

$$\frac{V(\omega)}{F_{ex}(\omega) + F_u(\omega)} = \frac{1}{Z_i(\omega)} \tag{2.40}$$

where $Z_i(\omega)$ is termed the intrinsic impedance of the system. In (2.40), $V(\omega)$, $F_{ex}(\omega)$ and $F_u(\omega)$ represent the Fourier transform of the velocity $v(t)$, excitation force $f_{ex}(t)$ and control force $f_{PTO}(t)$, respectively. Unless stated otherwise, the Fourier transform of time-domain signals or functions will be denoted by the corresponding capital letter, namely $X(\omega) \triangleq \mathcal{F}\{x(t)\}$.

The intrinsic impedance $Z_i(\omega)$ of the model in (2.40) is specified as (see [93]) for the full derivation:

$$Z_i(\omega) = B_r(\omega) + \omega\left(M + M_a(\omega) - \frac{K_b}{\omega^2}\right) \tag{2.41}$$

where $B_r(\omega)$ is the radiation resistance (real and even) and $M_a(\omega)$ is the frequency-dependent added mass, often replaced by its high-frequency asymptote m_∞.

The model in (2.40) allows the derivation of conditions for optimal energy absorption and the intuitive design of the energy-maximizing controller in the frequency-domain [93] as

$$Z_{PTO}(\omega) = Z_i^*(\omega) \tag{2.42}$$

where $()^*$ denotes the complex conjugate. The choice of Z_{PTO} as in (2.42) is referred to as complex conjugate control, but many (especially electrical) engineers will recognize this choice of Z_{PTO} as the solution to the impedance-matching problem represented by Figure 2.42. The result in (2.42) has a number of important implications:

- The result is frequency dependent, implying that there is a different optimal impedance for each frequency, which raises the question of how to specify for irregular seas containing a mixture of frequencies.
- Since $h_r(\tau)$ is causal, $h_c(t) = \mathcal{F}^{-1}\{Z_{PTO}(\omega)\}$ is anticausal, requiring future knowledge of the excitation force. While this knowledge is straightforward for the monochromatic case (single sinusoid), it is more problematic for irregular seas.
- Since force and velocity can have opposite signs in Figure 2.42, the PTO may need to *supply* power for some parts of the sinusoidal cycle, which is akin to reactive power in electrical power systems. Such a phenomenon places

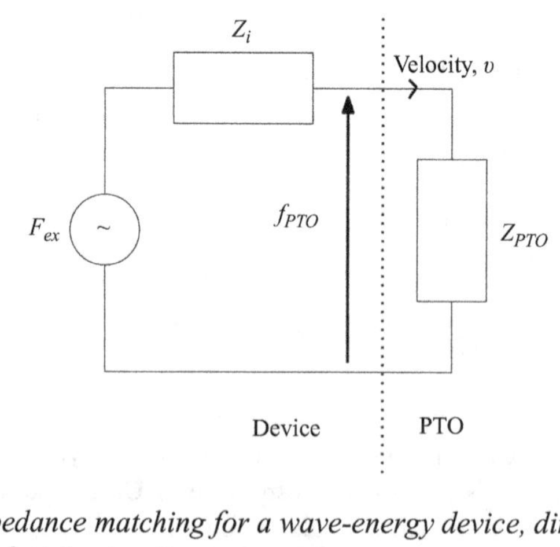

Figure 2.42 Impedance matching for a wave-energy device, directly analogous to its electric circuit counterpart

particular demands on PTO systems, not only in terms of the need to facilitate bidirectional power flow but also that the peak reactive power can be significantly greater than active power [146]. The optimal *passive* PTO is provided by $R_{PTO} = |Z_i(\omega)|$, which avoids the need for the PTO to supply power but results in a suboptimal control.

- The optimal control in (2.42) takes no account of physical constraints in the WEC/PTO, where there are likely to be limitations on displacement or relative displacement, and the PTO force, and there may be external constraints imposed by electrical grid regulations.

The condition in (2.42) can alternatively be expressed in terms of an optimal velocity profile as

$$V^{opt}(\omega) = \frac{F_{ex}(\omega)}{2R_i(\omega)} \tag{2.43}$$

where $R_i = 1/2(Z_i + Z_i^*)$ is the real part of Z_i. The condition in (2.43) is a condition on the amplitude of $V^{opt}(\omega)$, with the restriction that $v_{opt}(t)$ be in phase with $f_{ex}(t)$, since R_i is a real (and even) function. This phase condition, considered separately, forms the basis for some simple WEC phase-control strategies, such as latching.

2.4.2.2 WEC control structures

Based on the fundamentals of Section 2.4.2.1, two general control structured have evolved, as depicted in Figure 2.43. The ACC controller presents a simple feedback structure; there the PTO force, F_{PTO}, is determined directly from the measured device velocity. While this structure works well for monochromatic (perfectly sinusoidal) waves, it has the drawback that a suboptimal formulation must be used to cater for irregular (realistic) waves [147]. In addition, the ACC structure makes it very difficult to ensure that physical limits on displacement and PTO force are respected. The AVT controller structure, of which there are many variants [148], allows physical constraints and irregular sea conditions to be handled by virtue of a numerical optimization procedure but is more complex computationally. In addition, $F_{ex}(t)$ must be estimated and forecast into the future [149]. The AVT structure also has better robustness to modelling errors than the AVT structure [150].

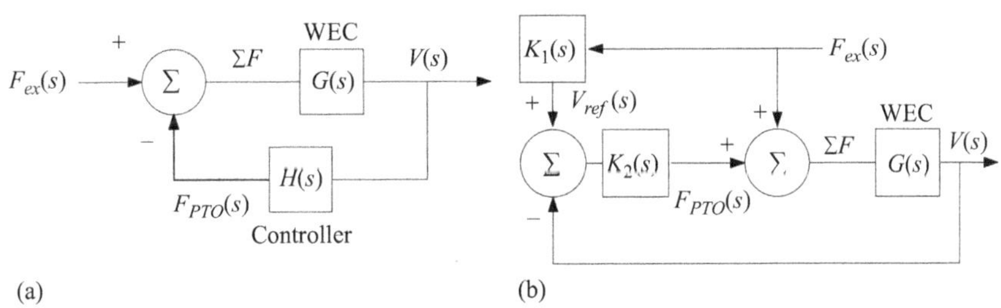

Figure 2.43 *(a) ACC and (b) AVT controller structures*

References

[1] Cornett AM. A global wave energy resource assessment. In: The Eighteenth International Offshore and Polar Engineering Conference, International Society of Offshore and Polar Engineers; 2008.

[2] Fusco F, Nolan G, and Ringwood JV. Variability reduction through optimal combination of wind/wave resources–An Irish case study. Energy, 2010; 35(1):314–325.

[3] Astariz S, Vazquez A, and Iglesias G. Evaluation and comparison of the levelized cost of tidal, wave, and offshore wind energy. Journal of Renewable and Sustainable Energy, 2015;7(5):053112.

[4] Koca K, Kortenhaus A, Oumeraci H, *et al.* Recent advances in the development of wave energy converters. In: Proc. 9th European Wave Tidal Energy Conf., Uppsala, Sweden; 2013.

[5] Young IR. Seasonal variability of the global ocean wind and wave climate. International Journal of Climatology, 1999;19:931–950.

[6] Hanley KE, Belcher SE, and Sullivan P. A global climatology of wind–wave interaction. Journal of Physical Oceanography, 2010;40:1263–1282. DOI: 10.1175/2010JPO4377.1.

[7] Vicinanza D, Ciardulli F, Buccino M, Calabrese M, and Koefed JP. Wave loadings acting on an innovative breakwater for energy production. Journal of Coastal Research, 2011;64:608–612.

[8] Besio G, Mentaschi L, and Mazzino A. Wave energy resource assessment in the Mediterranean Sea on the basis of a 35-year hindcast. Energy, 2016; 94:50–63.

[9] Liberti L, Carillo A, and Sannino G. Wave energy resource assessment in the Mediterranean, the Italian perspective. Renewable Energy, 2013;50: 938–949.

[10] Günther H, and Behrens A. *The WAM Model Validation Document Version 4.5.3, Tech. Rep.* Institute of Coastal Research Helmholtz-Zentrum Geesthacht (HZG); 2011.

[11] Liberti L, Carillo A, and Sannino G. Wave energy potential in the Mediterranean, the case of Pantelleria. In: Proceeding of OWEMES (Offshore Wind and Other Marine Renewable Energies), Rome; 2012.

[12] Datawell BV. *History of Datawell.* Haarlem: Datawell BV, B.13.01; 2001, pp. 1–4.

[13] Pinkel R, and Smith JA. Open ocean surface wave measurement using Doppler sonar. Journal of Geophysical Research, 1987;92:12967–12973.

[14] Korde UA, Robinett RD, Wilson DG, Bacelli G, and Abdelkhalik OO. Wave-by-wave control of a wave energy converter with deterministic wave prediction. In: Proceedings of EWTEC 13th European Wave and Tidal Energy Conference, Cork; August 2017, pp. 673-1–673-10.

[15] Li G, Weiss G, Mueller M, Townleys S, and Belmont M. Wave energy converter control by wave prediction and dynamic programming. Renewable Energy, 2012;48:392–403.

[16] BSI. *Marine Energy – Wave, Tidal and Other Water Current Converters, Part 101: Wave Energy Resource Assessment and Characterization*. London: The British Standards Institution; 2015. PD IEC/TS 62600-101:2015.
[17] Vögler A, Venugopal V, and Armstrong D. Wave sensor observations during a severe storm event at a marine energy development site. In: Proceedings of EWTEC 12th European Wave and Tidal Energy Conference, Nantes; 2015.
[18] Hasselmann K, Barnett TP, Bouws E, *et al.* Measurements of wind-wave growth and swell decay during the Joint North Sea Wave Project (JONSWAP). Ergänzungsheft zur Deutschen Hydrographischen Zeitschrift Reihe, A, 1973;80(12):1–95.
[19] Pierson WJJR, and Moskowitz LA. Proposed spectral form for fully developed wind seas based on the similarity theory of S. A. Kitaigorodskii. Journal of Geophysical Research, 1964;69:5181–5190.
[20] Adcock TAA, Taylor PH, Yan S, Ma QW, and Janssen PAEM. Did the Draupner wave occur in a crossing sea?. Proceedings of the Royal Society A, 2011;467:3004–3021.
[21] Sorensen R. *Basic Wave Mechanics: For Coastal and Ocean Engineers*. New York: John Wiley & Sons, Inc.; 1993, p. 180.
[22] Greenwood C, Vogler A, Morrison J, and Murray A. The approximation of a sea surface using a shore mounted X-band radar with low grazing angle. Remote Sensing of Environment, 2018;204:439–447.
[23] Pozzi N, Bracco G, Passione B, Sirigu SA, and Mattiazzo G. PeWEC: Experimental validation of wave to PTO numerical model. Ocean Engineering, 2018;167:114–129.
[24] Bracco G, Giorcelli E, and Mattiazzo G. ISWEC: A gyroscopic mechanism for wave power exploitation. Mechanism and Machine Theory, 2011;46(10): 1411–1424.
[25] Salter SH. The use of gyros as a reference frame in wave energy converters. In: Wave Energy Utilization, Trondheim, Norway; 22–24 June 1982.
[26] Kanki H, Arii S, Furusawa T, and Otoyo T. Development of advanced wave power generation system by applying gyroscopic moment.
[27] Salcedo F, Ruiz Minguela P, Rodriguez Arias R, Ricci P, and Santos-Mugica M. OCEANTEC: Sea trials of a quarter scale prototype. In: Proceedings of 8th European Wave Tidal Energy Conference, Uppsala, Sweden; January 2009.
[28] Falnes J. *Ocean Waves and Oscillating Systems*. Cambridge: Cambridge University Press; 2010.
[29] Vissio G, Valério D, Bracco G, Beiráo P, Pozzi N, and Mattiazzo G. ISWEC linear quadratic regulator oscillating control. Renewable Energy, 2017.
[30] Vissio G. ISWEC toward the Sea – Development, Optimization and Testing of the Device Control Architecture, PhD Thesis, Politecnico di Torino; 2017, pp. 164–214.
[31] Sirigu SA, Bracco G, Bonfanti M, Dafnakis P, and Mattiazzo G. On-board sea state estimation method validation based on measured floater motion. In: Proceedings of the IFAC-CAMS 2018, Opatija, Croatia; September 2018.

[32] The European Marine Energy Centre. *Tank Testing of Wave Energy Conversion Systems*. The European Marine Energy Centre; 2009.
[33] Passione B, Pozzi N, Sirigu SA, Bracco G, Brizzolara S, and Mattiazzo G. Numerical and experimental analysis of oscillating fluid tanks. In: Proceedings of ISOPE, San Francisco, USA; 2017.
[34] Contestabile P, and Vicinanza D. Coastal defence integrating wave-energy-based desalination: A case study in Madagascar. Journal of Marine Science and Engineering, 2018;6(2):64.
[35] Contestabile P, Di Lauro E, Buccino M, and Vicinanza D. Economic assessment of Overtopping BReakwater for Energy Conversion (OBREC): A case study in Western Australia. Sustainability, 2017a;9(1):51.
[36] Falcáo AFdO. Wave energy utilization: A review of the technologies. Renewable and Sustainable Energy Reviews, 2010;14:899–918.
[37] Evans DV. Wave-power absorption by system of oscillation surface pressure distribution. Journal of Fluid Mechanism, 1982;114:481–499.
[38] Takahashi S. A study on design of a wave power extracting caisson breakwater. Wave Power Laboratory. Yokosuka, Japan: Port and Harbour Research Institute, 1988.
[39] Müller GU, and Whittaker TJT. An investigation of breaking wave pressures on inclined walls. Ocean Engineering, 1993;20(4):349–358.
[40] Müller GU, and Whittaker TJT. Field measurements of breaking wave loads on a shoreline wave power station. In: ICE Proc. Water Maritime and Energy; September 1995, pp. 187–197. ISBN: 0965-0946.
[41] Jayakumar VS. Wave Force on Oscillating Water Column Type Wave Energy Caisson: An Experiment Study, PhD Thesis, Department of Ocean Engineering, Indian Institute of Technology, Madras, India; 1994.
[42] Boccotti P. On a new wave energy absorber. Ocean Engineering, 2003;30: 1191–1200.
[43] Boccotti P, Filianoti P, Fiamma V, and Arena F. Caisson breakwaters embodying an OWC with a small opening—Part II: A small-scale field experiment. Ocean Engineering, 2007b;34(5–6):820–841.
[44] Strati FM, Malara G, and Arena F. Performance optimization of a U-Oscillating-Water-Column wave energy harvester. Renewable Energy, 2016;99:1019–1028.
[45] Malara G, Gomes RPF, Arena F, Henriques JCC, Gato LMC, and Falcáo AFO. The influence of three-dimensional effects on the performance of U-type oscillating water column wave energy harvesters. Renewable Energy, 2017a;111:506–522.
[46] Malara G, Romolo A, Fiamma V, and Arena F. On the modelling of water column oscillations in U-OWC energy harvesters. Renewable Energy, 2017b;101:964–972.
[47] Hull P, and Müller G. An investigation of breaker heights, shapes and pressures. Ocean Engineering, 2002;29:59–79.
[48] Thiruvenkatasamy K, Neelamani S, and Sato M. Nonbreaking wave forces on multiresonant oscillating water column wave power caisson breakwater.

Journal of Waterway, Port, Coastal, and Ocean Engineering, 2005;131(2): 77–84.

[49] Preen S, and Robertshaw G. Development of a generic caisson design for an oscillating water column power generator. In: Proceedings of 9th International Conference on Coasts, Marine Structures and Breakwaters, Thomas Telford Ltd; 2009, pp. 2–266.

[50] Huang Y, Shi H, Liu D, and Liu Z. Study on the breakwater caisson as oscillating water column facility. Journal of Ocean University of China (Oceanic and Coastal Sea Research), 2010;9(3):244–250.

[51] Liu Y, Shi H, Liu Z, and Ma Z. Experiment study on a new designed OWC caisson breakwater. In: Proceedings of the 2011 Asia-Pacific Power and Energy Engineering Conference, Wuhan, China; 2011 pp. 1–5.

[52] Kuo YS, Lin CS, Chung CY, and Wang YK. Wave loading distribution of oscillating water column caisson breakwaters under non-breaking wave forces. Journal of Marine Science and Technology, 2015;23(1):78–87.

[53] Allsop W, Bruce T, Alderson J, *et al.* Large scale tests on a generalised oscillating water column wave energy converter. In: Proc. HYDRALAB IV Joint User Meeting, Lisbon; 2014.

[54] Viviano A, Naty S, Foti E, Bruce T, Allsop W, and Vicinanza D. Large-scale experiments on the behaviour of a generalised Oscillating Water Column under random waves. Renewable Energy, 2016;99:875–887.

[55] Ashlin SJ, Sundar V, and Sannasiraj SA. Pressures and forces on an oscillating water column–type wave energy caisson breakwater. Journal of Waterway, Port, Coastal, and Ocean Engineering, 2017;143(5):04017020.

[56] Tseng RS, Wu RH, and Huang CC. Model study of a shoreline wave-power system. Ocean Engineering, 1999;27:801–821.

[57] Bruce T, and O'Callaghan J. Oscillating water column wave energy converters at the coast: A review and forward look. In: Coastal Structures 2011: (In 2 Volumes); 2013, pp. 345–356.

[58] Aguado A, and Sánchez-Naverac V. Nuevo tipo de sección para diques en talud con espaldón. Revista de obras publicas, 1978:353–359.

[59] Aminti P, and Franco L. Performance of overspill basin on top of breakwaters. In: International Conference in Ocean Engineering, Madras, India; 2001, pp. 41–46.

[60] Geeraerts J, De Rouck J, Beels C, Gysens S, and De Wolf P. Reduction of wave overtopping at seadikes: Stilling Wave Basin (SWB). In: Coastal Engineering 2006: (In 5 Volumes); 2007, pp. 4680–4691.

[61] Burcharth HF, and Andersen TL. Overtopping of rubble mound breakwaters with front reservoir. In: Coastal Engineering 2006: (In 5 Volumes); 2007, pp. 4605–4615.

[62] Cappietti L, and Aminti PL. Laboratory investigation on the effectiveness of an overspill basin in reducing wave overtopping on marina breakwaters. Coastal Engineering Proceedings, 2012;1(33):20.

[63] Kofoed JP, Vicinanza D, and Osaland E. Estimation of design wave loads on the SSG WEC pilot plant based on 3-D model tests. In: The Sixteenth

International Offshore and Polar Engineering Conference, International Society of Offshore and Polar Engineers; January 2006.

[64] Margheritini L, Frigaard P, and Vicinanza D. Sea Slot Cone Generator overtopping performance in 3D conditions. In: The Eighteenth International Offshore and Polar Engineering Conference, International Society of Offshore and Polar Engineers; January 2008.

[65] Vicinanza D, and Frigaard P. Wave pressure acting on a seawave slot-cone generator. Coastal Engineering, 2008;55(6):553–568.

[66] Oumeraci H, Kortenhaus A, Allsop W, *et al. Probabilistic Design Tools for Vertical Breakwaters*. Rotterdam: CRC Press; 2001.

[67] Takahashi S, Tanimoto K, and Shimosako K. A proposal of impulsive pressure coefficient for design of composite breakwaters. In: Proc. of International Conference on Hydro-technical Eng. for Port and Harbour Construction, Port and Harbour Res. Inst.; 1994.

[68] Vicinanza D, Nørgaard JH, Contestabile P, and Andersen TL. Wave loadings acting on overtopping breakwater for energy conversion. Journal of Coastal Research, 2013;65(sp2):1669–1674.

[69] Vicinanza D, Contestabile P, Nørgaard J, and Lykke-Andersen T. Innovative rubble mound breakwaters for overtopping wave energy conversion. Coastal Engineering, 2014;88:154–170.

[70] Vicinanza D, Stagonas D, Müller G, Nørgaard JH, and Andersen TL. Innovative breakwaters design for wave energy conversion. Coastal Engineering Proceedings, 2012;1(33):1.

[71] Iuppa C, Contestabile P, Cavallaro L, Foti E, and Vicinanza D. Hydraulic performance of an innovative breakwater for overtopping wave energy conversion. Sustainability, 2016;8(12):1226.

[72] Goda Y, Nakada H, Ohneda H, Suzuki M, Takahashi S, and Shikamori M. Results of field experiment of a wave power extracting caisson breakwater. In: Proceedings of Civil Engineering in the Ocean, vol. 7, Japan Society of Civil Engineers; 1991, pp. 143–148.

[73] Nakada H, Ohneda H, Takahashi S, Shikamori M, and Nakazono T. Field experiments of a wave power converter with caisson breakwater. In: *Renewable Energy, Technology and the Environment*, 3rd ed., Reading: University of Reading; 1992, pp. 2530–2535.

[74] Takahashi S, Nakada H, Ohneda H, and Shikamori M. Wave power conversion by a prototype wave power extracting caisson in Sakata port. In: Proc. of 23rd International Conference on Coastal Engineering, ASCE, New York; 1992, pp. 3440–3453.

[75] Funakoshi H, Katoh H, Saito T, Takahashi S, and Oikawa K. Development of a wave power extracting caisson breakwater. Doboku Gakkai Ronbunshu, 1993;474:37–39.

[76] Suzuki M, Arakawa C, and Takahashi S. Performance of wave power gene rating system installed in breakwater at Sakata port in Japan. In: Proc. of ISOPE; 2004, pp. 202–209.

[77] Heath T, Whittaker TJT, and Boake CB. The design, construction and operation of the LIMPET wave energy converter. Wave Energy, 2000;2000.

[78] Boake CB, Whittaker TJT, Folley M, and Ellen H. Overview and initial operational experience of the LIMPET wave energy plant. In: 12th International Offshore and Polar Engineering Conference, vol. 1, Kyushu, Japan; 2002, pp. 586–594.

[79] McStay P. An Experimental Study of Cylindrical Oscillating Water Column Wave Power Devices, MSc Thesis, The Queen's University of Belfast; 1995.

[80] Brito-Melo A, Neuman F, and Sarmento AJNA. Full-scale data assessment in OWC Pico plant. International Journal of Offshore and Polar Engineering, 2008;18(01):1053–5381.

[81] Arup Energy, The Carbon Trust – Marine Energy Challenge. *Oscillating Water Column Wave Energy Converter Evaluation Report*. The Carbon Trust; 2005.

[82] Neumann F, and Le Crom I. Pico OWC – The Frog Prince of Wave Energy? Recent autonomous operational experience and plans for an open real-sea test centre in semi-controlled environment. In: Proceedings of the 9th European Wave and Tidal Energy Conference (EWTEC 2011), Southampton, UK; 5–9 September 2011.

[83] Falcào AF, and Henriques JC. Oscillating-water-column wave energy converters and air turbines: A review. Renewable Energy, 2015;85:1391–1424.

[84] Torre-Enciso Y, Ortubia I, Lopez de Aguileta LI, and Marques J. Mutriku wave power plant: From the thinking out to the reality. In: Proc., 8th European Wave Tidal Energy Conf., Uppsala, Sweden; 2009, pp. 319–329.

[85] Ibarra-Berastegi G, Sáenz J, Ulazia A, Serras P, Esnaola G, and Garcia-Soto C. Electricity production, capacity factor, and plant efficiency index at the Mutriku wave farm (2014–2016). Ocean Engineering, 2018;147:20–29.

[86] Arena F, Romolo A, Malara G, and Ascanelli A. On design and building of a U-OWC wave energy converter in the Mediterranean Sea: A case study. In: Proceedings of the 32nd International Conference on Ocean, Offshore and Arctic Engineering OMAE2013, Nantes, France; 2013.

[87] Arena F, Romolo A, Malara G, Fiamma V, and Laface V. The first worldwide application at full-scale of the REWEC3 device in the Port of Civitavecchia: Initial energetic performances. In: Progress in Renewable Energies Offshore: Proceedings of the 2nd International Conference on Renewable Energies Offshore (RENEW2016), CRC Press, Lisbon, Portugal; 24–26 October 2016, p. 303.

[88] Arena F. Un impianto REWEC3 per la produzione di energia elettrica da moto ondoso: dall'invenzione del Prof. Paolo Boccotti alla costruzione del primo prototipo. In: Workshop Energia dal Mare le Nuove Tecnologie per i Mari Italiani, Rome; 2014. Available at: http://www.enea.it/it/enea_informa/events/energia-dal-mare/Arena1.pdf. Accessed 20.09.2018.

[89] Contestabile P, Ferrante V, Di Lauro E, and Vicinanza D. Prototype overtopping breakwater for wave energy conversion at port of Naples. In: The 26th International Ocean and Polar Engineering Conference, International Society of Offshore and Polar Engineers, Rhodes, Greece; 26 June–2 July 2016.

[90] Falnes J. A review of wave-energy extraction. Marine Structures, 2007; 20:185–201.

[91] Margheritini L, Vicinanza D, and Frigaard P. SSG wave energy converter: Design, reliability and hydraulic performance of an innovative overtopping device. Renewable Energy, 2009;34:1371–1380.
[92] Stefano P, Jens Peter K, and Erik F-M. Experimental update of the overtopping model used for the wave dragon wave energy converter. Energies, 2013;6:1961–1992.
[93] Falnes J. *Ocean Waves and Oscillating Systems*. Cambridge, UK: Cambridge University Press; 2002.
[94] McCormick ME. *Ocean Wave Energy Conversion*. Mineola, New York: Dover; 2007.
[95] Hager R, Fernandez N, and Teng MH. Experimental study seeking optimal geometry of a heaving body for improved power absorption efficiency. In: Proceedings of the 22nd International Offshore and Polar Engineering Conference (ISOPE); 2012.
[96] Vantorre M, Banasiak R, and Verhoeven R. Modelling of hydraulic performance and wave energy extraction by a point absorber in heave. Applied Ocean Research, 2004:61–72.
[97] Hardisty J. Experiments with point absorbers for wave energy conversion. Journal of Marine Engineering & Technology, 2012;11:51–62.
[98] Ionescu D, and Ngwenya BS. Innovative design of a sea wave energy harnesser. In: International Conference on Renewable Energies and Power Quality (ICREPQ'14), Cordoba, Spain; 2014.
[99] Marquis L, Kramer M, and Frigaard P. First power production figures from the wave star Roshage wave energy converter. In: 3rd International Conference on Ocean Energy, ICOE, Bilbao; 2010.
[100] Coiro DP, Troise G, Calise G, and Bizzarrini N. Wave energy conversion through a point pivoted absorber: Numerical and experimental tests on a scaled model. Renewable Energy, 2015.
[101] Coiro DP, Troise G, and Bizzarrini N. Experiences in developing tidal current and wave energy devices for Mediterranean Sea. Frontiers in Energy Research, 2018;6:136. DOI: 10.3389/fenrg.2018.00136.
[102] Faltinsen OM. *Sea Loads on Ships and Offshore Structures*. Cambridge: Cambridge University Press; 1990.
[103] Coiro DP, Troise G, Calise G, and Bizzarrini N. Experimental test and numerical shape optimization of a point pivoted absorber for wave energy. In: VI International Conference on Computational Methods in Marine Engineering, Marine 2015, Rome, Italy; 2015.
[104] Castellini L, Martini M, and Alessandri G. Development and testing of a ballscrew electromechanical generator (EMG) for wave energy conversion. In: European Wave and Tidal Energy Conference (EWTEC); 2017.
[105] The Specialist Committee on Waves. Final report and recommendations to the 23rd ITTC. In: Proceedings of the 23rd ITTC – Volume II, Venice, Italy; 2002.

[106] Ringwood JV, Bacelli G, and Fusco F. Energy-maximizing control of wave energy converters: The development of control system technology to optimize their operation. IEEE Control Systems, 2014;34(5):30–55.

[107] IEA-OES. *International Levelised Cost of Energy for Ocean Energy Technologies, Tech. Rep.*; May 2015.

[108] SI Ocean Project. *Wave and Tidal Energy Market Deployment Strategy for Europe, Tech. Rep.*; June 2014.

[109] SI Ocean Project. *Ocean Energy: Cost of Energy and Cost Reduction Opportunities, Tech. Rep.*; May 2013.

[110] Tetu A. Chapter 8 – Power-take-off systems for WECs. In: A. Pecher, and P. Kofoed (Eds.), *Handbook of Ocean Wave Energy*. Cham: Springer; December 2016.

[111] Grimwade J, Hails D, Robles E, *et al. MARINET Deliverable 2.3 – Review of relevant PTO systems, Tech. Rep.*; 2012.

[112] Wave Energy Scotland. *International Search for Innovative Power Take-off Systems for Wave Energy Results in £7m Award to Technology Innovators*; 2015.

[113] Alves M. Chapter 10 – Wave-to-wire modelling of WECs. In: A. Pecher, and P. Kofoed (Eds.), *Handbook of Ocean Wave Energy*. Springer; December 2016.

[114] Tlali P, Wang R-J, and Gerber S. Magnetic gear technologies: A review. In: Electrical Machines (ICEM), 2014 International Conference on; 2014, pp. 544–550.

[115] Atallah K, and Howe D. A novel high-performance magnetic gear. IEEE Transactions on Magnetics, 2001;37(41):2844–2846.

[116] Kouhshahi MB, and Bird JZ. Analysis of a magnetically geared lead screw. In: Energy Conversion Congress and Exposition (ECCE), Milwaukee, IEEE; 2016.

[117] Frank NW, and Toliyat HA. Analysis and design of the trans-rotary magnetic gear. In: Energy Conversion Congress and Exposition (ECCE), 2012 IEEE; 2012, pp. 3340–3347.

[118] Emec.org.uk. *Technology Readiness Levels*. EMEC: European Marine Energy Centre.

[119] McGilton B, Mueller M, and Crozier R. Review of magnetic gear technologies and their applications in marine energy. In: Renewable Power Generation; 2016.

[120] Brochu P. *Dielectric Elastomers for Actuation and Energy Harvesting*. Los Angeles: University of California; 2012.

[121] Andritsch T, Morshuis PHF, Smit JJ, *et al.* Challenges of using electroactive polymers in large-scale wave energy converters. In: Conference on Electrical Insulation and Dielectric Phenomena; 2012.

[122] Wave Energy Scotland. *Ballscrew-based PTO Stage 2 – Public Report*; 2016.

[123] Penalba M, and Ringwood JV. A review of wave-to-wire models for wave energy converters. Energies, 2016;9(7):506.

[124] Penalba M, Davidson J, Windt C, *et al.* A high-fidelity wave-to-wire simulation platform for wave energy converters: Coupled numerical wave tank and power take-off models. Applied Energy, 2018;226:655–669.

[125] Bacelli G, Gilloteaux JC, and Ringwood J. State space model of a hydraulic power take off unit for wave energy conversion employing bondgraphs. In: Proc. World Renewable Energy Congress; 2008.

[126] Kurniawan A, Pedersen E, and Moan T. Bond graph modelling of a wave energy conversion system with hydraulic power take-off. Renewable Energy. 2012;38(1):234–244.

[127] Davidson J, and Ringwood JV. Mathematical modelling of mooring systems for wave Energy converters – A review. Energies, 2017;10(5):666.

[128] Folley M. *Numerical Modelling of Wave Energy Converters: State-of-the-Art Techniques for Single Devices and Arrays*. London: Academic Press; 2016.

[129] Giorgi G, and Ringwood JV. Computationally efficient nonlinear Froude–Krylov force calculations for heaving axisymmetric wave energy point absorbers. Journal of Ocean Engineering and Marine Energy, 2017; 3(1):21–33.

[130] Colagrossi A, and Landrini M. Numerical simulation of interfacial flows by smoothed particle hydrodynamics. Journal of Computational Physics, 2003;191(2):448–475.

[131] Currie IG, and Currie I. *Fundamental Mechanics of Fluids*. CRC Press; 2002.

[132] Yeung RW. Added mass and damping of a vertical cylinder in finite-depth waters. Applied Ocean Research, 1981;3(3):119–133.

[133] Cummins W. The impulse response function and ship motions. Schiffstechnik, 1962;9:101–109.

[134] WAMIT. *WAMIT User Manual*. Chestnut Hill, MA, USA: WAMIT Inc.; 2002.

[135] Clement AH. Using differential properties of the Green function in seakeeping computational codes. In: Proc. Euro. Control Conf.; 2009, pp. 3755–3760.

[136] Bacelli G, and Ringwood JV. A geometric tool for the analysis of position and force constraints in wave energy converters. Ocean Engineering, 2013; 65:10–18.

[137] Perez T, and Fossen T. Time-domain models of marine surface vessels for simulation and control design based on seakeeping computations. In: Proc. Of the 7th IFAC Conf. on Manoeuvring and Control of Marine Craft (MCMC); 2007.

[138] Faedo N, Peña-Sanchez Y, and Ringwood JV. Finite-order hydrodynamic model determination for wave energy applications using moment-matching. Ocean Engineering, 2018;163:251–263.

[139] Giorgi G, and Ringwood JV. Consistency of viscous drag identification tests for wave energy applications. In: Proceedings of the 12th European Wave and Tidal Energy Conference (EWTEC). Cork; 2017, pp. 643.1–643.8.

[140] Prony R. Essai Experimental et Analytique: Sur les lois de la Dilatabilite des fluides elastiques et sur celles de la Force expensiver de la vapeur del'eau et de la vapeur de l'alkool, a differentes temperatures. Paris J l'Ecola Polytechnique, 1795;1:24–76.

[141] Morison JR, O'Brien MP, Johnson JW, *et al.* The Force Exerted by Surface Waves on Piles. AIME Petroleum Transactions, 1950;189:149–154.

[142] Bhinder MA, Babarit A, Gentaz L, *et al.* Effect of viscous forces on the performance of a surging wave energy converter. In: Proc. Conf. of the Intl. Society of Offshore and Polar Engineers (ISOPE); 2012, pp. 545–550.

[143] Bacelli G, Balitsky P, and Ringwood JV. Coordinated control of arrays of wave energy devices – Benefits over independent control. IEEE Transactions on Sustainable Energy, 2013;4:1091–1099.

[144] Giorgi G, and Ringwood JV. Analytical representation of nonlinear Froude–Krylov forces for 3-DoF point absorbing wave energy devices. Ocean Engineering, 2018;164:749–759.

[145] Gilloteaux JC. *Mouvements de grande amplitude d'un corps flottant en fluide parfait. Application à la récupération de l'énergie des vagues*. Ecole Centrale de Nantes (ECN); Université de Nantes; 2007.

[146] Shek J, Macpherson D, and Mueller M. Phase and amplitude control of a linear generator for wave energy conversion. In: 4th IET Conf. on Power Electronics, Machines and Drives (PEMD); 2008, pp. 66–70.

[147] Nielsen SRK, Zhou Q, Kramer MM, *et al.* Optimal control of nonlinear wave energy point converters. Ocean Engineering, 2013;72:176–187.

[148] Faedo N, Olaya S, and Ringwood JV. Optimal control, MPC and MPC-like algorithms for wave energy systems: An overview. IFAC Journal of Systems and Control, 2017;1:37–56.

[149] Garcia-Abril M, Paperella F, and Ringwood JV. Excitation force estimation and forecasting for wave energy applications. In: Proceedings of the 20th IFAC World Congress, Toulouse; 2017, pp. 15257–15262.

[150] Ringwood JV, Merigaud A, Faedo N, *et al.* Wave energy control systems: Robustness issues. In: Proceedings of the 11th IFAC Conference on Control Applications in Marine Systems, Robotics, and Vehicles (CAMS), Opatija; 2018.

Chapter 3

Tidal and current energy

Brian Kirke[a] *and Domenico P. Coiro*[b]

3.1 Introduction

Electrical energy can be produced from various offshore sources including waves, offshore wind, ocean thermal energy conversion, ocean currents and tides. Tidal energy can be harnessed either from the potential energy in tidal rise and fall, or from the kinetic energy in flowing water. The energy in rivers can also be harnessed either from potential energy (conventional hydro) or from the kinetic energy in flowing water (hydrokinetic energy). This chapter deals with both tidal rise and fall energy (Section 3.3) and hydrokinetic energy from tides, ocean currents and rivers. (River energy is included because the proposed technologies share some characteristics with those proposed for tidal currents.) Out of the many devices proposed, some have been built, but most of these have disappeared as a result of either lack of financial support, poor design or poor management. The aim of this chapter is not to catalogue every device past or present, but rather to summarize the main types of technology and the most interesting and promising.

3.1.1 Characteristics of tidal current energy

3.1.1.1 Predictability

The big advantage of tidal current energy exploitation is that hourly, daily and yearly tidal flows can be predicted very accurately years ahead, unlike wind and solar power. Figure 3.1 shows predicted tidal flow velocities for a year at Cape Jervis, South Australia, provided in advance by the Australian National Tidal Facility (NTF) [1].

3.1.1.2 Daily and two-weekly variation in tidal current energy

Although predictable, flow velocities change constantly, typically dropping briefly to zero at slack water twice daily, with peak flows varying on a two-weekly cycle corresponding to full moon and new moon, and there is also some variation on a yearly cycle. Peak flow velocities are also highly site-specific.

[a]Barbara Hardy Institute, University of Australia, Adelaide, Australia
[b]Department of Industrial Engineering, University of Naples Federico II, Naples, Italy

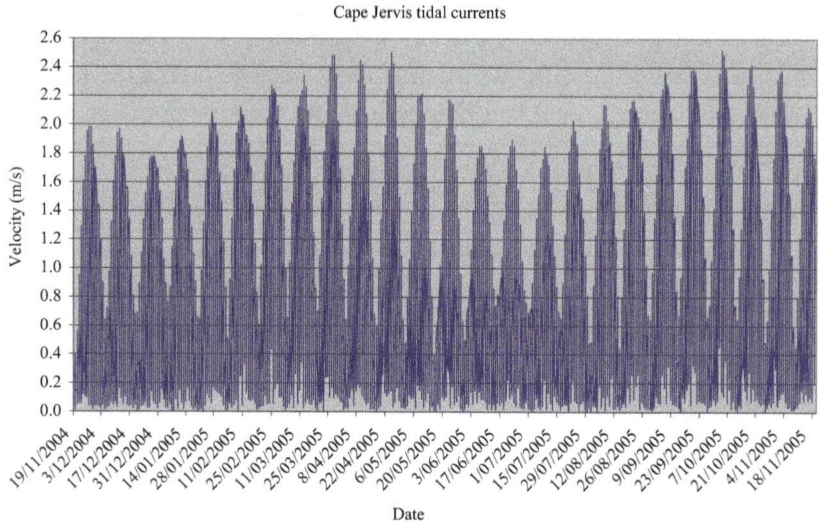

Figure 3.1 Predicted tidal flows at Cape Jervis, South Australia over a year (courtesy NTF)

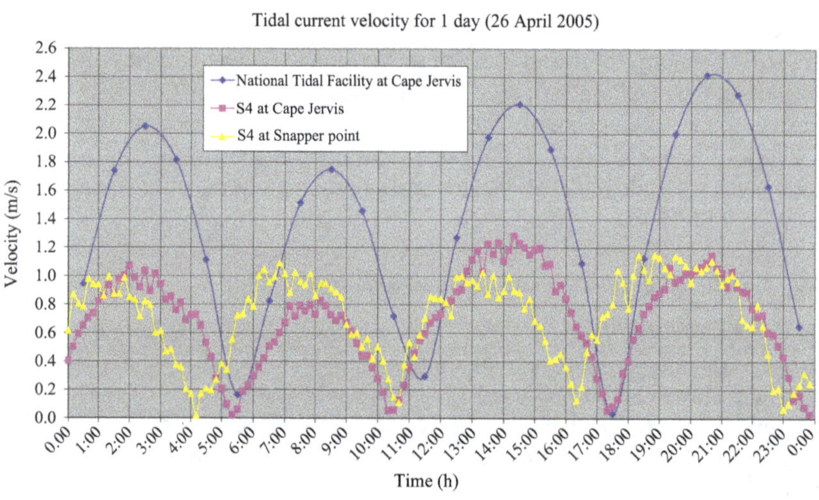

Figure 3.2 Predicted and measured tidal flows at Cape Jervis, over a 24-h period [2] (Bachmann, Tidal Current Energy in Backstairs Passage)

Figure 3.2 shows predicted velocities at Cape Jervis for a 24-h period as predicted by NTF, along with actual values recorded at two nearby sites. It will be seen that velocity typically peaks four times daily, but the peak value can vary substantially within a short distance, so calibration of models with actual measurements at the proposed site are essential.

Figure 3.3 shows velocity contours modelled for the area, and Figure 3.4 shows (i) actual locations of the two S4 current meters (green triangles), (ii) the point at

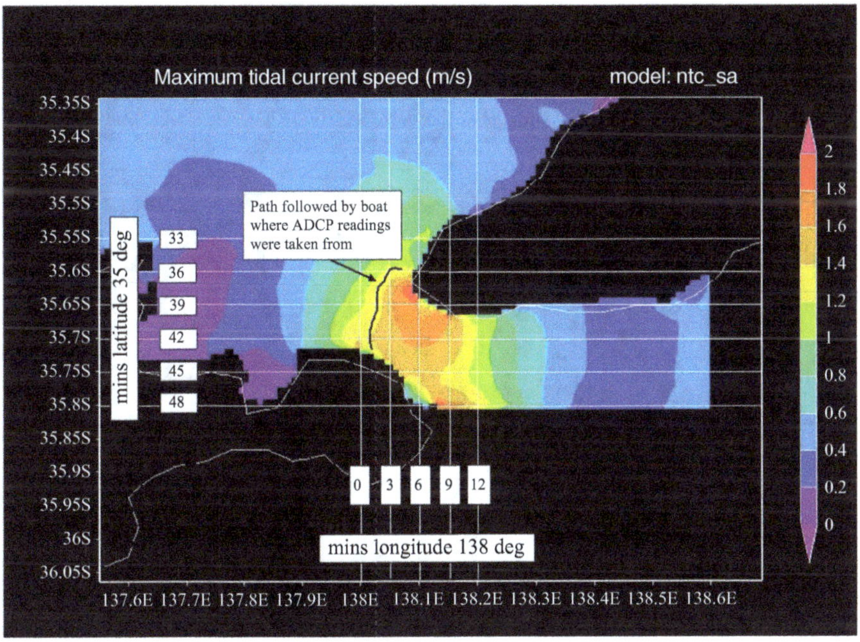

Figure 3.3 Modelled maximum tidal flows at Cape Jervis [1]

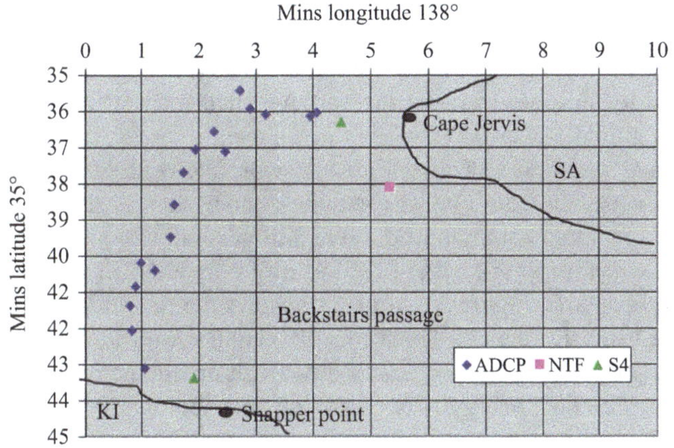

Figure 3.4 Locations of (i) two S4 current meters (green triangles), (ii) the point at which the values in Figures 3.1 and 3.2 were predicted by NTF (pink square) and (iii) the path followed by a boat taking readings with an ADCP (blue diamonds) [2]

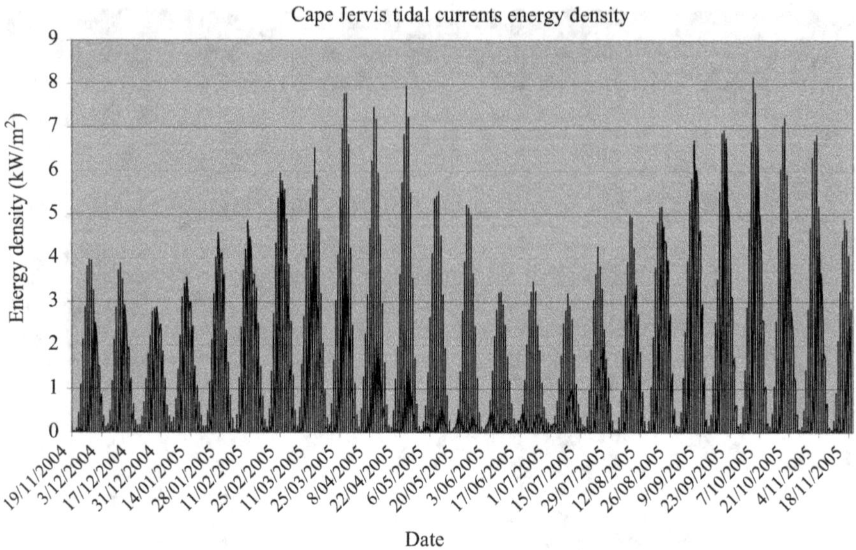

Figure 3.5 Variation in current power density

which the values in Figures 3.1 and 3.2 were predicted by NTF (red square) and (iii) the path followed by a boat taking readings with an acoustic Doppler current profiler (ADCP) (blue diamonds).

Current energy or power density varies with the cube of velocity, so the variations in available power vary even more than those of current velocity, as shown in Figure 3.5. The variations at this particular site are greater than those at most sites due to the peculiarities of the local coastline, and these figures are presented simply to illustrate some of the variables that must be taken into account when assessing tidal power potential at a given site.

The diurnal variation in available power with time is problematic for a single site without energy storage, but where two or more sites located some distance apart feed power into a common grid, there will generally be some power coming from one site when it is slack water (i.e. no flow) at another. Examples include the south coast of England. Figure 3.6, after [3], shows the time difference as the tide moves up the English Channel. However, the variations in flow corresponding to moon phases (Figure 3.5) are less easily dealt with unless the energy is fed into a large grid in which tidal energy is only a small contributor.

3.1.1.3 Energy production estimation

To capture the maximum possible energy per unit of turbine swept area at a given site, turbines would be designed without any governing mechanism. But the turbines and their mounting systems would then need to be strong enough to accept the maximum current velocity, which may only occur once or twice in a year, as shown in Figure 3.5. The system would then be very expensive, and it may be more

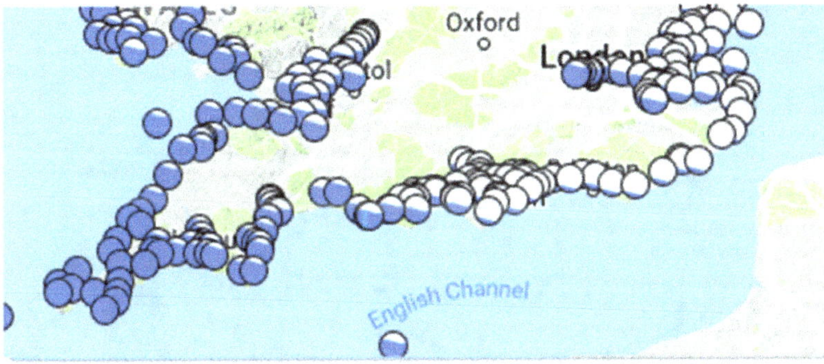

Figure 3.6 Time difference as the tide moves up the English Channel [3]

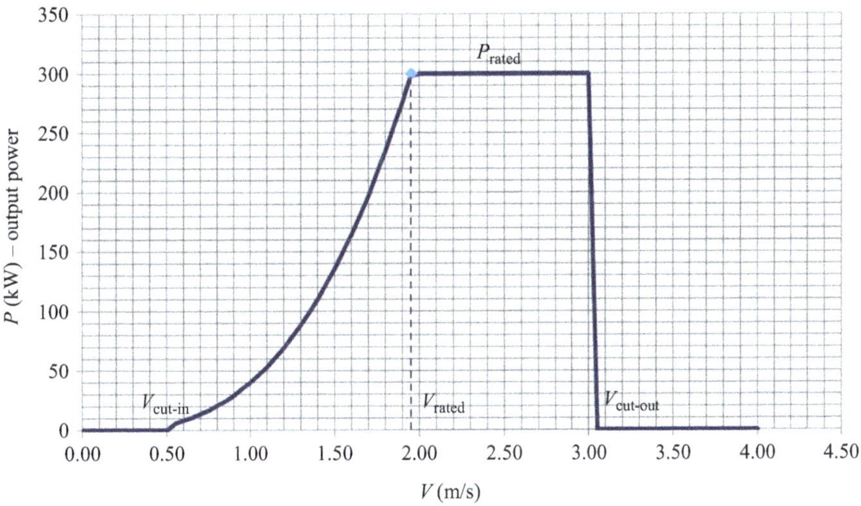

Figure 3.7 Typical power curve for a hypothetical 300-kW device

cost-effective to design less expensive systems with turbines that govern at some velocity below the absolute maximum and deliver only slightly less energy.

Given a typical power curve for a hypothetical 300-kW device delivering 300 kW rated power at 1.95 m/s, governing up to cut out at 3 m/s as shown in Figure 3.7, power production will vary with time as shown in Figure 3.8.

The total area under the power curve in Figure 3.8 as a function of the time is proportional to the total energy yield. The generated output power is zero below cut-in speed (0.5 m/s in this case), whereas, above the rated speed V_{rated}, power output is limited to the maximum generator power. If velocity increases beyond the cut-out speed, turbine operation is stopped and output power is zero. In Figure 3.8,

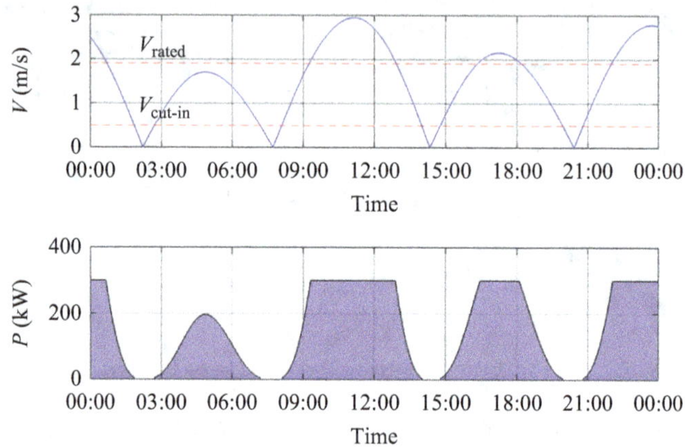

Figure 3.8 Typical daily trend of current speed in a semidiurnal tide site (interpolation of peak data) and related tidal turbine power output (power curve in Figure 3.7)

the velocity trend is obtained by interpolation of the sequence of peaks and slacks using a sinusoidal interpolation law. It has to be noted that this approach may introduce a significant approximation in current speed and power estimation, due to the real fluctuating nature of velocity variations, so care must be paid in choosing the best fitting curve of real data.

For a single device plant, the energy production assessment is based on the current temporal profile $V(t)$ and the power curve $P(V)$ of the system under consideration. Starting from the tidal time series at a site and in a certain period $\Delta T = t_f - t_i$, and assuming an almost steady behaviour of the conversion system, it is possible to assess the available energy, E_a, integrating power over time:

$$E_a = \int_{t_i}^{t_f} P(V(t))dt \tag{3.1}$$

where the available power P is estimated by the assumed device power curve.

3.1.2 Basic physics of tidal energy

Tidal rise and fall converters harness potential energy. Potential energy PE is measured in joules, megajoules, etc.

$$PE = mgh$$

where m is the mass in kg, g acceleration due to gravity (9.8 m/s^2) (approximately, varying slightly from place to place) and h the height in metres.

Energy and work: The gravitational force or weight = mg on a 1-kg mass is $1 \times 9.8 = 9.8$ N. Thus 1 L of water weighing 1 kg, falling through a height of 1 m, gives up 1 kg × 9.8 m/s^2 × 1 m = 9.8 N m = 9.8 joules (J). It can be considered to have done 9.8 J of work (work = force times distance).

Power: A flow $Q = 1$ L/s falling 1 m gives up 9.8 W (1 W = 1 N m/s) and a turbine with an efficiency η of 80% would generate $0.8 \times 9.8 = 7.84$ W of power, i.e. continuous flow of energy.

A flow of 100 m^3/s = 10^5 L/s falling 10 m gives up 9.8 MW and a turbine with an efficiency η of 90% would generate $0.9 \times 9.8 = 8.82$ MW.

Kinetic energy: Hydrokinetic turbines (HKTs) in open flow harness some of the kinetic energy in flowing water, in the same way that wind turbines capture kinetic energy from flowing air, i.e. wind. The power P captured is given by the following equation:

$$P = C_P ½ \rho A V^3$$

where C_P is the 'coefficient of performance' or 'coefficient of power'; ρ the fluid density = 1,000 kg/m^3 for freshwater, about 3% more for saltwater; A the 'swept area', i.e. the flow area occupied by the turbine – for example, a 1-m diameter axial flow (propeller type) turbine has a swept area of $\pi r^2 = 3.14 \times 0.5^2 = 0.785$ m^2; and V the flow velocity in m/s. The derivation of this equation can be found in any text on wind turbines.

Thus, for example, a 2-m diameter axial flow turbine in a 1.5-m/s flow of seawater with $C_P = 0.3$ would produce:

$P = C_P ½ \rho A V^3 = 0.3 \times ½ \times 1{,}030 \times \pi(1)^2 \times 1.5^3 = 1{,}638$ W or 1.63 kW of mechanical shaft power. Driving a generator with efficiency 80%, it would generate $1.63 \times 0.8 = 1.31$ kW electrical power.

There is a theoretical limit on C_P of 16/27 = 59.3 – i.e. in open flow it is impossible to capture more than 59.3% of the kinetic energy of the flow in the area A. This limit, called the Betz limit, is because the flow will simply go around the turbine rather than through it if there is too much resistance.

A diffuser can reduce the pressure downstream of a turbine, sucking in more flow and increasing the effective swept area A by a factor of up to about 4 (see Section 3.8).

In restricted flows like canals, small rivers and narrow tidal channels where the swept area A occupies more than about 10% of the total flow area, the Betz limit no longer applies because it becomes harder for the water to flow around the turbine or turbines, and with very high blockage up to say 60%, the apparent C_P can exceed 100%, which of course is impossible if only kinetic energy is being harnessed. But in fact some potential energy is being harnessed. This is discussed further in Section 3.5.7.

3.2 The resource

The total energy contained in tides and rivers is very large, but estimates of how much of this energy can realistically be converted to electricity vary widely, depending on the assumptions made. According to [4], the global hydrokinetic potential of ocean and tidal currents near shorelines is 7,800 TWh/year or 890 GW (gigawatts, billion watts) on average, roughly 40% of the world's total electricity output in 2013. According to [5], the total energy contained in tides worldwide is 3,000 GW. Estimates of how much of that energy is available for power generation by tidal barrages range from 120 to 400 GW, and one estimate for tidal streams is 434 GW [5]. No estimates have been found for global river hydrokinetic energy potential, but [6] estimated the technically recoverable resource for the continental United States at 120 TWh/year. Although these figures seem large, they are not large compared with wind power, whose actual global installed capacity has been growing exponentially for decades, passing 500 GW in June 2017, with nearly 800 GW forecast by 2020 [7], and solar, which is also growing exponentially and passed 300 GW in 2016 [8]. The potential of wind and solar are virtually limitless in comparison to even the most optimistic estimates of tidal and river hydrokinetic power. Thus the potential contribution of tidal, ocean current and river hydrokinetic energy to global energy production may be significant, but will always be small compared with wind and solar.

3.3 Tidal rise and fall concepts

3.3.1 Tidal barrages

Tidal barrages harness potential energy by blocking an estuary with a large tidal rise and fall so water flows inwards from the ocean through low head hydro turbines when the tide is high and/or the reverse when the tide is low, depending on whether the turbines are bidirectional or unidirectional. La Rance, the world's first major tidal power plant, was constructed in the mid-1960s and has a capacity of 240 MW. A detailed description is given in [9]. The 254-MW Sihwa Lake plant in South Korea was completed in 2011 [10], and there have been proposals for further barrages in the Severn Estuary in the UK, where several options have been proposed, the largest of which would generate 8.6 GW during flow and 2 GW on average [11]. This ambitious scheme was rejected in 2013 due to capital cost and ecological concerns [12].

3.3.2 Tidal lagoons

These would comprise U-shaped breakwaters and low head turbines similar to barrages but enclosing a large area with a large tidal range without completely blocking an estuary. A 320-MW proof of concept lagoon was proposed in Swansea Bay, Wales, but in June 2018 the project was abandoned [13].

3.3.3 Dynamic tidal power barriers

Proposed by [14], these would run out 20 km or more from a shore without a large tidal range but with a strong longshore tidal current, such that there is enough difference in level across the barrier to drive turbines contained in the barrier. Figure 3.9 shows a schematic plan view of such a barrier and the difference in head

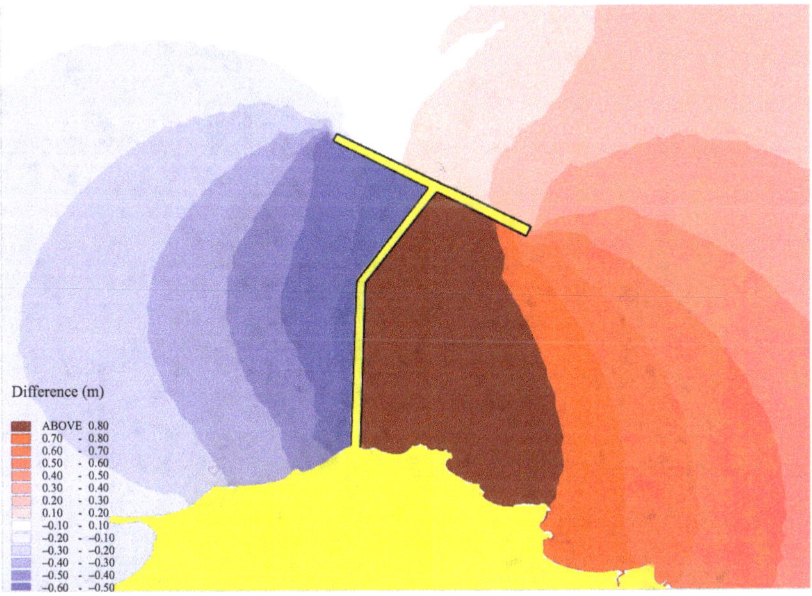

Figure 3.9 Dynamic Tidal Power barrier concept (image courtesy Kees Hulsbergen, Dynamic Tidal Power)

created. It has been estimated that a very large Dynamic Tidal Power barrier could generate up to 15 GW.

3.3.4 Turbines for tidal rise and fall schemes

Tidal barrage and lagoon concepts typically involve low head Kaplan hydro turbines of the type used in conventional low-head hydro plants, where the flow is unidirectional and turbines are designed for optimum performance with flow in one direction only. But according to [9], the La Rance turbines and alternators, although primarily intended for unidirectional flow, were designed to be able to operate in both directions, and they can also pump water. This flexibility is an advantage for tidal plants. Dynamic Tidal Power's bi-directional turbine concept is shown in Figure 3.10, after [15].

A totally different type of turbine, shown in Figure 3.11 and based on the old-fashioned waterwheel, has been proposed by [16], who argues that such a turbine would overcome the drawbacks of Kaplan turbines. It is efficient, bi-directional – it can block the flow without the need for sluice gates and can even act as a pump, an advantage at some stages of the tidal cycle. Because it turns slowly, there is no danger to marine life and there is no need for the long, tapered entry and exit ducts fitted to bidirectional Kaplan turbines to minimize the turbine diameter and recover velocity head as in Figure 3.10, thereby reducing the civil works required.

Figure 3.10 *Bidirectional turbine concept in the proposed Dynamic Tidal Power barrier (image courtesy Kees Hulsbergen, Dynamic Tidal Power)*

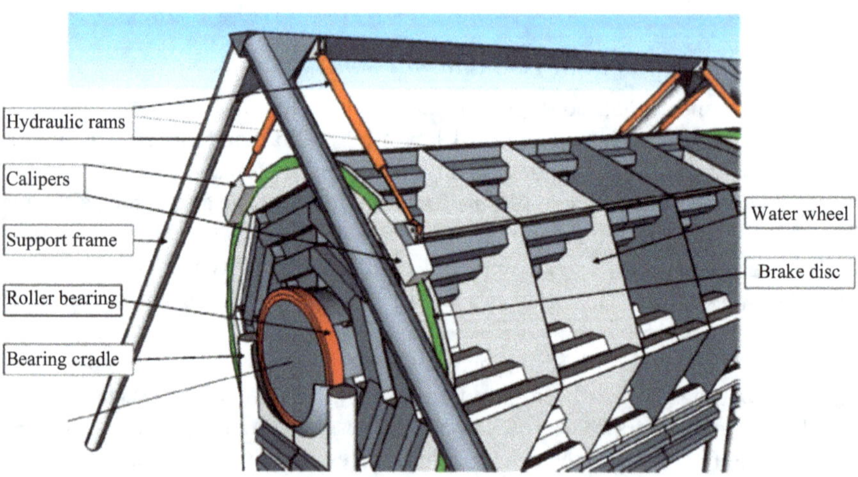

Figure 3.11 *Breast-shot waterwheel concept proposed by [16] (image courtesy R. Rainey)*

The large empty hub would make it buoyant, so it could be easily floated into place. According to [16], power transmission to a generator

> can be accomplished with a variable displacement hydraulic transmission, which acts like a gearbox with an infinitely variable gearing ratio. By coupling it to a synchronous generator, which runs at a constant speed set by the grid frequency, a prescribed volume flow rate can thus be achieved by suitable choice of the transmission gearing ratio, which will vary continuously over the tidal cycle.

Like other barrage and lagoon proposals, these turbines could generate up to several gigawatts, depending on the site.

3.4 Tidal bridges and fences

Like a tidal barrage across an estuary or enclosing a lagoon, a tidal bridge would comprise a barrier containing turbines, but the turbines could be hydrokinetic, i.e. driven primarily by flow velocity with very little static head, so it is debatable whether they should be classed under 'rise and fall' or 'hydrokinetic'. Bluenergy Canada and its predecessor Nova Energy pioneered the concept, developing ducted vertical axis turbine prototypes from 1982 to 1987 (Ref. [17] and Figure 3.12(a)) as

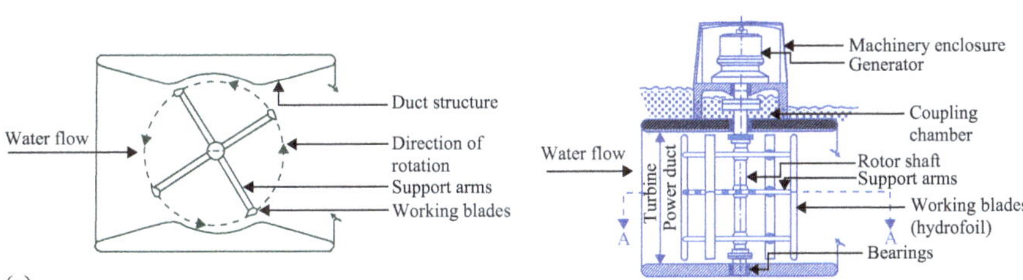

(a)

(b)

Figure 3.12 (a) Bluenergy Canada's original ducted vertical axis turbine prototype [17]. (b) Bluenergy Canada's tidal fence concept using banks of ducted vertical axis turbines [17]

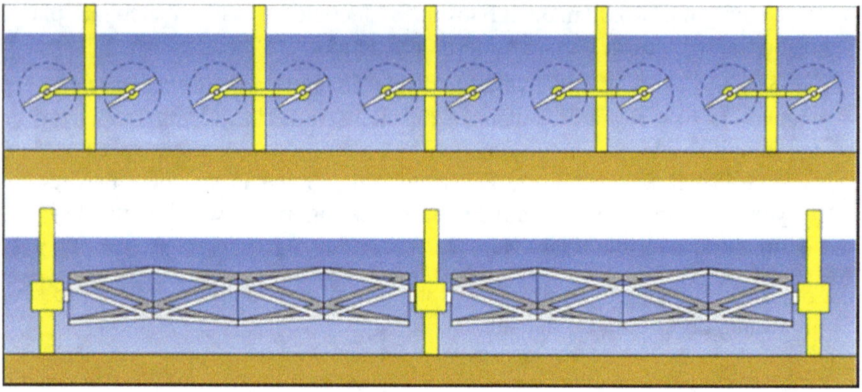

Figure 3.13 Kepler Energy's tidal fence concept (lower image) with horizontal axis crossflow turbines able to form a continuous barrier, contrasted with axial flow turbines (image courtesy Peter Dixon, Kepler Energy)

forerunners to the proposed tidal bridge concept (Ref. [17] and Figure 3.12(b)). It is significant that a slight drop in surface level is shown in Figure 3.12(a), showing that the Bluenergy designers were well aware that the 'zero head' HKTs referred to in references such as [18] are a myth.

Bluenergy Canada have proposed a 120-km tidal energy generating bridge across the Bohai Strait in China [19]. It was estimated that this ambitious project would not only generate over 70 GW but would shorten the drive between Dalian and Yantai to about 1 h from 7 to 8 h. However, a tunnel is now being proposed instead of a bridge.

A different type of tidal bridge, comprising a series of axial flow turbines mounted on a floating truss structure, is reportedly going ahead on the Indonesian island Flores [20], and Kepler Energy [21] has proposed a tidal fence concept comprising a series of horizontal axis crossflow turbines which can be mounted end to end to form a continuous high blockage barrier, unlike axial flow turbines, as shown in Figure 3.13.

3.5 Tidal, ocean current and river HKTs

HKTs capture kinetic energy from flowing water in a manner similar to wind turbines capturing kinetic energy from flowing air, i.e. wind, rather than potential energy from a drop in surface level. These have attracted increasing attention in recent years because they offer the promise of tidal, ocean current and river energy with minimal environmental impact, unlike tidal barrages which can cause environmental problems and conventional hydropower which involves damming a river, which may inundate fertile river valleys, displace communities and deprive downstream users of water, fish and river transport.

Figure 3.14 A 1.5-MW, 18-m diameter tidal turbine (image courtesy Atlantis Resources Ltd)

Large tidal farms are planned, comprising arrays of turbines along similar lines to wind farms. At the time of writing (January 2019), four 1.5-MW turbines have been deployed in the Pentland Firth off the north coast of Scotland, part of a planned 86 MW (possibly eventually 400 MW) array. At 18-m diameter, these are the largest single rotor tidal turbines so far developed, and Figure 3.14 shows their size.

3.6 Differences between hydrokinetic and wind energy

At first sight, the basic concept of hydrokinetic energy may appear the same as that of wind turbines, i.e. capturing kinetic energy from a moving fluid. This is generally true for isolated turbines in large tidal flows, but as more turbines are deployed in a given flow and they start to offer significant blockage to flow, major differences become apparent. These can favour turbine designs radically different from the conventional axial flow turbine on a tower, as discussed in the following.

3.6.1 Limited depth suitable for turbines

At some sites, turbines may need to be several metres under water to avoid shipping. Several metres submergence may also be needed to avoid increased stress from surface waves, and several metres clearance from the bottom may be needed because of low velocity near the bottom of the water column. There may be considerable turbulence around rocky outcrops at the bottom of some sites, or danger of scouring of sandy bottoms. All of these factors may limit the diameter of axial flow turbine rotors. The SR2000, a floating hull supporting twin 1-MW rotors (Figure 3.15), is the largest single unit at the time of writing [22] (January 2019), while the largest single rotor units are 18-m diameter, with 1.5-MW rated output [23].

It is unlikely that individual axial flow HKTs will get much bigger, so the rated capacity of single axial flow rotors will be limited to about 2 MW, unlike wind turbines whose size and capacity continues to grow, with recent announcements of a 12-MW, 260-m high wind turbine to be available in 2021 [24]. Larger wind turbines can access stronger winds at greater altitude, unlike HKTs, and economies of scale also favour small numbers of larger turbines with fewer electrical connections. This applies to wind turbines but perhaps even more to HKTs because of the added difficulty of subsea connections, so larger turbines would be an advantage. Although there is also a diseconomy of scale because power output scales with diameter squared, while bending moments on the blades scale with diameter cubed, this appears to be outweighed by the other factors.

Salter [25] has taken a completely different approach, proposing a close-packed array of very large diameter vertical axis turbines, each of which could generate tens of megawatts. Each turbine would have several layers of short, variable pitch blades as shown schematically in Figure 3.16, blade pitch being

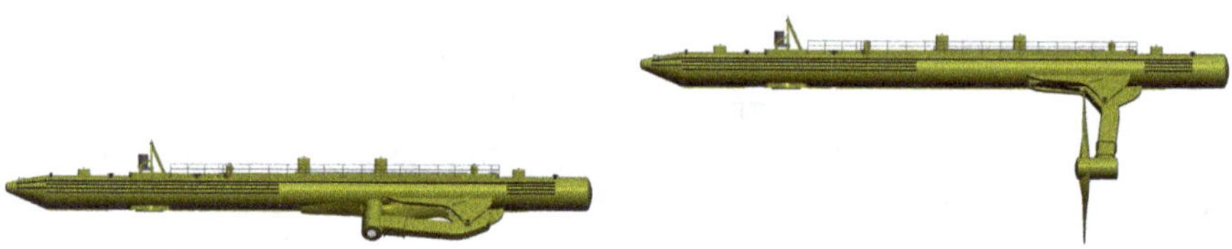

Figure 3.15 The SR2000 floating turbine with twin 1-MW rotors (image courtesy Orbital Marine Power)

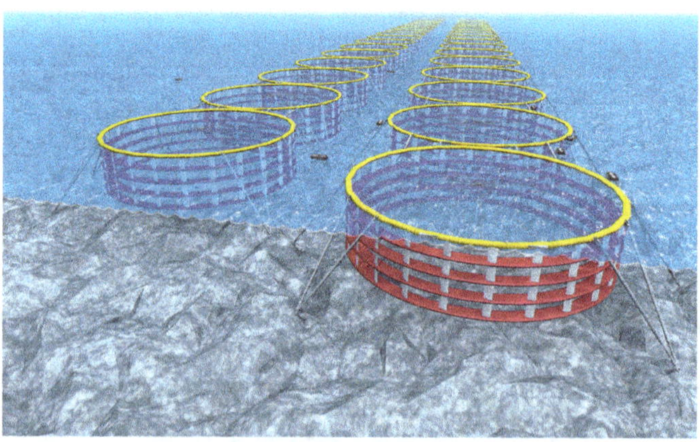

Figure 3.16 Salter's proposed very large diameter vertical axis turbine concept (image courtesy John MacNeill)

controlled by an algorithm to control the pitch so as to give an evenly distributed head, i.e. drop in downstream surface level. This would minimize downstream turbulence losses, which are considerable for isolated turbines and reduce the amount of useful energy that can be extracted from a given site. According to [26], 'when the free-stream mixes with the wake, significant energy is lost to turbulence. This constitutes an energy loss in excess of the loss due to extraction by the rotor'. Thus it is claimed that Salter's design would enable considerably more useful energy to be extracted from a given site.

3.6.2 Tidal and ocean current flow velocities are lower than wind

Tidal flow velocities are much lower than wind speeds, seldom exceeding about 3–4 m/s and typically nearer 2 m/s. Ocean currents, although they contain enormous amounts of energy, are too slow for economic exploitation at present. The Gulf Stream off Eastern Florida is among the fastest ocean currents on the planet, but according to [27], the average ocean current speed there is about 1.5 m/s (presumably in the most favourable location). Wind turbines are typically designed to develop rated power at wind speeds ranging from 12 to 17 m/s [28], i.e. several times that of tidal and ocean currents. Although power density (kW/m^2) is typically higher in strong tidal currents than in wind, power is basically the product of force and change of velocity across the turbine. Therefore the lower the flow velocity, the greater the force necessary to produce a given power output, so for the same power output, the force on an HKT would be several times that on a wind turbine, and for this reason HKTs are only being deployed in relatively fast flows.

Even in the fastest tidal flows such as the Pentland Firth and the Bay of Fundy, tidal HKTs are at present far from commercial viability, so it may be a long time before costs come down to the point where conventional HKTs are viable in ocean currents. However, a new 'flying' turbine concept is being developed by Minesto [29], similar in principle to the Makani flying wind turbine [30], which will be discussed in Section 3.9.

3.6.3 Much higher fluid density

The density of water is about 830 times higher than that of air (wind), so even with much lower velocities, power densities are typically higher. A smaller swept area is therefore required for the same power output, although as explained in Section 3.6.2 above, greater force is necessary to produce a given power output, so mounting structures must be stronger.

3.6.4 Cavitation

Unlike wind turbine blades, HKT blade velocities are limited by the danger of cavitation if the Cavitation Number σ is below 1. $\sigma = (P_T - P_V)/(\frac{1}{2}\rho V^2)$ where P_T is the pressure at the depth of the turbine, P_V is the vapour pressure, ρ is the water density and V is the velocity of the water relative to the blade [31]. For practical purposes, this means that blade tip speed should be kept below about 12 m/s.

3.6.5 Predictability

Tidal flow velocities, unlike wind, are generally predictable, but this is not always the case, especially where the bathymetry is irregular and flows are turbulent. For example, a 1-MW Open Hydro turbine failed in the Bay of Fundy in 2009 due to higher than expected maximum current velocities [32]: 'all 12 turbine rotor blades were destroyed by tidal flows that were two and a half times stronger than for what the turbine was designed'.

3.6.6 Bidirectional flow

Tidal flow, unlike wind, is bidirectional at most but not all sites. According to [33], writing of the Pentland Firth, 'ebb and flood streams did not fully overlap, and ... the tidal streams were more complicated, turbulent, and variable than existing models suggest'.

3.6.7 Limited flow field

Wind turbines operate in what is effectively an infinite flow field, and according to Betz's law, the upper limit of power extraction for a turbine in open flow is 16/27 or 59.3% of the kinetic energy flux in a flow area equal to the swept area of the turbine. A turbine's performance coefficient C_P is the ratio of actual power delivered to kinetic energy flux and it has an upper limit of 59.3% in open flow. But at some tidal and most river sites, channel cross-sectional area is limited, so that an array of turbines may block a significant portion of the flow area. In these cases, the assumptions on which the Betz (also known as Lanchester-Betz) Limit are based no longer apply, and numerous authors have stated that the efficiency of individual turbines can exceed this limit. Nishino and Willden [34] have predicted a C_P limit of 79.8% for a large number of devices arrayed across a part of an infinitely wide channel, and [35] and [36] measured an apparent C_P well in excess of 1 with a model crossflow turbine similar to that shown in Figure 3.13 except that the blades were straight, blocking 62.5% of the flow area.

As shown in Figure 3.17, after [35], the modelled velocity immediately downstream of the turbine was lower than that upstream, indicating some extraction of kinetic energy, but the velocity near the bottom and at the surface downstream

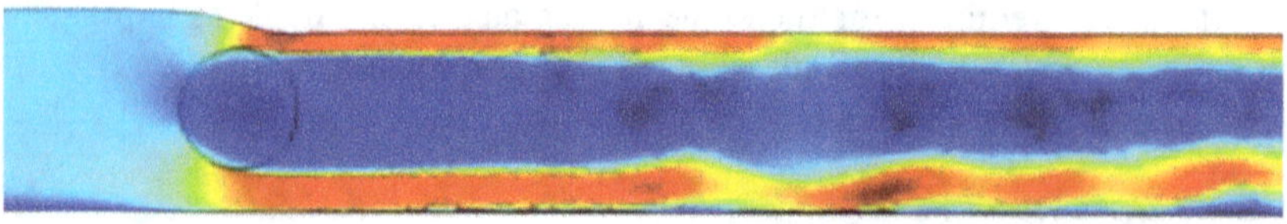

Figure 3.17 Modelled velocity contour plot with a blockage $B = 0.625$ and free surface deformation. Flow from left to right. High-speed flow regions are shown in red and low-speed regions in dark blue. The turbine is represented by the circle on the left (image courtesy Ross Mcadam, Oxford University [31])

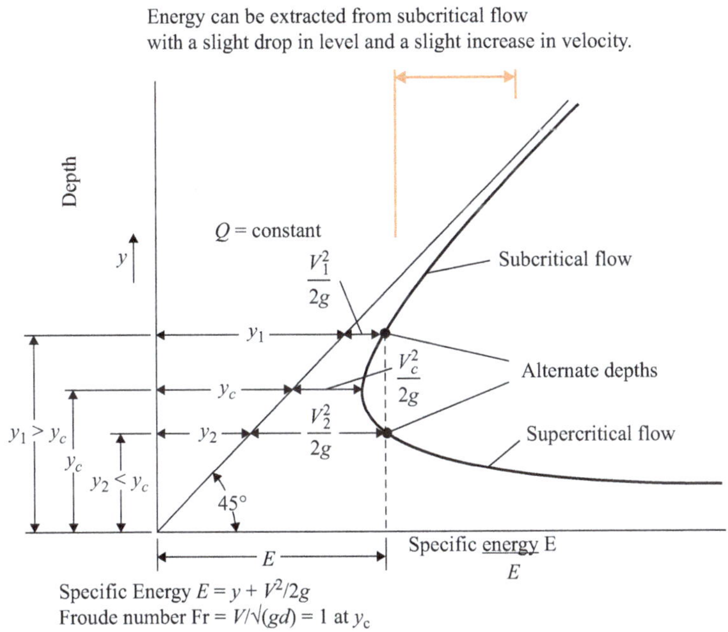

Figure 3.18 Energy diagram showing how the total specific energy (horizontal axis) is made up of depth y (potential energy) plus velocity head or kinetic energy $V^2/2g$

was considerably greater and there was a small drop in the surface-level downstream of the turbine, indicating that the average velocity and therefore the total kinetic energy flux had increased, so most of the power must have come from potential energy. The fact that the so-called 'hydrokinetic' turbines at high blockage extract potential energy while kinetic energy increases shows that the distinction between tidal rise and fall turbines and HKTs can be somewhat blurred. Figure 3.18 shows how the total specific energy is made up of depth y (potential energy) plus velocity head or kinetic energy $V^2/2g$. Moving to the left in Figure 3.18, potential energy can be extracted from subcritical flow (Froude number Fr < 1) with a drop in elevation accompanied by an increase in kinetic energy.

3.6.8 Floating debris

Clogging of turbine blades and/or damage from floating objects, ranging from jungle vines and floating logs in tropical rivers to chunks of ice in Alaska, has been identified by [37] as a major problem in some waterways. Screens upstream of turbines have been proposed, but these become clogged and need frequent cleaning. One possible solution proposed by [37] is to design blades to fold back when they become tangled or impact floating obstacles so that they are not producing forward torque, then swing forward as they clear the obstacle. More research is needed to assess the effectiveness of these measures.

3.7 Axial flow turbines

The three-blade axial flow turbine on a monopole support structure dominates wind turbine technology, and HKT technology has until recently followed suit, possibly because risk-averse investors feel more comfortable with familiar, proven geometries. But floating, tethered turbines are now gaining favour due to ease of deployment and access for maintenance. The 2 × 1-MW twin rotor tethered floating turbine mentioned in Section 3.6.1 was reportedly performing well off Northern Scotland in 2017 [38].

Other axial flow HKT concepts include a 2-MW Open Hydro ducted turbine which was deployed in the Bay of Fundy in Eastern Canada after some problems with unpredictable currents [32], and an array of small Schottel turbines which was to be deployed in the Bay of Fundy in late 2017 but has been delayed [39]. According to [39], a 2008 plan to deploy a 2.5-MW platform equipped with 40 Schottel turbines, each rated at 54–70 kW depending on flow velocity, has now been scaled down to 280 kW. Other axial flow turbine designs are being developed, such as the GEM twin ducted rotor 'ocean's kite' shown in Figure 3.19, after [40] and described in [41], rated at 20 kW.

3.8 Crossflow turbines

Although axial flow turbines currently (2019) dominate the field, it is by no means clear that this design is inherently better than any other. The blades are generally cantilevered from the hub, so they are subject to very high bending moments and the inner part of each blade is mainly structural, contributing little to power output.

Figure 3.19 The GEM 'Ocean's Kite' (image courtesy D. Coiro, University of Naples)

Figure 3.20 The Kobold turbine (image courtesy D. Coiro, University of Naples)

The generator is normally located under water, making it vulnerable to leakage and hard to access for maintenance.

Crossflow turbines may be either vertical axis as shown in Figure 3.20 or horizontal axis as shown in Figure 3.13. The generator of a vertical axis HKT can easily be placed above water level where it is readily accessible, as shown in Figure 3.20, the Italian Kobold floating turbine, in which the gearbox and generator are located in the hull. Being insensitive to flow direction, there is no need for a yaw mechanism to follow the reversing tidal flows, and the rectangular swept area allows them to be mounted in close-packed arrays, unlike axial flow turbines with their circular swept area, as shown in Figure 3.13.

Disadvantages of the basic straight blade, fixed pitch Darrieus turbine, include lack of starting torque, shaking and torque ripple, but variable pitch can improve all of these factors [42], and helical or inclined blades as shown in Figure 3.13, or multiple blades as shown in Figure 3.16 can almost eliminate shaking and torque ripple.

3.9 Ducts and diffusers

A duct around a turbine or a diffuser downstream of a turbine in open flow can increase its power output significantly, both by increasing the pressure drop across the turbine and by increasing the flow through it. This should not be confused with what happens in a venturi in a closed pipe, in which the velocity (and kinetic energy) increases but pressure decreases in the converging upstream section, while the reverse occurs in the diverging downstream section. Energy is simply converted from pressure (potential) to velocity (kinetic) and back again, with only a small loss of total energy provided it is designed correctly so there is no flow separation. To extract energy from a venturi, there must be a pressure drop.

Radial flow along the blades of an axial flow turbine can be prevented by a duct, and higher conversion efficiencies can be achieved. Figure 3.19 shows a pair of axial flow turbines with one-way ducts, which were shown in laboratory tests to approximately double the power output [41], but the thrust also doubled, necessitating stronger mountings. Rawlings [43] has shown a 73% power increase in a vertical axis turbine with a symmetrical two-way duct, which if repeated across a channel as shown schematically in Figure 3.21 could serve also as structural columns supporting a tidal bridge like that proposed by Bluenergy Canada.

Gilbert and Foreman [44] reported a power augmentation factor of 4.25 for a laboratory wind turbine with a diffuser; i.e. it produced 4.25 times as much power as the same turbine in open flow. They used slots in the diffuser to draw in high energy flow from outside the diffuser for boundary layer control and were able to achieve this result with a short, wide-angle slotted diffuser, a much more economical arrangement than the long diffusers studied by a number of other investigators. Smart Hydro [45] report success with a 1-m diameter ducted axial flow turbine which can be fitted with a slotted diffuser capable of delivering up to 5 kW.

Kirke [46] applied the same principle to a vertical axis HKT and measured approximately threefold increase in power with a rectangular slatted diffuser shown in Figure 3.22. However, the increased load on the turbine and mounting structure and the cost of the duct or diffuser offset any gain in power output and it remains to be seen whether ducts and diffusers will improve cost-effectiveness.

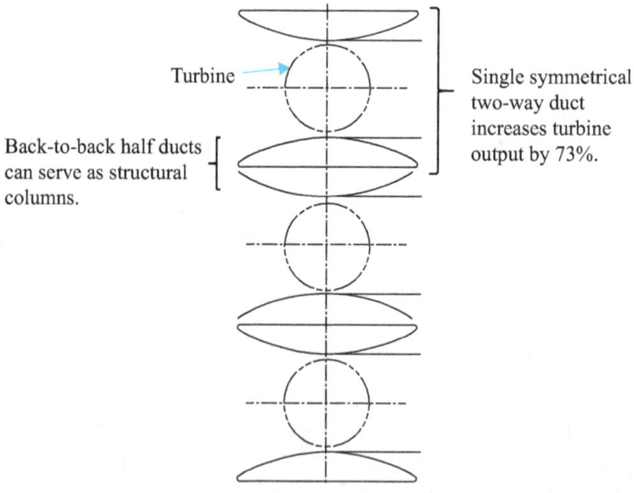

Figure 3.21 Symmetrical ducts around vertical axis turbines could increase power output while serving as structural columns. Adapted from [43] with permission

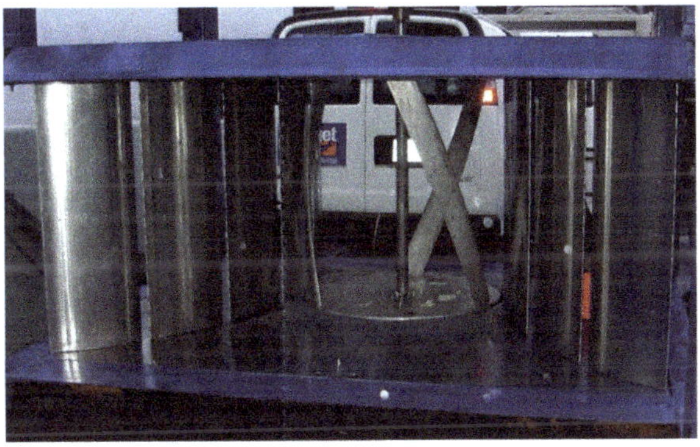

Figure 3.22 A helical crossflow turbine in slatted duct (photo by B. Kirke)

Figure 3.23 Minesto's 'flying' turbine: (a) components and (b) swept area (image courtesy Magnus Matsson, Minesto)

3.10 'Flying' turbines

Instead of making the water flow faster through a stationary turbine, Minesto's radical turbine shown in Figure 3.23(a), after [29], comprises a turbine attached to a tethered wing which 'flies' through the water in a figure 8 pattern as shown in Figure 3.23(b), like an underwater kite at speeds many times that of the current, so that a small, inexpensive turbine mounted on the wing experiences high relative velocities and can generate correspondingly high power. The energy is transmitted via the tether to the bottom. This may offer better economics in deep, slow flows. Makani, a wind turbine using a similar concept, is also being developed [30].

3.11 Oscillating foils

These have been proposed by [47–49] as an alternative to rotating blades. This concept is shown in Figure 3.24. Like the horizontal axis crossflow turbine, the

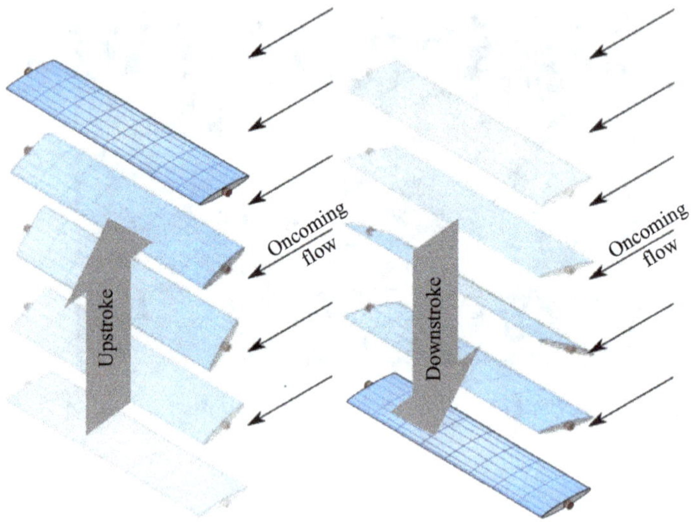

Figure 3.24 Oscillating hydrofoils [49] (image courtesy Jennifer Franck University of Wisconsin)

rectangular swept area lends itself to shallow locations where long slender foils can be deployed, but maximum C_P to date has been only about 22%, and fluctuations in power output are very large.

3.12 Vortex shedding

Another oscillating technology, the VIVACE Converter [50], uses forces exerted on a cylinder by vortex shedding. According to [50], multiple cylinders can work synergistically and achieve higher power-to-volume density than conventional HKTs, and high power conversion efficiency, estimated at 88.6% of the Betz limit ($C_P = 52\%$) at a flow speed of 1.08 m/s, which is superior to the peak efficiency of an isolated wind or tidal turbine.

3.13 Tidal sails

Another concept suitable for wide, shallow sites which uses linear rather than rotary motion is Tidal Sails, being developed by [51], in which multiple hydrofoils attached to a continuous belt 'sail' in a straight line across the current as shown in Figure 3.25, so that they can maintain a fairly constant optimum angle of attack.

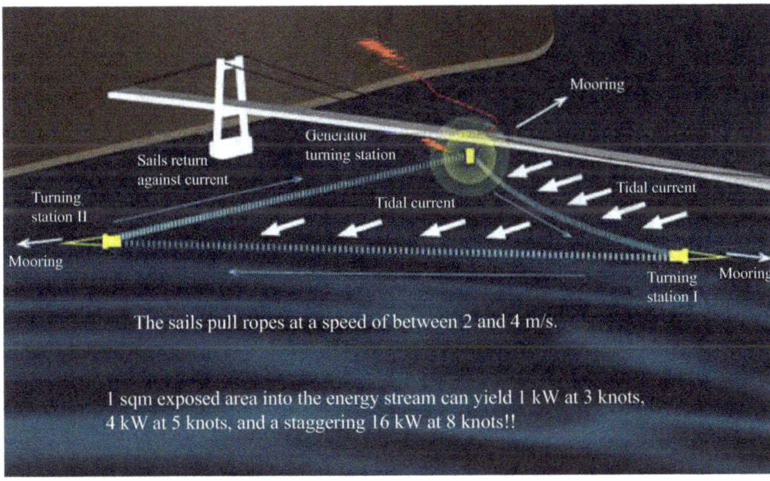

Figure 3.25 Tidal sails [51] (image courtesy Are Børgesen, Tidal Sails)

3.14 Arrays

It is likely that single turbine rotors will be limited to about 2 MW unless the extremely large vertical axis turbines proposed by [24] become a reality, which unfortunately seems unlikely due to the radical design, the large financial outlay that would be needed and the competing interests of conventional wind farm thinking. So to make a significant contribution to grid power, it is assumed that there is a need for arrays of turbines or 'tidal farms' analogous to wind farms.

3.14.1 Tidal farm layout

Despite the important difference between wind and tidal power, wind farm thinking continues to influence array design. Because wind can come from any direction, wind turbines are arranged in a grid with enough spacing between turbines in any direction to minimize wake effects on downwind turbines. A common assumption seems to be that tidal farms should follow the same pattern, but because tidal flows are typically bidirectional, a better arrangement may be to place the turbines close together across the flow, a bit like a barrage, and this has been shown to be the case in theory.

An optimization routine was carried out by [52] and the process has rearranged the turbines from an initial rectangular layout (Figure 3.26(a)) to 'barrages' perpendicular to the flow (Figure 3.26(b)). It also added small barrages of turbines on each side that force the water to flow through the perpendicular 'barrages' as shown in Figure 3.26 (right). It was found that the predicted power production of the optimized layout is 74% higher than the initial rectangular layout. This arrangement would also considerably reduce the amount of underwater cabling.

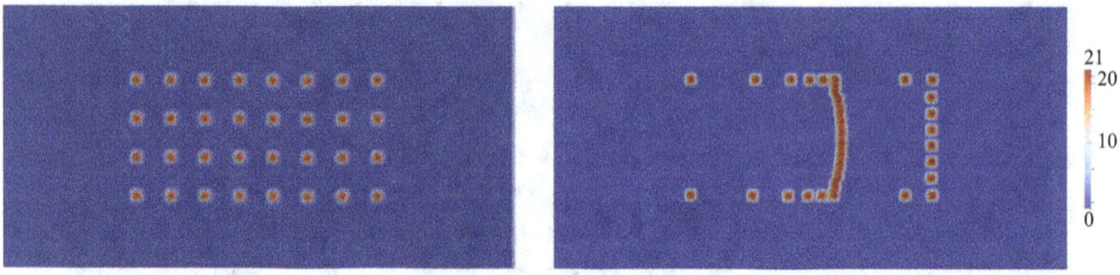

Figure 3.26 *(a) initial rectangular grid of 32 turbines and (b) optimized layout produces 74% more power from the same 32 turbines [48] (images courtesy Matthew Piggott, Imperial College)*

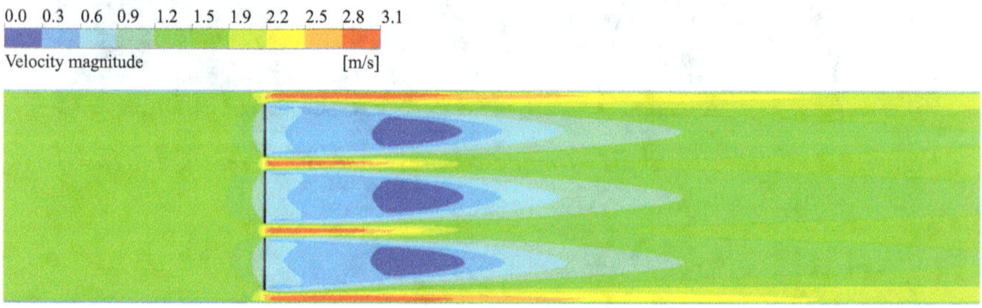

Figure 3.27 *Top view of velocity magnitude plots on a plane parallel to the channel bed passing through the centre of discs [53] (image courtesy Philip C. Malte, University of Washington)*

3.14.2 Mixing and wake recovery – streamwise and lateral spacing

As with wind farms, turbines must be spaced far enough apart to minimize wake effects on downstream turbines, but because flow is generally bidirectional, tidal farm turbines can be closely spaced across the flow, as shown in Fig. 3.26(b), unlike wind farms.

Figure 3.27, after [53], shows flow velocity reduced immediately upstream and downstream of the turbine array, increasing between turbines directly downstream, but evening out further downstream. It will be seen that the velocity almost reaches the free stream velocity 10–15 diameters downstream, where it will be about 90% of the freestream velocity. As power is a function of velocity cubed, this means the power output of downstream turbines will be about 30% less than those upstream, which is generally considered acceptable.

3.15 Economics

According to [54], ‘the capital costs would have to be less than €510,000 per MW installed which is currently an unrealistically low capital cost. Thus, it is concluded that tidal generation is not a viable option for the case system at the present time’.

This article was published in May 2009 and the economics may have improved since then.

According to [55],

> Although Tidal barrage power schemes have a high capital cost, they also have a very low running cost. Whilst a tidal power scheme may not produce returns for many years, once it moves into surplus it is likely to go on providing a profit for an indefinite period. Because of the long-term nature of returns, private investors may be reluctant to participate, and central governments may have to lead the way.

The cost of the turbine itself is only a small part of the complete installation, which includes gearing, generator, mounting structure, anchoring, electrical cabling, inverter, etc. It is difficult to get cost data from manufacturers but as far as this author has been able to determine, small complete systems of about 5-kW rated capacity in 3 m/s flows can cost anywhere from US $10,000 to $50,000; $10,000 for 5 kW 24/7 in a river compares favourably with solar and wind, but 3 m/s is rare, and even at a site with peak tidal flows of 3 m/s, the average power density will be much lower. So at most sites, the installed cost at present is prohibitive for all but one-off demonstrations. Hopefully the cost will come down as experience is gained, as has been the case with wind turbines. Yet there may be a need for a radical re-think of some design features.

3.16 Actual progress to date on large-scale grid-connected hydrokinetic power

There have been numerous scientific papers predicting turbine and array performance, and a few papers reporting actual performance of real prototypes. But despite optimistic predictions for large-scale grid-connected hydrokinetic power, actual progress is slow and potential for large-scale tidal farms may be less than previously believed. According to [55], 'only around twenty sites in the world have been identified as being ideal locations for large scale tidal power arrays'. To date the UK leads the world, but 'Of eight sites identified around the UK, only 2 are hydrokinetic: the Pentland Firth, possibly the best place in the world, and Islay'. In 2010 it was announced that a 400-MW farm would be installed in the Pentland Firth, to be completed by 2020 [56]. At the time of writing (late 2018), Stage 1 is complete, with four 1.5-MW turbines installed, but in July 2018 two of the turbines had to be retrieved for unscheduled repairs [57]. Judging by progress so far, it seems unlikely that 400-MW farm will be completed by 2020.

Another 400-MW Tidal Farm has been proposed off the west coast of Islay: 'The offshore construction works of the project are expected to start by 2019. The initial proposal is for an installed capacity of 30 MW by 2019' [58].

According to [59],

> Normandie Hydrolienne has been established with the intention of eventually harnessing up to 2 GW of power from the Alderney Race, the

eight-mile strait that runs between Alderney and La Hague, France, as well as more than 1 GW of resource from adjacent concessions under the control of the States of Alderney.

Elsewhere there are plans but little in the way of actual turbines in the water. According to [60], it's reported that the 12 MW stage 1 of a 150-MW scheme in, Indonesia has been approved, to be completed by 2020, but there is no mention of work actually starting.

It appears that there are two reasons for this slow progress. One is the high cost, as outlined above. The other is the fact that turbine designers want to avoid reports of embarrassing failures, such as the headline 'Failed tidal turbine explained at symposium' [32] and comments like 'The problem with those machines was that they kept breaking'. So they design for high velocity flows, and manufacturers want to present their products in a favourable light, so they generally quote rated power in flow velocities around 3 m/s. But such flow velocities are very rare.

3.17 Case study – development of GEM, a tidal stream energy system[1]

A specific case study will be presented with reference to the GEM system. The choice of this system, based on authors' experience, allows highlighting some particular features and problems potentially observed in tidal flow energy conversion systems. GEM development studies will be briefly described with particular reference to the experimental test campaigns performed in laboratory and in operating environment.

3.17.1 GEM system configuration

The GEM system is composed by a tethered floating structure, supporting two HKTs, with the ability of self-alignment with current stream. It may be equipped with a self-towing winch, which is capable of setting the desired operating depth. Operating underwater, it has a limited impact on navigation. The system is moored at a single point on seabed, allowing the rotation of the floating structure in response to current direction change and may reduce maintenance cost and simplify deployment operations: by releasing the winch mooring cable, the system may be raised at surface for easier maintenance.

A computer-aided design (CAD) drawing of the GEM schematic configuration in operating conditions is shown in Figure 3.28.

The system is sustained by the buoyancy provided by a streamlined axial-symmetric floating body, placed at the top of the structure. Stability and oscillation damping are improved by means of tail fins mounted on the floating body. Two generators are installed on board and mounted on the turbines shafts through a

[1]Written with cooperation of Giancarlo Troise, Seapower Scrl Research Consortium

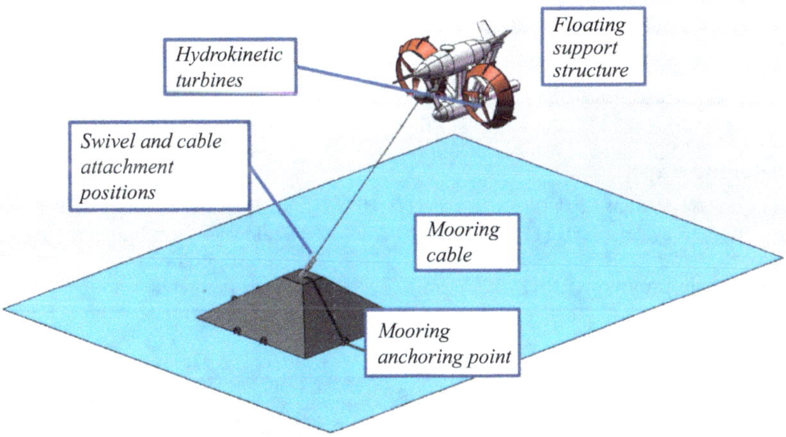

Figure 3.28 Schematic view of GEM configuration

gearbox. Each generator is electrically controlled by an inverter, both for grid connection purposes and in order to attain optimal working conditions at different current speed. Suitable control logic is also needed to pursue optimal operating conditions. The power connection is provided by means of a power cable, passing along the mooring cable and extended up to an onshore grid connection point.

The three bladed turbines have been developed at the University of Naples and designed in order to reach a high efficiency in a relatively wide operating conditions range, using a properly designed airfoil section shape. The turbines have been intensely tested both in wind tunnel and towing tank experimental campaigns. In a possible configuration a diffuser, also developed at the University of Naples, is placed around each turbine with the aim of enhancing the energy conversion process by increasing the mass flowing though the turbine swept area. After further investigations, a solution with bare turbines (without diffusers) may be considered for effective implementation on the system.

The development of the system started in 2005 and, after preliminary numerical analyses and design work, many experimental tests have been performed on different systems characterized by different scales as summarized in Table 3.1.

The main results of the experimental test campaigns, performed at different project stages, are reported in the following paragraphs [61].

3.17.2 Turbine and diffuser experimental tests (small-scale models)

In order to characterize the behaviour of the system in different operating conditions, experimental tests were performed in the towing tank on isolated turbine as well as on two different scaled model of the full GEM system. A first experimental

Table 3.1 Experimental tests on different scale prototypes and components

#	Tests	Dimensions and scale	Notes
1	Bare turbine (2005)	1.20 m	Rotor diameter
2	Bare and Shrouded turbines (2008–2010)	0.6 m	Rotor diameter
3	Complete system (2010)	(scale)1:10	Scale based on buoyancy
4	Complete system (2011)	(scale) 1:8	Scale based on buoyancy (different configuration)
5	Full-scale prototype (2012)	3 m	Rotor diameter

Figure 3.29 Towing tank test facility

campaign was carried out on a single isolated, reduced-scale hydro-turbine – which is the main component of the GEM system – with and without a diffuser. After a set of preliminary tests performed in air in the wind tunnel facility of the University of Naples, a second main session of experimental tests has been completed in the towing tank of the Industrial Engineering Department. A picture of the towing tank, with the dynamometric carriage, is reported in Figure 3.29.

3.17.2.1 General definitions

Prior to presenting tests data, some useful definitions of dimensionless quantities are recalled here. These coefficients are widely used to represent the power generation characteristics of wind and HKTs.

In general, a turbine is a device that works at a given rotational speed Ω when immersed in a fluid stream of speed V. The desired effect is the establishment of a torque τ that keeps the blades in rotation and brings about the generation of a mechanical power:

$$P = \tau\Omega \quad (3.2)$$

An important state variable of the turbine is the tip speed ratio $TSR = \Omega R/V$, i.e. the tip speed, ΩR, non-dimensionalized with respect to the current speed V. The power coefficient (which may be used as representation of conversion efficiency) is defined as

$$C_P = P/\left(0.5\rho V^3 A\right) \quad (3.3)$$

where P is the power generated (measured or estimated), V is the asymptotic speed of the fluid stream (in the case of towing tank tests, the steady-state travelling carriage speed), A is the area of a chosen reference surface and ρ is the water density. Similarly, the torque coefficient is defined as

$$C_\tau = \tau/\left(0.5\rho V^2 AR\right) \quad (3.4)$$

where τ is the torque (measured or estimated), at the turbine hub. Finally, the thrust coefficient has the following expression:

$$C_T = T/\left(0.5\rho V^2 A\right) \quad (3.5)$$

where T is the turbine thrust, i.e. the axial force developed by the device immersed in the flow and functioning at a given TSR. The area A in the above definitions is always taken coincident with the bare rotor disc surface area, both for the bare turbine and for the diffuser-augmented one. This is important when comparing performance of different solutions in terms of efficiency.

3.17.2.2 Model set-up

Support structure and actuation system

An initial test campaign has been performed on a single isolated turbine to characterize the behaviour of the power conversion system prior to the installation on the floating system. The turbine was mounted on a partially submersed pole connected to the towing tank carriage by means of a support structure capable of placing the turbine shaft at the desired depth. During tests, the towing cart was operated at a constant speed simulating a given relative current, and the turbine shaft was actuated by a controlled motor in order to achieve the desired rotational speed and operating conditions.

The mounting assembly consists of a hollow tubular steel mast with length of 2.46 m, whose circular section has an external diameter of 114 mm and inner diameter of 108 mm and it could rotate around its axis in order to test the device with a fixed yaw angle with respect to the oncoming water current. A frame built by several welded tubular square sections fixes the tubular mast in a vertical position.

Measurements have been made using a torque meter of 226 N m (2,000 lbf-in) of full-scale range, with an accuracy of ±0.1% full scale (FS). Also, a load cell with

5 kN FS was installed at the end of the transmission shaft to measure compressive or traction loads produced by the rotor.

A ducted turbine configuration, i.e. with a diffuser (shroud) placed around the turbine, has been developed and tested in order to study the possible increase in conversion efficiency. Basically, the shroud is an annular diffuser with a streamlined cross section. A high lift airfoil has been used for the cross-section shape of the shroud.

A controlled motor is set up to control the desired rotational speed, a torque meter measures the turbine torque and load cells measure separately the axial forces experienced by the turbine and by the diffuser immersed in the flow.

Figure 3.30 represents a picture of the test set-up installed in towing tank, while Figure 3.31 shows a CAD detail view of the support structure with measuring equipment.

Diffuser and turbine geometry

The diffuser is realized as an axially symmetric revolution body generated by an airfoil shape, set at a proper inclination angle. A characteristic angle, associated to the diffuser geometry, is the angle ϕ formed by the chord of the airfoil-shaped annular cross section of the shroud (taken with a radial plane) and the turbine axis. Two geometries of streamlined diffusers have been tested; these geometries had the same annular cross section, i.e. the same airfoil shape, but with different airfoil chord angles, $\phi = 23°$ and $\phi = 26°$, respectively. Figure 3.32 illustrates the turbine and shroud principal geometric features.

3.17.2.3 Tests on single turbine

Power coefficient

The main results of the tests on the turbine are reported in terms of non-dimensional coefficients. Preliminary numerical and experimental results show that, between

Figure 3.30 Single turbine test system mounted on towing tank carriage

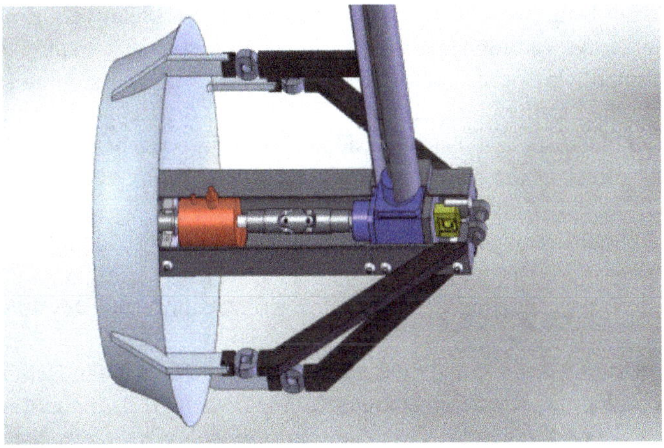

Figure 3.31 Detail of the diffuser on the support structure and the arrangement of measurement devices

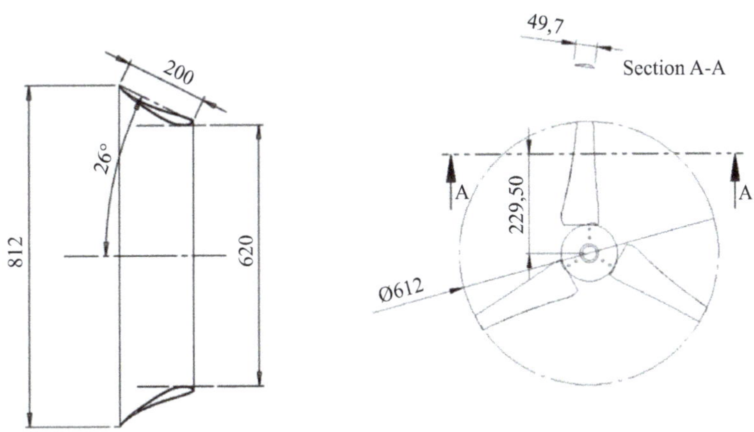

Figure 3.32 Geometric data of the turbine test system

the two considered shroud geometries, the most effective in terms of power coefficient increase is represented by the configuration with the higher section angle $\phi = 26^\circ$, at least for the examined case. Such shape has been considered for further analyses and tests.

It is worth to specify that the blade pitch angle has been kept always fixed without any attempt to optimize its value in order to increase the maximum efficiency, since the main goal of the tests was the investigation on diffuser installation effect on bare turbine performances.

The Reynolds number, based on airfoil chord and on relative airfoil velocity, shows a significant variation across the operating test range, mainly due to the variation of rotational speed. For a flow velocity of 2.0 m/s, assuming a reference chord length equal to the blade chord at 75% radius, the Reynolds number of the representative blade section spans the approximate interval between 120,000 and 730,000. The effect of the diffuser on turbine apparent efficiency over the whole range of *TSR*s is shown in Figure 3.33.

The maximum power coefficient reaches an almost double value with respect to the case of bare turbine. Figure 3.34 reports the effect of yaw misalignment on C_P, clearly showing a reduction of conversion efficiency at increasing yaw angle.

It may be noted that the apparent value of the C_P exceeds the Betz limit, with the assumed value of the reference area (turbine disc area). If the exit area of the diffuser is used as reference for the definition of power coefficient, the maximum value is limited to a lower value.

Thrust coefficient

A comparison of thrust coefficient plots versus TSR is reported in Figure 3.35. Reported data are relative to three different turbine configurations (bare rotor, ducted rotor with $\phi = 23°$ and ducted rotor with $\phi = 26°$).

The values of C_T are relative to the thrust acting on the rotor only. These data are consistently evaluated by taking the reference area A as the frontal area of the bare rotor disc. It is observed that, in presence of the diffuser, the rotor features a

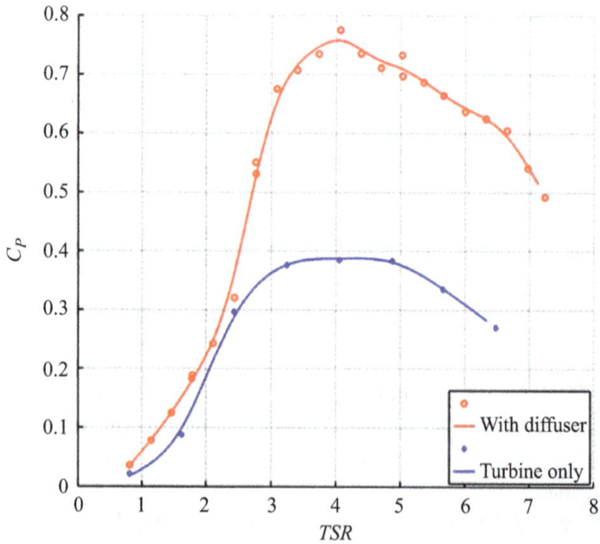

Figure 3.33 Power coefficient for single isolated turbine. Comparison between bare and ducted turbine ($\phi = 26°$). Relative speed, 2.0 m/s

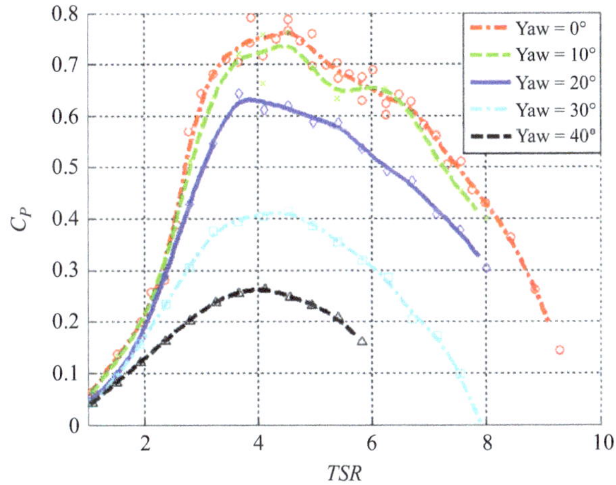

Figure 3.34 Effect of yaw misalignment on ducted turbine ($\phi = 26°$). Relative current speed, 1.5 m/s

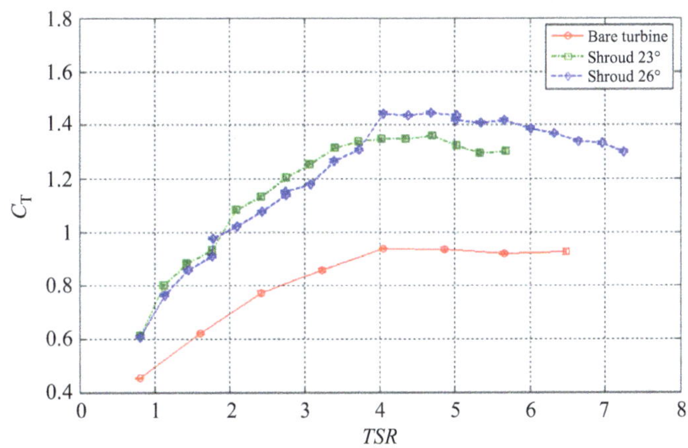

Figure 3.35 Comparison of rotor thrust coefficients for three different configurations (bare rotor, ducted rotor with $\phi = 23°$ and ducted rotor with $\phi = 26°$). Relative current speed, 2.0 m/s

sensibly higher thrust coefficient with respect to the bare rotor case. This experimental result does not agree with the hypothesis made in some studies reported in literature; see, e.g., Van Bussel [62], who claims that the thrust on the rotor remains the same even if it is ducted, but the observed power performance increment, on a ducted turbine, is essentially due to the extra mass flow through the rotor. However, further investigations and validations of numerical and experimental analyses may be useful on this topic.

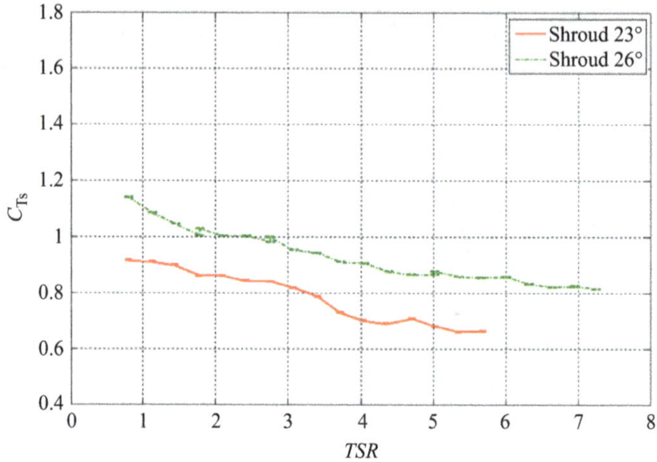

Figure 3.36 Diffuser thrust coefficient for two shrouded turbines with different diffuser geometries ($\phi = 23°$, $\phi = 26°$). Relative current speed, 2.0 m/s

For the design of GEM system, the estimation of thrust acting on the diffuser supports has also an important role. In Figure 3.36, the diffuser-only thrust coefficients (C_{Ts}), referred to the swept area of the rotor ($C_{Ts} = T_s/(0.5\ \rho V^2 A)$), are compared as functions of *TSR*, for two shrouded turbines with different diffuser geometries ($\phi = 23°$, $\phi = 26°$).

It is observed that the diffuser thrust coefficient reduces with increasing *TSR*. Such behaviour may be due to a flow reattachment on the diffuser inner side promoted by the turbine induced flow.

An important observation from the design point of view is related to the significant increase in thrust on the turbine–diffuser assembly with respect to the turbine only configuration.

3.17.3 Tests on full model in small scale

3.17.3.1 GEM tethered model set-up

In this section, the main results of towing-tank tests on a reduced-scale model of the entire GEM system are presented. These tests were mainly aimed at characterizing the steady-state conditions and the related power production performance of the overall floating system and to study its stability and dynamic behaviour in different possible operating conditions; some off-design conditions were also examined, such as those caused in real situations by an abnormal shut-off of one turbine.

In the GEM system design, two counter-rotating turbines are mounted at the sides of the main floating body. The model used in the tests, scaled by eight times with respect to a possible real-scale installation of the GEM, is shown in Figure 3.37.

It consists of a floating submerged system made up of two bodies: the upper body serves to produce the necessary buoyancy force, while the lower body accommodates instruments and auxiliary devices. The upper body is also equipped

Figure 3.37 The GEM reduced-scale model tested in the towing tank. Scale 1:8

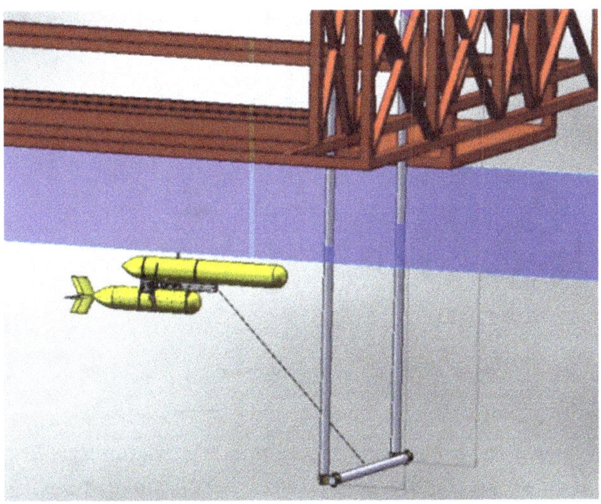

Figure 3.38 Arrangement of the towing-tank trials on reduced-scale GEM model. The test article is linked to a submerged structure, which is fixed to the running carriage

with two fins forming an angle of 45° with the longitudinal plane of symmetry. A 'V tail' configuration has been used for the tail planes with the aim to improve the necessary lateral stability of the system and to allow the regulation of the body pitching trim at stabilized advancing speeds, while reducing the possible interference with turbine wake. Different tail configurations have also been tested throughout the campaign.

A schematic CAD view of the GEM model towed at a steady speed is shown in Figure 3.38. The main dimensions of the GEM scaled model are reported in Table 3.2. Table 3.3 reports the main geometric characteristics of the floating model configuration.

Table 3.2 Turbine model geometry (bare turbine tests)

Turbine diameter	0.612 m
Support pole height	2.46 m
Diffuser throat diameter	0.620 m
Diffuser exit diameter ($\phi = 26°$ airfoil chord angle)	0.812 m
Diffuser section chord	0.200 m
Section chord at 75% radius (0.2295 m)	~0.0497 m

Table 3.3 GEM reduced-scale model main geometric data

Turbine diameters	0.61 m
Distance between turbine axes	1.13 m
GEM overall length	3.55 m
GEM overall height	1.84 m

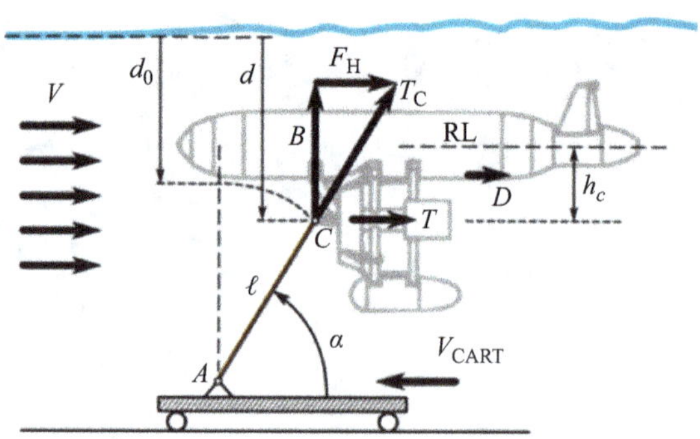

Figure 3.39 Schematic of the GEM system in a steady-state condition, immersed in a current V (due to the advancing carriage speed V); the equilibrium involves the cable tension force T_C, the buoyancy B and the total horizontal force F_H

At zero carriage speed, the turbine axis was set at 1.65 m below water surface, while the final axis depth is dependent on carriage speed; a schematic of the forces acting on the system is illustrated in Figure 3.39. The tethering cable arrangement in steady flow is an important aspect of the tested device. In the towing tank tests, an immersed steel frame structure, fixed to the running carriage, simulates the seabed single-point anchoring point (represented by point *A* in Figure 3.39). The model is connected through the cable to the running anchorage point. A winch located in point *A* allows to release or to tighten the cable and, consequently, to position the model at the desired depth in the tank. Geometric data of the mooring arrangement, for the tested tethered GEM model, are reported in Table 3.4.

Table 3.4 Geometric data of GEM scaled model tethering system

Initial depth d_0 of attachment point C	1.65 m
Depth d_A of towing attachment point A	3.27 m
Distance h_C of attachment point C from upper body reference line (RL)	0.75 m
Cable length, ℓ	1.62 m

3.17.3.2 Characterization of the submerged tethered system

During the tests, the following data are monitored and measured:

- torque and angular velocity of the left-hand-side turbine, from which the total output power is estimated, assuming symmetrical operating conditions;
- mechanical tension on the tethering cable, measured by a dedicated load cell;
- trim of the GEM system with respect to a reference frame fixed to the tank floor, by means of an inertial platform.

As seen from the schematic shown in Figure 3.39, in steady-state condition, the equilibrium is guaranteed by the cable tension force T_C, the net buoyancy B (buoyancy force minus dry weight) and the total horizontal force F_H.

Based on the cable tension load and on the knowledge of the buoyancy, the total horizontal drag developed by the whole system may be estimated, which in non-dimensional form is expressed by the total resistance coefficient:

$$C_D = \frac{F_H}{\frac{1}{2}\rho V^2 S_{\text{ref}}} \tag{3.6}$$

where and S_{ref} is a properly chosen reference area. For this application, the assumed reference area is $S_{\text{ref}} = 2\ A$, i.e. twice the turbine disc area. The cable traction T_C balances the vector sum of total drag F_H and the buoyancy B.

Experimental values of power and torque coefficients plotted versus *TSR* are shown in Figures 3.40 and 3.41, respectively. While the data reported in Figure 3.33 were related to a single isolated turbine, the latter graphs are referred to the complete small-scale GEM model.

In the same figures, fitting cubic splines of C_P and C_Q are also reported in order to highlight the overall trend of the coefficients. The scattered values of C_P and C_Q are estimated by averaging the power and torque measured signals, sampled at a frequency of 500 Hz, over different steady-state conditions with constant rotational speed of the turbine rotors and at a fixed towing carriage speed, $V = 1.5$ m/s. It has to be noted that a certain degree of scattering has been observed during measurements. The uncertainty levels, indicated on plots in Figures 3.40 and 3.41, are taken as three times the standard deviation of the sampled values. The two plots show that the maximum non-dimensional torque, $C_{Q,\text{max}} = 0.22$, occurs at a value of $TSR = 3.15$, which, as expected, is not the same *TSR* at which the maximum C_P occurs. It has to be noted that, due to data scattering, some difficulties are encountered in the estimation of the optimal C_P and *TSR*.

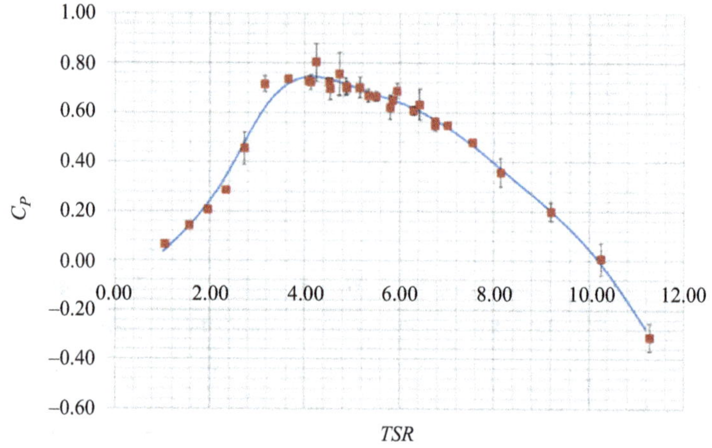

Figure 3.40 Power coefficient averaged values. Turbines with diffusers, upright V-tail, V = 1.5 m/s

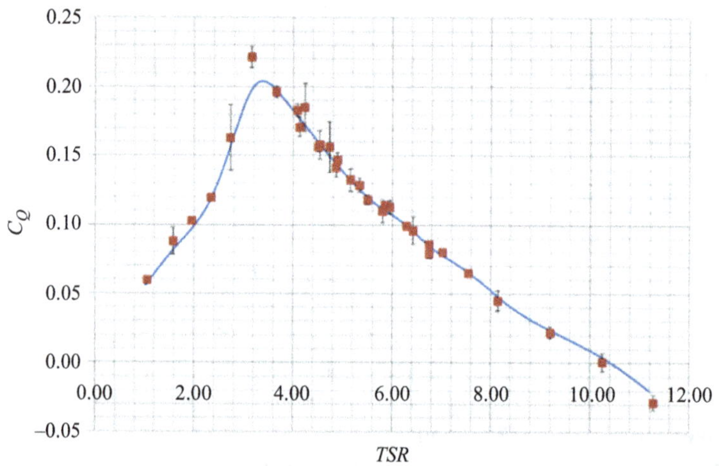

Figure 3.41 Torque coefficient averaged values. Turbines with diffusers, upright V-tail, V = 1.5 m/s

In Table 3.5, the maximum efficiencies and optimal *TSR* values, estimated on the reduced-scale GEM model, are compared for the two cases of shrouded and bare turbines.

Although a high scattering of test data may be observed, such results seem to confirm the results observed for the isolated turbine; that is, the presence of diffusers nearly doubles the maximum power coefficient referred to rotor disc area.

Table 3.5 GEM efficiencies with and without diffusers (scaled model tests)

	$C_{P,\max}$	TSR @ $C_{P,\max}$
Turbines w/o diffusers (from isolated turbine test data)	0.40	4.0
Ducted turbines	0.74	4.17

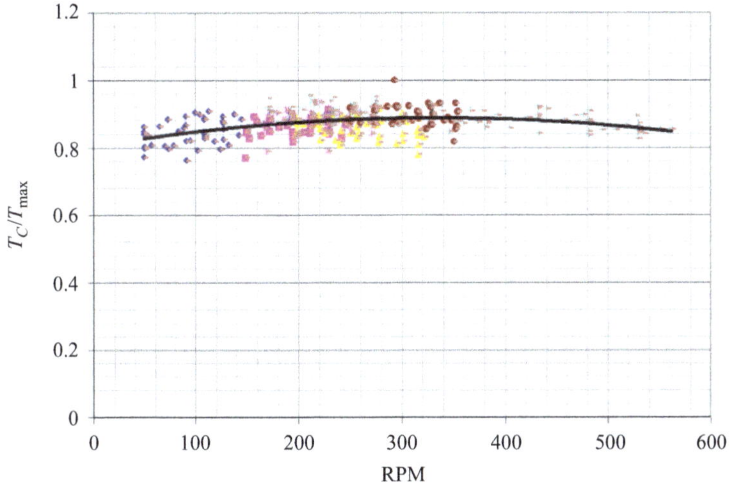

Figure 3.42 Tension force at the connection tether, normalized with maximum observed tension (turbines with diffusers, upright V-tail configuration, V = 1.5 m/s)

3.17.3.3 Tether force and trim

Figure 3.42 shows typical measured values of the tension force T_C in the tether cable for different values of the turbine angular speeds; these tension force values are normalized with respect to the maximum measured tension force. The plot contains all experimental measurements of a single carriage run, including those in transient-state conditions (variable Ω), at the imposed current speed V. It can be observed, besides a large scatter in estimated data, a difficulty in determining a clear trend.

A simplified evaluation of the tether angle α and the total horizontal force F_H is reported. Assuming a rectilinear shape of the cable in nominal conditions, knowing the net buoyancy force B (which is easily evaluated in static conditions), and having measured T_C, the angle α is given by

$$\alpha = \frac{\pi}{2} - \cos^{-1}(B/T_c) \tag{3.7}$$

All experimental tests were carried out with a fixed value $B = 2{,}570$ N of the buoyancy force.

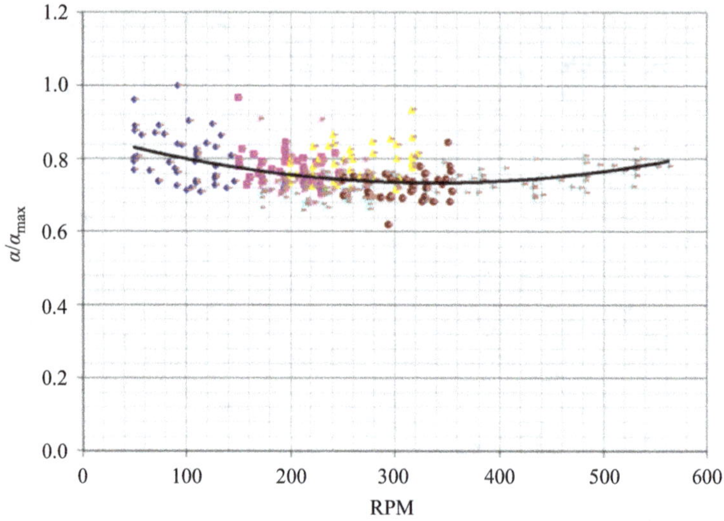

Figure 3.43 Normalized tether angle with respect to the horizontal plane. Turbines with diffusers, upright V-tail, $V = 1.5$ m/s

Figure 3.43 shows the normalized tether angles α, formed by the GEM connection cable with the horizontal, plotted versus the angular speed Ω, for a fixed steady-state speed $V = 1.5$ m/s, with the same consideration about large scattering as reported for tension force. The value of the horizontal thrust is also estimated using the measured cable tension and the calculated cable angle, as $F_H = T_C \cos \alpha$.

The variation of the estimated total thrust coefficient with *TSR* is also reported in Figure 3.44, for the 26° shroud configuration. A value of approximately 2.5 with very large scatter may be observed. Results are based on 1.5 m/s tests. Results show, anyway, a very large scatter, probably related to the oscillations of the floating body, resulting in cable instantaneous tension variations.

Tests have also been performed in some off-design conditions, such as in asymmetric conditions with one turbine inoperative, showing an overall stable behaviour.

3.17.4 Full-scale prototype tests

Experimental tests were carried out on a full-scale GEM prototype, designed to produce 20 kW of power at a nominal current speed of 1.5 m/s. During the field tests, GEM off-design operating conditions have been measured as well. This prototype has been developed, built and installed in a test site near Venice, Italy. The test campaign was supported partially by a consortium of companies operating in the Italian Veneto Region and partially by the Veneto Regional Authority.

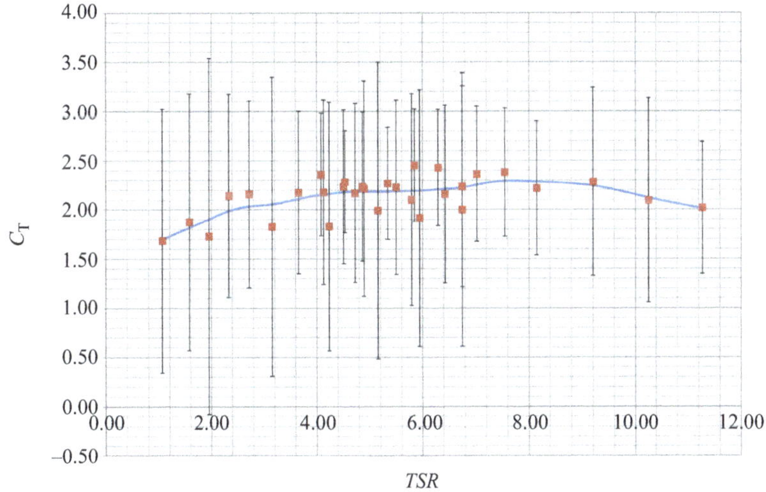

Figure 3.44 Thrust coefficient, averaged values. Turbines with diffusers, upright V-tail, V = 1.5 m/s

Figure 3.45 GEM large-scale prototype

3.17.4.1 Test plant configuration

Prototype general data

The first prototype has been installed in the Venice lagoon, near Forte Sant'Andrea, with a seabed depth of about 25 m. In Figure 3.45, a picture of the large-scale prototype is reported. Operational characteristics and other features of the system are summarized in Table 3.6.

Table 3.6 GEM prototype geometrical data

Turbine diameter	$D_t = 3.08$ m
Diffuser throat diameter	$D_g = 3.10$ m
Diffuser exit area diameter	$D_e = 4.08$ m
Overall length	$L = 9.2$ m
Overall height	$H = 5.2$ m
Overall width	$S = 10.4$ m
Weight (overall weight including ballast weight)	$W = 16{,}100$ kg
Weight (steel/composite structure only)	$W = 10{,}700$ kg
Net buoyancy force	$B = 51{,}000$ N
Horizontal thrust (on the overall system estimated at 1.5 m/s)	$T = 45{,}000$ N
Working rotational speed (turbine shaft at 1.5 m/s)	$\Omega = 38$ rpm

Starting from the information gathered in the preceding research steps, a prototype plant was designed and built-up with the objective to operate in a site with 1.5 m/s speed, generating a nominal total power of 20 kW, with a depth operating range between about 9.8 m (without current) and about 15 m. The adopted constructive solution uses carbon fibre for the turbine blades and of glass fibre for the diffuser and for the tail planes, while steel structures are used for the floating body and the connection frame. Turbine blades and diffuser design are based on the results of previous aerodynamic research studies.

Measurement equipment and main observed data

The GEM system full-scale prototype has been equipped with a set of devices for measuring several operating parameters, mainly, system trim and generator power production data. The data have been recorded on-board and transmitted via radio to a remote server. An ADCP was installed in proximity of the GEM mooring point in order to measure current speed (the mooring point is about 25 m below the free water surface and its horizontal distance from the GEM depends on the current speed). This device allowed the reconstruction of the whole velocity profile of the current, from seabed to water surface.

A typical example of time histories during a day of observation of the main state variables is shown in Figure 3.46. The plots show the GEM sideslip angle β, i.e. the rotation angle around the GEM vertical axis, the current direction δ with respect to the North and the averaged actual current speed measured by ADCP. It may be observed that the actual current shows a typical reversing pattern, characteristic of tidal currents. Figure 3.47 reports the measured power output for the left and right turbines.

Power production results

The power production has been measured for both installed turbines. In Figure 3.48, the measured power is reported versus the measured speed, in order to determine the device power curve. The data in Figure 3.48 are non-dimensionalized using the maximum value of velocity observed during a single cycle (V_{max} is about 1.3 m/s, slightly lower than the device rated velocity of 1.5 m/s) and the maximum measured

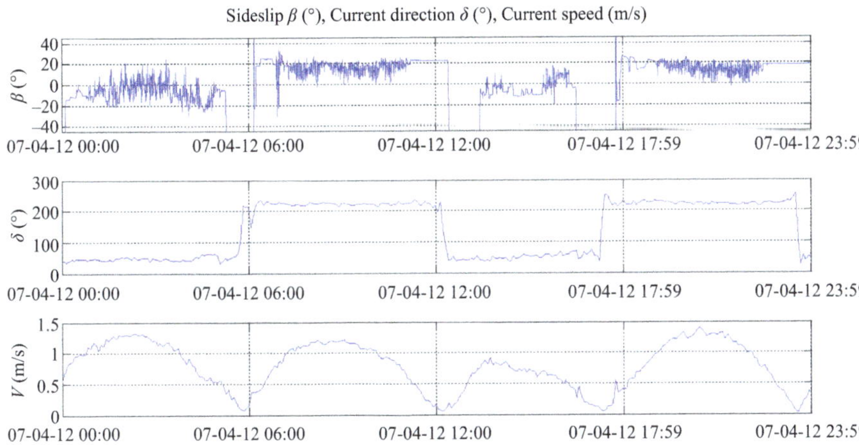

Figure 3.46 GEM large-scale prototype typical attitude and speed data measurements for a day-long measurement

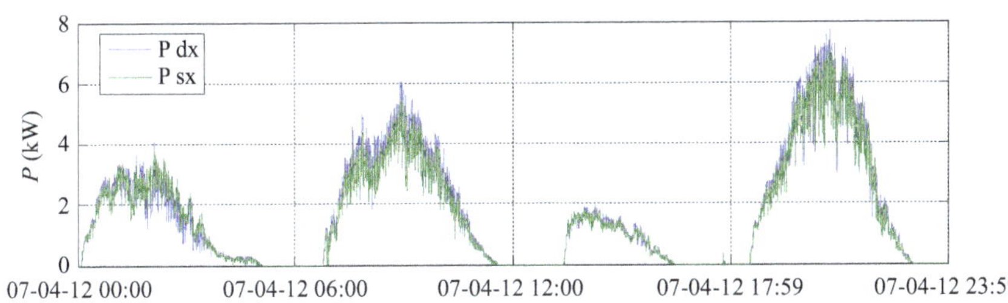

Figure 3.47 GEM large-scale prototype generator output power data measurements (left and right turbines) for a day-long measurement

power for a single turbine (P_{max} is about 7 kW). The test campaign presented many challenges and the data show a rather wide dispersion. Data dispersion is also partially due to a number of trials performed to set up and optimize the maximum power tracking control procedure.

A sort of binning procedure has been used in order to better analyse the power performance: the binning procedure consists in dividing velocity and power measurements into small intervals (bins), for each bin an average value of velocity and power can be extracted. Thus, power production data are grouped into velocity bins over which an arithmetic mean has been performed. This is the same procedure used for determining wind turbine power curve.

The power curves after the binning procedure, for the right turbine, are reported in Figure 3.49. In the figure, estimated shaft power is also compared with the measured electrical power. As a reference, in the same figure, the dashed lines represent the theoretical cubic power curves for a given constant power coefficient.

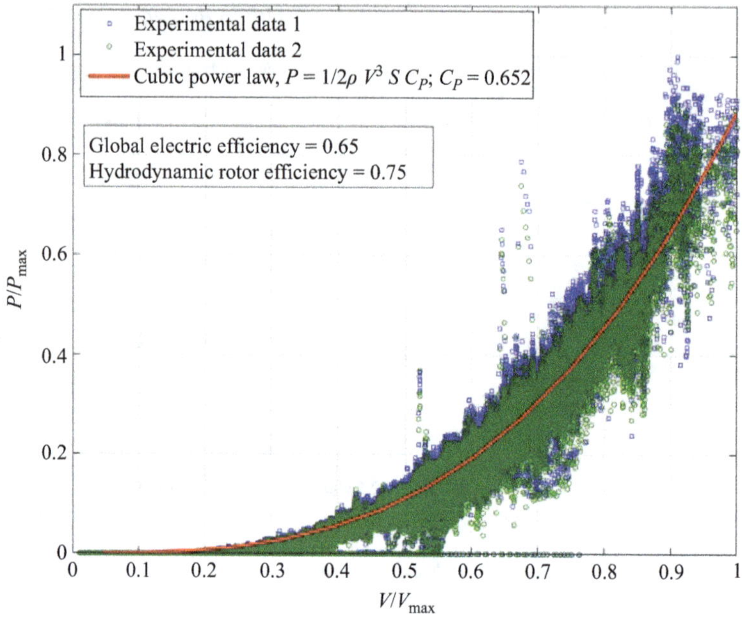

Figure 3.48 GEM large-scale prototype generator output power (normalized with respect to maximum measured power) versus current speed data measurements (left and right turbines)

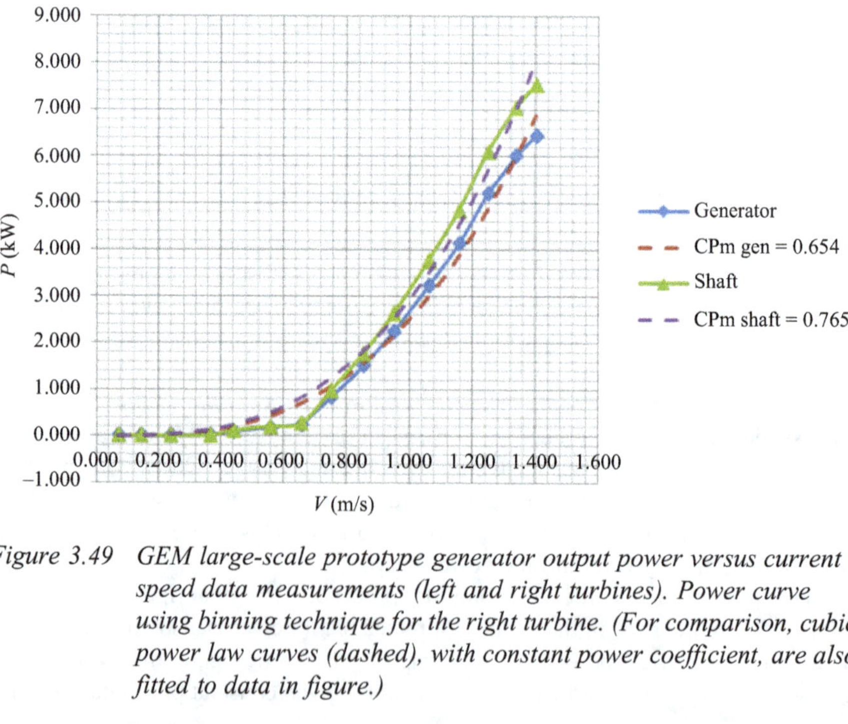

Figure 3.49 GEM large-scale prototype generator output power versus current speed data measurements (left and right turbines). Power curve using binning technique for the right turbine. (For comparison, cubic power law curves (dashed), with constant power coefficient, are also fitted to data in figure.)

A rough estimation of power coefficient may be obtained by comparison between cubic laws and binned curves: the maximum overall power coefficient, $C_{P\,\max}$, is within the range:

$$C_{P\max} = 0.6 \div 0.65 \tag{3.8}$$

also including mechanical and electrical efficiencies.

Consequently, supposing a generator efficiency of approximately 0.9 and a mechanical gearing efficiency of about 0.9, it can be claimed that, approximately, the shaft power coefficient is within the range $0.74 \div 0.8$, which is in good agreement with previous model towing tank testing.

Tests have proved the suitability of the system to operate in nominal conditions and the capability to convert efficiently the energy of the flow in mechanical and electrical energy. In particular, the use of ducted diffusers with rotor shroud of optimized shapes has proven to augment power generation capability with respect to the simple bare turbine solution. Nonetheless, the increased structural complexity and higher thrust levels, due to the installation of the diffusers, have to be accounted for in the overall evaluation of the shrouded configuration.

The full-scale tests on a GEM prototype seem to prove the feasibility, reliability, stability and efficiency of the system. Based on the experiences gained with GEM system, further research studies are undergoing for the development of the second generation device, named GEMSTAR, where improvement of some technological aspects are being performed aiming to obtain global cost reduction and to enhance the ratio cost/benefit to make the device appetible at commercial scale.

Symbols used

A	swept area, i.e. the flow area occupied by a turbine
B	blockage
C_P	coefficient of performance or coefficient of power
C_Q	torque coefficient
TSR	tip speed ratio
Fr	Froude number $= V/\sqrt{(gd)}$
g	acceleration due to gravity
GW	gigawatts, billion watts
h	height
HKT	hydrokinetic turbine
J	Joule
KE	kinetic energy
L	litre
MJ	megajoule
m	mass

N	newtons
NTF	National Tidal Facility of Australia operated by Flinders University of South Australia until December 2003; now called the National Tidal Centre, which commenced operation in January 2004 by the Australian Government Bureau of Meteorology
P	power
PE	potential energy
Q	volumetric flow or discharge
T	turbine thrust
TWh	terawatt hours (1 terawatt = 10^{12} watts)
V	velocity
W	watt
η	efficiency
ρ	fluid density
σ	Cavitation Number
τ	shaft torque
Ω	rotational speed

References

[1] Australian Government Bureau of Meteorology. Tide predictions for Australia, South Pacific and Antarctica. http://www.bom.gov.au/australia/tides/.

[2] Bachmann, R. (2005). Tidal current energy in backstairs passage. Final Year Project, University of South Australia School of Advanced Manufacturing and Mechanical Engineering.

[3] UK Tide Times. https://www.tidetimes.org.uk/.

[4] Hydro Quebec. Hydro power. http://www.hydroquebec.com/data/developpement-durable/pdf/file-hydrokinetic.pdf. Accessed 26 January 2019.

[5] Selin, N.E. Tidal power. https://www.britannica.com/science/tidal-power Accessed 26 January 2019.

[6] Electric Power Research Institute, U.S. Department of Energy (DOE). Assessment and mapping of the riverine hydrokinetic resource in the continental United States. https://digital.library.unt.edu/ark:/67531/metadc844286/#description-content-main. Accessed 26 January 2019.

[7] Steel, W. (2018). Bigger turbines, better economics, more digitization on deck for 2018 wind power market. *Renewable Energy World*, 21(1). 2017. Accessed 26 January 2019.

[8] Developing nations are driving record growth in solar power. https://reneweconomy.com.au/developing-nations-driving-record-growth-solar-power-94939/. Accessed 4 May 2019.

[9] La Rance tidal power plant: a world first. https://www.edf.fr/en/the-edf-group/industrial-provider/renewable-energies/marine-energy/tidal-power. Accessed 26 January 2019.

[10] Sihwa Lake Tidal Power Station. https://en.wikipedia.org/wiki/Sihwa_Lake_Tidal_Power_Station. Accessed 5 May 2019.

[11] Severn Barrage Tidal Power. http://www.reuk.co.uk/wordpress/tidal/severn-barrage-tidal-power/. Accessed 13 January 2019.

[12] Severn Barrage proposal rejected. https://www.renewableenergymagazine.com/ocean_energy/severn-barrage-proposal-rejected-20130611. Accessed 13 January 2019.

[13] McIntyre, F. (2018). After Swansea are tidal lagoons dead in the water? New Civil Engineer, 27 June 2018. https://www.newcivilengineer.com/latest/insight-after-swansea-are-tidal-lagoons-dead-in-the-water/10032529.article. Accessed 13 Jan 2019.

[14] Dynamic Tidal Power. https://en.wikipedia.org/wiki/Dynamic_tidal_power. Accessed 13 January 2019.

[15] Steijn, R. (2015). Dynamic Tidal Power technology advances. http://www.renewableenergyworld.com/articles/2015/01/dynamic-tidal-power-technology-advances.html. Accessed 13 January 2019.

[16] Rainey RCT. (2018). Lightweight steel tidal power barrages with minimal environmental impact: application to the Severn Barrage. *Proc. R. Soc. A*, 474:20170653. http://dx.doi.org/10.1098/rspa.2017.0653.

[17] Blue Energy Vertical Axis Hydro Turbine. http://www.bluenergy.com/technology_method_vaht.html. Accessed 13 January 2019.

[18] Olatunji, O.A.S., Raphael, A.T. and Yomi, I.T. (2018). Hydrokinetic energy opportunity for rural electrification in Nigeria. *Int. Journal of Renewable Energy Development*, 7(2), 183–190. http://dx.doi.org/10.14710/ijred.7.2.183-190

[19] Blue Energy – The Bo Hai Tidal Bridge Project. https://www.youtube.com/watch?v=uABv1mQcuEM. Accessed 6 December 2018.

[20] BAM takes share in tidal bridge scheme as Strukton opts out.https://marineenergy.biz/2018/05/28/bam-takes-share-in-tidal-bridge-scheme-as-strukton-opts-out/ Accessed 13 January 2019.

[21] Kepler Energy. http://www.bluenergy.com.

[22] SR2000 – Orbital Marine Power. https://orbitalmarine.com/technology-development/sr2000. Accessed 26 January 2019.

[23] The Bo Hai Tidal Bridge Project By Blue Energy. http://francisrousseau-emr.blogspot.com/2012/10/the-bo-hai-tidal-bridge-project-by-blue.html.

[24] GE unveils massive new 12MW wind turbine. https://reneweconomy.com.au/ge-unveils-massive-new-12mw-wind-turbine-19016/. Accessed 26 January 2019.

[25] Salter, S.H. and Taylor, J.R.M. (2007). Vertical-axis tidal-current generators and the Pentland Firth. *Proc. IMechE Vol. 221 Part A: J. Power and Energy*, Special Issue Paper 181. DOI: 10.1243/09576509JPE295.

[26] Polagye, B.L. (2009). Hydrodynamic effects of kinetic power extraction by in-stream tidal turbines. PhD Thesis, University of Washington.

[27] Sniderman, D. (2012). Energy from Florida's ocean currents. https://www.asme.org/engineering-topics/articles/renewable-energy/energy-from-florida-s-ocean-currents. Accessed 26 January 2019.

[28] http://www.wind-power-program.com/turbine_characteristics.htm. Accessed 27 January 2019.
[29] Minesto. https://minesto.com/our-technology. Accessed 27 January 2019.
[30] https://x.company/projects/makani/. Accessed 27 January 2019.
[31] Batten, W.M.J., Bahaj, A.S., Molland, A.F. and Chaplin, J.R. (2006). Hydrodynamics of marine current turbines. *Renewable Energy*, 31(2006), 249–256.
[32] Failed tidal turbine explained at symposium. http://www.cbc.ca/news/canada/nova-scotia/failed-tidal-turbine-explained-at-symposium-1.1075510. Accessed 28 January 2019.
[33] Goddijn-Murphy, L., Woolf, D.K. and Easton, M.C. Current patterns in the inner sound (Pentland Firth) from underway ADCP data. *Journal of Atmospheric and Oceanic Technology*, 30, 96–111.
[34] Nishino, T. and Willden, R.H.J. (2012). The efficiency of an array of tidal turbines partially blocking a wide channel. *J. Fluid Mech.*, 708, 596–606. and blockage correction for tidal turbines. (2009). *J. Fluid Mech.*, 624, 281–291.
[35] McAdam, R.A., Houlsby, G.T. and Oldfield, M.L.G. Structural and hydrodynamic model testing of the transverse horizontal axis water turbine. Proc 9th European Wave and Tidal Energy Conference, University of Southampton, 5–9 September 2011.
[36] McAdam, R.A., Houlsby, G.T. and Oldfield, M.L.G. (2013). Experimental measurements of the hydrodynamic performance and structural loading of the Transverse Horizontal Axis Water Turbine: Part 1. *Renewable Energy*, 59(2013), 105–114.
[37] Anyi, M. and Kirke, B.K. (2015). Tests on a non-clogging hydrokinetic turbine. *Energy for Sustainable Development*, 25(2015), 50–55.
[38] Whitlock, R. (2017). Scotrenewables Tidal Power SR2000 floating tidal turbine generates 116 MWh in less than a week of continuous generation. https://www.renewableenergymagazine.com/ocean_energy/scotrenewables-tidal-power-sr2000-floating-tidal-turbine-20170913. Accessed 14 March 2018.
[39] Schottel Hydro scales down Bay of Fundy tidal ambitions. https://tidalenergytoday.com/2017/12/15/schottel-hydro-scales-down-bay-of-fundy-tidal-ambitions/. Accessed 14 March 2018.
[40] University of Naples Aircraft Design and Aeroflight Dynamics Group. http://www.adag.unina.it/english/research/renewable/gem.html. Accessed 14 March 2018.
[41] Coiro, D.P., Troise, G., Scherillo, F., De Marco, A., Calise, G. and Bizzarrini, N. (2017) Development, deployment and experimental test on the novel tethered system GEM for tidal current energy exploitation. *Renewable Energy*, 114(2017), 323–336.
[42] Lazauskas, L. and Kirke, B.K. Modeling passive variable pitch cross flow hydrokinetic turbines to maximize performance and smooth operation. *Renewable Energy*, 45, 41–50.

[43] Rawlings, G.W. (2008). Parametric characterization of an experimental vertical axis hydro turbine. BASc, University of British Columbia, 2005. https://circle.ubc.ca/bitstream/handle/2429/14706/ubc_2008_spring_rawlings_george.pdf?seque.

[44] Gilbert, B.L. and Foreman, K.M. (1983). Experiments with a diffuser-augmented model wind turbine. *J. Energy Resources Technology, Trans ASME*, 105, 46–53.

[45] Smart Hydro Power. https://www.smart-hydro.de/

[46] Kirke, B.K. (2003). Developments in ducted water current turbines. https://www.researchgate.net/publication/237809015.

[47] Oscillating hydrofoil http://www.esru.strath.ac.uk/EandE/Web_sites/05-06/marine_renewables/technology/oschydro.htm

[48] Huxham, G.H, Cochard, S. and Patterson, J. Experimental parametric investigation of an oscillating hydrofoil tidal stream energy converter. 18th Australasian Fluid Mechanics Conference Launceston, Australia, 3–7 December 2012.

[49] Oscillating Foil Technology. http://leadingedge.engin.brown.edu/wordpress/?page_id=353.

[50] Kim, E.S. and Bernitsas, M.M. (2016). Performance prediction of horizontal hydrokinetic energy converter using synergy of multiple cylinders in flow induced motion. *Applied Energy*, 170(2016), 92–100.

[51] Tidal Sails. http://tidalsails.com/.

[52] Funke, S.W., Farrell P.E. and Piggott, M.D. (2014). Tidal turbine array optimisation using the adjoint approach. *Renewable Energy*, 63(2014), 658e673.

[53] Hoseyni-Chime, A. and Malte, P.C. Hydrokinetic turbines at high blockage ratio. Proceedings of 2nd Marine Energy Technology Symposium METS2014, 15–18 April 2014, Seattle, WA.

[54] Denny, E. (2009). The economics of tidal energy. *Energy Policy*, 37(5): 1914–1924. DOI 10.1016/j.enpol.2009.01.009.

[55] Tidal Energy. http://www.tidalenergy.eu/tidal_energy_uk.html.

[56] MeyGen to develop 400 MW UK tidal power site. http://www.renewableenergyfocus.com/view/13491/meygen-to-develop-400-mw-uk-tidal-power-site/. Accessed 20 December 2018.

[57] Repairs needed at Flagship Scottish tidal energy project. https://www.insider.co.uk/news/meygen-tidal-energy-repairs-needed-12892338. Accessed 20 December 2018.

[58] West Islay Tidal Farm. https://www.power-technology.com/projects/west-islay-tidal-farm-scotland/

[59] Ingram, E. (2018). SIMEC Atlantis Energy plans to launch Europe's 'largest planned tidal power project' in 2021. https://www.hydroworld.com/articles/2018/11/simec-atlantis-energy-plant-to-launch-europe-s-largest-planned-tidal-power-project-in-2021.html. Accessed 27 January 2019.

[60] Final investment decision made for tidal project. http://www.4coffshore.com/windfarms/final-investment-decision-made-for-tidal-project-nid6511.html. Accessed 27 January 2019.

[61] Coiro, D.P., Troise, G. and Scherillo, F. (2012). Design, towing tank test and deployment of full scale GEM, a novel tethered system for harnessing tidal energy. Proceeding of Asian Wave and Tidal Energy Conference (AWTEC 2012). Jeju (South Korea).

[62] Van Bussel, G.J. (2007). The science of making more torque from wind: diffuser experiments and theory revisited. *Journal of Physics: Conference Series*, 75:012010.

Chapter 4
Thermal and salinity gradient systems

Gianfranco Rizzo[a] *and Francesco Antonio Tiano*[a]

This chapter presents thermal and salinity gradient energy (SGE) systems. First, a description of both energy resources and the determination of energy potential is given. Then, power plants that convert the thermal gradient potential as well as the salinity gradient are discussed. Environmental and economic aspects associated with these technologies are also considered.

4.1 Energy resources

4.1.1 Oceanic thermal gradients

The oceanic thermal gradient is the temperature difference between the sea surface water (warm) and the deep water (cold) which is greater in the equatorial zone [1]. In the area between the Tropic of Cancer and that of Capricorn, the average daily solar irradiation is about 250 W/m^2 and guarantees the constant presence of sea surface water temperature of the order of 27 °C [300 K]. A low temperature water resource is formed by a global natural phenomenon called thermohaline circulation [2]. The thermohaline circulation represents the set of slow ocean currents that carry the deep water, which is generated at high latitudes, through the major ocean basins [3]. It is the component of the global oceanic circulation caused by the variation of temperature and density of the masses of water. At high latitudes, the water sinks, both due to the low temperature and due to the high salinity caused by the formation of the ice pack [4]. Once the ocean floor is reached, the cold water with higher density moves toward the equator. Mixing with surface water is minimal and a thermally stratified structure is generated [5]. Figure 4.1 shows a diagram of the thermohaline circulation, where salinity is expressed with a practical salinity scale [4].

4.1.1.1 Optimal characteristics

The exploitation of thermal gradient resources by an energy conversion plant requires that some important parameters are met. The fundamental one is the

[a]Department of Industrial Engineering, University of Salerno, Fisciano, Italy

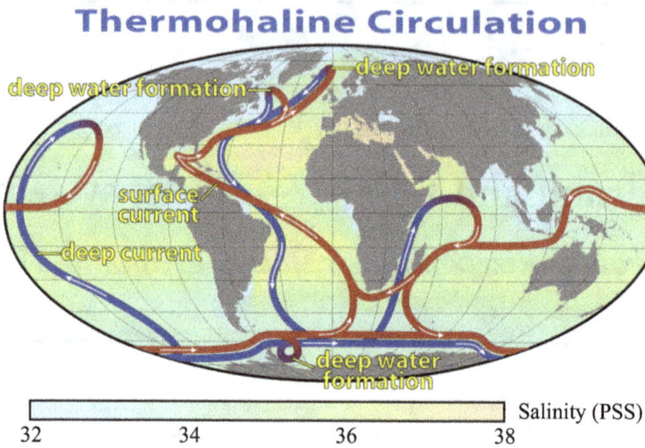

Figure 4.1 Thermohaline circulation

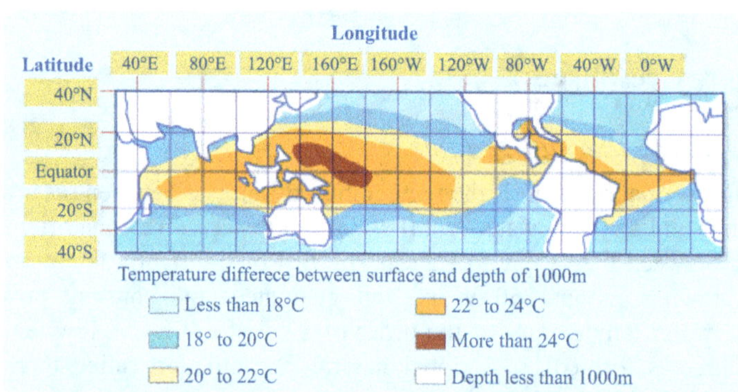

Figure 4.2 Temperature difference between surface and depth of 1000 m

existence of a temperature difference of at least 20 °C in the water column at a depth in the range of 0–1,000 m [6]. The stratification of the water column is intense in the tropical zone, where the thermocline layer is closer to the surface [7]. The thermocline layer, as shown in Figure 4.3, is the transition zone between the superficial layer of the ocean waters and the deep layer, where the temperature decreases rapidly from the value present on the surface to the one of deep water [8]. Figure 4.2 shows the value of the temperature difference at the depth of 1,000 m in the various ocean regions [9], while Figure 4.3 shows the thermocline [10].

The annual temperature variations must be equal to a few degrees and about 1° in the day–night cycle [5]. By that, the plant is allowed to operate with a non-oscillating power [6]. The temperature of the deep water must be about 4 °C–7 °C [277 K–280 K] [11]. The plant distance from the land coast, which has to be at most

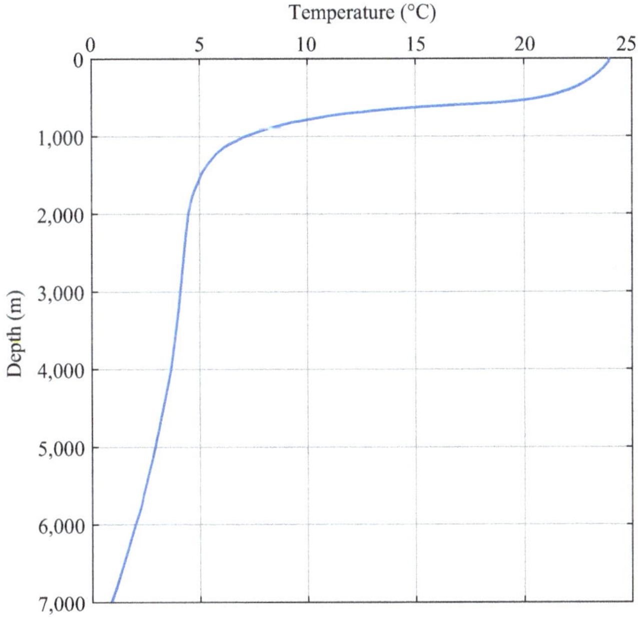

Figure 4.3 Thermocline

10 km, is a characteristic of the thermal gradient resources linked to the techno-economic aspects [5]. If the plant is located on the coast, the bathymetric characteristics must include the presence of a steep and uniform slope of the seabed toward the sea. The climatic and seismic conditions of the area of interest must not be extreme; in particular, the waves must not exceed 6 m, the wind must not exceed the speed of 20 m/s, the sea currents must not exceed 1.5 m/s and the expected earthquakes must be of low intensity [6].

4.1.1.2 Estimation of energy potential

The estimation of the power obtainable from oceanic thermal resources has been addressed in the literature by several methods. These studies use the incident solar power absorbed by the oceans to quantify the potential. Starting from the absorbed solar power, approximately equal to 240 W/m^2, the evaporation flow is chosen as reference. This flow is an example of the natural conversion of solar energy into latent heat and is equal to about 95 W/m^2. The Carnot cycle efficiency sets a theoretical limit for the operating efficiency of plants exploiting this source. Considering a temperature of the deep water equal to 4 °C [277 K] and a temperature difference between surface and deep water equal to 24 °C, the theoretical efficiency of a plant is 7.97%. The conversion into useful power by a real plant has an efficiency of about 3% and considering the total ocean surface area is around 370 million km^2, the estimated achievable power from the thermal resources would be around 1,000 TW. This estimation does not consider the limitations that may exist in using this renewable resource; in fact, the flows of water that can be taken from the environment must be of the same order of magnitude as the rate of formation of deep cold water. The exploitation of thermal resources must be done in order to adequately space and dimension the power plants. Previous assessments of the resource potential tend to overestimate the power that can be obtained, and for this reason, more studies have been carried out that take into account the vertical and horizontal ocean structure and a diffusion of conversion plants over the entire available area [12].

The oceanic structure has been simulated with different models:

- a one-dimensional model of oceanic temperature [13];
- a numerical model of the general ocean circulation MITgcm (Massachusetts Institute of Technology general circulation model) with a horizontal resolution of $4° \times 4°$ (longitude and latitude) and a vertical stratification in 15 zones [14];
- a numerical model of the MITgcm oceanic general circulation with a horizontal resolution of $1° \times 1°$ and a vertical stratification in 23 zones [3];
- a model identical to the previous one with the addition of the atmospheric feedback [2].

The one-dimensional model considers the overall effects of the conversion plant operations on the vertical water column, while the phenomena of horizontal transport are neglected. The extractable power from the resources is limited due to the perturbation of the vertical structure. The simulation of the model shows that there will be heating of the tropical deep cold water resource due to the flows of water necessary for the operations of the plants in a global scenario. The maximum power available according to this simulation is 5 TW [13].

The three-dimensional model that uses MITgcm, with a vertical grid of $4° \times 4°$ and a vertical stratification of 15 zones, simulates atmospheric and oceanic phenomena with greater accuracy. The phenomena of horizontal transport are higher by some orders of magnitude compared to vertical phenomena; consequently, the power limit that can be extracted with this approach increases, reaching 30 TW. The water flow, related to the maximum power value, significantly increases the thermohaline circulation and causes changes to ocean temperatures [14].

The model that uses a $1° \times 1°$ ocean grid implements a more elaborate description of oceanic mechanisms. The difference of this model compared to the previous one lies in the fact that the massive use of conversion plants has a greater impact on the thermal resources; in fact, the maximum value of extractable power calculated with this model reduces to 14 TW. The simulation is allowed to analyze a scenario where the use of thermal resources is more limited and the power extracted is 7 TW. Therefore, the impact on the environment is negligible [3].

The model with atmospheric feedback modifies the thermal condition at the ocean/atmosphere boundary, adding an approximate relation that corresponds to the long wave atmospheric radiation to the space and an additional term that expresses horizontal atmospheric transport. In this way the atmospheric temperature can vary according to changes in the oceanic one. Simulations showed that the maximum achievable power is in the range of 8–10.2 TW [2].

All models have been simulated for a minimum period of 1,000-year thermal resource exploitation [2,3,13,14].

4.1.2 Salinity gradients

The salinity gradient represents the difference in salt concentration between two masses of water. For instance, the mixing of fresh water and sea water at the mouth of a river can be framed as an energy resource [15]; in fact, when a river flows into the ocean, the spontaneous and irreversible mixing of fresh and salt water increases the entropy of the system. The change in entropy can be converted into electricity [16]. Energy from the salinity gradient is also known as *osmotic power* [17] or *blue energy* [18]. This resource is highly decentralized and is mainly situated at the mouths of the rivers where human settlements are dense and the energy demand is high [19].

4.1.2.1 Optimal characteristics

The exploitation of salinity gradient resources by a conversion plant requires that they meet some important requirements. The main characteristics concern the depth and stability over time of the salinity gradient. A marked stratification between fresh and salt water is fundamental. This depends on the buoyancy of the water with the lower concentration of salt discharged into the sea and the mixing by the tides. A strong stratification is present in rivers with a high or medium flow rate and a medium-low tide intensity [19]. The natural phenomenon of stratification of water with a different degree of salinity is called a *halocline*. In estuaries, fresh water is stratified at the top and the salty water at the bottom. In the tropical zone, an instability of the phenomenon can be born because the greater evaporation of the superficial layers causes the descent of the latter as a consequence of the increase in density. A mixing is generated instead of a well-defined stratification [20].

The salinity of sea water recorded at the mouth of a river is about 35 g/l (this value varies during the year), while that of fresh water is 1 g/l [15]. The salinity close to the river mouth is lower than that calculated using the mean value of global ocean waters [19]. The vertical distance between the high concentration (HC) water source, and the low salty water source should be as small as possible, thus allowing shorter pipelines to be used in the conversion plant and to limit losses [15]. The tide height mentioned above, must be less than 1.2 m [19]. This allows the presence of a stable gradient at a shorter distance from the river mouth [21]. The abundant presence of a fresh water discharge is also important [21]. Figures 4.4 and 4.5 show the river mouths and the salinity and temperature measurement points [22–24]. Figure 4.6 shows coastal areas with a tidal height of less than 1.2 m [19].

4.1.2.2 Energy potential

The quantification of the theoretical energy potential of SGE has been addressed in several studies since 1970. The values were found to range from 1.4 to 2.6 TW. More recent studies speak of 0.23, 3.13 and 1.724 TW. Only the last evaluation takes into account the salinity of the water in the vicinity of the river mouths [25], instead of the global mean values.

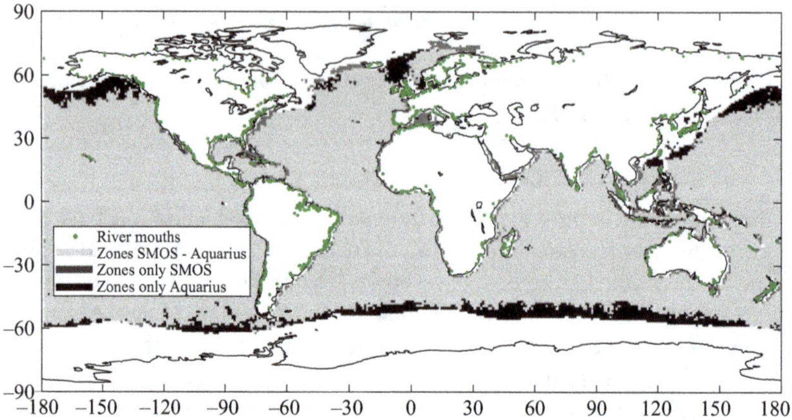

Figure 4.4 Coverage areas of salinity gradient data [19]

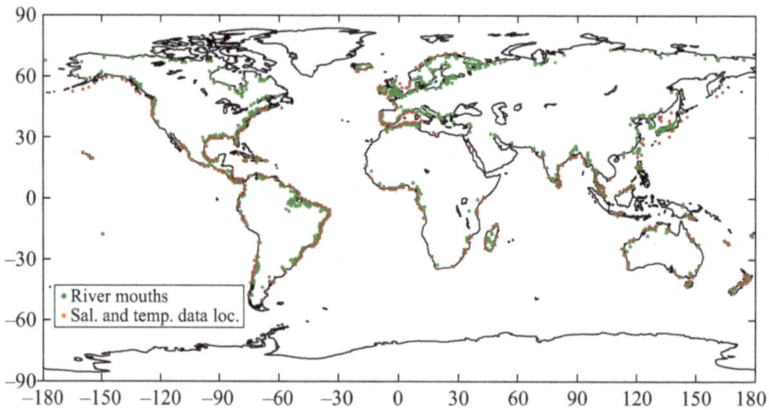

Figure 4.5 Location of sea surface salinity and sea surface temperature data for each river mouth [19]

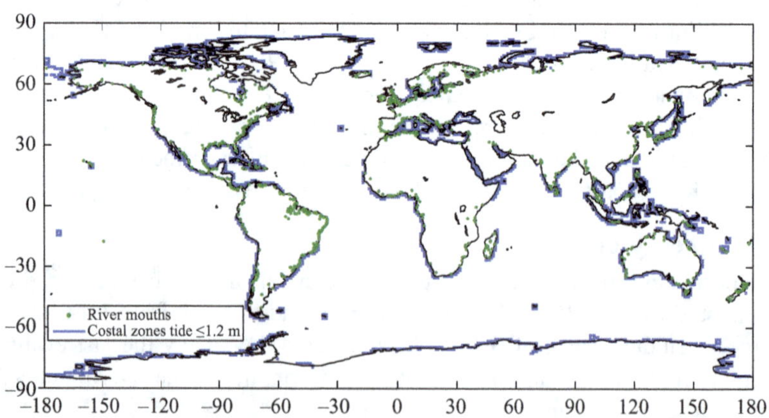

Figure 4.6 Coastal regions with mean tidal range ≤1.2 m [19]

A correct assessment must take into account that not all river mouths in the world are suitable. In fact, areas with reduced salinity gradients, low water quality and not permanently accessible resources must be neglected [19]. Water with a low quality contains a large number of colloidal particles, organic and inorganic materials that can cause encrustations on the components of the conversion plant, reducing the power generated [26]. The estimate of the potential is also useful in choosing the optimal site. The actually exploitable potential (EE, extractable potential) can be derived from the theoretical potential (TP), reducing the extraction factor (EF) and the capacity factor (CF). The EF takes into account the impossibility of exploiting all the river discharge, and the CF takes into account the seasonal variability of river flow and the saline stability of the aquatic environment near the mouth [19]. To simulate the variability in time in the estimation of the potential, one can use a model that describes the hydrodynamics of the estuaries and the ocean, for example, ELCOM (Estuary, Lake and Coastal Ocean Model) [15]. It is possible to write a formula to quantify the TP reduction by the two EF and CF factors:

$$EE = \sum_{k=1}^{sm} (TP_k \cdot EF_k \cdot CF_k) \tag{4.1}$$

In the sum, *sm* represents the number of river mouths suitable for extraction. The determination of the *TP*, resulting from the mixing of sea and river water at a given k – mouth, is carried out considering the chemical potential difference before mixing the chemical potential after mixing:

$$TP_k = (G_{s_k} + G_{r_k}) - G_{b_k} \tag{4.2}$$

where G_s, G_r and G_b are the Gibbs free energies of the sea water (s), fresh water (r) and the solution after mixing (b). The free energy of each electrolyte i ($i = s, r, b$) depends on the volume, salinity and temperature of the solutions and is given by

$$G_i = T_i \cdot Q_i \cdot m_i \cdot R \cdot [x_i \cdot \ln(x_i) + y_i \cdot \ln(y_i)] \tag{4.3}$$

where T (in Kelvin) is the absolute temperature, Q (in m^3/s) is the volumetric water flow ($Q_b = Q_s + Q_r$), m (in mol/m^3) is the total number of moles per unit of volume, R is the universal constant of gases (8.314 J/(mol K)), x and y are, respectively, the molar fractions of ions (Na^+ and Cl^-) and water. The global TP is therefore identified by summing the potential of all river mouths in data K:

$$TP = \sum_{k=1}^{K} TP_k \tag{4.4}$$

Sea and fresh water flows are considered in a ratio of 1:1 ($Q_s = Q_r$). Temperatures are assumed equal to the surface temperature near the river mouths ($T_r = T_s = T_b$). The TP is greater than the energy required for the regular operation of the plant only where the river mouths have the characteristics previously addressed, in particular stable stratification of salinity and a short distance between the layers [19].

The EF can be calculated from the variability of the natural flow of the river Q_N, which is the greatest limitation in the design of an SGE plant. The EF is defined as the relationship between the design flow Q_D of the plant (river water flow that can be extracted to generate design power) and the average river flow $\bar{Q}$ [19]:

$$EF = \frac{Q_N}{\bar{Q}} \tag{4.5}$$

The design flow, as anticipated, is limited by environmental factors. A residual flow rate Q_R can be defined:

$$Q_R = Q_N - Q_D \tag{4.6}$$

The residual flow must not be lower than the so-called environmental flow Q_E. The environmental flow has been attributed a value equal to 30% [19] of the average flow of the river:

$$Q_E = 0.3 \cdot \bar{Q} \tag{4.7}$$

The condition $Q_R \geq Q_E$ involves periods where the extraction of fresh water will be less than the design flow, and therefore the plant will not operate at full load. It is possible to identify three operating conditions:

- $Q_N > (Q_D + Q_E)$, the plant operates at full load;
- $Q_E < Q_N < (Q_D + Q_E)$, the plant operates at partial load;
- $Q_N < Q_E$, the system is stationary [19].

The CF is the ratio between the effective yield of the plant over the year and the theoretical annual generation. This factor is calculated, for each river mouth, as the ratio between the operating flow Q_{OP} (determined by the previous limit conditions) and the design flow Q_D during the year:

$$CF = \frac{\int_0^T Q_{OP}\ dt}{Q_D \cdot T} \tag{4.8}$$

The CF depends on the EF, the design flow and the environmental flow. To evaluate the extractable energy, the CF can be expressed as a percentage of the full load hours per year, multiplying it by the total hours present in a year (8,760 h/year) [19]. The determination of the design flow requires a compromise analysis. The cost of electricity generation is lowered if the conversion plant operates with a high CF and a relatively low design flow, but this would cause a low output power. On the contrary, if the plant operates with a high design flow, the installation costs per unit of power are lowered, thanks to economies of scale. The increase in this flow increases the installed power and the generation costs but decreases the CF and the plant will operate at full load for a shorter time during the year. While the design flow for hydroelectric power plants is approximately equal to the average available flow of the river, for SGE plants it cannot have that value because it would compromise the salinity structure and the ecosystem at the mouth of the river [21].

The global average TP has been estimated in 1,183 GW [19]. Considering only the mouths of the suitable rivers, this potential reduces to 412 GW. If the EF is equal to 0.2, the value of the potential is 82.5 GW. The CF is around 0.84, which corresponds to 7,358 h/year. The CF is greater than the estimated value for energy generation from waves, tides, wind and photovoltaics (PVs). The actually exploitable potential (*EE*) is estimated at 625 TW h/year [19]. The global distribution of exploitable potential and the energy density is shown in Figures 4.7 and 4.8.

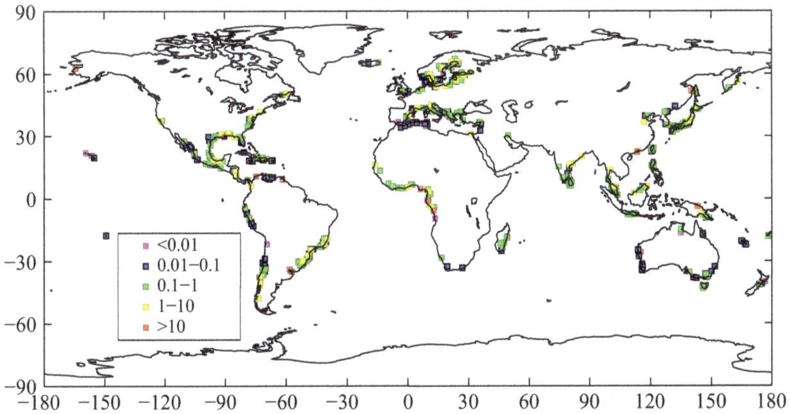

Figure 4.7 Global map of extractable salinity gradient energy resources. Extractable energy (TW h/year) [19]

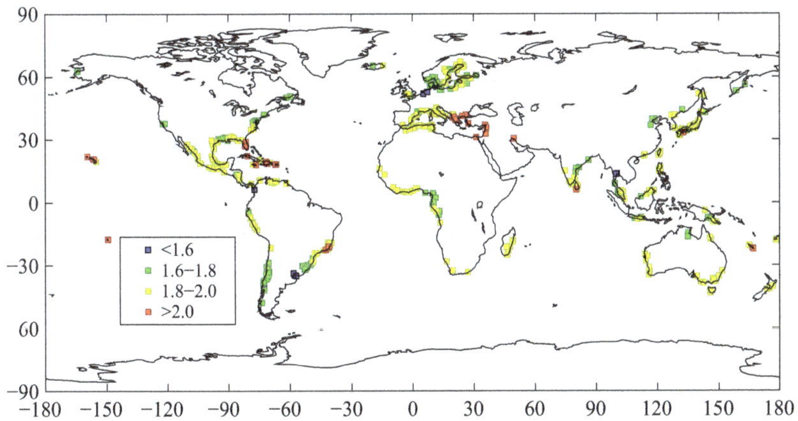

Figure 4.8 Global map of extractable salinity gradient energy resources. Energy density (MJ/(m³/s)) [19]

4.2 Ocean thermal energy conversion (OTEC) plants

The oceanic thermal gradient energy resources can be exploited by the OTEC (ocean thermal energy conversion) systems, which are able to convert thermal energy into electrical energy. The conversion of oceanic thermal energy into electricity was discovered by the French physicist Jacques Arsen d'Arsonval in 1881. In 1930, one of his students, Georges Claude, built the first open cycle OTEC plant in Cuba [11]. In 1935, another power plant was built in Brazil, but it did not manage to produce a positive net power. The progress of the OTEC plants was stopped until 1979, when a pilot plant was built at the Natural Energy Laboratory of Hawaii Agency (NELHA), Hawaii, which had a net power of 18 kW. A demonstration plant with a net capacity of 30 kW was built in Tokyo. In 1993, the NELHA built a 103 kW net open loop facility on the coast. In 2012, the estimated budget for a 10 MW plant was around 300 million euros [27]. Later, in 2015, Makay Ocean Engineering inaugurated the first closed-cycle OTEC onshore facility in North Kona, Hawaii [17]. Figure 4.9 shows the current and the future status of OTEC plants in the world [28].

OTEC systems operate a thermodynamic cycle. Part of the thermal energy of sea level warm water is converted into electricity. Deep cold sea water is used as a low temperature source and is taken from submerged ducts. Carnot efficiency (i.e., the maximum limit of the thermodynamic cycle) depends on the temperature difference. As described in Section 4.1.1.1, this thermal gradient is small and therefore the theoretical efficiency of the cycle will be in the range of 6%–8%. Considering the irreversibility and a real system, the efficiency will further drop to around 3%–5% [29]. Although the resource used is practically free, low thermodynamic efficiency has negative effects on the economic feasibility and quantity of electricity produced [30]. The OTEC system can have different configurations, as schematically represented in Figure 4.10.

4.2.1 Open-cycle plants

Open-cycle OTEC systems use seawater as the working fluid and consist of a flash evaporator, a steam expansion turbine, a generator, a steam condenser, a non-condensable gases ejector and a deaerator. The system configuration is shown in Figure 4.11.

The plant is based on a simple Rankine cycle model. The surface warm water is deaerated (i.e., the removal of oxygen and other dissolved gases from water which can cause serious corrosion damages in evaporators and turbines), then is passed through the flash evaporation chamber, where a fraction of seawater is converted into low-pressure steam [5]. Only 0.5% of the mass of water entering the chamber evaporates [30]. The steam is passed into the turbine for conversion into electrical energy and then sent into a condenser. Gases from the deaerator and the condenser are ejected in the atmosphere via specific gas ejectors. The cycle name derives from the fact that the working fluid does not return to the evaporator. Instead, the steam coming out of the turbine can be passed through a surface heat

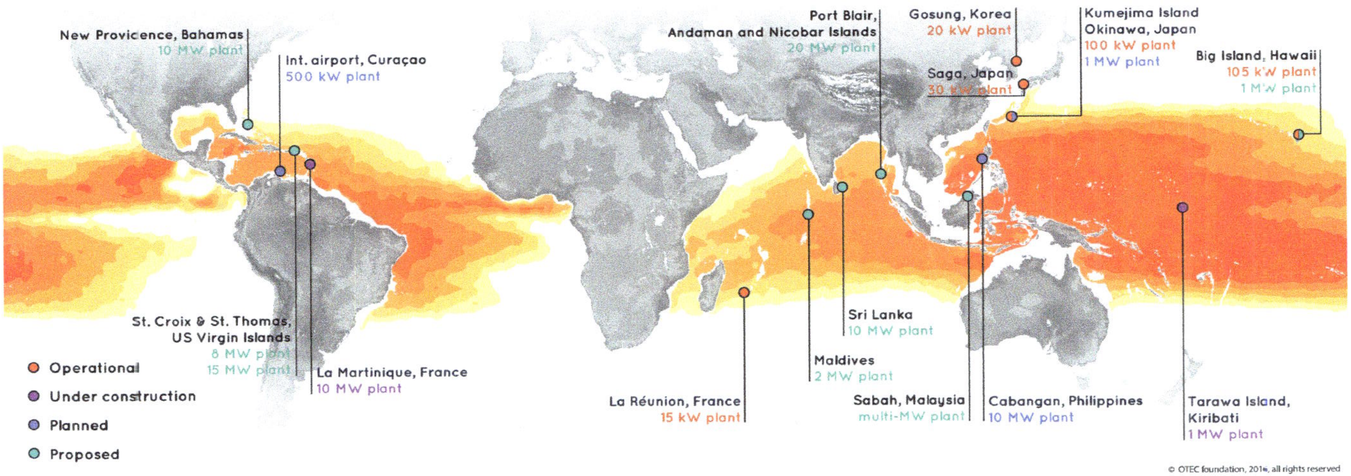

Figure 4.9 Current and future status of OTEC plants in the world [28]

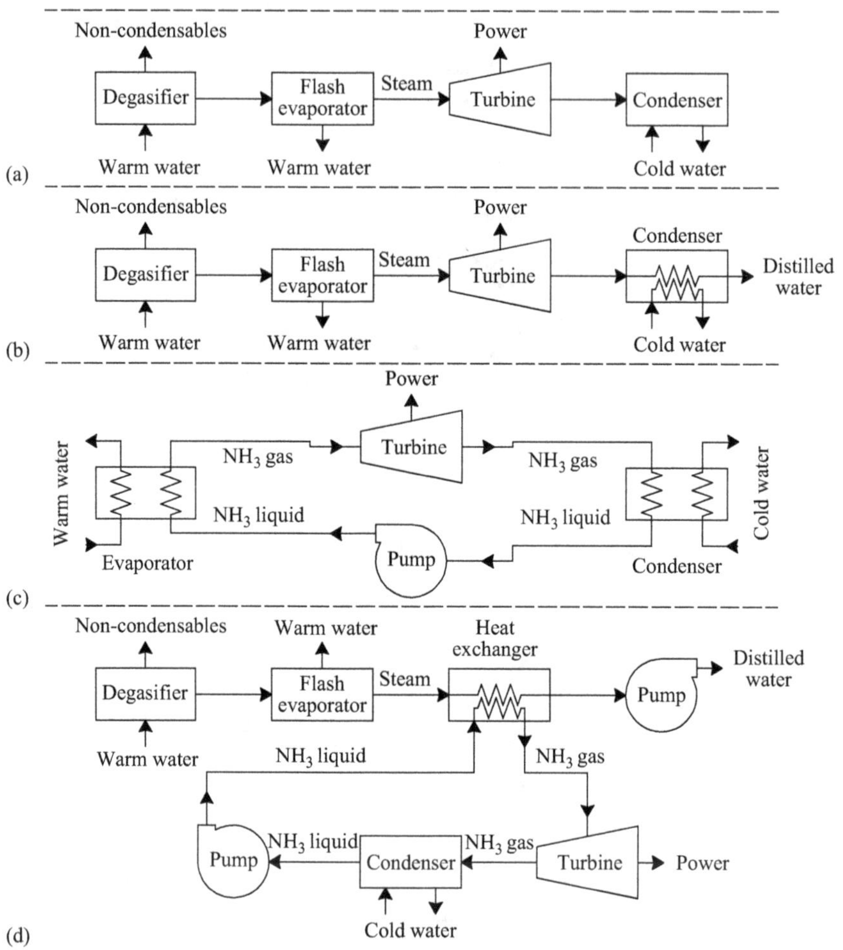

Figure 4.10 OTEC configurations: open-cycle type without distilled water production (a), the open-cycle type with distilled water recovery (b), the closed-cycle (c) and the hybrid-cycle (d)

exchanger to obtain desalinated water as a secondary product. Alternatively, it is sent to a direct contact heat exchanger to mix steam and deep cold water and drain the product into the ocean. The open cycle can be represented on the $T - S$ diagram as shown in Figure 4.12.

State 1 corresponds to the surface warm seawater at a temperature of about 27 °C (300 K) and atmospheric pressure. Sea water flows to the entrance of the deaerator in state 2, where the pressure is suddenly reduced to a value slightly higher than the saturation pressure at the corresponding temperature. Gases dissolved in the water are removed in this stage. The deaerated seawater in state 2 is sent to the flash evaporator, where the value of the ambient pressure changes to

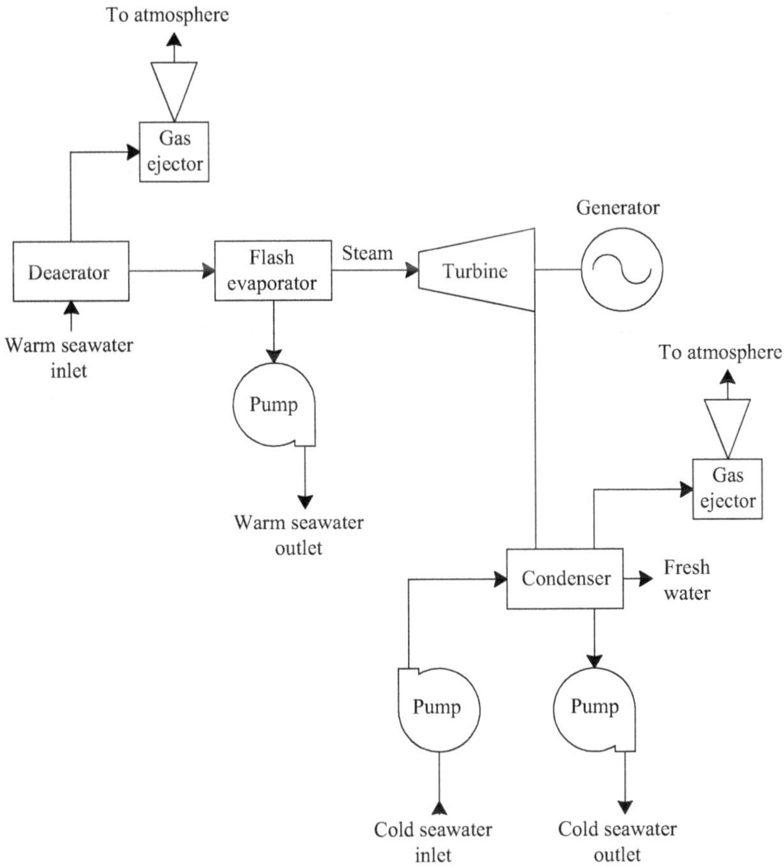

Figure 4.11 Open-cycle OTEC schematic diagram

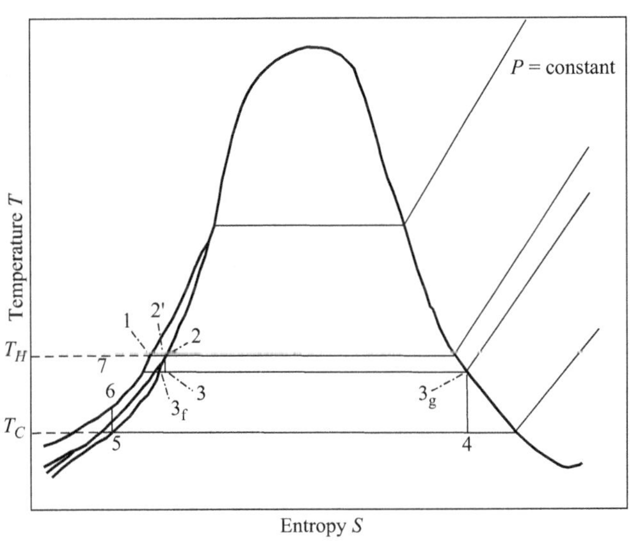

Figure 4.12 Open-cycle OTEC T–S diagram

the vapor saturation pressure value, equivalent to that of the outlet steam temperature. The pressure drop is the driving force for evaporation. A small part of the working fluid is immediately transformed into steam. The flashdown temperature drop corresponds to the required latent evaporation heat. The reference thermodynamic state in the flash evaporator is represented by state 3 [5]. The water reaches the evaporation chamber through nozzles, designed to maximize the heat exchange surface [30]. The vapor phase of the fluid, corresponding to the 3_g state, has a mass flow rate of $x\ \dot{m}$ (the x is the mass fraction of steam and $\dot{m}$ is the total mass flow of sea water) and expands in the turbine, where the thermal energy is converted into mechanical work. The remaining fraction of seawater, with a mass flow rate of $(1-x)\dot{m}$, is in the 3_f state and is pumped into the ocean reaching state 7. It is at a lower water temperature at the plant inlet, and this temperature difference is called flashdown. The turbine exhaust at state 4 is sent to the condenser, where it is cooled by deep cold sea water. At the output of the condenser, it is in state 5 and subsequently a pump brings it to state 6 for removal. If the processed seawater is discharged directly into the ocean, it will follow steps 6, 7 to state 1, absorbing the solar thermal energy. The mass and energy balances on the flash evaporator, the turbine and the surface condenser allow to quantify the exchanged heat and power output of the OTEC open cycle system:

$$\begin{aligned} \dot{Q} &= \dot{m}_H c_p (T_H - T_W) \\ P_g &= \eta_T \dot{Q}_{in}\left(1 - \frac{T_c}{T_H}\right) \\ \dot{Q}_{out} &= \dot{Q}_{in} - P_g \\ P_n &= P_g - P_{CSW} - P_{WSW} - P_{misc} \\ \eta &= \frac{P_n}{\dot{Q}_{in}} \end{aligned} \tag{4.9}$$

where $\dot{m}_H$ is the mass flow rate of warm water, T_H is the temperature of the incoming warm water, T_W is the temperature of the warm water at the evaporator outlet, T_c is the condensation temperature of the steam, c_p is the specific heat, $\dot{Q}_{in}$ is the thermal power entering into the system, $\dot{Q}_{out}$ is the thermal power discharged by the system, η is the efficiency of the system, η_T is the turbine efficiency, P_n is the net power, P_g is the total power, P_{CSW}, P_{WSW} and P_{misc} are, respectively, the parasitic powers absorbed by the systems for cold water, warm water and auxiliary systems [5].

The mist lift and foam lift cycles are variants of the open cycle and use a hydraulic turbine instead of a steam turbine. They directly exploit sea water to produce power. Warm sea water is evaporated in a flash mode in a two-phase mixture of vapor and liquid. The mist cycle consists of liquid droplets suspended in the vapor, in the foam cycle the vapor bubbles are enclosed in a liquid phase. The mixtures work against gravity and in the process the thermal energy of the steam is transformed into gravitational potential energy. The flow can be passed through the hydraulic turbine before or after the ascension process. An advantage of these cycles is that they operate with hydraulic turbines which are cheaper than steam turbines [30].

4.2.2 Closed-cycle plants

Closed-cycle OTEC systems have a similar configuration to standard Rankine-cycle-based systems. The main components of the closed-cycle system are a steam turbine, a condenser, a working fluid pump, an evaporator and auxiliary pumps for warm and cold water. Ammonia can be used as a working fluid. R134 refrigerant may also be used. Figure 4.13 shows a typical configuration of a closed-cycle OTEC system. The thermodynamic diagrams are shown in Figure 4.14, in (a) the plane *p–h*, in (b) the plane *p–v* and in (c) the plane *T–s*.

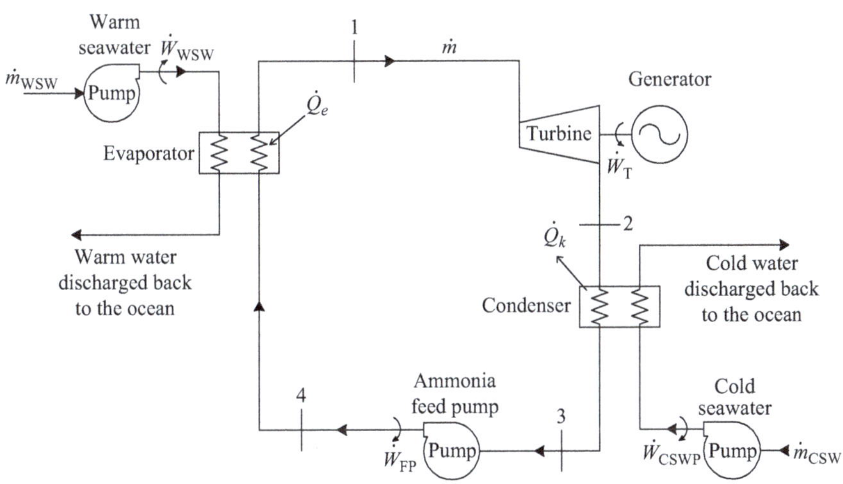

Figure 4.13 Closed-cycle OTEC schematic diagram

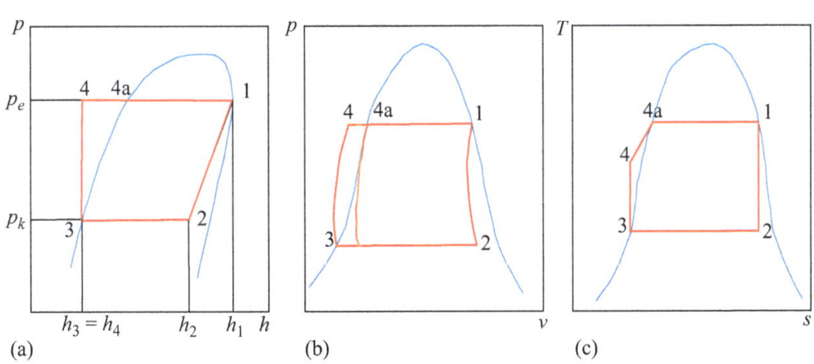

Figure 4.14 Phase diagrams for the closed-cycle OTEC system: (a) the p–h diagram, (b) the p–v diagram and (c) the T–s diagram

Ammonia-saturated steam is generated in the evaporator, where sea-level warm water exchanges heat with liquid ammonia at pressure p_e. The saturated vapor of ammonia enters the turbine that drives an electrical generator to produce electricity. The steam exits the turbine at a pressure, p_k, from where it is diverted into the condenser. The heat exchange with cold water, taken at 1,000 m depth, allows ammonia to condense. The liquid ammonia, which is at a low temperature and pressure, flows to the pump which increases the pressure and sends it to the evaporator, repeating the cycle. According to the diagrams in Figure 4.14, the added heat, q_A, in the 4–1 transformation is

$$q_A = h_4 - h_1 \tag{4.10}$$

where h_i is the specific enthalpy of the working fluid in the state i. The work of the turbine, in 1–2, is

$$w_T = h_1 - h_2 \tag{4.11}$$

The heat released in transformation 2–3 is

$$|q_R| = h_2 - h_3 \tag{4.12}$$

The work of the pump, in process 3–4, is

$$|w_P| = h_4 - h_3 \approx v_3(p_4 - p_3) = v_{kf}(p_e - p_k) = v_{kf}\Delta p \tag{4.13}$$

where v is the specific volume and p is the absolute pressure of the working fluid. Subscripts e, k and kf refer, respectively, to the pressure in the evaporator, in the condenser and at the pressure p_k on the curve of the saturated liquid (f), shown in Figure 4.14(a). The network can be expressed as

$$w_{net} = w_T - |w_P| = h_1 - h_2 - h_4 + h_3 \tag{4.14}$$

The thermal efficiency of the Rankine cycle is

$$\eta_{th} = \frac{P_{net}}{\dot{Q}_A} = \frac{w_{net}}{q_A} = 1 - \frac{h_2 - h_3}{h_1 - h_4} \tag{4.15}$$

where $\dot{Q}_A$ and P_{net} are, respectively, the added thermal power and the produced net power. The net power of a Rankine cycle is calculated as

$$P_{net} = \dot{m}\, w_{net} = \dot{m}(h_1 - h_2 + h_3 - h_4) \tag{4.16}$$

where $\dot{m}$ is the mass flow rate of the working fluid through the system.

In an ideal system, with the assumption of perfect heat exchangers, the warm water enters the evaporator at the temperature T_b and leaves at the temperature T_o with a volumetric flow rate $\dot{V}$. In this case, the thermal power transferred from the warm water to the OTEC system is

$$\dot{Q}_A = \dot{m}\, c_p\, \Delta T_A = \rho_{ww}\, \dot{V}_{ww}\, c_p\, \Delta T_A \tag{4.17}$$

where $\Delta T_A = T_b - T_o$ is the temperature difference in the evaporator, $\dot{V}_{ww}$ is the warm water volumetric flow rate and ρ_{ww} its density. Given the thermal efficiency (η_{th}) and the electric generator efficiency ($\eta_{el,gr}$), the net power is calculated as

$$\begin{aligned} P_{net} &= \rho \dot{V}\, c_p\, \eta_{th}\, \eta_{el,gr}\, \Delta T \\ \eta_{el,gr} &= \eta_t \cdot \eta_{gr} \end{aligned} \tag{4.18}$$

The thermal power given in the condenser is

$$\dot{Q}_R = \dot{m}\, c_p\, \Delta T_R = \rho_{cw}\, \dot{V}_{cw}\, c_p\, \Delta T_R \tag{4.19}$$

The warm and cold water feed pumps require work both for moving the water and for overcoming friction. The head losses in a section of the ducts are

$$\Delta H_{f,i} = \frac{2f\, L_i\, v_i^2}{D_i\, g} \tag{4.20}$$

Pressure drops can be expressed as

$$\Delta p_{f,i} = \rho\, g\, \Delta H_{f,i} \tag{4.21}$$

where L_i and D_i are, respectively, the length and the diameter of ducts, f is the friction coefficient, $i = CW, WW$ refers to the cold and warm water, respectively. v_i is the average speed of sea water and is calculated as

$$v_i = \frac{\dot{V}_i}{A_i} \tag{4.22}$$

Head losses in feed and discharge ducts, in heat exchangers and in other zones of the plant, are assumed equal to ΔH_f.

Consequently, the overall losses in the water feed system are

$$\Delta H_t = 4\Delta H_f \tag{4.23}$$

where the multiplicative factor 4 is equal to the water-feeding zones of the plant.

The required power of the feed pumps can be expressed as

$$P_{P,i} = \frac{\rho\, g\, \dot{V}_i\, \Delta H_t}{\eta_{p,i}} \tag{4.24}$$

where η_p is the pump efficiency.

The net power generated from the OTEC plant, considering also the sea water withdrawal operations, is given by:

$$\begin{aligned} P &= P_T - P_P - P_{CW} - P_{WW} \\ P &= P_{net} - P_{CW} - P_{WW} \end{aligned} \tag{4.25}$$

The heat-exchange surfaces represent one of the major limitations for this type of system. In fact, for a 130 MW closed-cycle OTEC, the surfaces of evaporator and condenser are equal to 5.02 and 4.28 km^2, respectively. The surface extension depends on the small available temperature difference [31].

The Rankine base cycle can be modified with the use of more complex systems that increase its efficiency. An example is the Kalina cycle (shown in Figure 4.15), which uses a mixture of ammonia and water. By the use of the Kalina cycle, it is possible to increase the power output from the plant, reducing the losses due to irreversibility in the cycle [32]. The Kalina cycle uses the mixture of water–ammonia in different concentrations in the different parts of the plant. An advantage lies in the fact that at a given pressure, a binary mixture condenses or evaporates within a temperature range, with respect to a pure substance which changes phase at a fixed temperature. This further degree of freedom reduces the irreversibility in the condenser and in the evaporator [30]. The Kalina cycle has been further modified to achieve greater thermal efficiency. The modified cycle, shown in Figure 4.16, is called the Uehara cycle. This cycle proposes the introduction of two turbines and the process of absorption and extraction [33].

Another way to improve thermal efficiency is to increase the available temperature difference. The temperature at the inlet of the turbine can be increased from 25 °C to 70 °C with a solar collector system, as shown in Figure 4.17 [29].

A closed-loop OTEC system can be coupled to a conventional thermal plant, such as a nuclear plant, to obtain a performance improvement of both. In this way, it is possible to use a flow of water at a higher temperature than that of sea surface water. The flow of condenser discharge water from a conventional thermoelectric

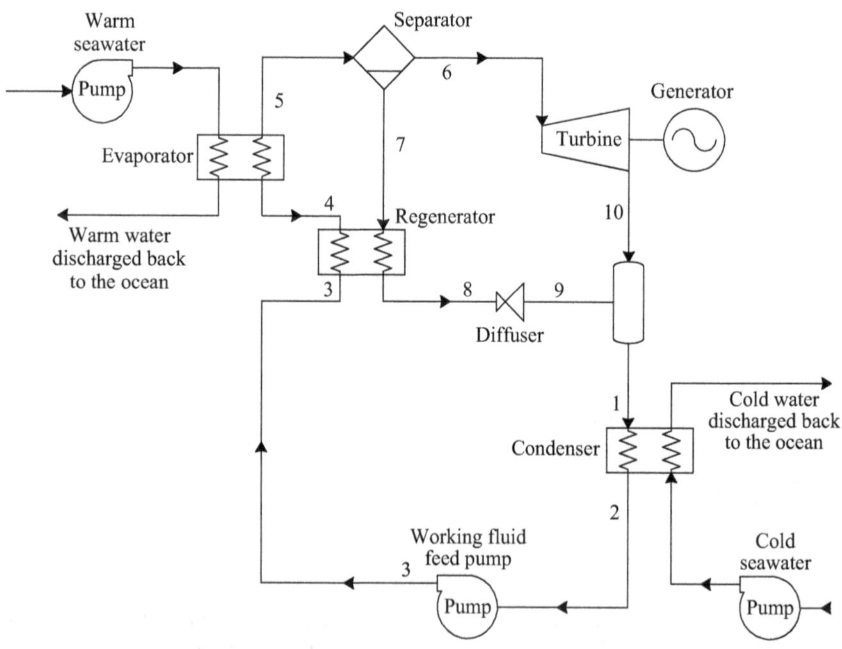

Figure 4.15 Kalina cycle flow

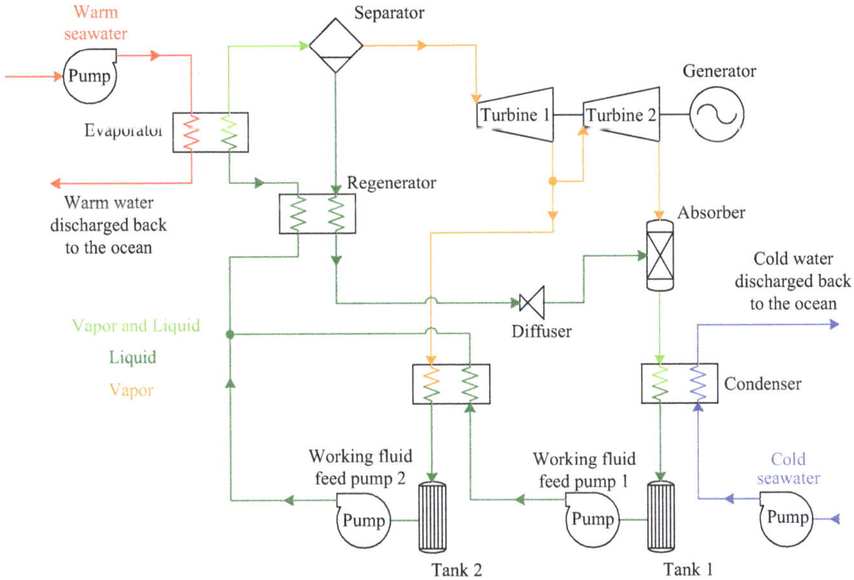

Figure 4.16 Uehara cycle flow

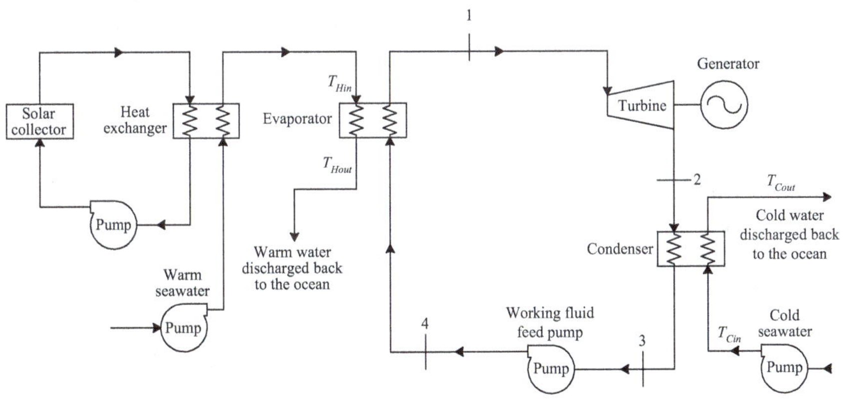

Figure 4.17 Closed-cycle OTEC coupled with solar collector

power plant can be sent to the evaporator of an OTEC system. By making this modification to the plant, it is possible to exploit the OTEC in areas of the world where the natural thermal gradient does not meet the minimum requirements. The proposed plant is schematized in Figure 4.18. It consists of a closed loop Rankine cycle (1-2-3-4), an open cycle of sea water taken from the thermoelectric power plant (5-6-11) and an open cycle of cold water taken deep (7-8-10). Part of the high-temperature seawater flow is desalinated into a flash evaporator.

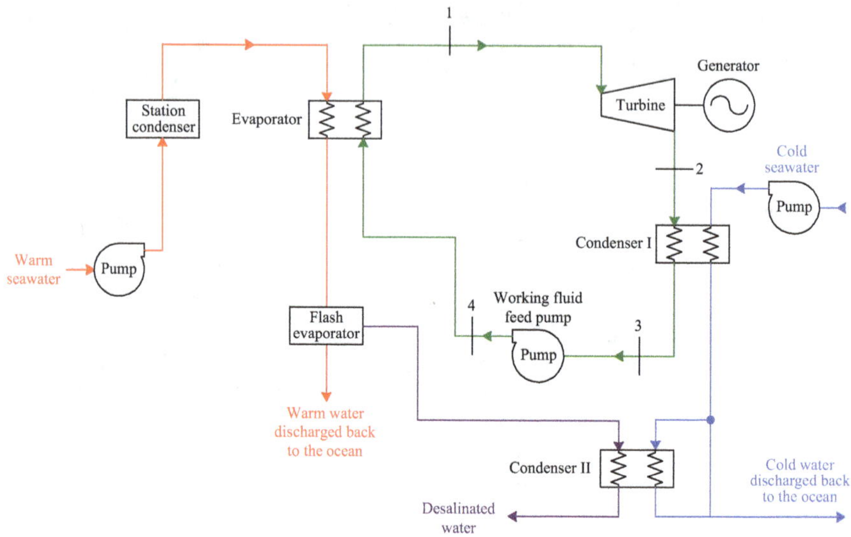

Figure 4.18 Closed-cycle OTEC coupled with a conventional thermal plant

Improvements made by this solution are an increase in the power generated by the combined plant without an increase in emissions; a reduction in the thermal pollution of the area, since the condenser discharge water is cooled by the OTEC system; and a production of desalinated water. The efficiency of the thermoelectric plant can increase by 1.3% [34].

4.2.3 Hybrid-cycle plants

Hybrid-cycle OTEC systems combine the characteristics of an open-cycle plant with those of a closed-cycle plant in order to simultaneously produce desalinated water and electrical power [35]. Although an open cycle plant can produce both products, a hybrid system combines the advantages of the two cycles. In the hybrid cycle, warm sea water is evaporated in flash mode with partial vacuum conditions, as shown seen in process 1-2 of Figure 4.19. The heat enclosed in the low pressure steam is transferred, through a heat exchanger, to the ammonia that flows through a Rankine cycle. The water vapor condenses in the ammonia evaporator as shown in process 3-4. The ammonia evaporates in process 5-6 and flows through the turbine in transformation 6-7 and is condensed in process 7-8 in a surface exchanger.

The deep cold sea water is heated in the condenser with the transformation 9-10. The scheme of the hybrid cycle plant is shown in Figure 4.20.

Warm water is taken at a depth of 10–15 m, and only the 0.5% is transformed in vapor in the flash evaporation process. The ammonia cycle is identical to that of the closed-cycle systems. The hybrid cycle uses a tube bundle heat exchanger for the condensation of water vapor and the generation of ammonia steam. The hybrid system overcomes some limitations of the open cycle plant: it is not necessary to use

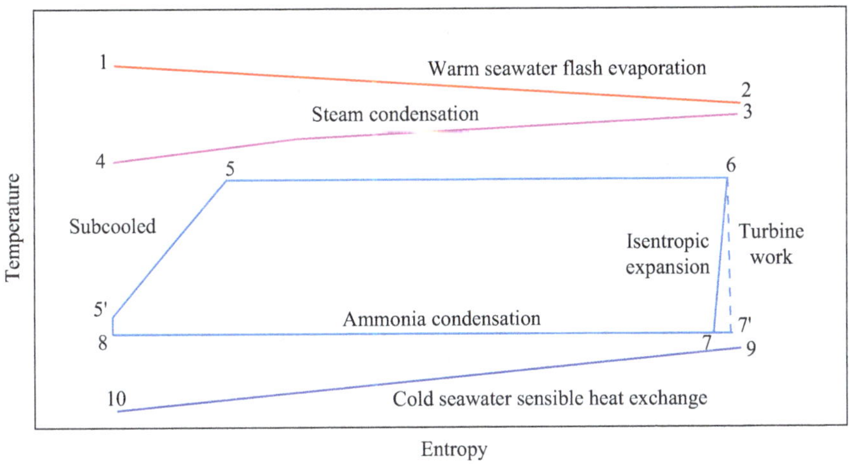

Figure 4.19 Hybrid-cycle OTEC T–S diagram

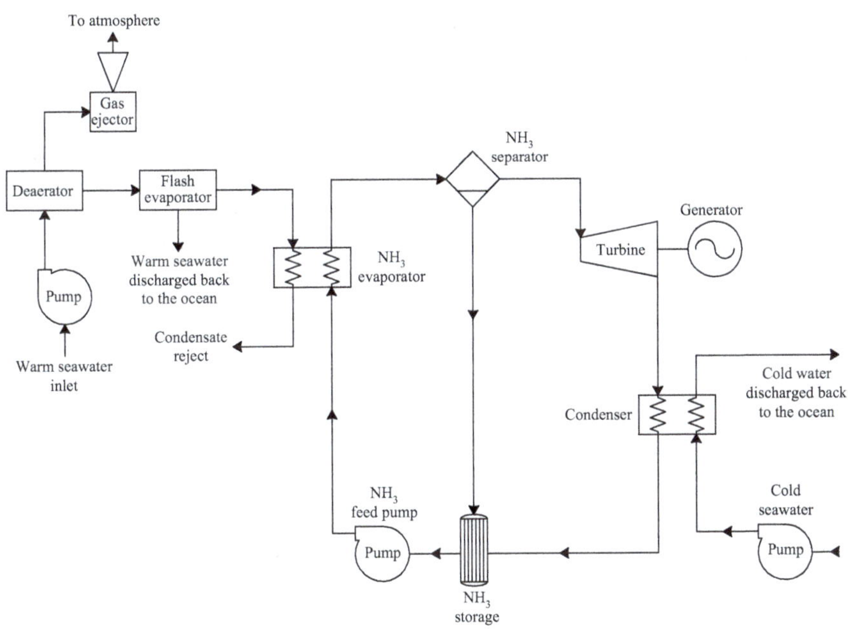

Figure 4.20 Hybrid-cycle OTEC schematic diagram

large diameter turbines; the void is present in a smaller and structurally less complex area; steam condensation occurs at a higher pressure (2.6 kPa) than in the open cycle (1.3 kPa), making it possible to manage more easily the non-condensable gases [5]. In fact, the power absorbed by the compression and discharge of non-condensable gases is lower than the other types of plant [30]. The estimation of the total heat exchange area for each component is made based on the respective exchange coefficients, the average logarithmic temperature difference and the thermal powers required by the hybrid system:

$$A_E = \frac{Q_E}{U_E\ (LMTD)_E}$$
$$A_C = \frac{Q_C}{U_C\ (LMTD)_C} \tag{4.26}$$

where A_E and A_C are the exchange surfaces of evaporator and condenser, respectively, $(LMTD)_E$ and $(LMTD)_C$ the logarithmic medium temperature differences between the ammonia and sea water in the condenser and in the evaporator.

The thermal power in the evaporator $\dot{Q}_E$ and in the condenser $\dot{Q}_C$ is determined from:

$$\dot{Q}_E = \dot{m}_{wf}\ c_p\ (T_3 - T_4) = \dot{m}_A\ (h_6 - h_5)$$
$$\dot{Q}_C = \dot{m}_c\ c_p\ (T_9 - T_{10}) = \dot{m}_A\ (h_7 - h_8) \tag{4.27}$$

where $\dot{m}_{wf}$, $\dot{m}_A$ and $\dot{m}_c$ are, respectively, the mass flow rates of warm water evaporated in flash mode, ammonia and cold sea water; c_p is the specific heat of sea water; T is the temperature of the sea water; h is the enthalpy of ammonia in the various states shown in Figure 4.19.

Deaerators are needed in the hybrid cycle. In fact, the presence of non-condensable gases such as oxygen can accelerate the corrosion process in the exchangers. Moreover, the heat exchange performances are reduced by the formation of a layer on the exchange surfaces by non-condensable gases.

The turbine and the ammonia pump are related to the thermodynamic states of Figure 4.19 by the following relationships:

$$P_T = \eta_T\ \dot{m}_a (h_6 - h_7)$$
$$P_P = \frac{\dot{m}_a (h_5 - h_8)}{\eta_P} \tag{4.28}$$

where P_T is the power produced by the turbine, P_P is the power required by the pump, η_T and η_P are, respectively, the efficiency of the turbine and the pump. The mass flow of water vapor generated in the evaporator is

$$\dot{m}_{WF} = \frac{\dot{m}_W\ c_p (T_1 - T_2)}{h_{fg} - c_p (T_2 - T_E)} \tag{4.29}$$

where $\dot{m}_W$ is the mass flow of the warm sea water, c_p the specific heat of the sea water, T_1 the temperature at the inlet of the warm sea water, T_2 the outlet temperature of the warm sea water, T_E is the saturation temperature at the pressure in the evaporator [5].

The hybrid cycle planting scheme can be modified, adding a solar collector and a thermal storage tank, as shown in Figure 4.21. The solar-hybrid system increases the benefits, generating desalinated water and increasing efficiency [35].

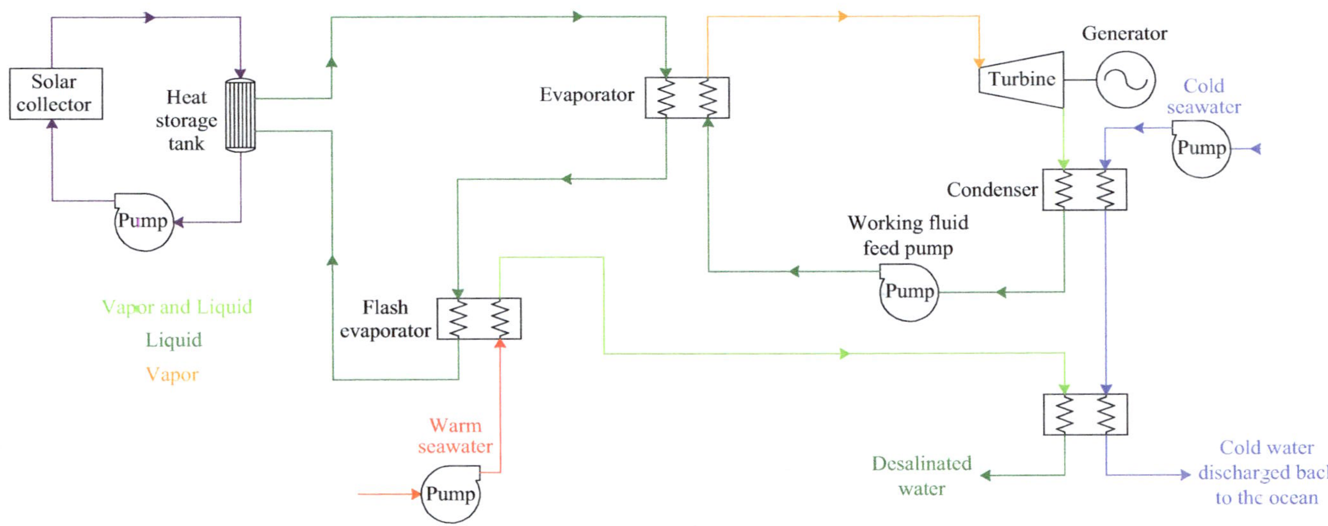

Figure 4.21 Solar-hybrid OTEC system

4.2.4 Feed-pump removal technique

In an OTEC system, the variation of 1 °C in the available temperature difference corresponds to a variation in output power of 15% [3]. OTEC systems have to deal with a rather low temperature difference between the warm and the cold water, so they need a very large flow of water to absorb the necessary heat. For example, for a system with a power rating of 100 MW, it is assumed that it takes about 750 kg/s of water, this implies a power absorbed by the feed pumps equal to 40% of the net generated power.

An innovative technique that eliminates the need for the pumping system has been proposed. This technique consists of increasing the salinity of surface water at high temperatures, cooling it in depth and bringing it back to the surface. The whole process is driven by the salt gradient, resulting from the deliberate salinization of surface waters. Two methods for increasing salinity and allowing natural descent are the intensification of solar evaporation, using a dedicated surface; addition of salt from a nearby source, such as a saline. The two techniques are illustrated in Figure 4.22. Water with higher concentration of salt descends and is cooled with an osmotic heat exchanger, which also decreases the salt concentration. In this way, the difference in salinity is again used in the ascent duct to bring the water back to the surface and use it as a low temperature source in the system. A simulation made it possible to have data relating to the exchange surface that the solar evaporator must possess and the amount of salt to be added.

Results show that for the evaporator, the field of use is reduced to plants with power not exceeding 1 MW (evaporator surface of 10^3–10^4 m^2), and for the technique of adding salt, the systems must not exceed 100 kW (consumption of 172,000 kg of salt per day) [36].

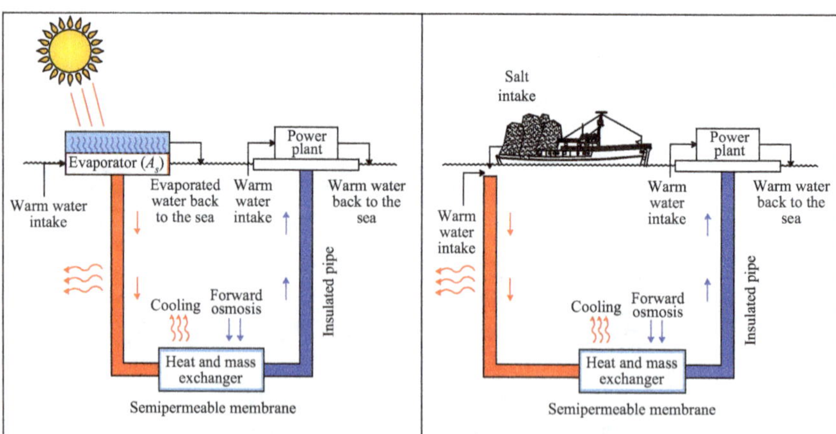

Figure 4.22 OTEC concept where salinization of seawater is generated by enhancing solar evaporation (left) or by dumping salt brought from onshore saline brine directly into the seawater (right)

4.2.5 OTEC and solar pond

Solar ponds (SPs) normally consist of artificial lakes having a small depth. In nature, when the sun rays fall on the surface of a lake or a pond, the water molecules absorb heat, thereby increasing their temperature. Therefore, the water in the bottom gets warmer, rising to the surface due to convection. The dissolution of salt in the bottom thin layer of the SP inhibits the convection; in this way, the molecules, although hotter, are too heavy to rise to the surface, producing high temperatures in the lower layer. Thus, this lower layer can be used as a source to provide heat in a continuous manner [37].

Conceptually, an SP can be used to hybridize OTEC systems, thus forming the so-called SPOTEC (solar pond ocean thermal energy conversion). This system was studied in the 1980s by researchers from the Department of Planning and Economic Development, State of Hawaii, for possible application at the NELHA, Ke-ahole Point, in order to provide power for the NELHA and the Ke-ahole Airport Hawaii [38]. Recently, SPOTEC plants were studied in 2008 by Straatman from the Department of Science, Technology and Society, Copernicus Institute for Sustainable Development and Innovation, Utrecht University, The Netherlands [23]. In the analysis of Straatman *et al.*, the proposed design is called OTEC offshore solar pond hybrid, and it has been optimized for the lower production costs.

In SPOTEC plants, the cold water taken from the ocean depths thanks to the presence of an OTEC tube can be used in the place of surface water of the pond (warmer) to condense the working fluid at the exit of the turbine. The Rankine power cycle configuration consists of an evaporator, a turbine, a condenser, a pump and an organic fluid which is used as the working fluid, exploiting the difference of temperature between the water on the bottom of a solar pond (about 80 °C–85 °C) and the deep cold water of the ocean (5 °C–10 °C) to generate electricity. In this way, a higher temperature difference between the Organic Rankine Cycle (ORC) cycle hot set (bottom of the pond) and cold set (ocean depths) is created that increases the Carnot cycle efficiency up to almost 24% [39]. The principle of SPOTEC is shown in Figure 4.23.

Figure 4.24 represents the four processes of an ideal ORC on the *T–S* diagram. The working fluid is pumped from low-to-high pressure (1-2) and follows an

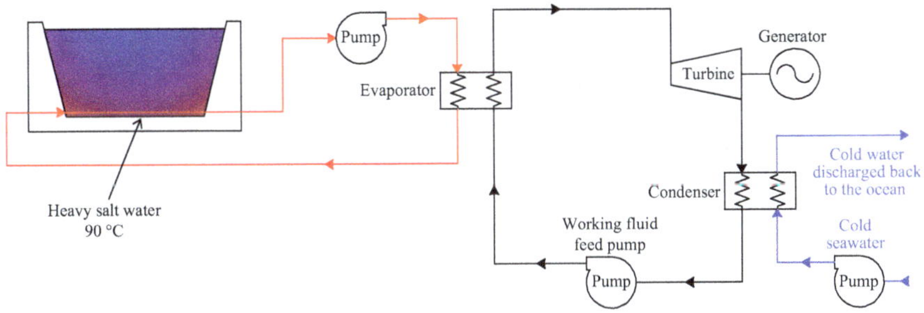

Figure 4.23 Scheme of solar pond integrated in the OTEC system

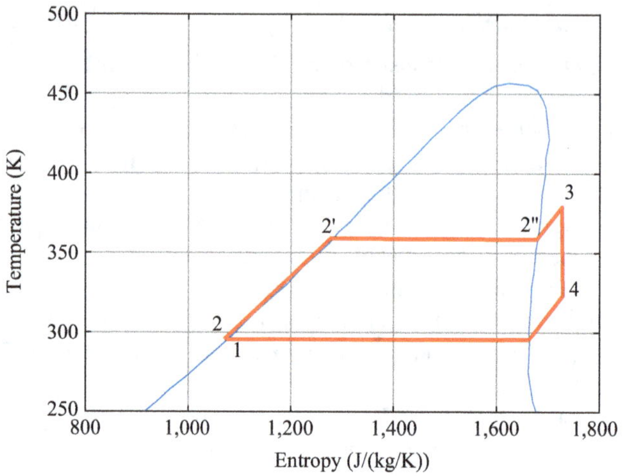

Figure 4.24 T–S diagram of an ideal ORC

evaporation process within the heat exchanger (2-2′-2″-3), which envisages a overheating in vapor phase (2″-3). Afterwards, it evolves as overheated vapor in a turbine (3-4) directly coupled to a generator to produce electrical energy. During the condensation process (4-1), there is an isobaric heat rejection that condenses the vapor back to a liquid [37].

Simulations on a real case show that a SPOTEC plant can reach an efficiency up to 8.5%. This is much higher than that of a standard OTEC system whose efficiency, as previously reported, is lower than 5%. From a comparison between a simulation of a SPOTEC system and a PV system on the same site placed in Malta emerged that, despite the higher energy efficiency and production of a PV system (i.e., 312 kW h/m^2/year) than the SPOTEC ones (i.e., 42 kW h/m^2/year), a SPOTEC system allows for reduced land requirements, thanks to the exploitations of potentially marginal sea surface [37].

4.2.6 OTEC systems characteristics

OTEC systems can be built on the coast (onshore) or offshore on floating platforms. The onshore type requires the use of very long ducts to provide the necessary sea water, a high initial-construction cost and a large amount of energy for pump operation. Positioning the plants on land can limit the availability of thermal resources and has greater impact on the environment. Onshore systems, however, are easier to repair and maintain, can provide ancillary products to nearby areas such as desalinated drinking water, air conditioning (seawater air conditioning), refrigeration systems and nutrients for mariculture. Offshore installations located on floating platforms do not need expensive coastal land. However, offshore systems have higher costs for energy transport and need large structures for anchoring and protection from wind, waves and adverse weather conditions. Onshore plants are suitable for tropical islands that need electricity and drinking water [6], while offshore plants are suitable for the production of hydrogen, ammonia or methanol [40]. The possibility of the development of mariculture is to be found in deep water at low temperature (deep ocean water, DOW), taken from an OTEC plant. These waters are rich in nutrients and have a low concentration (LC) of pathogens. DOW can also be used to create an artificial microclimate and cultivate products that would be impossible to find in tropical areas, such as strawberries or spring flowers [30]. Air conditioning obtained from an OTEC system requires less energy expenditure than a conventional air conditioner [6].

4.2.7 Environmental impact

The construction of an OTEC plant can have effects on the atmosphere, terrestrial environment, marine ecosystem and on human activities near the site where the plant operations take place.

In the case of plants operating on the mainland, the DOW ducts lie on the sea floor and cover a significant distance before reaching the designated depth. These surfaces alter the ecosystem of the seabed, creating an artificial habitat for different organisms. The artificial reef that is created generates the same effects as an offshore installation of an oil rig or an offshore wind farm. The extraction and discharge operations of the plant at different depths alter the concentration of nutrients in the waters near the plant, the fish production and the plankton population density [6]. The dumping water of OTEC systems can be at a lower temperature than surface water. Furthermore, they can contain an HC of nutrients and have a different salinity. Therefore, it is necessary to choose an optimal discharge depth to minimize the environmental impact on the marine ecosystem. Analyzing the outflow, the optimal depth must be between 50 and 100 m.

OTEC systems that use toxic substances as the working fluid, such as ammonia, pose risks associated with accidental spillage. However, in normal operation, the only emissions are the gas dissolved in the processed water [30]. It is important to carefully regulate the use of substances harmful to the environment in these facilities in close contact with the ocean [40].

A worldwide development of OTEC systems can cause a variation in the thermohaline circulation. The consequences are the lowering of the ocean surface temperature in tropical regions and an increase at high latitudes. Simulations carried out, through the aforementioned MITgcm model, show that heat penetrates into the ocean and a new steady state is reached. Considering a volume flow rate of the deep seawater intake per unit (latitude–longitude) area equal to 20 m/year, the global temperature changes from 3.65 °C to 3.42 °C with a 500-year simulation time. The yearly averaged temperature variation of surface and deep seawater in the simulation time is up to −4 °C and −2.5 °C at the equator and up to 7 °C and 10 °C, respectively [3]. The withdrawal of deep cold water, in a global scenario, must be of the same order of magnitude as the thermohaline circulation [12].

The development of OTEC plants can limit the release of greenhouse gases into the atmosphere and increase the absorption of CO_2 from marine organisms brought to the surface [30]. Offshore plants can have less effect on the environment if they are used to produce an energy carrier (ammonia, hydrogen or methanol). In fact, if an offshore OTEC generates electricity, it needs a seafloor cable to transport the energy to the shore. Seafloor cables are a well-developed but costly technology and impact the environment since they disrupt seafloor communities [40].

Onshore systems can potentially be coupled with under-water compressed air energy storage (UWCAES) systems. UWCAES system store compressed air in submerged distensible air accumulators, placed at or near the bed of lakes or oceans, in order to use the hydrostatic pressure resulting from the water column. Consequently, it would be feasible to simultaneously build an onshore OTEC plant and a UWCAES system in order to make the most of construction works. UWCAES system can be used to store electricity produced by renewable energy sources during off-peak periods. On the other hand, when there is more energy demand, the compressed air is taken from the submerged accumulators in order to produce the additional electricity required. In this way, it is possible to maximize the production of renewables and, therefore, reduce the environmental impact and the emission of pollutants into the atmosphere. In fact, it has been shown that the introduction of UWCAES technology integrated with other renewable energy power plants, such as PV and wind, can reduce the overall CO_2 emissions of sea-related industry up to 54% [41].

4.2.8 Economic aspects

As shown in Figure 4.9, there are only five operative OTEC power plants and three currently in the construction phase [28]. For this reason, there is limited actual project cost data available for OTEC. Instead, most cost references are based on feasibility studies from a limited number of sources [38,39,42–45].

The economic variables affecting the design and deployment of an OTEC plant range from macro-factors, such as economic environment of the plant location and market for OTEC generated electricity and by-products, to micro-factors such as the scale of the plant, the cost components including capital costs, operations and maintenance cost and CF. The most important economic variables, but also the easiest to quantify, include

- *Capital cost*. Since OTEC technology requires no fuel for plant operation, its major cost component is the capital investment. At over $10,000/kW, the capital cost of OTEC plants is almost one order of magnitude higher than that of oil-fired plants.
- *Operation and maintenance (O&M) costs*. O&M costs for well-developed oil-fired plants are low (typically 3% of capital cost per annum), whereas for low-efficiency OTEC plants, they will be relatively high. For floating OTEC plants, including inspection and crew costs, they will be about 5% of capital cost per annum, but may be considerably less for land-based plants.
- *Fuel cost*. Although OTEC fuel is free, the price of fossil (and nuclear) fuel still inherently determines the competitiveness of OTEC against conventional power generation.

- *Size and economy of scale*. Depending on the plant configuration, unit-generating costs will increase significantly below a critical size.
- *Economy of replication*. In predicting the production cost of OTEC hardware, it is necessary to keep in mind various cost-reduction measures applied to industrial processes. For example, industry distinguishes between the cost of early units of a production line, and the asymptotic cost after a given number of units has been produced and learning has been achieved.
- *Knock-on effect*. In addition to possible direct cost reduction, technological refinements may also have indirect effects on the overall design cost.
- *CF*. The CF may be separated into two components:
  - *availability*, which is the percentage of time the plant is ready to deliver output;
  - *environmental factor*, which is the percentage of full output based on the seasonal variation of the thermal resource. The CF for oil and coal fired plants is around 70% and may reach higher values up to 95%. For OTEC plants, with routine preventive maintenance, an availability factor of 90% is possible.
- *Operating life*. The 30-year design life of OTEC plants is similar to that of fossil fuel power stations, and 5 years longer than that of nuclear plants.
- *Temperature difference*. In order to be able to produce the rated power at all times, the design ΔT must be the minimum available at the plant location. There is also a strong dependence of capital cost on design ΔT. In fact, it has been shown that $Cost \propto \Delta T^k$, where $2 < k < 2.5$.
- *Insurance*. Potential accidents include accidental release of hazardous chemicals, possible interference with shipping and with defense submarines by the mooring line, the cable and the cold-water pipe. An annual value of 2% of the capital cost is typical.
- *Operating conditions*. Early OTEC experiments drew on the DOW as a key element of the energy system. Now, applications for the cold water will play a key role in improving the economic appeal of the total OTEC system, and in purely economic terms, there is substantial logic in choosing a combination plant of OTEC, desalination and aquaculture providing power, water and a product to sell.
- *Fresh water*. In regions where both electric power and fresh water are valuable, it may be economically advantageous to produce both because the desalinated water will offset the high cost of electricity produced from the system. However, there is a trade-off situation between the generation of power and the production of fresh water, because one is produced at the expense of the other.
- *Aquaculture*. To the aquaculture user, such as SeaWater GreenHouse, OTEC plants provide a significant supply of electric power to run the sea-water pumps and other operations. In such circumstances, the costs of the OTEC plant can become quite competitive.
- *DOW applications*. Although the concepts of temperature control in refrigeration or air-conditioning systems and irrigation have been proven, their application on a large scale still has to be demonstrated.

- *Other options*. The remaining options consist of using OTEC electrical power in situ for the generation of energy intensive products such as hydrogen and ammonia. However, the economic viability of these processes is not yet established [44].

To summarize, though OTEC does not have major fuel-cost implications (it uses abundant ocean water), the relationship between the scale of the plant and the corresponding capital costs have made OTEC plants very unattractive compared to fossil fuel plants, as well as some of the renewable technologies such as solar and wind. But an important feature of OTEC technology that improves its financial viability is the option to colocate production of various by-products with electricity generation [46].

Figure 4.25 shows the trend of capital cost of an OTEC plant with the plant size (in MW) according to Muralidharan's analysis. The power regression line follows the equation:

$$y = a \cdot x^{-b} \tag{4.30}$$

where y is capital cost of the plant ($/kW) and x is the plant size (MW). The intercept is $a = 39{,}900$ and the exponent is $b = 0.35$. Therefore, the cost reduction with plant size follows the trend line:

$$y = 39{,}900\ x^{-0.35} \tag{4.31}$$

This trend line indicates that as plant size doubles, the costs/kW of installed costs decreases approximately by 22%. A one-fifth reduction in capital costs/kW for every doubling of plant output is a significant reduction and, if the projection holds true, can make the technology attractive at large-scale outputs [46].

Another study performed by Vega [38] and shown in Figure 4.26 obtained a least squares curve from which stands over that OTEC capital cost is a strong

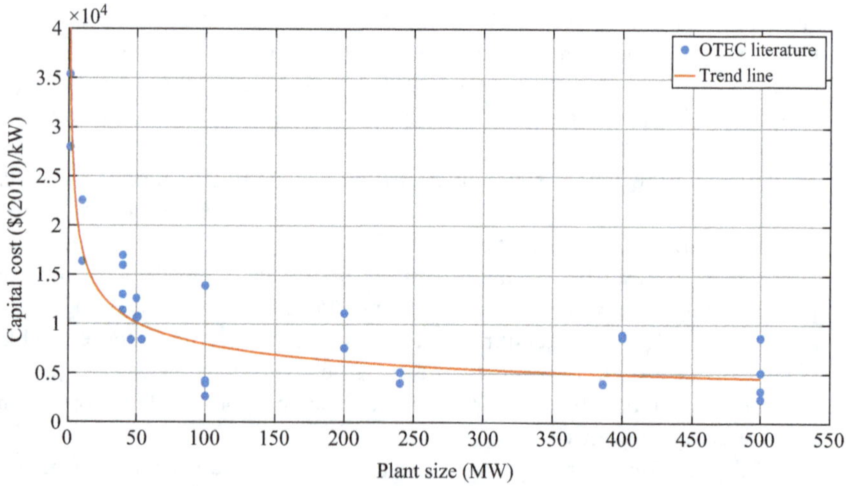

Figure 4.25 Capital cost as a function of plant size

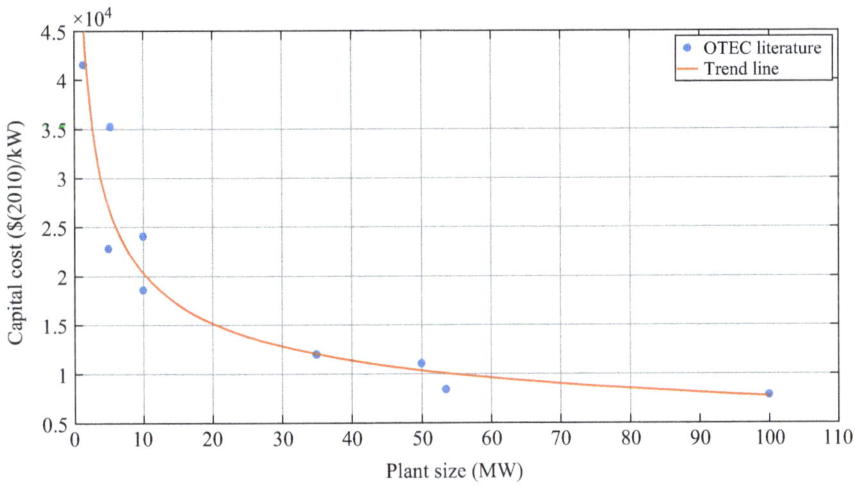

Figure 4.26 Capital cost as a function of plant size

function of plant size. The least squares curve is provided by the following equation:

$$y = 53{,}160\ x^{-0.418} \tag{4.32}$$

The evaluation of cost for an OTEC power plant should pass through cost drivers for each component. Six major cost categories are considered:

1. *Platforms*: The major cost driver for platforms (or land-based containment system, in the case of a land-based plant) is the customization costs associated with modifying the conventional design from other industries to execute OTEC plants. This might also impact the operation and maintenance costs of the platform and allied platform services.
2. *Heat exchangers*: The operating conditions of low temperature and pressure require heat exchangers which are good for moderate strength conditions, compared to which are used in conventional power plants. This can lower the cost of installation and other supporting systems.
3. *Power generation system*: Costs associated with turbines and pumps in an OTEC plant are predictable as the commercially available design of these components that can be easily matched for OTEC applications. OTEC systems will also require feed pumps and recycle pumps which are commercially available with a low acquisition cost but a significant maintenance cost.
4. *Energy transfer system*: The energy transfer costs are influenced by the costs of the power cable system which depend on special equipment designed for unique local sea conditions and seafloor characteristics.
5. *Water ducting system*: The process of fabrication and deployment of the cold-water pipe can significantly influence the installed capital cost of an OTEC plant.

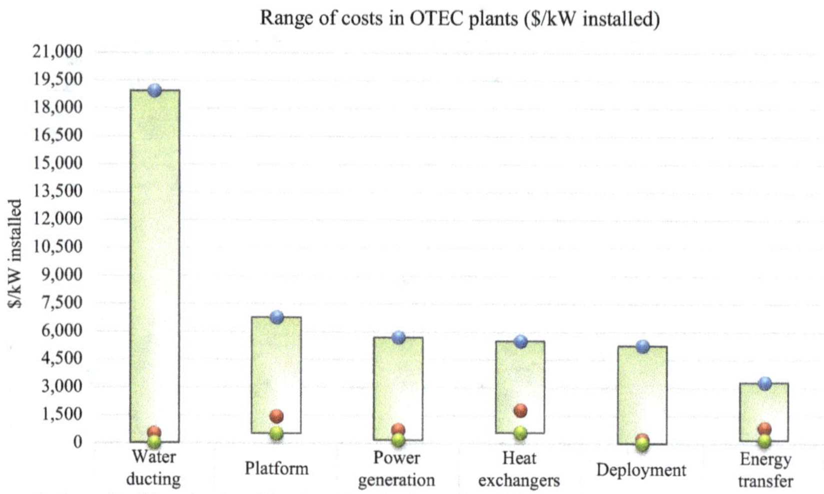

	Water ducting	Platform	Power generation	Heat exchangers	Deployment	Energy transfer
Max	18,942	6,776	5,698	5,501	5,250	3,300
Median	512	1,436	707	1,797	219	834
Min	30	530	184	586	13	202

OTEC cost components

Figure 4.27 OTEC components cost

6. *Deployment and installation processes*: Deployment costs include costs for installing OTEC components which are either built and assembled onsite or fabricated offsite and deployed on location.

The platform structure and the heat exchangers are the major cost contributors to an OTEC installation cost. The most cost-effective heat exchangers are made of aluminum. The cost of the platform will depend on the design configuration of the plant. The cost of heat exchangers per kW increases linearly with scale of the plant. Using concrete platforms instead of steel has the potential of reducing cost by around 30%. Another significant cost is the cold-water pipe where the cost per unit volume of handling cold water decreases with an increase in the cold-water pipe diameter. Conversely, deployment and operation and maintenance costs per kW do not increase significantly with increasing scale of the plant. Figure 4.27 shows the variation in each of the cost components across all plant sizes. Water-ducting systems have the most variation in prices and this is potentially an important lever for cost reduction [46].

4.3 Salinity gradient energy (SGE) plants

The potential energy of the difference in salinity between a fresh water source and a salty water source can be converted into electricity, placing an appropriate device between the two water flows [16].

The concept of salinity gradient power has been known for a long time; the first studies date back to 1953 [47]. In 1974, Norman published a scientific paper entitled "Water Salination: A Source of Energy" [48] in which he stated that techniques for water desalination can be used in reverse to generate power from the salt gradient [49]. Such research indicates that when a volume V of pure solvent is mixed with a larger quantity of a solution, whose osmotic pressure is p, the free energy lost is equal to $p\ V$. The osmotic sea water pressure is about 20 atm, the theoretical free energy released in mixing with fresh water is about the same as that obtained by making the water make a jump of 200 m. SGE is based on the difference in chemical potential [47]. The technologies that allow to carry out the conversion are

- pressure-retarded osmosis (PRO),
- reverse electrodialysis (RED),
- electric double-layer (EDL) capacitor,
- Faradaic pseudo-capacitor and
- devices that exploit the pressure difference of vapor.

PRO and RED technology are based on membranes. These are the most suitable for large-scale applications [16]. Membrane processes can introduce a considerable process intensification in the extraction of energy from renewable sources. It consists in developing new equipment and techniques that can lead to significant improvements in production and processes. A substantial reduction in equipment for production, energy consumption and waste generation may be achieved, making this technology cheaper and more sustainable [49].

Plants with PRO technology have been considered economically unfavorable for a long time. Only in recent years they have rediscovered interest thanks to the development of membrane technology. The first pilot plant was built in Norway in 2009 by the Statkraft company. New PRO plants are under development in Japan, Korea, Canada, Iran and Australia [50].

RED technology is gaining increasing interest and plants that exploit it are trying to move from the prototyping to the industrialization phase. The first RED pilot plant was developed in the Netherlands and was inaugurated in 2014. It is located on Afsluitdijk, a 32 km long dam that separates Lake Ijssel from the Wadden Sea. In 2014, a demonstration plant was built in Italy by the REAPower project group, coordinated by the University of Palermo. The plant is located in the Ettore–Infersa brine basins near the city of Marsala (Figure 4.28) [51].

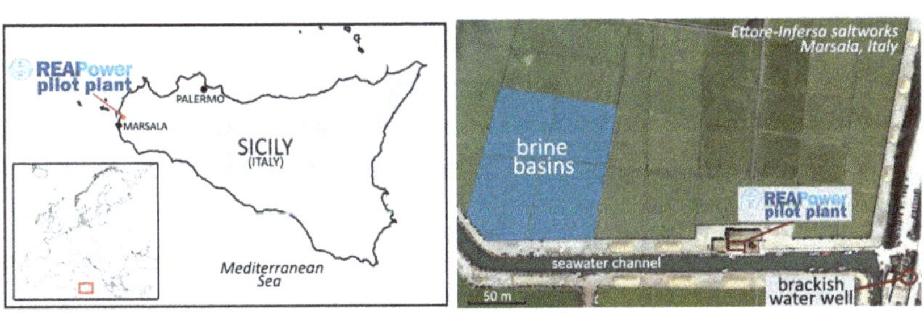

(a) (b)

Figure 4.28 REAPower pilot plant in Marsala (Italy): (a) location and (b) satellite image [51]

4.3.1 Pressure-retarded osmosis membrane (PRO) plants

The technology used in a plant that works with PRO membranes was developed by Sidney Loeb in the early 1970s. The conversion of the energy of the salt gradient occurs, thanks to the use of semi-permeable membranes, which allow the passage of water from the low salt concentration zone (fresh water) to the HC area (sea water) [49]. The two water flows flow into adjacent ducts, separated by a membrane, as shown in Figure 4.29. The difference in concentration between the two sides of the membrane allows fresh water to spread to the side with salt water.

The accumulation of water volume in the region with the highest concentration generates an increase in pressure. The pressure difference drives a hydraulic turbine, producing electrical energy [49]. Before entering the turbine, part of the water flow is sent to a pressure exchanger to pressurize the incoming salt water flow [16]. In the pressure exchanger, the incoming seawater is brought to the pressure of 10–13 bar, about half of the osmotic pressure. In the membrane module, the river water permeates against the pressure difference thanks to the osmotic effect [21]. The osmotic pressure π is calculated with the Van't Hoff equation:

$$\pi = 2R\,C\,T \tag{4.33}$$

where C is the salt concentration in water, R is the constant of ideal gases and T is the water temperature.

The difference in osmotic pressure $\Delta\pi$ between two masses of water is expressed as follows:

$$\Delta\pi = 2R(C_m\,T_m - C_r\,T_r) \tag{4.34}$$

where subscripts m and r refer to marine and river water, respectively.

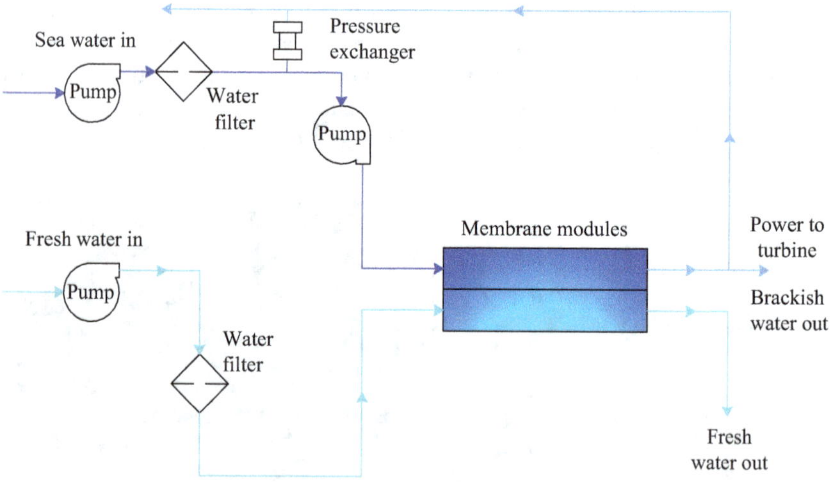

Figure 4.29 Simplified process layout for a typical osmotic power plant

The potential energy E depends on the osmotic pressure difference and on the average flow of the river Q_m:

$$E = \Delta\pi \, Q_m \tag{4.35}$$

Technical limitations in the conversion process decrease the value of this potential. A coefficient can be used to express these limitations and take into account energy and pressure losses in the real process and the overall system efficiency [8]. The power density of the osmotic cell W is calculated using the relation:

$$W = J_w \, \Delta\pi \tag{4.36}$$

where J_w is the flux through the membrane (m^3/m^2 s) [26].

Key components of the system are the membrane and the pressure exchanger. The process is limited by the properties of the membrane. In fact, the membranes are not completely impermeable to the solute, so a part of energy is lost due to the inevitable migration of a quantity of salt into the LC zone which causes a decrease of the available osmotic pressure [16]. PRO membranes are generally formed by a thin film composite (TFC) membrane and by a hydrophilic (permeable) support layer [26].

An important challenge to successfully realize a conversion with PRO membranes is the design of a membrane that is effectively resistant to fouling [49]. The fouling of osmotic membranes can be avoided by lowering the silt density index of the water drawn to a value lower than 3. If this index exceeds the value of 4 or 5, a pretreatment of the water entering the system is needed. Pretreatment must limit the presence of turbidity, reduce water hardness and carbon particles. Following some text results, it has been shown that ultra-filtration can improve the efficiency of PRO membranes. The disadvantage of the ultra-filtration technique is its high cost [26]. The membrane support layer causes a high internal concentration polarization (ICP) [16].

In order to be economically advantageous, PRO membranes need a power density of about 3–5 W/m^2 (the surface refers to the extension of the membrane) [26]. Cellulose-acetate membranes designed specifically for direct osmosis have reduced the ICP effect. The energy density is increased from 1–1.7 to 4–5.1 W/m^2.

A new type of composite polyamide thin film membranes can increase the power density up to 5.7–10 W/m^2 [16]. The Statkraft pilot plant in Norway achieved values of power density of 1.3 W/m^2 with cellulose-acetate membrane, while with TFC membranes the value of the power density was 3.5 W/m^2 [26]. The plant is shown in Figure 4.30. The company has announced that it will not continue to develop PRO technology because, from their point of view, it will not be sufficiently developed to become competitive within the foreseeable future [49,52].

The use of high efficiency membranes and pressure exchangers allows a PRO plant to generate 1 $MW/m^3/s$ of fresh water through the membranes [16]. Following a thermodynamic analysis, considering the intrinsic inefficiencies due to the entropic production and the unused energy due to the discontinuity during the passage of water in the membrane, the maximum efficiency at constant pressure

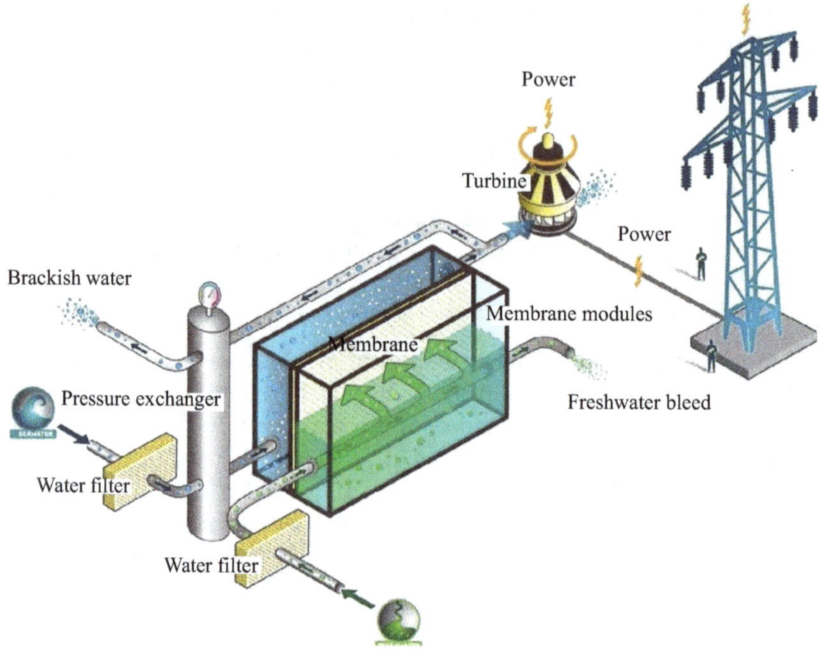

Figure 4.30 Schematic diagram of the pilot PRO plant constructed by Statkraft [16]

can reach up to 91.1%. Losses in the other components were neglected [15]. The largest losses in PRO systems occur in pumps, in the pressure exchanger, in the filtration system and in the turbine [21].

The estimated cost for the generation of energy using this technology is about \$50–100/MW h and is comparable with that of offshore wind and even better than resources that exploit the waves and tides [49].

4.3.2 Reverse electrodialysis (RED) plants

Power plants using RED technology do not generate pollutants and directly convert into electrical power the free energy produced by mixing two solutions with different concentrations, avoiding the use of a hydraulic turbine [16]. RED technology uses ion transport membranes as shown in Figure 4.31 [49]. Sea water with an HC of salt and water with LC are pumped into arrays or stacks of membranes arranged in such a way as to alternate those of anion exchange (A) with cation exchange (C). The difference in concentration, present between the two solutions, and the ion-selectivity of the membranes forms an electrochemical potential. This potential difference allows the movement of the ions in one direction and of the electrons in the other, generating an electric current in the stack [16].

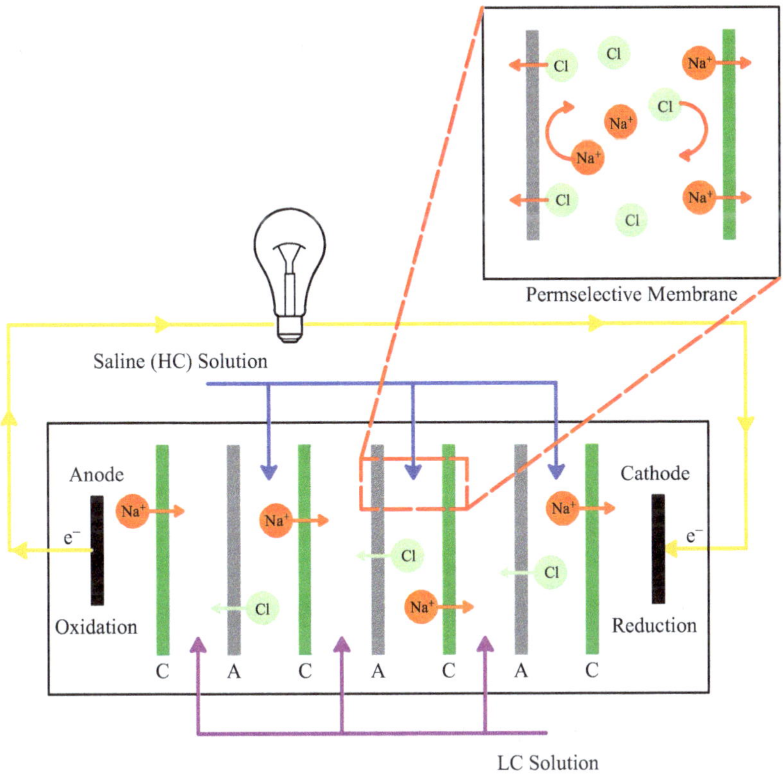

Figure 4.31 Schematization of reverse electrodialysis (RED) [49]

To calculate the TP of the ΔV_{theo} membrane, using an aqueous monovalent electrolyte (NaCl), the Nerst equation may be applied:

$$\Delta V = \frac{2\alpha_{avg}\, R\, T}{z\, F} \ln \frac{a_c}{a_d} \tag{4.37}$$

where α_{avg} is the membrane average perm-selectivity (which indicates a limitation of membrane permeability), R is the ideal gas constant (8.314 J/mol/K), T is the absolute temperature (K), z is the electrochemical valence, F is the Faraday constant (96,485 C/mol) and a_c and a_d are, respectively, the concentrated and diluted solution activity (mol/l).

The global potential of the system with N pairs of membranes is given by the sum of the potential differences:

$$V^0 = N\, \Delta V_{theo} \tag{4.38}$$

The global resistance is defined as the sum of the resistances of each individual component of the stack:

$$R_{stack} = \frac{N}{A}\left(R_{aem} + R_{cem} + \frac{d_c}{k_c} + \frac{d_d}{k_d}\right) + R_{el} \tag{4.39}$$

where N is the number of membrane pairs, A is the effective membrane area (m^2), R_{aem} and R_{cem} are, respectively, the resistance of the anion-exchange membrane and of the cation-exchange membrane (Ωm^2), d_c and d_d are, respectively, the thickness of the membrane HC and of the LC compartment (m), k_c and k_d are, respectively, the conductivity of the HC and of the LC compartment (S/m) and R_{el} is the electrode resistance (Ω).

The total output power P of the entire RED stack can be defined by Kirchhoff's law:

$$P = I^2 R_{load} = \frac{(V^0)^2 R_{load}}{(R_{stack} + R_{load})^2} \tag{4.40}$$

where I is the current and R_{load} is the load resistance (Ω).

To get the maximum output power $W_{\max}$, R_{load} must be equal to R_{stack}. The equation that binds the open circuit potential, the value of the maximum output power and the stack resistance can be written as

$$P_{\max} = \frac{(V^0)^2}{4R_{stack}} \tag{4.41}$$

The combination of the previous equations leads to express the value of the maximum power according to the contributions of the individual components [16]:

$$P_{\max} = N\,A\,\frac{((2\alpha_{av}\,R\,T/z\,F\,\ln(a_c/a_d)^2)}{R_{aem} + R_{cem} + (d_c/k_c) + (d_d/k_d)) + R_{el}} \tag{4.42}$$

The pilot plant located in Marsala (Italy) at the Ettore–Infersa salt marshes is an example of a system that uses a real flow of water at HC (brine), coming from the salt pans, and a flow of water with lower concentration (brackish water), coming from a well on the coastline. The system diagram is shown in Figure 4.32. The water suction ducts are about 200 m long and connect to the two water sources. There are also two tanks (each with a 2 m^3 capacity) to carry out experiments with artificial solutions of NaCl. The two brackish water and brine flows are first sent to a filtering system, then to a buffer tank and then to the RED unit for power generation. The solutions at the output of the RED module are the slightly diluted frost and the slightly concentrated brackish water. The brine is recycled directly into the salt marshes where the original concentration is restored thanks to the evaporation process and the wind action [51].

The use of slightly salted water, such as brackish water, instead of fresh water, allows the lowering of the electrical resistance of the diluted compartment, increasing the output power [49]. The RED module that allows the energy conversion, installed in the Italian pilot plant, has a total cell pair area equal to 44×44 cm^2 and is equipped with 125 pairs of cells (48 m^2, membranes total area). The unit is designed for cross flow, it has four $Ru - Ir$ (ruthenium–iridium) electrodes coated with a Ti (Titanium) mesh, and it is installed in an external compartment. The electrode rinse solution has been specifically chosen to minimize the impact that it could have if it leaks into the environment [51]. Before the Marsala

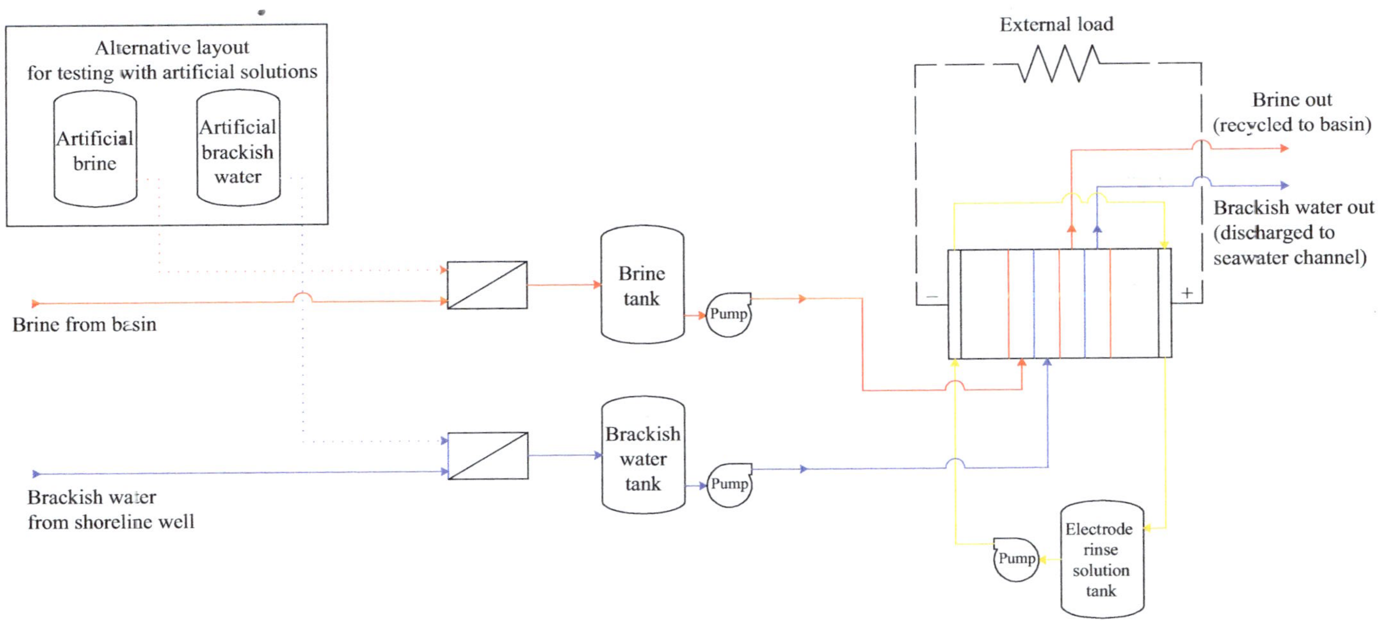

Figure 4.32 Simplified scheme of the Marsala (Italy) plant layout

experimentation, the highest power density for a stack was 1.18 W/m^2, neglecting losses [16]. The Italian demonstration plant managed to achieve, in the 5-month observation period, an average output power of 40 W (i.e., 1.6 W/m^2) using real water flows. Using artificial solutions, a peak of 65 W has been reached. The reduction of the maximum output power of the real solution is attributable to the presence of salts other than NaCl, especially Mg^{2+} (magnesium ions) [51]. The laboratory experiments of the RED modules allowed to achieve 80% conversion efficiency, considering only the ion transport losses. Energy required by the auxiliary systems (estimated in 50 kJ/m^3) has not been taken into account in the assessment of the efficiency [15]. The maximum energy value that a RED plant can produce depends on the salinity of the water source. The cost item with the highest weight, for this type of system, is that of ion-exchange membranes. The increase in the energy density of RED systems also involves the development of new spacers, structures that guarantee the mechanical support of the membranes. Conventional spacers interfere with ion transport, while new spacers have a greater permeability to this flow [49]. Future work in the Ettore–Infersa saltworks will focus on the installation and testing of two larger RED units, each consisting of 500 pairs of cells with a total of 400 m^2. The plant will have a target power of 1 kW [51].

4.3.3 Electric Double-Layer Capacitors plants

Capacitive energy extraction is a new and interesting method for converting free energy into electrical energy. The technique is based on the EDL technology. This is described by the Gouy–Chapman–Stern theory. The EDL condenser consists of an activated carbon electrode. When it is immersed in sea water, it stores the charge on its surface. Subsequently, fresh water flows into the EDL condenser. The charged ions spread far from the electrodes, acting against the electrostatic force. The electrostatic energy of the whole system increases and the surplus of energy can be extracted from the condenser, converting the difference in salinity into electrical energy. The cell that allows conversion has two electrodes, as shown in Figure 4.33 and it is crossed by the high salinity solution and the low salinity

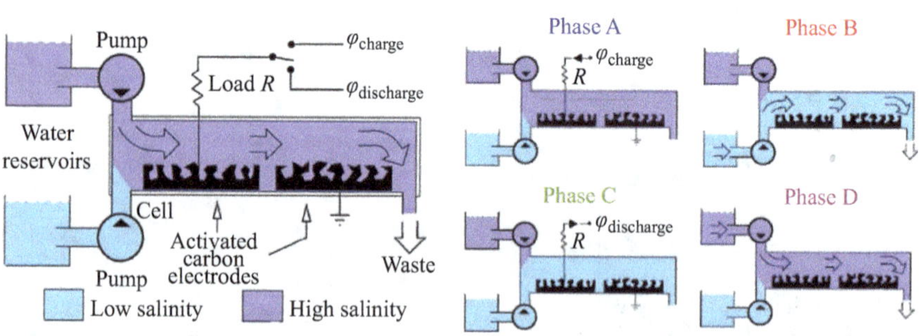

Figure 4.33 EDL systems: (left) capacitor scheme and (right) cycle for extracting energy from salinity difference [16]

solution alternately. The solution with an LC of NaCl may be river water, while the solution with an HC may be seawater. The two tanks are connected to the cell by means of two pumps that are activated in sequence, changing the concentration value. The carbon electrodes are extremely porous conductors and have the shape of discs with a diameter and thickness of 2 and 0.1 mm, respectively. A switch can temporarily connect the electrodes to the charge and discharge circuit.

The load is represented by a resistor with $R \approx 1\ \text{k}\Omega$ through which the charge and discharge current flows. The energy conversion takes place in a four-step cycle, shown in Figure 4.33.

The relationship between the surface charge density σ and the potential difference φ between the electrode and the bulk solution is

$$\varphi = \frac{2k_B T}{e} \sinh^{-1}\left(\frac{\sigma}{\sqrt{8C N_A \varepsilon_0 \varepsilon_r k_B T}}\right) \tag{4.43}$$

where k_B is the Boltzmann constant, T the temperature, e the electron charge, N_A the Avogadro number, ε_0 the electrical constant, ε_r the relative dielectric constant and C the electrolyte concentration. The equation is valid for a monovalent electrolyte such as NaCl. The decrease of the electrolyte concentration leads to an increase in potential.

EDL technology is still in the laboratory testing phase, but it can provide a solid basis for further development. In particular, an increase in the cycle frequency and a better engineering of the porous structure of the electrode material can allow an increase in performance, making this technology competitive with those based on membranes [16].

4.3.4 *Faradaic pseudo-capacitor plants*

Systems that exploit the pseudo-capacitor technology are based on the storage of the Faraday charge of the capacitor electrodes. The pseudo-condenser consists of a pair of reversible electrochemical systems where the reagents are the salt present in the electrolyte and the electrodes. Ions are stored in the electrodes. This system has a higher energy density and a simpler structure than EDL. The pseudo-capacitor is constructed according to the following chemical reaction:

$$5MnO_2 + 2Ag + 2NaCl \leftrightarrow Na_2Mn_5O_{10} + 2AgCl \tag{4.44}$$

Two electrodes are used in this device: a silver anode, which interacts selectively with Cl^- and a MnO_2 cathode, which selectively interacts with Na^+. In this electrochemical system, two separate reactions occur:

$$\begin{aligned} &5MnO_{2(\alpha)} + 2Na_\varepsilon^+ + 2e^- \leftrightarrow Na_2Mn_5O_{10(\alpha)} \\ &2AgCl_{(\beta')} + 2e^- \leftrightarrow 2Ag_{(\alpha')} + 2Cl_\varepsilon^- \end{aligned} \tag{4.45}$$

where α is the phase of $Na_{2-x}Mn_5O_{10}$, ε is the electrolyte phase, β' is the AgCl phase, α' is the silver phase.

The potential of the two reactions can be written as

$$E_{+} = E_{+,0} + \frac{RT}{F} \ln\left[\frac{\alpha_{\mathrm{Na},\varepsilon}}{\alpha_{\mathrm{Na},\alpha}}\right]$$
$$E_{-} = E_{-,0} - \frac{RT}{F} \ln\left[\alpha_{\mathrm{Cl},\varepsilon}\right] \tag{4.46}$$

where E_{+} and E_{-} are the electrodes potentials, $E_{+,0}$ and $E_{-,0}$ are the electrodes potential standards, $\alpha_{\mathrm{Na},\alpha}$ is the activity of the solid phase of sodium α, $\alpha_{\mathrm{Na},\varepsilon}$ is the activity of the electrodes sodium ions present in the electrolyte, $\alpha_{\mathrm{Cl},\varepsilon}$ is the activity of chlorine ions in the electrolyte. The difference between the two potentials is ΔE.

If the activity of the solid sodium phase is fixed, it can be written as

$$\Delta E = \Delta E_0(\alpha_{\mathrm{Na},\alpha}) + 2\frac{RT}{F} \ln[C_{\mathrm{NaCl}}] + 2\frac{RT}{F} \ln[\gamma_{\mathrm{NaCl}}] \tag{4.47}$$

where ΔE_0 is the standard voltage of the cell, C_{NaCl} is the concentration of NaCl and γ_{NaCl} is the average activity coefficient of NaCl.

The dependence of γ_{NaCl} with respect to C_{NaCl} is described by Hückel's law:

$$\ln[\gamma_{\mathrm{NaCl}}] = -\frac{A\sqrt{C_{\mathrm{NaCl}}}}{1 + B\sqrt{C_{\mathrm{NaCl}}}} \tag{4.48}$$

where A and B are constant. ΔE increases as the concentration of NaCl increases.

Electrodes are loaded into a solution with a low ionic power (river water) and Na^{+} and Cl^{-} ions are removed from the respective electrodes as shown in step 1 of Figure 4.34.

In step 2, the low-concentration electrolyte is replaced with the high-concentration electrolyte causing the increase of the open-circuit potential of the cell. In step 3, the pseudo-condenser is discharged, the anions and cations are reincorporated into the respective electrodes. In step 4, the seawater is removed and replaced by the more diluted solution (river water), in this way, the potential difference between the electrodes decreases. During steps 2 and 4, no energy is consumed or produced in the cycle. During step 1, the capacitor needs energy to expel ions from the crystalline structure, while during step 3, the capacitor produces energy by storing the ions. The energy gain is given by the integral on the voltage cycle as a function of the charge:

$$W = -\oint_C \Delta E \, dq \tag{4.49}$$

In tests carried out in the laboratory, the geometric surface of the electrode in contact with the solution is 1 cm^2, the distance between the positive and negative electrodes is 1 cm. In the experiment, the energy produced, using sea water (0.6 M NaCl) and river water (0.024 M NaCl) is about 29 mJ/cm^2 (power density 105 mW/m^2).

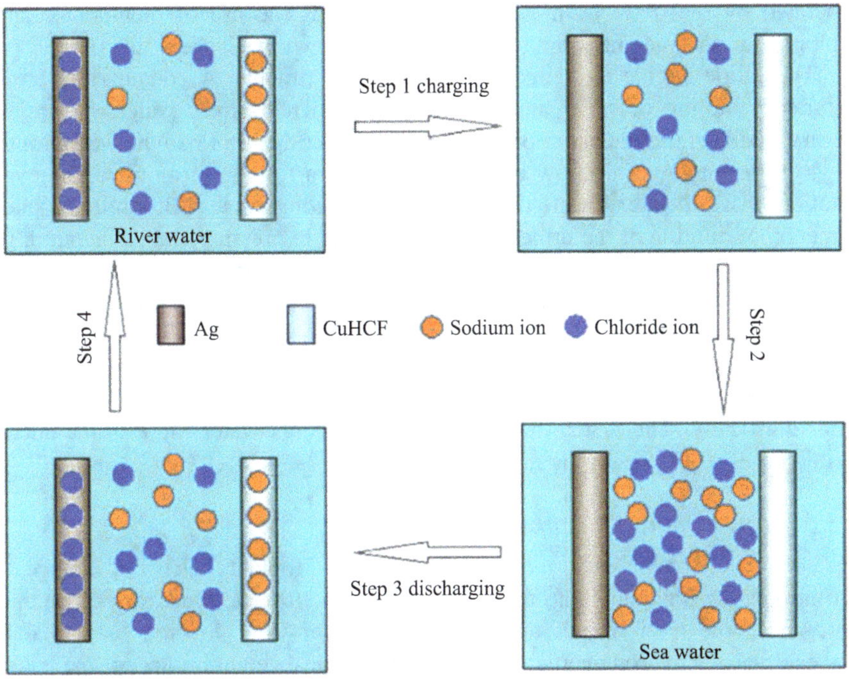

Figure 4.34 Schematic representation of the working principle of the pseudo-capacitor [16]

The technology that exploits the pseudo-capacitors is far from being a practically feasible solution but offers a promising future and can significantly contribute to the development of energy production from renewable sources. The process can also be reversed to produce desalinated water [16].

4.3.5 SGE systems characteristics

A comparison, under the same conditions, between the potential performances of the PRO and RED technologies showed that each of them has its own field of application. PRO technology seems to be more suited to power generation by exploiting a saturated salt solution, while RED technology is more appropriate for power generation using sea and river water [49]. At present and using real solutions, PRO technology has a power density of 2.3 W/m^2, higher than that obtained from a RED system which does not exceed 1.6 W/m^2 [51]. These technologies suffer from the high cost of membranes. In fact, it is 80% of the capital cost of these plants, and the membranes specifically designed for energy conversion from the salt gradient are 2–3 times more expensive than commercially available membranes [49]. The time-dependent changes of the salinity gradient structure have effects on the salinity of the incoming water and therefore they alter the power generated by the plant [21]. A system that avoids the use of membranes and exploits the salt gradient is the saline engine as shown in Figure 4.35.

The saline engine is composed of two compartments: A contains the brine (saturated solution of water and salt) and in B there is fresh water. Assuming thermal equilibrium, the vapor pressure in A is lower than in B, so the water vapor flows from B to A. This flow can be used to set a turbine in motion. The evaporation of fresh water lowers the temperature of compartment B, while the phenomenon of dilution of the brine causes an increase in pressure and temperature in A. Once the equilibrium is reached, the turbine stops. By making adiabatic the partition wall of the two compartments, the condensation heat flows from A to B. In this way, the process will theoretically stop when all the water in B will be passed into compartment A. The partition wall is a thin conductive sheet. The brine can also be replaced by sea water and fresh water with river water, but the output power drops drastically. The system requires that there is only water vapor in the ducts, so it is necessary to degas the inlet water [53].

4.3.6 Environmental impact

The energy conversion performed with the use of SGE plants has the advantage of not introducing pollutants into the environment [16,19]. The process of extracting water from the river cannot fully exploit the river flow, in order not to alter the ecological, hydrodynamic and sedimentological equilibrium at the river mouths [19]. Studies carried out for the river Leon, in Colombia, have shown that using a 20% EF, and considering an "ecological flow" equal to 12% of the average river discharge (minimum flow of the river after water withdrawal by the plant), the impact on the salt structure in the mouth is tolerable. The recorded effect is a slight movement toward the coast of the mixing area, mainly during the rainy season [15]. The discharge of waste water to a depth of 5 m has no detectable

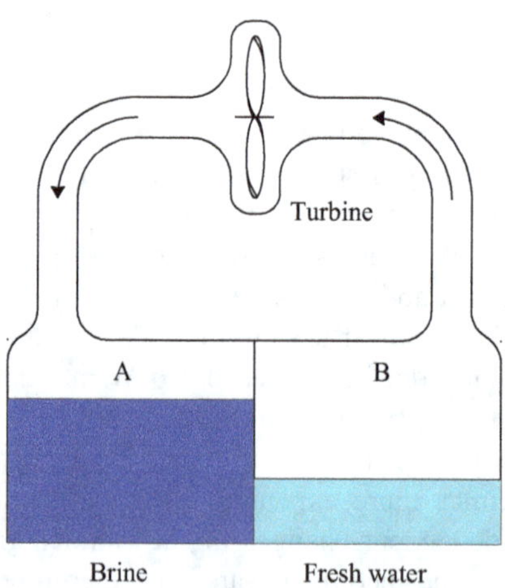

Figure 4.35 A membraneless salination engine

impact on the salt structure and does not negatively affect the operations of the plant [21]. Oxidation-reduction reactions in the RED system electrodes are a limitation for these implants as they can have a strongly negative environmental impact [49]. The reactions in RED electrodes can produce Cl_2 on the anode side and H_2 on the cathode side. Hydrogen, especially in tight spaces, is explosive, while Cl_2 is highly toxic [54].

4.3.7 Economic aspects

At present, a cost estimation for PRO and RED systems is difficult to obtain due to the lack of large-operational plants.

Regarding PRO systems, capital cost contributes for a fraction of over 60%, while operation and maintenance costs would be a smaller fraction. The capital cost is high since the power density of membrane is low as seen in the previous sections. Currently, membranes for desalinization plant range from \$20 to \$40/m^2. Consequently, considering the power density of the Statkraft pilot plant (1.3 W/m^2), a 20 MW capacity power plant would cost at least \$15,384/kW. This analyses did not consider the costs relative to water intake and outfall, and streams pretreatment. Including these cost components, the capital cost for PRO system would range from \$29,200 to \$32,000/kW, assuming a membrane with a 2.4 W/m^2 power density, which is much higher than other renewable energy power plants [55]. In fact, offshore wind power has an average total cost equal to \$4,487/kW, while for onshore wind power, the average total cost lowers to \$1,477/kW. Solar PV has a total cost range from \$1,050 and \$4,550/kW. Several cost ranges are available for concentrated solar depending on the energy storage condition. Concentrated solar power (CSP) for the “no storage” configuration has a cost range equal to \$2,550–\$7,000/kW. CSP with up to 4 h storage show a range of total costs between \$3,500 and \$9,000/kW. In the case of CSP with 4–8 h storage, total costs rises to a range from \$6,000 to \$12,600/kW. CSP with more than 8 h storage has a total cost range equal to \$7,300–\$11,300/kW. Regarding hydropower projects, their typical range of total costs is equal to \$500–\$4,500/kW. Bioenergy-fired power generation technologies have a wide range of total costs depending on the selected technology. Biomass combustion technologies show a range of total costs from \$450 to \$2,500/kW. Biomass projects using steam cycle have a total costs in the range \$500–\$2,000/kW, while fixed bed gasifiers show a cost range from \$2,000 to \$7,000/kW. Most projects do not disclose the technology used, and their total costs range from \$500 to \$8,000/kW [56].

Analogously to PRO system, the lack of commercially available RED systems makes difficult an economic analysis of plants exploiting this technology. As perceivable, membranes represent the component with the highest capital cost, contributing to about 80%. The actual prices for membranes (about \$55/m^2) suggests that RED power plants are not economically viable [55].

Lastly, EDL and Faradaic pseudo-capacitor systems are at laboratory stage. Consequently, there are no data for the performing of an economic analysis at this moment.

Acknowledgments

Sincere gratitude is hereby expressed to Eng. Michele Sparano and Eng. Assunta Valentina Aprile, bachelor graduated students in Mechanical Engineering and Management Engineering at the University of Salerno, respectively, for their valuable work in the bibliography research support for the writing of this chapter.

List of acronyms

CF	capacity factor
CSP	concentrated solar power
DOW	deep ocean water
EDL	electric double-layers
EE	extractable potential
EF	extraction factor
elcom	Estuary, Lake and Coastal Ocean Model
ICP	internal concentration polarization
MITgcm	Massachusetts Institute of Technology general circulation model
NELHA	Natural Energy Laboratory of Hawaii Agency
O&M	operation and maintenance
OTEC	ocean thermal energy conversion
PI	process identification
PRO	pressure retarder osmosis
RED	reverse electrodialysis
SGE	salinity gradient energy
SP	solar pond
SPOTEC	solar pond ocean thermal energy conversion
TFC	thin film composite
TP	theoretical potential
UWCAES	under-water compressed air energy storage

References

[1] García Huante A, Rodríguez Cueto Y, Silva R, *et al.* Determination of the potential thermal gradient for the Mexican Pacific Ocean. Journal of Marine Science and Engineering. 2018;6(1):20.

[2] Jia Y, Nihous GC, and Rajagopalan K. An evaluation of the large-scale implementation of Ocean Thermal Energy Conversion (OTEC) using an ocean general circulation model with low-complexity atmospheric feedback effects. Journal of Marine Science and Engineering. 2018;6(1):12.

[3] Rajagopalan K, and Nihous GC. An assessment of global ocean thermal energy conversion resources with a high-resolution ocean general circulation model. Journal of Energy Resources Technology. 2013;135(4):041202.

[4] NASA Earth Observatory. Explaining Rapid Climate Change: Tales from the Ice; 2006. Viewed on December 18, 2018. Available from: https://earthobservatory.nasa.gov/Features/Paleoclimatology_Evidence/paleoclimatology_evidence_2.php.

[5] Avery WH, and Wu C. Renewable energy from the ocean – a guide to OTEC. Johns Hopkins University Applied Physics Laboratories Series in Science and Engineering. New York, United States: Oxford University Press; 1994.

[6] Devis-Morales A, Montoya-Sánchez RA, Osorio AF, *et al.* Ocean thermal energy resources in Colombia. Renewable Energy. 2014;66:759–769.

[7] Osorio A, Ortega S, and Arango-Aramburo S. Assessment of the marine power potential in Colombia. Renewable and Sustainable Energy Reviews. 2016;53:966–977.

[8] Hogan CM. Analysis of highway noise. Water, Air, and Soil Pollution. 1973;2(3):387–392.

[9] Burman K, and Walker A. Ocean Energy Technology Overview. National Renewable Energy Laboratory (NREL); 2009.

[10] University Corporation for Atmospheric Research (UCAR). Transfer and Storage of Heat in the Oceans; 2018. Viewed on December 18, 2018. Available from: https://scied.ucar.edu/longcontent/transfer-and-storage-heat-oceans.

[11] Etemadi A, Emdadi A, Asef-Afshar O, *et al.* Electricity generation by the ocean thermal energy. Energy Procedia. 2011;12:936–943.

[12] Nihous GC. An order-of-magnitude estimate of ocean thermal energy conversion resources. Journal of Energy Resources Technology. 2005;127(4): 328–333.

[13] Nihous GC. A preliminary assessment of ocean thermal energy conversion resources. Journal of Energy Resources Technology. 2007;129(1):10–17.

[14] Rajagopalan K, and Nihous GC. Estimates of global Ocean Thermal Energy Conversion (OTEC) resources using an ocean general circulation model. Renewable Energy. 2013;50:532–540.

[15] Alvarez-Silva O, and Osorio AF. Salinity gradient energy potential in Colombia considering site specific constraints. Renewable Energy. 2015;74:737–748.

[16] Jia Z, Wang B, Song S, *et al.* Blue energy: current technologies for sustainable power generation from water salinity gradient. Renewable and Sustainable Energy Reviews. 2014;31:91–100.

[17] Melikoglu M. Current status and future of ocean energy sources: A global review. Ocean Engineering. 2018;148:563–573.

[18] Post JW. Blue Energy: Electricity Production from Salinity Gradients by Reverse Electrodialysis; 2009.

[19] Alvarez-Silva O, Osorio A, and Winter C. Practical global salinity gradient energy potential. Renewable and Sustainable Energy Reviews. 2016;60: 1387–1395.

[20] National Centers for Environmental Information (NOAA). The Global Temperature and Salinity Profile Programme (GTSPP). Viewed on December 18, 2018. Available from: https://www.nodc.noaa.gov/GTSPP/.

[21] Ortega S, Stenzel P, Alvarez-Silva O, *et al.* Site-specific potential analysis for pressure retarded osmosis (PRO) power plants – The León River example. Renewable Energy. 2014;68:466–474.

[22] Dai A, and Trenberth KE. Global River Flow and Continental Discharge Dataset; 2014. Viewed on December 18, 2018. Available from: http://www.cgd.ucar.edu/cas/catalog/surface/dai-runoff/.

[23] NASA Jet Propulsion Laboratory and California Institute for Technology (CalTech). Physical Oceanography Distributed Active Archive Center (PODAAC); 2004. Viewed on December 18, 2018. Available from: http://podaac-ftp.jpl.nasa.gov/allData/aquarius.

[24] University of Hamburg. SMOS Ocean Surface Salinity Data; 2013. Viewed on December 18, 2018. Available from: https://icdc.cen.uni-hamburg.de/1/daten/ocean/smos-sss/.

[25] National Oceanographic Data Center (NODC). World Ocean Database 2005; 2005. Viewed on December 18, 2018. Available from: http://www.nodc.noaa.gov/OC5/WOD05/pr_wod05.html.

[26] Abbasi-Garravand E, Mulligan CN, Laflamme CB, *et al.* Role of two different pretreatment methods in osmotic power (salinity gradient energy) generation. Renewable Energy. 2016;96:98–119.

[27] Martins M, Sinama F, and Lucas F. Equivalent Gibbs systems for modelling an onshore OTEC experimental plant on Reunion Island. International Journal of Energy Research. 2013;37(9):1112–1121.

[28] Ocean Energy Systems (OES). Status of OTEC Power Plants; 2017. Viewed on December 18, 2018. Available from: https://www.ocean-energy-systems.org/oes-projects/task-11-status-of-otec-and-its-resource-assessment/.

[29] Khan N, Kalair A, Abas N, *et al.* Review of ocean tidal, wave and thermal energy technologies. Renewable and Sustainable Energy Reviews. 2017;72:590–604.

[30] Masutani S, and Takahashi P. Ocean thermal energy conversion (OTEC). Oceanography. 2001;22:625.

[31] Oko C, and Obeneme W. Thermo-economic analysis of an ocean thermal power plant for a Nigerian coastal region. International Journal of Ambient Energy. 2018;39(6):562–572.

[32] Ikegami Y, Yasunaga T, and Morisaki T. Ocean thermal energy conversion using double-stage Rankine cycle. Journal of Marine Science and Engineering. 2018;6(1):21.

[33] Matsuda Y, Yoshitake T, Sugi T, *et al.* Construction of a static model for power generation of OTEC plant using Uehara cycle based on experimental data. Journal of Marine Science and Engineering. 2018;6(1):18.

[34] Soto R, and Vergara J. Thermal power plant efficiency enhancement with ocean thermal energy conversion. Applied Thermal Engineering. 2014;62(1):105–112.

[35] Park S, Chun W, and Kim N. Simulated production of electric power and desalination using Solar-OTEC hybrid system. International Journal of Energy Research. 2017;41(5):637–649.
[36] Arias FJ. Ocean thermal energy conversion by deliberate seawater salinization. International Journal of Energy Research. 2018;42(2):499–507.
[37] Ciccone MF, Cricchio A, Marra D, *et al.* A model for an offshore solar pond with organic Rankine cycle power plant. In: Proceedings of the 3rd Offshore Energy & Storage Symposium (OSES 2016). Valletta, Malta; 2016.
[38] Vega LA. Economics of ocean thermal energy conversion (OTEC): an update. In: Offshore Technology Conference. Offshore Technology Conference; 2010.
[39] Vega LA. Ocean thermal energy conversion primer. Marine Technology Society Journal. 2002;36(4):25–35.
[40] Pelc R, and Fujita RM. Renewable energy from the ocean. Marine Policy. 2002;26(6):471–479.
[41] Tiano F, Rizzo G, and Feo N. Study and optimization of an under-water compressed air energy storage system coupled with renewable energy plants. In: Proceedings of the 3rd Offshore Energy & Storage Symposium (OSES 2016). Valletta, Malta; 2016.
[42] Martel L, Smith P, Rizea S, *et al.* Ocean Thermal Energy Conversion Life Cycle Cost Assessment, Final Technical Report, 30 May 2012. Manassas, VA, United States: Lockheed Martin; 2012.
[43] Vega LA. Ocean thermal energy conversion. In: Encyclopedia of Sustainability Science and Technology. New York: Springer; 2012, pp. 7296–7328.
[44] Cavrot D. Economics of ocean thermal energy conversion (OTEC). Renewable Energy. 1993;3(8):891–896.
[45] Vega LA, and Michaelis D. First generation 50 MW OTEC plantship for the production of electricity and desalinated water. In: Offshore Technology Conference. Offshore Technology Conference; 2010.
[46] Muralidharan S. Assessment of Ocean Thermal Energy Conversion. Massachusetts Institute of Technology; 2012.
[47] Brauns E. Towards a worldwide sustainable and simultaneous large-scale production of renewable energy and potable water through salinity gradient power by combining reversed electrodialysis and solar power?. Desalination. 2008;219(1–3):312–323.
[48] Norman RS. Water salination: a source of energy. Science. 1974;186(4161): 350–352.
[49] Buonomenna M, and Bae J. Membrane processes and renewable energies. Renewable and Sustainable Energy Reviews. 2015;43:1343–1398.
[50] Kim J, Jeong K, Park MJ, *et al.* Recent advances in osmotic energy generation via pressure-retarded osmosis (PRO): a review. Energies. 2015; 8(10):11821–11845.
[51] Tedesco M, Scalici C, Vaccari D, *et al.* Performance of the first reverse electrodialysis pilot plant for power production from saline waters and concentrated brines. Journal of Membrane Science. 2016;500:33–45.

[52] Statkraft. Statkraft Halts Osmotic Power Investments; 2013. Viewed on December 18, 2018. Available from: https://www.statkraft.com/media/news/News-archive/2013/Statkraft-halts-osmotic-power-investments/.

[53] Da Rosa AV. Fundamentals of Renewable Energy Processes. London: Elsevier Academic Press; 2012.

[54] Veerman J, Saakes M, Metz SJ, *et al.* Reverse electrodialysis: evaluation of suitable electrode systems. Journal of Applied Electrochemistry. 2010; 40(8):1461–1474.

[55] Le NL, and Nunes SP. Materials and membrane technologies for water and energy sustainability. Sustainable Materials and Technologies. 2016;7:1–28.

[56] IRENA. Renewable Power Generation Costs in 2017. Abu Dhabi: International Renewable Energy Agency; 2018.

Chapter 5

Offshore wind energy

Graeme McCann[a]

Today offshore wind power arguably represents the poster child of ocean energy technologies. It provides the most tangible and sizable evidence to the public in general (and policymakers in particular) of ocean energy's ability – as opposed to mere potential – to supply modern economies with electrical power at industrial scale. While one British politician notoriously remarked in 2013 that a modern wind turbine could not be relied upon to '... pull the skin off a rice pudding... ' [1], the objective reality is that wind power in the UK supplied 14% of the nation's total electricity demand in 2017, with almost half of this coming from offshore wind [2]. Furthermore, the UK now boasts the two largest offshore wind farms in the world, with final construction of Walney in the Irish Sea having reached completion in the summer of 2018. Combined with the previous incumbent – London Array – these two giant wind farms alone provide over 1 GW of electrical power, equivalent to meeting the needs of over one million households for the duration of their operating lifetimes. And yet Walney and London Array – whilst the largest examples globally today of industrial-scale ocean-based energy – are by no means isolated examples. The bigger picture is one where installed offshore wind power in Europe has now reached almost 16 GW as of the end of 2017 [3], delivered from around one hundred separate projects. The majority of these projects are sited in the UK, Germany and Denmark, with the Netherlands, Belgium, France and Sweden also making significant investments. The 25% year on year growth rate in Europe in 2017 is also noteworthy [3]. While the bare statistics of offshore wind installations outside of Europe may not appear so impressive (totalling 2 GW in Asia and the US combined at the end of 2017 [4]), it is nevertheless noteworthy that China now boasts 15% of total global capacity (recently overtaking early adopters such as Denmark) – most of it having been installed in just the past few years. Likewise, the first offshore wind farm in the US went operational in 2016, with a pipeline of further projects now emerging [4].

For engineering professionals working in the offshore wind industry since the turn of the millennium, the rise and success of the technology has been nothing short of remarkable and was considered by no means guaranteed. Early arguments

[a]Goldwind Denmark A/S, Bristol, UK

made for investment in and development of the technology focused on wider societal concerns pertaining to decarbonisation, security of supply and local job creation. Tacitly, the assumption was held that on a pure 'cost of energy' basis, offshore wind would continue to be relatively expensive, and require ongoing taxpayer subsidy. The remarkable cost curve that offshore wind power has traversed over the past 20 years has brought it today to a point where 'subsidy-free' bids are being placed in national auctions for new projects, coming in well below 100 euro/MWh [5]. Such costs, if realised, place offshore wind competitively against technologies like nuclear – a commonly used benchmark given its own twin challenge of achieving lower costs and societal acceptable.

What has driven this dramatic reduction in the cost of offshore wind over the past two decades? Undoubtedly, the policy environments within a number of key industrial countries (namely Denmark, Germany and the UK) have helped to foster sufficient investment certainty to stimulate development of technologies and projects. These developments in turn have enabled the industry to traverse learning curves (some painful!) such that efficiency gains have been leveraged in the project development process, the supply chain and in the operations phase. While this 'industrialisation' can be questioned in terms of its scale (many argue that offshore wind will not attain the levels of industrialisation and associated economies of scale of, e.g., the automotive industry unless it moves to smaller, higher-volume multi-rotor design concepts [6]), nevertheless it is clear that offshore wind power's move beyond small-scale pilot projects to full-scale commercial deployment has stimulated intensive learning and cost reduction in all phases of the asset life cycle.

Alongside supportive policies, the other clear driver in rapid cost reduction has been technology innovation. It is towards the technology that the present chapter will primarily focus.

A final word on offshore wind's privileged position today in the vanguard of ocean energy technologies is, what lessons and warnings does it have for other technologies seeking to join the party? Clearly, as illustrated above, the history of offshore wind power development provides succour to other less established technologies as they aim to demonstrate their own efficacy (ability to produce predictable electrical power at scale), reliability and survivability (ability to do so with high availability levels within a harsh operating environment), affordability (ability to provide value for investors and wider society) and safety (ability to mitigate inordinate risks to the industry, the public and the wider environment). One major advantage offshore wind power has had relative to many other ocean energy technologies is that it could evolve from its onshore wind 'cousin' – and so unleash previous decades' worth of learning and innovation made on dry land prior to 'dipping its toes'. This option is not so readily open to many other ocean technologies where a greater leap of faith has been (or will be) required. One area where such prior experience has proved invaluable for offshore wind is that of survivability. For many ocean energy technologies, the fundamental challenge is producing a design that can effectively generate electrical power during the relatively benign, typical environmental conditions (be it wind, wave or tidal states) while also being equipped to withstand the incredibly large forces experienced during the rare, but extreme, environmental states. This extreme survival requirement applies also for onshore wind turbines and offshore wind has been fortunate to find that many (not all) of the techniques developed to mitigate damage to onshore wind turbines have also found application offshore. The challenge remains for other technologies, however – can they be made strong enough to survive extreme ocean conditions, while ensuring that the designs nevertheless remain light and cheap enough to deliver a competitive cost of energy? Solving this conundrum will largely dictate whether other technologies can join offshore wind power in delivering power at utility scale in future.

Offshore wind power, then, provides tangible evidence of ocean energies' ability to contribute to our future energy mix. However, the continuing challenges faced by offshore wind should also serve as a warning to other less-developed ocean energy technologies. Stakeholders should be clear-eyed in appreciating, e.g., that, while offshore wind power's cost reduction over the past decade has been impressive, it has nevertheless paled in comparison with that of solar power [7]. While the installation rate of offshore wind power over the past decade represents significant growth and investment, it is nevertheless eclipsed by the scale of *onshore* wind power development over the same period [4]. By 2050, installed capacity of onshore wind globally is predicted to outweigh offshore by a factor of five to one [8]. The motivation to state these statistics is not to burst ocean energy technologies' bubble but rather to quell some of the hype around the field and ensure that the expectations placed on these technologies are both ambitious *and* realistic. Ultimately, bar some unforeseen and dramatic leap in some other technology area, it is likely that offshore wind power will continue to act as the flag-bearer for ocean energy technologies for the coming decades, decades in which onshore wind and solar will continue to boast lower costs and higher installation rates globally. This does not mean offshore wind power as an industry cannot or will not thrive. Rather it means that, to do so, it needs to understand the conditions under which it can co-exist with ultra-cheap onshore wind and solar. For many countries, the unique selling point of offshore wind is the opportunity it holds out to develop relatively remote ocean spaces rather than already over-subscribed land, while keeping energy supply indigenous. Offshore wind's challenge is then to stay close enough to the onshore wind and solar cost point in the coming decade to make the additional cost premium it entails still palatable to society. Other ocean technologies face this challenge as well, but are starting further back and hence are under even greater pressure to drive their cost reduction while identifying the additional niche 'selling points' they can offer society to bridge the gap with solar and onshore wind, both of which by 2030 will be considered as ubiquitous, conventional, low-cost energy sources worldwide.

5.1 Offshore wind characterisation

The power output from a wind turbine is given by the following well-known equation:

$$P = \frac{1}{2} C_P \rho A U^3 \tag{5.1}$$

where ρ is the density of the air, C_P is the power coefficient, A is the swept area of the rotor and U is the incident free stream wind speed. The form of this equation and derivation of the power coefficient parameter C_P will be considered presently. For now, we note that the power is a function of the incident wind speed *cubed*, indicating that even a small increase in speed can drive a large increase in power capture. Similarly, we observe that the power is a function of the rotor diameter *squared* – i.e., a doubling in rotor diameter (e.g., an increase from 75 to 150 m) leads to a factor of four increase in maximum power capture, assuming all other factors remain constant.

These two fundamental relationships:

$$P \propto U^3 \tag{5.2}$$

$$P \propto D^2 \tag{5.3}$$

where D is the rotor diameter, explain two of the often-stated driving motivations to site wind farms offshore: (i) the generally improved wind resource (higher mean wind speeds) compared to onshore sites and (ii) the relaxation of turbine tip-height restrictions offshore enabling developers to install machines with larger rotors to maximise energy capture. Of course, these factors both come with limiting characteristics – the higher mean wind speeds which drive more power capture also increase loading on the wind turbine and its support structure offshore, hence a more expensive structural design is generally required to withstand such exposure. Similarly while pursuit of larger rotors to capture more power seems like an obviously beneficial design evolution, the underlying physics is not so accommodating. It turns out that, while the square-law governing the relationship between rotor size and power capture is strong, it is not as strong as the *cubic* law governing the relationship between rotor size and material mass, and hence cost. Therefore, the offshore wind turbine design engineer, in pushing rotor growth to drive energy yield improvement, eventually finds that the growth in material requirements (and therefore cost) quickly begins to outweigh the improved power output. As with all engineering optimisation problems, the cost function of interest must include a consideration of *economic* cost, not just technological constraints. We will return to this question of optimal turbine size later in the chapter.

Based on a consideration of levelised cost of energy (LCoE), it is easily demonstrated that (today at least) the generally higher wind speeds found offshore compared to onshore or nearshore sites are still not enough to offset the much higher capital cost (capex) and operational cost (opex) associated with offshore wind, particularly if the water depth is anything other than very shallow (<10 m). A modern, well-sited onshore wind farm can boast an LCoE profile less than half of that of the most competitive offshore projects planned for the coming years [7]. Wind speed, we may conclude, is important, but not supremely so. Other influencing factors, such as proximity to electrical load centres (e.g., US Eastern seaboard, mainland Chinese coast and Thames Estuary), onshore siting restrictions, political and societal factors, are ultimately decisive in promoting the adoption of offshore wind.

Nevertheless, once a decision has been made (for whatever reason) to develop an offshore wind project, it generally makes sense to do so where the mean wind speed is high. Simply stated, higher wind speeds mean higher energy yield which generally means a much high revenue stream and more robust investment case. With wind speed being of such fundamental importance in determining the economic success of an offshore wind project, it is not surprising that the industry devotes so much effort to understand the fundamental behaviour and characteristics of the wind.

5.1.1 The random nature of wind

Equation (5.1) applied naively could imply that the wind is deterministic. Our everyday experience tells us this is not the case. The opposite error committed by many is to assume that the wind is entirely unpredictable. In fact, meteorologists and wind engineers alike reject both descriptions; instead the wind energy industry adopts the language of randomness, variability and intermittency to describe the behaviour of wind flow [9]. While this may seem like splitting semantic hairs, the precision of language is important. Consider more closely the term unpredictable and we find that, on reflection, it does not align well with our own everyday experiences of the wind; we intuitively predict that we will have more windy days in winter than in summer, we expect that when we go for a walk on a hill or mountain side that the winds will be stronger at the top than in the sheltered valley below; we assume that, when the evening weather forecast tells us to expect high winds tomorrow, there is a fair chance high winds will indeed follow. In other words, we acknowledge through daily life that the wind is in some sense predictable. Scientists and engineers agree, as do bank and fund managers. No bank or investment fund would consider financing an offshore wind farm development if the wind was truly unpredictable – investment confidence amongst the finance community relies on sufficient certainty (or conversely, the ability to quantify the uncertainty) in the wind behaviour at the particular project site in question.

To state that the wind is random and intermittent is to say that, while it is not completely unpredictable (as we have seen), it is nevertheless not *entirely* determined. Rather, our approach to characterise the behaviour of the wind must adopt the mathematical language of probability. Probability theory is applicable whenever we need to say something useful about a process or system for which we have limited knowledge, or which contains random features, or both [9].

The variability of wind (both onshore and offshore) applies in both temporal and spatial sense and over a range of timescale and length scale. Starting at the macro scale, we know that wind speeds at a given location can exhibit dramatic changes from one year to the next and even more so over decades. Such large temporal scale effects can be very difficult to understand or quantify; for example, an active and pertinent research question being pursued today relates to the impact that climate change is having on long-term global wind speeds, e.g., [10]. On timescales of the order of months, seasonal wind speed fluctuations are observed, and are generally better understood and quantified than the inter-annual variations (IAVs). Moving to yet smaller temporal scales, we observe 'synoptic' variations measured on a timescale of days, which relate to the movement of weather systems through the geographic location of interest. Going smaller still, we see variations on a daily scale (diurnal variation) which are often related to thermal effects driving changes in the atmospheric stability [9]. Finally, we come down to the very small-scale variations observed on the scale of minutes and seconds, commonly referred to as turbulence. Turbulence of the wind has both temporal and spatial components; for example – considering that a modern offshore wind turbine may have a rotor diameter in excess of 160 m – the wind speed experienced at the tip of one blade will vary significantly from that at the other two tips located 150 m ($80 + 80 \times \cos 30°$) away. This temporal and spatial turbulent variation of the wind can create dramatic fluctuations in electrical power (if not suitably controlled), as well as generate high aerodynamic loading on the structure. We will return to look at these loading effects in more detail later in the chapter. For now, we simply summarise by observing that the frequency spectrum of the wind is broadband and – as van der Hoven illustrated in the 1950s – generally composed of two distinct peaks (one corresponding to the low-frequency synoptic and diurnal effects and the other the higher-frequency turbulent energy in the flow which tends to follow Kolmogorov's famous turbulent cascade of energy moving from larger length scales (eddies) to smaller [11]). Between these two distinct peaks, one observes a valley or 'spectral gap' in the frequency spectrum at a timescale between 10 min and 2 h. It is, incidentally, due to the presence of this spectral gap that time-domain turbulent wind simulations of the dynamic and loading response of a wind turbine are recommended to run for at least 10 min, in order that the full turbulent energy spectrum is sampled in the simulation. In the context of offshore wind turbine dynamic analysis, a similar argument is applied to justify a minimum 3-h simulation length – now to adequately sample the typical broadband energy spectrum of the sea state (waves). Clearly, for integrated wind and wave time-domain simulation, care needs to be taken that the reference periods for these two stochastic processes (i.e., wind and waves) are applied consistently.

5.1.2 Long-term offshore wind speed characteristics

As mentioned above, average wind speeds at a given location (onshore or offshore) are unlikely to be stationary over very long timescales (say, decades) and this non-stationarity may become even more pronounced as climate change effects accelerate through this century. Nevertheless – largely due to the extreme challenge in quantifying these very long-term trends looking forward – it is common in the practical project development phase of an offshore wind farm to *assume* long-term (decadal) stationarity but then monitor the assumption through the operational lifetime. IAV is considered by correlating the wind speeds measured at the site with a suitable long-term reference location [12]. Ideally at least 2 years of site measurement data should be used, though 1 year may be tolerable. Depending on the region in which the offshore wind farm is being developed, available long-term reference sources will vary (e.g., offshore platforms and coastal met stations) and are becoming increasingly rare as sites move further offshore. Fortunately, global reanalysis data sets (e.g., MERRA2 and ERA5) have been found to correlate very well offshore. For European markets, the coefficient of variation representing IAV is typically in the range of 4.5%–5.5%.

The annual synoptic and diurnal variations are represented in both energy yield and structural loading calculations using probability distributions. Concretely, the Weibull distribution has been shown to give reasonable representation of the 10-min average wind speed variation within a given year for many sites, including offshore. The Weibull distribution is expressed as

$$F(U) = e^{\left(-\left(\frac{U}{c}\right)^k\right)} \tag{5.4}$$

where $F(U)$ is the probability that the wind speed U is exceeded, c is the 'scale' parameter which governs the average mean wind speed implied by the distribution and k is the 'shape' parameter which influences the distribution's skewedness. Higher values of k indicate lower variance relative to the average wind speed at the site.

Equation (5.4) only deals with wind *speed.* In reality, it is also important to quantify the variation in wind *direction* [13]. The directionality is important for at least two reasons: first, for a given wind farm layout, the wind rose (directional probability distribution) will dictate how much time turbines spend operating in the wake of neighbouring machines. This so-called wake-affected operating condition has a double-negative impact: reduced energy capture *and* increased fatigue loading. An appreciation of the long-term directional variance of the wind is necessary to mitigate these effects during project development and micro-siting. Knowledge of the expected wind direction distribution is also important for offshore wind farms as it greatly influences the design, and hence cost-effectiveness, of the offshore support structures [14]. The design of onshore wind foundations typically makes the tolerable assumption that the wind is unidirectional in terms of ascribing the internal load path through the tower and foundation. For offshore wind farms, however, such an assumption can lead to unnecessarily heavy and expensive support structures and should be avoided by use of the *joint* probability distribution of wind speed and direction during design. Indeed, for a rigorous treatment, the designer should also consider long-term correlation of wind speed and sea state (wave heights and periods), and wind and wave directional misalignment [15].

5.1.3 Turbulence

Turbulent variations in the wind, measured on the timescale of seconds and minutes, are driven by two principle effects; surface roughness or 'friction' at the fluid-surface boundary and thermal effects. A range of descriptors are available to characterise these variations in time and space. Gust factors are used to give the ratio of peak (say, 3-s sustained) wind speed to the average 'background' level. Turbulence 'intensity' is defined as the standard deviation σ of the wind speed normalised by the mean wind speed U and gives a handy measure of the overall level of turbulence. Mathematically turbulence intensity is defined as

$$TI = \frac{\sigma}{U} \tag{5.5}$$

Spectral models of turbulence (see, e.g., [16]) build on these basic concepts by quantifying the frequency content of the wind speed variations at given spatial points (typically assumed to have Gaussian form). In turn these point representations can be correlated in space through use of empirically parameterised coherence models which represent the decay in point-to-point correlation as the distance between the points increases, over a range of frequencies. It is then a relatively simple step to use efficient fast Fourier transformation (FFT) algorithms to re-cast these spectral models of turbulence and coherence in the time domain for subsequent use in time-domain dynamic simulation (see Figure 5.1).

Due to the link between turbulence intensity in a neutral atmosphere (i.e., where stability effects are not dominating) and surface roughness, it is generally the case – owing to the smoother surface of the sea compared to land – that turbulence intensity is less severe offshore. This greatly benefits offshore wind turbine design as turbulence intensity is the primary source of fatigue loading (and potentially extreme loading) for the structure, more important even than the average wind speed level.

A number of standard turbulence spectral models are available to the wind industry and are applicable for use in the context of offshore sites. The international

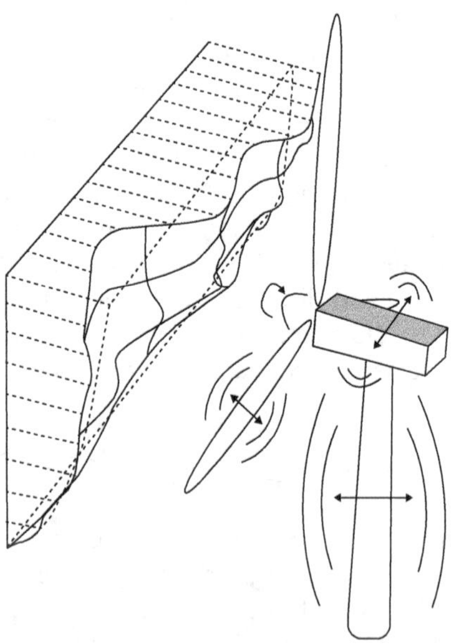

Figure 5.1 Time-domain numerical 'wind field' synthesised from a spectral model of turbulence, including a spatial and frequency coherence model

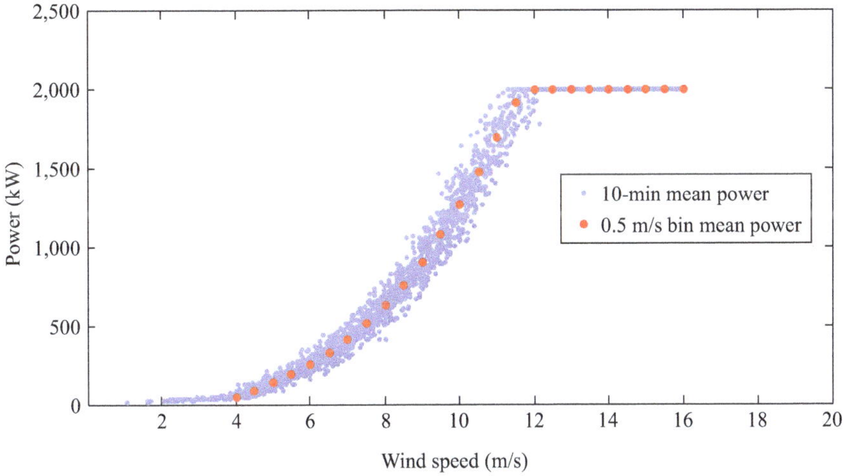

Figure 5.2 Ten-minute averaged power measurements (blue) and data re-binned in 0.5 m/s increments

design standard for offshore wind turbines International Electro-technical Commission (IEC) 61400-3 [17] (itself referencing [16]) gives more details on the application of this models in design.

Finally, it is noted that time-domain turbulence modelling is the state of the art for offshore wind turbine loading and design analysis but is not common for energy yield analysis. Rather the effect of turbulence on energy capture calculated at a farm level is typically included in the so-called dynamic power curve for the wind turbine (see Figure 5.2). 'Dynamic' power curve refers to the fact that the turbine's response to wind speed variations experienced on the timescale of seconds and minutes is included *implicitly* in the 10-min averaged power output per mean wind speed. Comparing the shape of the 'steady-state' power curve of Figure 5.3 with the dynamic power curve in Figure 5.2, we note that the principal effect of turbulence is to smooth out the knee of the power curve around the rated wind speed. It also tends to reduce the average output in the below-rated regime.

5.1.4 Wake flow effects

The mechanism by which a wind turbine extracts energy from the wind (see Section 5.2) generates a wake which propagates downstream of the rotor. This wake has a number of defining characteristics; it contains on average lower wind speeds than the free stream surrounding it, it has a low-frequency rotational component due to the flow's reaction to the rotating turbine blades, it has additional high-frequency turbulence components caused by vorticity shedding at the blade tips and hub location and finally it has an additional low-frequency turbulent component, due to directional 'meandering' of the wake under the influence of the $> 1D$ length-scales in the background atmospheric turbulence (where D is the rotor diameter). Collectively,

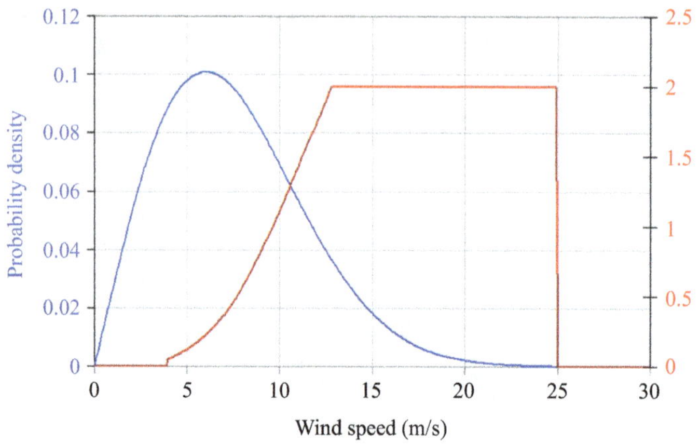

Figure 5.3 Weibull distribution of wind speed (blue line) and turbine steady power curve (red line). 'Steady' indicates that the effect of turbulent wind variations has not been included. Integrating the blue and red curves would provide an estimate of the annual energy yield

these effects create a complex physical flow environment downstream of a wind turbine, and the complexity is compounded in a farm environment where many turbine wakes are interacting and mixing. Much research effort has been devoted by the wind industry over the past three decades to better understand the physics governing wind turbine wakes not least because, as already mentioned, they have a dramatic (detrimental) effect on both energy capture and structural loading.

As the wake propagates downstream, it recovers, in the sense that the size of the wind speed deficits contained within the wake diminishes. This recovery is principally caused by energy mixing at the wake-free stream boundary which re-energises the latter. Onshore, a rule of thumb is that full wake recovery is complete at a propagation distance of $10D$ downstream. However, this assumption appears less secure in the offshore context. Recalling that wake recovery is driven by turbulent energy mixing between the wake and the free stream, with typically lower turbulence intensity levels at offshore sites, one should expect wakes to persist for greater distances downstream. This prediction is indeed supported by actual wake measurements in offshore wind farms where wake structures have been detected even up to $18D$ downstream of the wind turbine from which they originally propagated [18]. Hence, the benefit in terms of energy capture and loading that we identified for offshore wind due to the lower ambient turbulence levels now turns against us – wake effects are generally much more severe offshore due to the higher level of wake persistence (and also because the number of turbines in an offshore wind farm tends to be greater).

Models to represent the 'added' turbulence in wind turbine wakes have been in development since the late 1980s. Quarton and Ainslie [19] examined a range of wake-affected turbulence measurement sets (both model and full scale) and from this proposed an empirical formulation (later updated by Hassan):

$$I_{\mathrm{add}} = 5.7 C_T^{0.7} I_0^{0.68} \left(\frac{x}{x_n}\right)^{-0.96} \tag{5.6}$$

where C_T is the turbine thrust coefficient (see Section 5.2), I_0 is the ambient turbulence intensity of the background flow and x_n is the length of the new wake region [20].

Frandsen [21] proposed a similar empirical approach but used more full-scale measurement sets from larger wind turbines and calibrated his model using the measured turbine fatigue loading (tower and blade load sensors) rather than just the wake turbulence intensity per se. This approach was codified in the international design standard for wind turbines [16,17] and has become state of the art within the industry for most turbine site loading calculations, both onshore and offshore. While it has the attractive features of simplicity and ease of application, it has some significant limitations which have motivated continued research into more advanced wake models. The primary limitation of the Frandsen model is that it does not explicitly attempt to capture much of the governing wake physics – this means that while it focuses on the prediction of fatigue loading in wakes, it does not reflect the corresponding impact on energy capture. Similarly, given the empirical formulation was focused on fatigue loading, it is questionable to what degree the model can be safely applied in the prediction of extreme wake-affected loading. Finally, the model validation was conducted well before the advent of today's super-sized offshore wind turbines. Again, one is justified in questioning whether the model retains its applicability when the scale of the machine has grown by a factor of four compared to those used in the original Frandsen model calibration. For all these reasons, higher-fidelity wake models are now making the transition from research into industrial application; e.g., the dynamic wake meandering (DWM) model developed by researchers at the Wind Energy department at the Danish Technical University (DTU) has been accepted for inclusion in the latest edition of the IEC 61400-1 international wind turbine design standard. For more information on the DWM approach, the reader is directed to [22].

5.1.5 Other offshore-specific conditions

As the offshore wind industry has grown up in northern Europe, it is understandable that standard representations of wind conditions have reflected observations typical of the North Sea and Baltic Sea areas. However, offshore wind is now a global industry with more and more projects planned across Asia and in the US (both coastal and on the Great Lakes). As offshore wind gains traction in global markets, it is confronted by a number of new environmental challenges. For example, in large parts of Asia (China, Taiwan, Japan and India) where offshore is forecasted to grow, as well as on the Eastern seaboard of the US, tropical cyclones (a.k.a. hurricanes in the US and typhoons in parts of Asia) are prevalent and constitute a significant risk to offshore wind development. In response, the industry has proposed models that attempt to simulate the wind conditions characteristic of a tropical cyclone event; see, e.g., [23]. While peak sustained wind speeds is important, it is not the only parameter of interest in a tropical cyclone – it is also necessary to consider extreme changes in wind *direction* as well as wind shear and turbulence levels. Ultimately the design of a wind turbine for siting in tropical cyclone-prone areas comes down to economic considerations and assessment of risk. To design the structure to *guarantee* survival may be technically possible and yet economically disastrous. As such, the risk may be better managed through a more nuanced combination of technical and insurance/financial mitigations.

As well as tropical cyclones, globalised offshore wind needs to consider sea ice effects (particularly in the Baltic Sea area), extreme breaking or rogue wave impacts, a range of challenging sea-bed geotechnical conditions and, as ocean spaces and users become more condensed, potential ship impacts.

5.2 Wind turbine technology development: a historical perspective

Ask almost anyone today – engineering professional or otherwise – to describe a modern wind turbine (onshore or offshore) and they will invariably paint the picture of a three-bladed, horizontal-axis machine. This ubiquitous architecture – often referred to as the 'Danish design' – can trace its early heritage back to a 200-kW, 24-m diameter wind turbine which was installed on the Danish island of Gedser in the mid-1950s [6]. It was not until the 1980s, however (following the global oil-crisis of 1973), that commercial wind turbine designs really began to converge upon today's common blueprint. Two broad innovation pathways were pursued at this time. First, an incremental approach in Denmark whereby *kW*-scale machines were rapidly designed, built and operated, with lessons quickly learnt and adopted in subsequent design iterations. (As an aside, one can reasonably reflect that – long before the phrase became popular – the Danes were applying 'agile thinking' in the pursuit of commercially viable wind turbine designs well over three decades ago.)

A parallel innovation drive in the US, the UK and Germany was also in evidence around this same time, characterised by (i) direct jumps to ambitious MW-scale prototypes, (ii) a strong focus on developing the underpinning science of wind energy and (iii) the backing of significant state funding. For example, the US in the 1970s and 1980s – driven by a desire for greater energy security following the events of 1973 – established state and federal funds for wind turbine research and development, and tax credit schemes aimed at stimulating wind farm developments. While the ambition of many of the MW-scale prototypes developed during this period is to be applauded, a simple reading of the history makes it difficult not to conclude that they were generally commercially unsuccessful. For a more comprehensive treatment of the history, the reader is directed to [24]. In the end, the boom in wind farm developments in California in the mid-1980s driven by the lucrative tax credit scheme created the environment whereby the two innovation pathways could in some sense converge – the incrementally developed, technically superior Danish designs secured a lucrative export market for Danish manufacturers and so enabled them to build the foundations of the industry we know today.

The taxonomy of modern offshore wind turbines represents a later branch in this evolutionary tree. Indeed, the first offshore wind turbines installed in Europe in 1991 were fundamentally re-purposed *on*shore designs. It was arguably not until the first offshore deployment of the (then called) REpower 5M 5 MW wind turbine in 2006 [25] that the industry really demonstrated it was serious about developing bespoke designs for the offshore market, opening up an innovation pathway in pursuit of very large machines that has precipitated the offshore 'behemoths' we see today.

5.3 Basic principles of wind turbine operation

5.3.1 Energy conversion and concentration

Fundamentally, the purpose of a modern wind turbine is to convert the kinetic energy of the wind to electrical energy. This electrical energy is then exported from the wind farm – in most applications – into the wider electrical grid. Going a step further, the wind turbine can be said to perform two fundamental transformations of the energy in the wind: *conversion* and *concentration*.

Considering conversion first, it turns out that it is overly simplistic to state that a wind turbine converts kinetic energy (of the wind) into electrical energy. Rather, there is an intermediate conversion step introduced by the rotor aerodynamics, which converts the kinetic energy of the wind into a pressure energy gradient at the rotor plane. The induced drop in pressure across the rotor plane corresponds to the kinetic energy being extracted from the flow and converted via the aerodynamics into rotational mechanical energy in the turbine drive train. This rotational energy in turn drives the rotor of an electrical generator within a stator, producing electrical energy.

So much for conversion of energy, wind turbines also act as highly effective concentrators of energy. Following the illuminating treatment of an onshore wind turbine by Jamieson [6] but updating it to consider the length scales typical of a modern offshore wind turbine, Table 5.1 shows the power concentration 'cascade'

Table 5.1 Wind turbine power concentration cascade, after [6]

Stage	Length scale (m)	Area (m^2)	Efficiency (.)	Power (MW)	Power density (kW/m^2)	Concentration ratio	Cumulative concentration
Wind	160	80,435	1.0	15.0	0.186	1.0	1.0
Blades	80	4,021	0.5	7.5	1.865	10	10
Drive train	1.7	9.08	0.94	7.0	770.9	413.4	4,134
Export cable	0.06	0.02	1.0	7.0	350,000	454.0	1,876,836

that occurs from the wind flowing through the rotor plane down to the electrical power flowing out from the export cable.

The reader's attention is drawn to the exponential growth in power density from the rotor (measured with the units of kW/m^2) to the electrical export cable (measured with units of GW/m^2). It is this ability to cost-effectively concentrate the relatively low-density energy of the wind that has made modern wind turbines such a success. As Table 5.1 demonstrates, the rotor blades play a key role in this regard. Modern offshore wind turbine rotors typically deploy blade designs with a so-called solidity ratio close to 5%. Solidity quantifies the fraction of the swept rotor area occupied by the projected area of the physical blade. With a solidity value of only 5% (representative of modern, slender, low mass and hence low-cost designs), the rotor is nevertheless able to extract energy from the entire swept area of its rotation with a C_P efficiency (as per (5.1)) of around 50%. Compared to an idealised 'actuator disc', the well-designed rotor can achieve extraction efficiencies to within 15% of the Betz theoretical limit, while using only 5% of material surface area compared with the Betz idealised (solid) disc. The material mass and cost savings of the slender, low-solidity rotor far outweigh the 15% loss in extraction efficiency, making the modern aerodynamic rotor design difficult to surpass. Section 5.3.2 considers these aerodynamic features of the rotor in more detail.

5.3.2 One-dimensional momentum and Betz

Significant insight into wind turbine aerodynamics can be derived from a relatively simple representation of the system fluid dynamics. The actuator disc concept had found application in helicopter rotor and ship propeller analysis before wind turbines, and much of the developed theory was readily transferable. Despite its simplicity and vintage, it nevertheless remains the state of the art within many wind turbine design calculations.

The actuator disc theory postulates an energy extraction device (e.g., a rotor) situated within an open flow, as depicted in Figure 5.4.

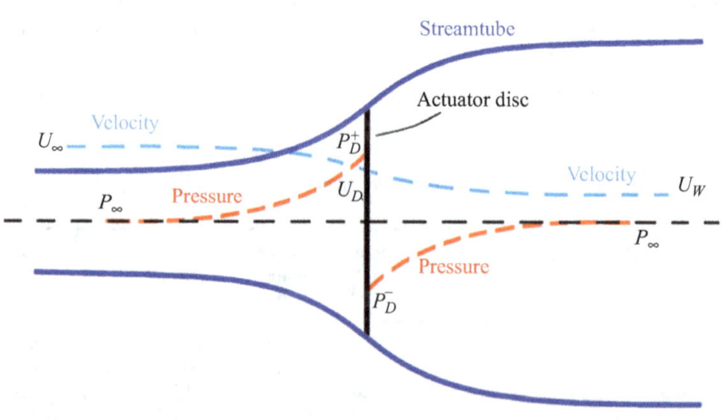

Figure 5.4 Actuator disc in an open flow with stream-tube development

The actuator disc approach makes several fundamental assumptions:

- The mass flow is steady and uniform.
- The flow is bound by a stream-tube, with no fluid interaction across the stream-tube boundary.
- The flow is inviscid (no energy dissipation).
- The flow is incompressible.

Using momentum theory, the flow momentum change across the plane of the actuator disc can be used to calculate the forces acting on the disc and the corresponding pressure energy that it extracts.

The actuator disc induces a velocity change (acceleration) in the flow which must be superimposed on the free stream velocity. The stream-wise component of this induced flow *at the rotor plane* is therefore given by $-aU_\infty$, where a is referred to as the axial induction factor (alternatively the inflow factor). At the rotor plane, the stream-wise velocity is therefore calculated as

$$U_D = U_\infty(1-a) \tag{5.7}$$

Re-arranging (5.7) gives:

$$a = \frac{U_\infty - U_D}{U_\infty} \tag{5.8}$$

Continuity of mass means that the mass flow rate must be constant at each location along the streamtube. Bernoulli's equation states that under steady conditions the total energy in the streamtube must remain constant assuming no work is done by or on the fluid. Finally, momentum balance considerations allow us to equate the force F acting on the actuator disc by the fluid (itself equal and opposite to the force exerted by the fluid by the disc) as the rate of change of momentum. Re-stating these relationships mathematically gives us the following:

Mass flow rate balance:

$$\dot{m} = \text{constant} \tag{5.9}$$

Momentum balance:

$$F = \dot{m}\Delta U \tag{5.10}$$

Energy balance per Bernoulli (neglecting gravitational potential energy), where ρ is the air density and p is the static air pressure:

$$\frac{1}{2}\rho U^2 + p = \text{constant} \tag{5.11}$$

Adopting the notation of Figure 5.5, (5.9) implies that

$$\rho A_\infty U_\infty = \rho A_D U_D = \rho A_W U_W \tag{5.12}$$

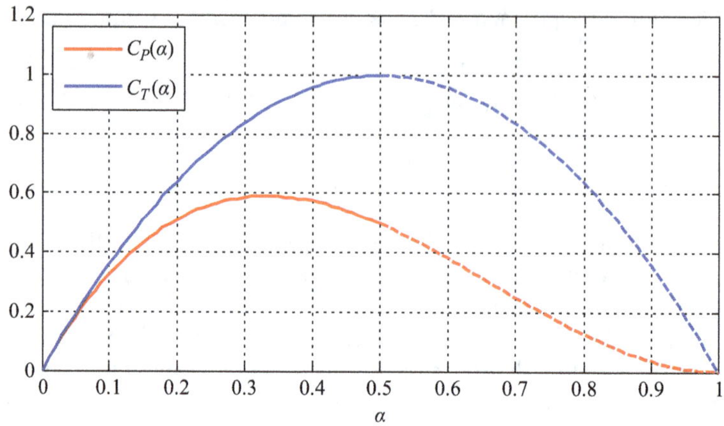

Figure 5.5 C_P and C_T as functions of axial induction factor a

The flow undergoes an overall change in velocity from the far upstream to far downstream states equivalent to

$$\Delta U = U_\infty - U_W \tag{5.13}$$

and a rate of change of momentum equal to this change in far-field velocity multiplied by the mass flow rate:

$$\dot{m}\Delta U = \rho A_D U_D (U_\infty - U_W) \tag{5.14}$$

This rate of change of momentum can be considered as the force F acting on the flow by the disc.

This force is entirely generated by the change (drop) in pressure either side of the actuator disc plane. Examining Figure 5.4, one notes the smooth increase in pressure as the flow approaches the disc plane followed by a discrete drop in pressure across the plane. Just behind the plane, the pressure drops below ambient atmospheric levels, from which it recovers smoothly back to atmospheric pressure in the far wake. The smooth rise in pressure upstream of the actuator disc is balanced by a corresponding drop in the flow velocity according to Bernouilli's equation (5.11) in order to keep overall energy at any point along an upstream streamline constant. A similar, but opposite, energy balance is observed in the downstream development with the pressure now increasing with a corresponding reduction in wake velocity asymptotically tending to the far-field velocity U_W.

This force can be expressed accordingly:

$$F = (p_D^+ - p_D^-)A_D = \rho A_D U_D (U_\infty - U_W) \tag{5.15}$$

Introducing (5.7) for the unknown U_D gives

$$F = (p_D^+ - p_D^-)A_D = \rho A_D U_\infty (1 - a)(U_\infty - U_W) \tag{5.16}$$

Similarly, we need to eliminate the unknown far-field wake velocity U_W from (5.16) and can do so by considering energy balance along streamlines in both the upstream and downstream locations.

Upstream, we observe:

$$\frac{1}{2}\rho U_\infty^2 + p_\infty = \frac{1}{2}\rho U_D^2 + p_D^+ \tag{5.17}$$

Similarly, downstream:

$$\frac{1}{2}\rho U_W^2 + p_\infty = \frac{1}{2}\rho U_D^2 + p_D^- \tag{5.18}$$

Taking the difference of (5.17) and (5.18) gives

$$\left(p_D^+ - p_D^-\right) = \frac{1}{2}\rho\left(U_\infty^2 - U_W^2\right) \tag{5.19}$$

Substituting (5.19) into (5.16) gives

$$\frac{1}{2}\rho\left(U_\infty^2 - U_W^2\right)A_D = \rho A_D U_\infty(1-a)(U_\infty - U_W) \tag{5.20}$$

hence:

$$U_W = (1-2a)U_\infty \tag{5.21}$$

Equation (5.21) implies that half of the flow velocity reduction occurs upstream and half downstream of the actuator disc.

Reintroducing (5.21) back into (5.16) allows us to re-cast the force acting on the flow from the actuator disc as

$$F = \left(p_D^+ - p_D^-\right)A_D = 2\rho A_D U_\infty^2 a(1-a) \tag{5.22}$$

The rate of work done by this force is FU_D, giving the following equation for power:

$$Power = 2\rho A_D U_\infty^3 a(1-a)^2 \tag{5.23}$$

We can now introduce the concept of a power coefficient C_P, as the ratio of the power extracted from the flow by the actuator disc (see (5.23)) to the *maximum* power in the free stream for the same reference disc area A_D:

$$C_P = \frac{Power}{0.5\rho A_D U_\infty^3} \tag{5.24}$$

Hence,

$$C_P = 4a(1-a)^2 \tag{5.25}$$

From basic calculus, we know the maximum value of C_P occurs when its derivative equals zero:

$$\frac{dC_P}{da} = 4(1-a)(1-3a) = 0 \tag{5.26}$$

Equation (5.26) has two solutions, $a = 1$ and $a = 1/3$. $a = 1$ corresponds to an entirely blocked flow. $a = 1/3$ provides the more interesting solution and defines the axial induction factor corresponding to maximum efficiency of the actuator disc. Physically, $a = 1/3$ can be understood to mean that this maximum efficiency is achieved whenever the flow velocity at the actuator disc plane drops to $2/3$ of the free stream velocity.

Introducing $a = 1/3$ back into (5.25) provides a numerical expression for the maximum C_P value, corresponding to the well-known Betz limit:

$$C_{P-\text{Betz}} = \frac{16}{27} = 0.593 \tag{5.27}$$

Similarly, the thrust coefficient C_T can be stated as

$$C_T = \frac{Thrust}{0.5\rho A_D U_\infty^2} = 4a(1-a) \tag{5.28}$$

Equations (5.25) and (5.28) are plotted in Figure 5.5 to illustrate how the power coefficient and thrust coefficient vary as a function of the axial induction factor a.

The actuator disc model breaks down for axial induction factor values greater than 0.5. In this case, the model predicts that the downstream wake velocity becomes zero or negative (reversed flow). Clearly this is not physically representative. In reality, a high axial induction factor (i.e., close to or above 0.5) corresponds to a heavily loaded flow state. Under these conditions, the wake becomes highly turbulent with significant mixing of flow at the stream tube boundary which re-energises the slow-moving flow through the rotor plane. The behaviour of the power and thrust coefficients at high-induction factor values can be determined empirically, e.g., [26]. In modern offshore wind turbine applications, the value of axial induction rarely exceeds 0.6 and for well-designed rotor blades, the factors will often stay very close to the theoretical optimum of 0.33 across their operational range.

5.3.3 1D momentum with rotational wake

The one-dimensional (1D) momentum theory applied in Section 5.3.2 can now be extended to include consideration of rotational effects in the flow. The exertion of a torque on the rotor disc created by the flow moving through it creates an equal and opposite torque on the fluid. The transfer of this rotational energy to the flow occurs across the finite thickness of the disc plane. The change in tangential velocity of the flow is defined in terms of a tangential induction factor a'.

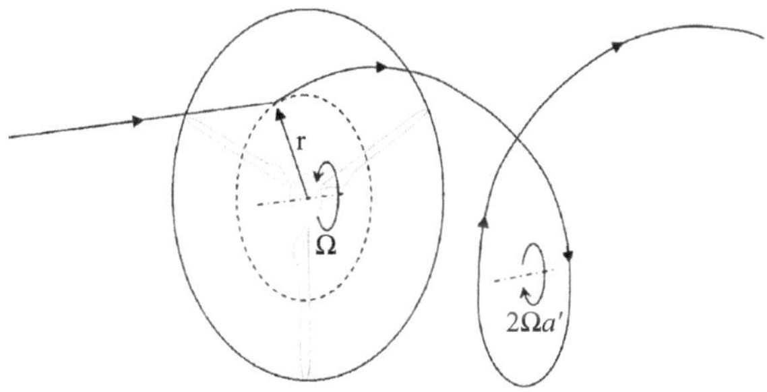

Figure 5.6 Tangential induction factor a' with a rotor rotation speed Ω

As illustrated in Figure 5.6, upstream of the rotor disc the tangential velocity of the flow is zero. At the disc plane, at a radial distance r from the rotational axis, the induced tangential velocity is $\Omega ra'$, where Ω is the rotational velocity of the rotor. Immediately downstream of the rotor plane, the tangential flow velocity is $2\Omega ra'$.

Applying a similar treatment as for the axial induction case, but now considering angular momentum instead of axial and torque instead of force, enables us to develop a complementary expression for rotor power extraction incorporating the tangential induction factor term applied at a discrete rotor annulus at radius r:

$$\delta Power = 2\rho\delta A_D U_\infty(1-a)\Omega^2 a' r^2 \tag{5.29}$$

Combining (5.29) with (5.23) leads to the following relationship:

$$a(1-a) = \lambda_r^2 a' \tag{5.30}$$

where $\lambda_r = \frac{\Omega r}{U_\infty}$ is the local speed ratio and $\lambda = \frac{\Omega R}{U_\infty}$ is the tip speed ratio for a rotor of radius R.

5.3.4 Basic aerofoil principles

As illustrated in Figure 5.7, two fundamental aerodynamic forces occur when an aerofoil is situated within an incident flow field; lift (perpendicular to the resultant flow direction) and drag (in line with the flow direction). Depending on where the reference 'aerodynamic centre' is located along the aerofoil section, an aerodynamic moment is also generated. By convention, the aerodynamic centre is considered to be located at 25% of the chord length c (where 0% corresponds to the leading edge and 100% the trailing edge).

The lift and drag (and moment) forces are functions of the flow density ρ, the resultant inflow air velocity W, the chord length c and the corresponding dimensionless aerodynamic coefficient. In turn, the aerodynamic coefficients (namely the life coefficient C_L, the drag coefficient C_D and the moment coefficient C_M)

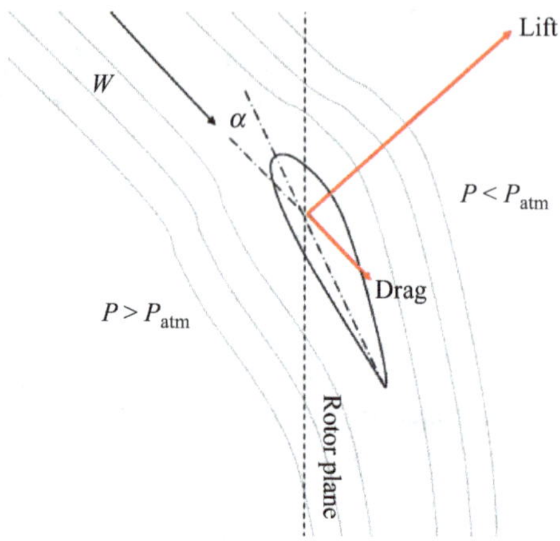

Figure 5.7 Flow past a wind turbine aerofoil

are functions of the angle of attack α. The aerodynamic forces are specified as follows:

Lift force per unit blade length:

$$L = \frac{1}{2}\rho c W^2 C_L(\alpha) \tag{5.31}$$

Drag force per unit blade length:

$$D = \frac{1}{2}\rho c W^2 C_D(\alpha) \tag{5.32}$$

Moment per unit blade length:

$$M = \frac{1}{2}\rho c^2 W^2 C_M(\alpha) \tag{5.33}$$

Offshore wind turbine aerofoils typically demonstrate linear behaviour at low values of α up to the so-called stall point between 10° and 15°. As α approaches and enters the stalled flow region, the steady-state lift coefficient flattens off and the drag coefficient increases dramatically. Earlier wind turbine designs intentionally exploited this stall behaviour as a technique to passively control the peak power output of the machine; however, it tended to also attract high structural loading and occasionally instability as a by-product. Today, the vast majority of utility-scale wind turbines, and all modern offshore machines, instead use active blade pitch control in order to ensure that the rotor blade *generally* remains in the linear pre-stall aerodynamic regime for most of its operational range, though complete avoidance of stall – even for a pitch-controlled machine – is not possible.

Today modern offshore wind turbines will use aerofoil profiles specifically designed and optimised for wind turbine application. Considerations applied in the design include their aerodynamic effectiveness (often measured as the ratio of the lift to drag coefficient, with values of between 100 and 150 common), the range of Reynolds numbers they will experience during operation, their robustness to dirt and other degradation factors, their acoustic emission profile and their behaviour in stall. The aerodynamicist also needs to compromise with the blade structural designer's requirements in order that the eventual design has sufficient strength to withstand the substantial aerodynamic loads produced during both normal operating and extreme (e.g., storm) conditions.

5.3.5 BEM theory

So far in our consideration of how energy is extracted from the wind by a wind turbine, we have been content to simply represent the rotor as an idealised actuator disc with no specific consideration of the discrete blades that comprise a real rotor. Using the fundamental principles of aerofoil behaviour described in Section 5.3.4, we can extend the basic momentum theory of Section 5.3.3 by stating that the rate of change of momentum of the airflow through a discrete annulus of the rotor plane must be produced by the forces generated at the blade element(s) which sweep out that same annulus. This is the basic premise of the blade element momentum (BEM) theory. Consequently, it clearly relies on the assumption that there is no flow interaction between adjacent radial annuli which – although not strictly true – is generally tolerable for most practical applications of the method.

Considering the geometry of the blade section with respect to the plane of rotation and hence the direction of free stream airflow, the elemental aerodynamic force acting on N blade elements in the axial direction is given by

$$L\cos\phi + D\sin\phi = \frac{1}{2}\rho W^2 Nc(C_L\cos\phi + C_D\sin\phi) \tag{5.34}$$

Similarly, the total elemental torque acting around the axis of rotation is

$$(L\sin\phi - D\cos\phi)r = \frac{1}{2}\rho W^2 Ncr(C_L\sin\phi - C_D\cos\phi) \tag{5.35}$$

where

$$W = \sqrt{U_\infty^2(1-a)^2 + (\Omega r)^2(1+a')^2} \tag{5.36}$$

$$\sin\phi = \frac{U_\infty}{W}(1-a) \tag{5.37}$$

$$\cos\phi = \frac{\Omega r}{W}(1+a') \tag{5.38}$$

Inspection of (5.34)–(5.38) shows that the blade elemental forces are functions of the axial and tangential induction factors previously determined by momentum theory. The blade element equations can now be solved together with the momentum equations using a numerical iterative procedure. Typically, the iteration is initiated with a random value (or zero) for the axial and tangential induction factors; using this initial set of induction factors, the elemental forces along the blades are calculated – these forces are then equated to the rate of change of momentum and new values for the induction factors are determined, which are then fed back into the blade element calculation. The iterative loop continues until some convergence criteria are met.

5.3.6 Basic corrections to BEM

Despite the relative simplicity of the BEM theory, it has achieved remarkable longevity by providing a reasonably descriptive model of wind turbine rotor aerodynamics with very low computational cost. Nevertheless, the approach does have important limitations that have motivated the development of a range of correction terms to the basic formulation – resulting in 'modified BEM' models. A brief summary of such corrections is provided in the following; for a more comprehensive survey, the reader is referred to [27].

5.3.6.1 Tip loss correction

Although commonly referred to as a 'tip loss correction', this factor is better considered as a technique to model the fact that real rotors have a finite number of blades. This is important because the basic BEM theory assumes that each infinitesimal 'packet' of air that passes through a given radial annulus of the rotor interacts with (i.e., has its momentum affected by) one of the rotor blades, and that all such air 'packets' are equally affected around the annulus. Due to the low solidity of modern wind turbine rotors (typically ~5%), it is clear however that, while the air passing close to a blade will indeed experience a fluid–structure interaction, many other air particles will pass straight through the rotor plane with little or no change to their momentum, given their relative large distance from the nearest passing blade. Based on the solidity of 5%, one might be tempted to estimate that close to 95% of the airflow passes through the rotor plane with little or no interaction with a blade; fortunately, this is not the case. Due to the relatively high tip speed ratio of the rotor (typically around ten), we instead find that the blade tips pass through the swept area an order of magnitude more quickly that the air which is flowing through the rotor plane. This greatly enhances the probability of a given blade element interacting with a given collection of air particles. Nevertheless, basic BEM theory does not include this 'finite blade' effect and so it must be added as a sub-model, the most well known of which is arguably the Prandtl model; see, e.g., [28]. Prandtl approximated the effect by introducing the concept of a tip-loss factor f_{tl}, which can then be used to modify the induction factor terms in the basic BEM equations (5.34)–(5.38). The Prandtl tip-loss factor f_{tl} can be expressed as a function of the normalised blade element radius μ as follows:

$$f_{tl} = \frac{2}{\pi}\cos^{-1}\left[e^{-\frac{N}{2}\left(\frac{1-\mu}{\mu}\right)\sqrt{1+\frac{(\lambda\mu)^2}{(1-a)^2}}}\right] \tag{5.39}$$

where N is the number of blades and λ is again the tip speed ratio. A similar correction, $f_{rl}(\mu)$, can be expressed for the blade root losses, leading to an overall rotor correction factor $f(\mu) = f_{tl}(\mu)f_{rl}(\mu)$.

It is because of this 'finite blade' effect that even the most optimised modern wind turbine rotor designs cannot achieve performance at the theoretical Betz limit which itself can be considered to represent the special case of a rotor with infinite blades. Clearly, a direct implication of the finite blade effect is that the corresponding efficiency loss will be more pronounced in lower tip speed rotor designs for a given number of blades. This is one (certainly not the only) reason why designers are interested in very high tip speed designs for offshore applications where acoustic noise constraints may be relaxed.

5.3.6.2 High-induction factor correction

As illustrated in Figure 5.5, theoretical momentum theory breaks down for high-induction factor values corresponding to highly loaded rotor conditions. An empirical correction can be applied for induction factors > 0.5, e.g., [28]:

$$C_T = C_T^1 - 4\left(\sqrt{C_T^1} - 1\right)(1 - a) \tag{5.40}$$

where C_T^1 is the empirically determined thrust coefficient for an induction factor $a = 1$ and is typically reported in the literature to lie in the range $1.6 \leq C_T^1 \leq 2$.

5.3.6.3 Corrections for skewed wake effects under yawed rotor conditions

Due to the turbulent nature of the wind, the rotor-averaged wind direction (as well as the speed) is continuously changing in time at a rate that cannot be perfectly tracked by the wind turbine's nacelle yaw system. Therefore, it is apparent that wind turbines spend a non-negligible proportion of time operating in misaligned inflow. Under such conditions, even if the wind speed were steady, the rotor blades would still experience an azimuthal- and time-dependent variation in their angle of attack, applied aerodynamic forces and induced local velocities. One of the basic assumptions of standard BEM theory is that the average induction factors calculated for a given rotor annulus is representative of *all* azimuthal positions around the annulus. Clearly, due to the periodic variation in angle of attack for blades operating in misaligned inflow, this assumption begins to look unsafe.

A range of engineering models are proposed in the literature to correct for wind turbine operation in yawed conditions including the effect of skewed wakes; see [27,28]. An early proposal from Glauert models the modified axial induced velocity $u_{a,skewed}$ with the equation:

$$u_{a,\text{skewed}} = u_a\left(1 + \left(\frac{r}{R}\right)K \sin\phi\right) \tag{5.41}$$

where ϕ is the blade azimuthal position, R is the rotor radius, r is the radius of a discrete blade element and K is a function of the rotor yaw angle with again various empirical formulations found in the literature. Sant in [27] describes these variants in more detail, as well as more recent extensions of Glauert's basic formulation calibrated with wind tunnel rotor measurements and higher-fidelity physics-based simulations of rotors in yawed flow. As with all empirical and semi-empirical models, the user must take care that the tuned parameters of such models have not been overfitted to the data and hence lose generality.

5.3.6.4 Corrections for dynamic inflow and dynamic stall

Dynamic (or unsteady) inflow describes the varying inflow states (resultant velocities and angles of attack) that a given blade element experiences as it rotates. Time-dependent variation in the angle of attack causes the aerodynamic coefficients for the aerofoil section to differ from those representatives of the 2D static case. Below the stall region, this dynamic behaviour generally corresponds to a gentle hysteresis still reasonably close to the linear, steady-state behaviour. As the mean operating point moves closer towards stall, however, the lag introduced by stall onset delay and subsequent delay in flow re-attachment can introduce highly pronounced hysteresis effects in the aerodynamic coefficients. These hysteresis loops in turn can result in negative aerodynamic damping of the blade structure. As blades are slender, highly flexible structures, this loss of damping, if not understood and mitigated, can result in catastrophic damage to the turbine.

5.3.6.5 Corrections for 3D stall delay

Early BEM models used 2D aerodynamic coefficients derived from static wind tunnel measurements as input to the blade load calculation and found that the resulting predictions did not match well with measurements from rotating rotors. It was discovered that the complex 3D flow induced by the rotating blades, especially at the blade extremities (tip and root), caused the lift and drag coefficients to vary significantly from those derived in the static scenario. At angles of attack in the linear region of the lift coefficient, rotating and non-rotating behaviours are very similar. As the angle of attack approaches the stall region, however, it is found that the onset of stall is delayed in the 3D (rotating) scenario compared with the static case, and the peak lift is increased. While this complex aerodynamic phenomenon has been acknowledged and subject to extensive research for many decades, it is still the case that a comprehensive physical understanding still eludes today's state of the art. Based on a combination of measurements taken from rotating rotors in wind tunnels and higher-fidelity fluid dynamics modelling, empirical corrections have again been proposed, such as that from Snel [29]:

$$C_{L\,3D} = C_{L\,2D} + 3(c/r)^2 \Delta C_L \tag{5.42}$$

where $C_{L\,2D}$ is the measured static lift coefficient, $C_{L\,3D}$ the corresponding lift coefficient for the rotating case, *c*/*r* the local solidity of the blade element and ΔC_L is the difference between a hypothetical linear extrapolation of the static lift curve beyond the static stall point and the actual static lift curve in the stalled region.

5.3.6.6 Summary of BEM correction terms

Many different engineering (approximate) models covering dynamic inflow, dynamic stall, skewed inflow and non-standard rotor geometries (e.g., highly coned rotors) have been proposed as corrections to the basic BEM model. The fact that these models are typically empirical or semi-empirical means that they are unable to capture all the underlying physics over the entire state space of interest and as such can still display significant shortcomings when validation comparisons are made with either model or full-scale measurements. For this reason, while 'modified-BEM' remains the industry's state-of-the-art aerodynamic model for wind turbine load and performance simulation, research continues to focus on improved, physics-based models.

5.4 Offshore wind turbine control systems

5.4.1 Control system fundamentals

If the rotor blades represent the 'wings' of a modern wind turbine, then the control system of the turbine surely represents its 'brain'. While the control hardware, functional actuators and underlying software algorithms are essentially hidden from view, they nevertheless comprise one of the key sub-systems of the turbine, and are vital in ensuring robust, safe and cost-optimised performance of the entire machine. Control system design and optimisation for a modern offshore wind turbine – the world's largest examples of rotating machinery – is challenging due to the combination of highly stochastic forcing conditions (e.g., wind loading) and the highly dynamic (flexible) structural response of the main turbine components (e.g., blades, drive train, tower and support structure). The rotation of the blades through a turbulent wind field also creates modulated forcing frequencies that – if care is not taken – can align with natural vibration modes of the structure, amplifying loading and damage. Developing a 'good' controller, then, is a difficult engineering task but, when achieved, can be highly effective in optimising power performance and reducing dynamic loading in the structure. It is fair to state that the growth in offshore wind turbine scale observed over the past two decades would simply not have been possible without the accompanying innovation in the underlying control technologies that regulate and optimise the machine behaviour. A good summary of modern wind turbine control methods is given in [30].

The schematic of Figure 5.8 shows the basic feedback nature of the controller. During wind turbine operation, the controller receives input measurement signals (represented by the blue arrow) providing information on a range of turbine observed states (e.g., power output, rotor speed and component internal loading) at discrete time steps (typically on the timescale of milliseconds). The control algorithm contained within the control hardware PLC (programmable logic controller) processes these signals and, depending on the policy reflected within the algorithm design, computes a number of 'demand' signals that are then sent to the appropriate actuators within the wind turbine (red arrow in Figure 5.8). These actuators may

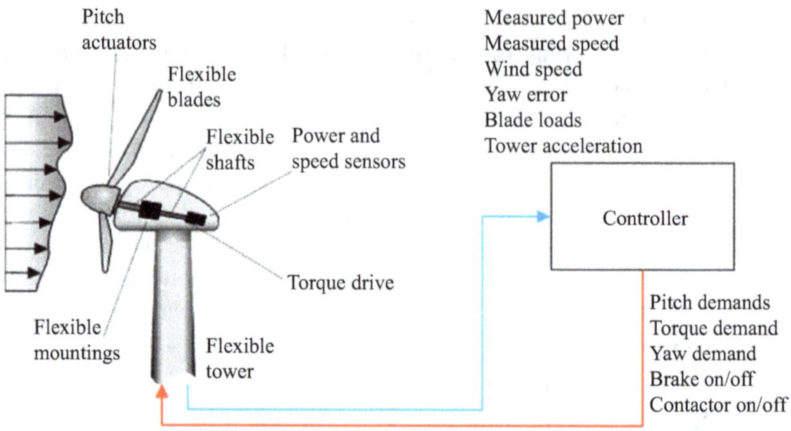

Figure 5.8 Basic wind turbine controller schematic

include pitch drives, yaw drives, the electrical generator and mechanical brake(s). Based on these new 'demand' signals, the actuators adjust their set points accordingly (assuming actuator limits are not violated). This change in actuator set point invokes a change in the physical state response of the wind turbine – e.g., a demanded change in pitch angle will cause the rotor blades to rotate which in turn will affect the aerodynamic loads on the blades. The latter will likely cause the rotor speed to change. Some of these changes in turbine state will be 'observed' by the measurement sensors in the turbine. These are fed back to the controller which in turn will update its actuator demand signals – and so the feedback loop continues. Due to the fluctuating wind conditions experienced in real operation, a wind turbine will be continually adjusting its actuator demands – i.e., fine changes will be happening multiple times every second of operation.

Fundamentally, the wind turbine controller is performing a multi-objective optimisation whereby it simultaneously attempts to regulate power capture and rotor speed while minimising structural fatigue and extreme loads. The controller also attempts to minimise the duty demanded from the system's actuators as far as technically possible. For some designs, the controller may also be tasked with regulating aerodynamic noise emissions. Many of these objectives are in direct or indirect conflict with each other and so require compromise to ensure acceptable overall behaviour. In theory, the controller optimisation goal can be re-framed as a requirement to minimise the LCoE for the turbine. While elegant in theory, this creates the challenge of representing each of the lower-level design objectives in the units of cost of energy (e.g., \$/MWh) which can introduce significant uncertainty terms into the underlying mathematics. Discontinuities and constraints further complicate the optimisation.

The function of a wind turbine controller can also be framed in terms of rejecting system disturbances to maintain stable operation. Particularly as modern offshore turbines get bigger and more flexible, it becomes a vital control task to ensure that the system remains stable (i.e., its dynamic response to random or harmonic forcing does not amplify over time).

Finally, it is appropriate to consider the architecture of a modern wind turbine control system as composed of discrete layers, often with complex interactions occurring between them. Without too much loss of fidelity, we can conceptually identify three distinct layers as follows:

- a 'power-production' closed-loop controller, responsible for optimising the behaviour of the machine under normal operating conditions;
- a 'supervisory' controller, responsible for controlling the transitions between discrete operational states of the turbine (e.g., the transition from 'power production' to 'shut-down', or from 'idling' to 'start-up' states);
- a 'safety system', responsible for ensuring that the turbine does not exceed safe thresholds (e.g., the rotor speed does not exceed its maximum permissible range).

In many modern controller architectures, it can be difficult to establish clear boundaries between the closed-loop and supervisory behaviour as they are typically coupled. The safety system, however, should be kept completely independent to reduce the probability of a systemic fault undermining the entire safety chain.

5.4.2 Steady-state control

Recalling the basic equation for wind turbine power capture (see (5.1)) but now with the power coefficient C_P given explicitly as a function of the tip speed ratio λ and the blade pitch angle β, we have

$$P = \frac{1}{2} C_P(\lambda, \beta) \rho A U^3 \tag{5.43}$$

We can calculate C_P for a range of tip speed ratios and blade pitch angles to understand how the power coefficient changes as a function of these parameters.

In the example of Figure 5.9 (with the pitch angle set at fine), we see that the peak C_P value occurs at a tip speed ratio of eight. Hence below the rated wind speed (the speed at which the controller starts to 'feather' the blades away from fine pitch angle to shed aerodynamic loading and hence maintain power at the 'rated' output level), the controller (notwithstanding any other constraints that may be imposed on it) will typically aim to maximise power capture which will be achieved by controlling the rotor speed to maintain this constant peak C_P value. A key point of note is that the controller is not here seeking to maintain constant *rotor speed* – rather, as the rotor-averaged wind speed fluctuates in time, the controller will alter the speed of the rotor accordingly such that $C_{P-\max}$ is maintained. So, assuming the wind speed is 6 m/s, to achieve optimal power capture, the controller will need to achieve a blade tip speed of 48 m/s. Should the rotor-averaged wind speed increase to, say, 8 m/s, then the rotor tip speed will need to increase to 64 m/s to maintain optimal power performance. The relationship is a simple linear one.

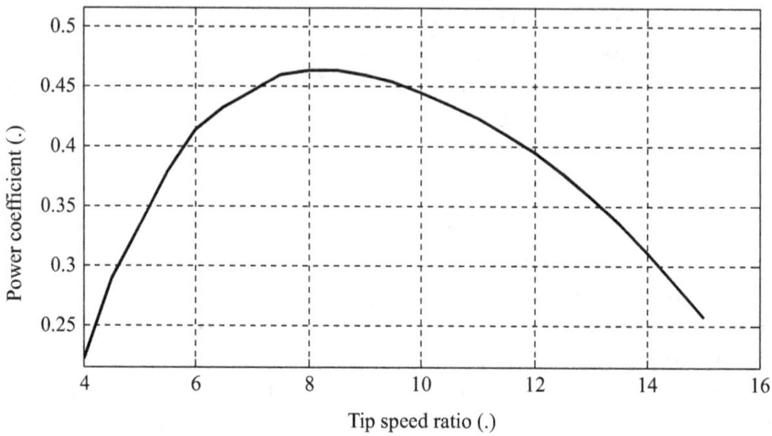

Figure 5.9 Power coefficient as a function of tip speed ratio, constant (fine) pitch angle

In practice, the wind turbine controller cannot demand a change in rotor speed directly, rather it needs to give the corresponding torque demand signal to the generator such that the requisite rotor speed is induced. The relationship between the demanded generator torque and drive train speed is given as follows, where we introduce the concept of a steady-state optimal mode gain for the wind turbine:

$$\lambda = \frac{\omega_r R}{U} = \frac{\omega_g R}{GU}, \quad U = \frac{\omega_g R}{G\lambda} \tag{5.44}$$

$$P = \frac{1}{2}\rho A C_P U^3 = \frac{1}{2}\rho\pi R^2 C_P\left(\frac{\omega_g R}{G\lambda}\right)^3 = \frac{\rho\pi C_P R^5 \omega_g^3}{2\lambda^3 G^3} \tag{5.45}$$

$$Q_d = \frac{P}{\omega_g} = \frac{\rho\pi C_P R^5}{2\lambda^3 G^3}\omega_g^2 = K_{\text{opt}}\omega_g^2 \tag{5.46}$$

where R is the rotor radius, ω_g is the generator speed, G is the gearbox ratio (1 for direct drive machines), Q_d is the demanded generator torque and K_{opt} is the optimal mode gain. With this relationship established, we can derive an optimal steady-state torque–speed operating curve for the generator, as shown in Figure 5.10.

Above the rated wind speed, the generator torque and speed are basically held constant (see Figure 5.11). Control of the rotor speed is now handed over to the blade pitch controller and actuators. In the steady state, as the rotor-averaged wind speed increases above the rated wind speed threshold, the controller demands an increase in blade pitch angle β, which results in shedding of aerodynamic loading and hence power capture at the rotor. Figure 5.12 illustrates this pitch control effect in the steady state. The speed and power are relatively insensitive to changes in pitch angle just above the rated wind speed; hence, the rate of change in pitch angle at this point in the operating envelope is high. As the wind speed increases, the aerodynamics alter such that smaller increases in pitch are needed to maintain

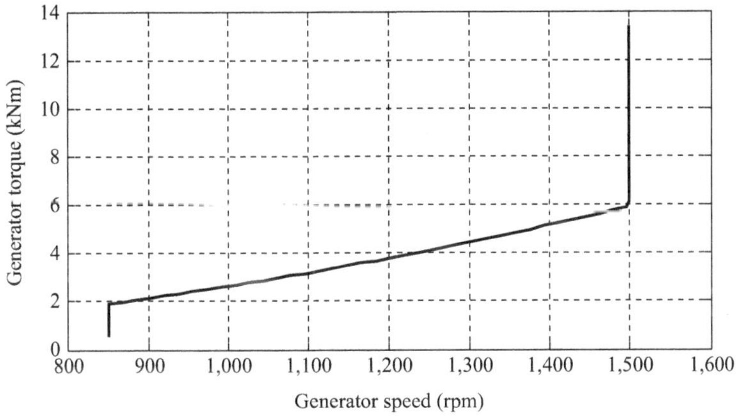

Figure 5.10 Optimal steady-state torque–speed curve for below rated wind speed operation with upper and lower generator speed limits

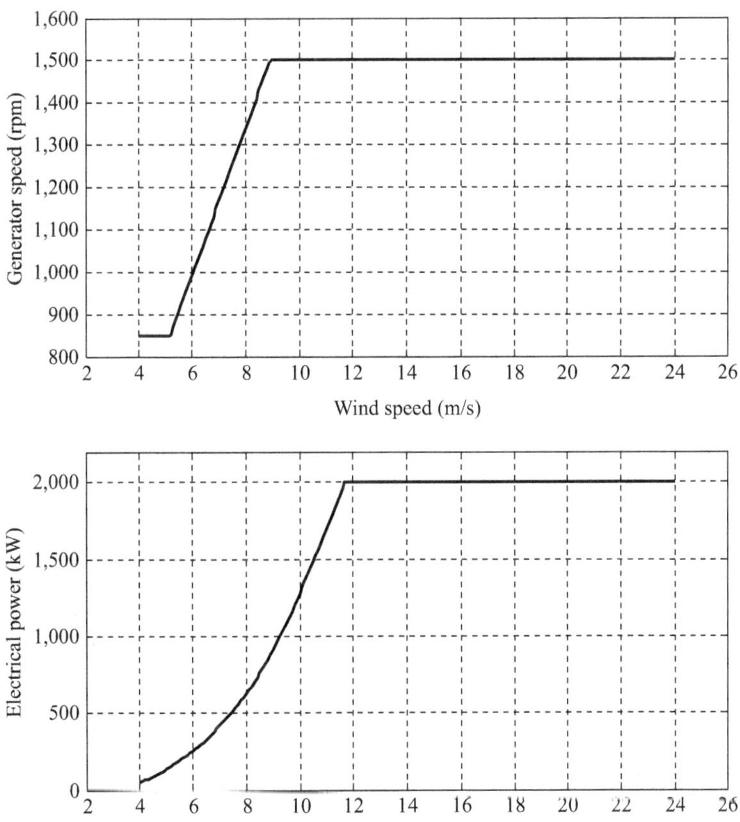

Figure 5.11 Steady-state generator speed (upper) and power output (lower) curves showing optimal power control below rated generator speed, with power limited to 2 MW above rated wind speed to reduce mechanical and electrical loading of the turbine. Note how rated generator speed and rated wind speed are not necessarily coincident

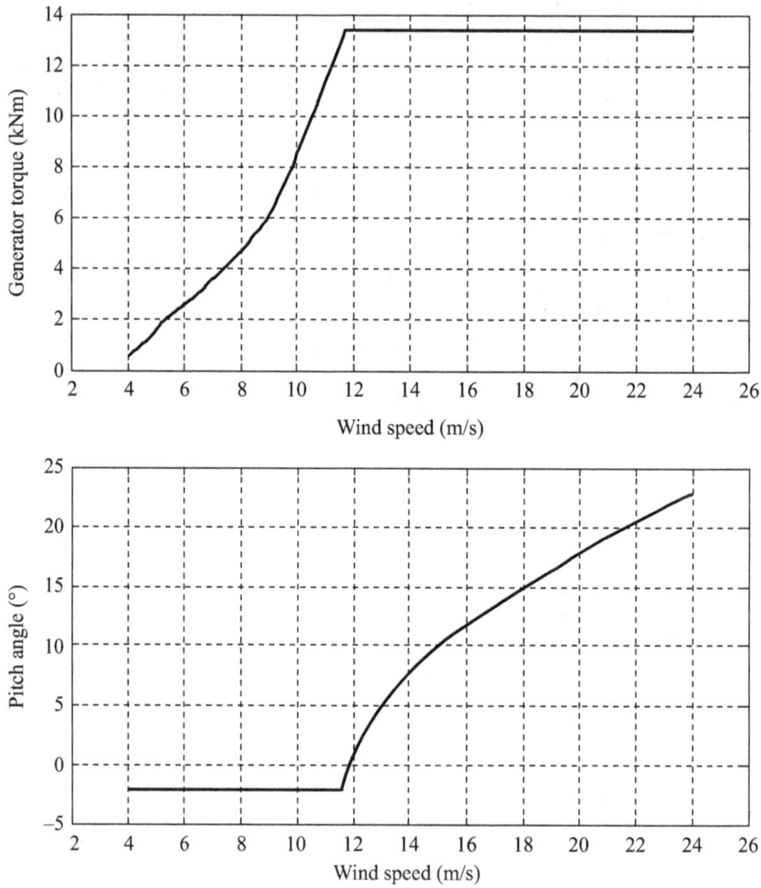

Figure 5.12 Steady-state generator torque (upper) and blade pitch angle (lower)

speed and power regulation. We see in the example of Figure 5.12 that just before the turbine reaches cut-out wind speed the blade pitch angle has feathered to 23°. At cut-out wind speed, the turbine's supervisory controller will take over and demand a blade pitch angle close to 90°, effectively applying aerodynamic braking. From maximum power production at 25 m/s wind speed to shut down of the machine with the rotor gently idling in the same high-wind conditions can take as little as 10–15 s, less if done under emergency conditions.

5.4.3 Dynamic control

The optimal steady-state torque–speed operating curve for a wind turbine between minimum and maximum generator speed is achieved with the simple optimal mode gain law from (5.47). At the extremities of this operating region, additional controller terms are required, as illustrated in Figure 5.13.

Operation in the range between ω_{min} and ω_{max} (green curve) can be governed by a simple look-up table. As the speed approaches these minimum and maximum

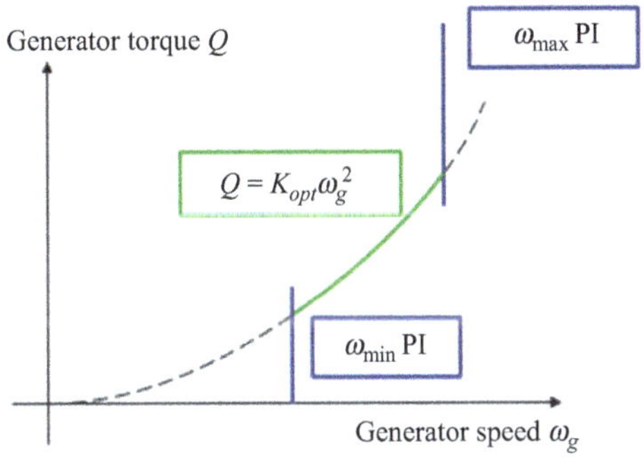

Figure 5.13 Optimal torque–speed curve with discontinuities at minimum (ω_{min}) and maximum (ω_{max}) generator speed thresholds (blue vertical lines)

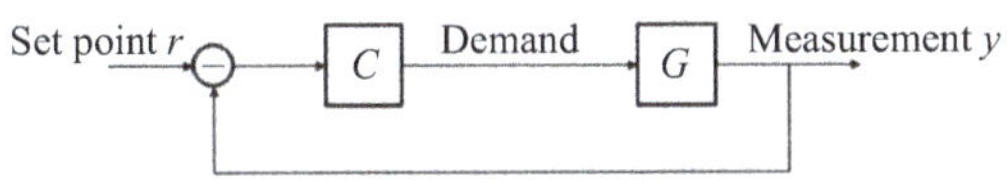

Figure 5.14 Control function C in series with a linear dynamic system G with feedback of measured signal y

limits, however, a PI (proportional, integral) controller is typically used to ensure that, under dynamic disturbance, the limits are not violated. As the wind speed around the rated value increases, the PI controller for the torque–speed envelope has to manage a handover in control to the pitch-speed PI(D) controller (where D represents an additional derivative term) – typically achieved through appropriate switching or blending functions.

Considering the well-known block diagram representing closed-loop control of a linear dynamic system G as depicted in Figure 5.14, we can represent the closed-loop dynamics of the system by

$$\frac{y}{r} = \frac{CG}{1 + CG} \tag{5.47}$$

The engineering challenge is to then design an appropriate control policy C for the dynamic wind turbine system G such that the error (r-y) is minimised whilst other constraints (e.g., loading, actuator duty) are respected. In practice, the basic linear PI(D) architecture ($C = K_P + \frac{K_I}{s} + K_D s$, where K_P = proportional gain, K_I = integral gain and K_D = derivative gain and s is the Laplace variable) is modified and extended with additional linear (e.g., notch and bandpass filters) and non-linear (e.g., gain scheduling) terms in order to provide suitable system response across the full range of operating points.

Assessment of a given control policy C is achieved through a range of techniques. In the frequency domain, linear system behaviour (e.g., stability) can be assessed using Bode diagrams. In the time domain a linearised representation of the actual non-linear system can be used to estimate how the system is likely to respond

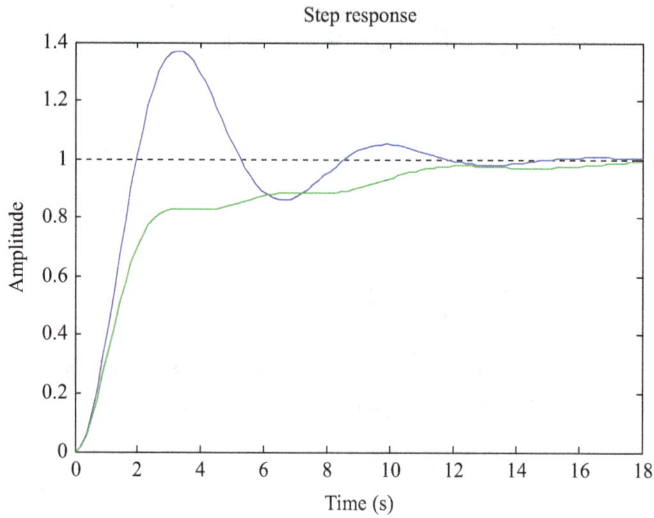

Figure 5.15 Time domain response of a linearised system to a step change in wind speed. The blue line represents an open loop response (no active control), and green line a closed-loop (PI) response which removes oscillations and peak overshoot

at given operating points to step changes in external (e.g., wind speed) or internal (e.g., pitch angle) states (see Figure 5.15). Finally, the controller C can be implemented within a non-linear time-domain simulation of the complete wind turbine system where the power performance, stability, loading and actuator duty can all be modelled. Typically, the controller design is an iterative process where the design team will repeat these assessment stages as they refine the design [30].

5.4.4 Advanced control for load suppression

The wind turbine dynamic controller described in Section 5.4.3 is an oversimplified version of the actual algorithms used in today's commercial turbines, including those offshore. Many of the control techniques applied in modern machines are closely protected intellectual property, given the competitive advantage that an optimised control system can confer. Nevertheless, we can identify some common themes in the area of advanced control design for wind turbines. Underpinning them all is the motivation to reduce structural loading (to make the structural design lighter and cheaper) without adversely affecting the power curve. They are also typically characterised by the use of additional measurement data as input to the controller in order to better infer the present and future operating state of the machine.

5.4.4.1 Load measurements as input for advanced control

Control systems in older wind turbines, in trying to minimise structural loading of key components, typically had to infer the loading from other proxy measurements such as nacelle acceleration and rotor speed variance. Today, many modern machines take advantage of direct measurements of loading in key components. A good example is the use of blade load sensor data to control the three blades independently rather than collectively – so-called individual pitch control (IPC) [31]. The principal motivation behind IPC is the fact that the wind speed varies considerably across the rotor plane – this is particularly true for large offshore machines. This variation in spatial wind speed across the rotor plane, in combination with rotation of the blades through it, results in a phenomenon called 'rotational sampling' whereby the turbulent energy in the ambient flow is sampled, or experienced, by the wind turbine at frequencies modulated by both the blade rotation frequency (1P, where P refers to the rotor frequency) and the blade-tower passing frequency (3P for a three-bladed machine). This rotational sampling effect means that fatigue damage on the structure due to turbulence and other spatial variance factors (like wind shear) is potentially greater than it would be for a static structure. IPC can alleviate some of this damage by reducing the magnitude of loading cycles experienced by each blade (and the hub) per rotation.

IPC can be applied with modulation of the blade pitch at frequencies of 1P and 2P, in order to target load reduction in different structural components (for which the loading is typically dominated by the response at these corresponding frequencies). Figure 5.16 provides an example of loading spectra for three key wind turbine components: the 'out of plane (OOP)' blade root bending moment $\left(Blade\ M_y\right)$ and the hub-centre asymmetric bending moment in both stationary and rotating reference frames.

1P IPC can be used to reduce the spectral load peaks observed at 1P in both the blade OOP bending moment (black) and the hub-asymmetric bending moment (green) in the rotating reference frame. It also reduces the peak hub moment in the stationary reference frame (red) observed at 0P. 2P IPC on the other hand can be

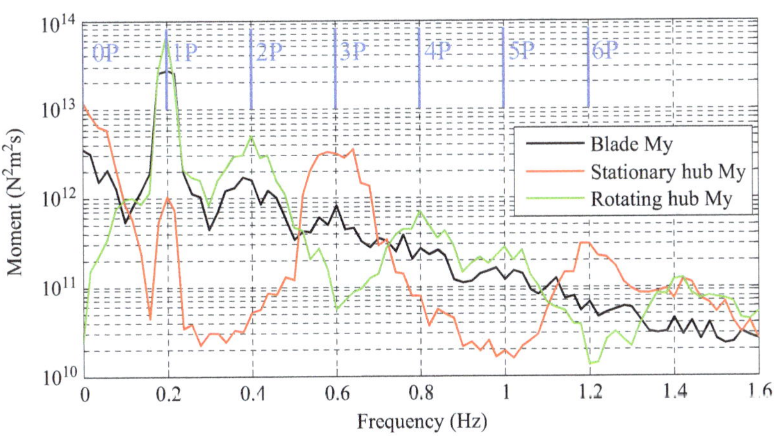

Figure 5.16 Wind turbine loading spectra for three key components (OOP blade bending moment (black) and hub-centre asymmetric bending moment in stationary (red) and rotating (green) reference frames)

applied to mitigate the load peaks at 2P frequency in the blade and rotating hub, as well as the peak in stationary hub load observed at 3P. Similarly, 2P IPC has been shown to be effective in reducing the torsional bending moment transferred into the tower and support structure. For offshore wind jacket-type support structures which are particularly sensitive to torsional loading, application of 2P IPC may enable significant weight and cost savings in the structure [14].

While IPC has been shown to be highly effective in reducing structural loading without incurring a penalty in terms of overall energy capture, it nevertheless presents some challenges for the offshore wind turbine designer. First, the additional blade pitching activity involved inevitably leads to greater pitch system duty requirements (i.e., pitch travel may increase by a factor of up to five compared to a collective pitch control policy). Second, the method requires additional load sensors (either in the blade themselves or in the tower top) to be installed and maintained throughout the turbine operating life. Particularly offshore, where maintenance is expensive and reliability of key systems becomes of paramount importance, designers and operators alike may be nervous about the robustness of control systems that require continuous data streams from load sensors in the machine. More sophisticated designs can manage this robustness challenge by running in 'safe' modes during periods when (for whatever reason) the load sensor signals are not available, ensuring that a complete shut down of the machine is avoided.

5.4.4.2 Wind measurements as input for advanced control

Although offshore wind turbines have anemometers installed on top of their nacelles to measure wind speed, they typically do not use this measurement of wind speed as input to the closed-loop power production control algorithm – instead the signal tends to be used to support supervisory (turbine state-transition) control actions. Why, if the turbine has a direct signal of wind speed, does it not use it second by second to optimise power capture? The answer is that the best wind speed measurement device installed on the wind turbine is typically not the anemometer (which provides only a point wind speed measurement located where the wind flow is highly disturbed behind the turbine blades) but rather the rotor itself (which is sampling the entire swept area of the flow). Hence the steady-state control law (see (5.46)) for optimal power production does not require any direct (anemometer) measurement of wind speed, only the speed of the drive train. That being the case in the steady state, it is nevertheless clear that in a dynamic context (i.e., when the turbulent wind speed is fluctuating in time) the state of the wind field 'observed' by the rotor is that which is directly incident on the rotor. By the time the turbine has 'observed' the incident state of the wind field and passed this 'observation' to the control algorithm, and control actuators have responded in return, some finite time has passed within which the state of the wind at the rotor plane has likely changed. The question is therefore posed – could a controller perform even better if it could observe the turbulent wind speeds *upstream* of the rotor plane before they became incident on the rotor itself. This is the motivation behind so-called 'LiDAR-assisted control' (LAC), where LiDAR itself is an acronym of '*li*ght *d*etection *a*nd *r*anging'.

LiDAR works by emitting a beam of light (either pulsed or continuous) which reflects and scatters as it interacts with aerosols (particles) located in the air along the beam. The frequency of the backscattered signal is measured and compared with the frequency of the originally emitted beam from which a Doppler frequency shift can be computed. This shift can in turn be used to quantify the wind speed at points along the line of the beam. By using multiple beams, or by scanning the beam in multiple directions, information on upstream wind speeds across the entire rotor plane can be inferred (see Figure 5.17).

This observed state of upstream wind speed assists the power production closed-loop controller by enabling it to *anticipate* disturbances – otherwise known as feed-forward control [31].

Numerical simulations of wind turbines with LAC need to consider the fact that the common 'frozen' turbulence hypothesis assumed in many spectral models of turbulence does not create an unrealistically favourable scenario for the controller. Concretely, frozen turbulence assumes that turbulent fluctuations propagate downstream at a constant speed corresponding to the mean speed of the entire flow field. Applying this approach when simulating LAC would mean that the wind speeds measured upstream by the LiDAR would be exactly those that are then incident at the rotor plane some finite time later – in other words the LiDAR would have complete advance knowledge of the flow. In real flows, turbulence is not 'frozen' in this manner but rather evolves as it propagates. Methods have therefore been developed to account for this evolution so that LAC can be simulated and designed under realistic conditions [32].

LAC has been demonstrated, both in numerical simulation and in full-scale testing, to facilitate improved rotor speed control and structural load reductions in key components without power capture penalties and with less actuator activity (blade pitching). A published study by DNV GL demonstrated that LAC applied in the design of a 7-MW offshore wind turbine could reduce fatigue loading in key components by up to 11%, with extreme load reduction potential of a similar order of magnitude [33]. Full-scale field tests of LAC are reported in [34] whereby measurements show significant improvements in rotor speed control and pitch activity.

While LAC makes intuitive good sense and has been demonstrated to achieve impressive results – both in simulation and in full-scale studies – it nevertheless must

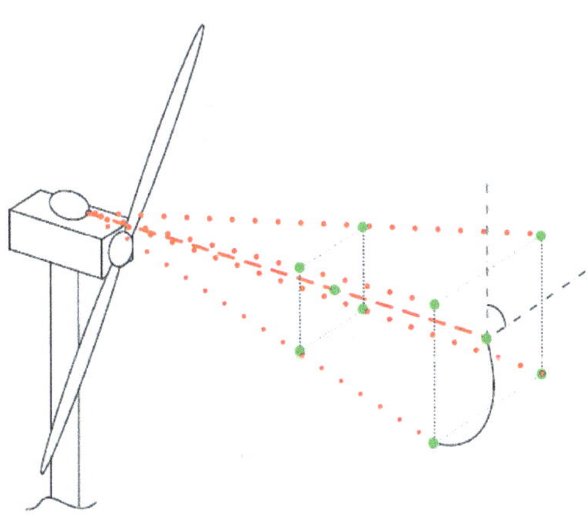

Figure 5.17 Wind turbine with nacelle-mounted LiDAR

overcome the robustness challenge if it is to become a standard component in future offshore wind turbine control systems. Robustness concerns relate to (i) the inclusion of an additional complex system that requires maintenance and (ii) the fact that LiDAR requires the presence of aerosols in the air to work, and raises the question of what to do when atmospheric conditions are not favourable for the LiDAR to function. In this case, the turbine may need to de-rate or even shut down entirely, both of which are unpalatable from an energy yield and project finance viewpoint.

5.4.4.3 Synthetic damping of wave-induced structural excitation

Offshore turbines are subject to additional excitation due to wave loading. It turns out that the wind turbine rotor and closed-loop control action can also be effective in mitigating this excitation [15]. The frequency spectrum of the sea-state is broadband with a non-negligible proportion of the energy typically located at frequencies close to the first natural frequency of the structure. While passive damping sources exist in the system (e.g., internal material damping of the structural steel, soil-pile interaction damping effects and hydrodynamic damping of the substructure), the magnitude of these damping sources tends to be small and, on their own, insufficient to prevent unwanted dynamic response in the presence of wave loading. The wind turbine rotor aerodynamics, however, can introduce additional damping of the system at an order of magnitude higher than these other sources. This aerodynamic damping exists in passive form but can also take on an active nature via nacelle-acceleration measurement feedback in the closed-loop controller. Simulation and measurements of offshore wind turbines subject to combined wind and wave loading demonstrate that the aerodynamic damping effect can contribute up to 4% of critical damping in the system.

Active damping of the system under wave loading can be achieved by taking advantage of the rotor aerodynamics in combination with nacelle-acceleration feedback in the closed-loop controller. This is achieved by developing an additional pitch rate demand term in the controller based on the acceleration signal. As this additional term affects the basic pitch-speed control loop, care needs to be taken in the overall controller design to ensure the desired rotor speed behaviour is not compromised [15].

Clearly, the application of fore–aft aerodynamic damping from the rotor to mitigate wave-induced dynamics depends on the wind and wave directions being sufficiently aligned. In cases where there is pronounced misalignment (or indeed when the rotor is idling), the wave-induced dynamics can be very difficult to damp actively – though techniques using, e.g., generator torque demand have been explored.

5.5 The future of offshore wind turbine technology

The rise of offshore wind power over the past two decades would appear to support the thesis that 'bigger is better'. The fact that the exponential increase in turbine size (both rotor diameter and power rating) has almost completely aligned with dramatic cost reductions over this period has created a direct causal link between the two in the minds of experts and non-experts alike. To a large degree, the inference is correct; the offshore wind industry's ability to reduce cost of energy *is* in large part due to the emergence of today's super-sized machines with rotor diameters over 160 m. However, the fundamental connection between turbine size and cost of energy is often missed. To understand this, we return to the basic physical relationships that govern the system explored in Section 5.1. Recall that rotor power capture and rotor mass (and hence cost) scale as square and cubic functions of the rotor diameter, respectively, i.e.,

$$P \propto D^2 \text{ and } mass \propto D^3$$

where D is the rotor diameter.

Hence, while larger rotors will indeed capture more power, this is in rising tension with the even greater increase in structural mass and cost that the larger rotor entails. At some point, the additional power capture is no longer enough to offset the increased mass and cost, and so the size of machine is constrained. This limit would have been reached long before the size of machines we see today if it were not for two principal reasons. First, significant technical innovation in areas like materials, controls and structural design have allowed wind turbine designers to, in effect, 'fight back' against the onerous $mass \propto D^3$ relationship. In other words, if we introduce a constant into the expression such that $Mass = Constant \times D^3$ then we can say that technology innovation over the past two decades has reduced the size of this constant term within the governing relationship. Second, the relative dominance of the balance of plant costs in an offshore wind farm (i.e., the offshore support structures, electrical infrastructure, etc.) means that it tends to make sense to maximise the power capture per unit foundation located in the farm.

The following question remains: will offshore wind turbines continue to grow in size and power capacity as we embark on the next two decades of innovation up to 2040? Certainly today, the industry is already discussing offshore wind turbine designs with rotor diameters in excess of 200 m. No final consolidation of turbine scale, it would seem, will be established in the near future. That being the case, it is also clear from detailed cost of energy modelling of the entire offshore wind farm system that the design space becomes rather flat around the 200-m rotor diameter point – in other words, going above this size does not seem to generate much additional benefit. However, such 'system' models of cost of offshore wind energy are inherently complex, uncertain and may not anticipate well future step-changes in technology. As such, the most certainty we can probably have regarding the question of optimal wind turbine size offshore is that it will continue to encourage ongoing debate and technical innovation.

Finally, it is interesting to note that some voices in the industry, when discussing optimal turbine scale offshore, argue that in future turbines will not grow bigger, but smaller. What could possibly justify such a trend-defying perspective? In fact, as we have already established, the fundamental $P \propto D^2$ and $mass \propto D^3$ relationships provide this argument with a sound theoretical basis – smaller rotors are fundamentally more cost-effective. The practical challenge then becomes, can the size of an offshore turbine rotor be kept relatively small *and* the balance of plant costs kept low? The 'multi-rotor' concept emerges as a possible solution to meet these twin design constraints, as (at least in theory) it allows multiple small, cost-efficient rotors to share a common support structure and electrical infrastructure.

Multi-rotors, then, have theoretical merit, but what of their practical viability? Here, the picture is nuanced and – given that very few multi-rotor prototypes exist today (for a recent example see Figure 5.18) – largely hypothetical. Nevertheless, it is still instructive to consider the practical challenges that such a design might face. Take, e.g., system reliability. A single multi-rotor turbine with n rotors will have around n-times as many components compared with a single-rotor turbine of the same power rating; hence, the probability of one of those components failing at any given time increases. However, one may counter that the system now has n-times greater redundancy so is less sensitive to sub-component failure and downtime. Furthermore, the factor n growth in the number of rotors means that the industry has the potential to stabilise and standardise the basic turbine design, reducing unit cost and increasing unit reliability through a process of volume-driven industrialisation [6].

In conclusion, multi-rotor wind turbines – while not a new idea – are attracting renewed interest for the reasons sketched above with some experts considering it probable that the industry will eventually go 'back to the future' – gathering together the major technology developments achieved over the past two decades and applying them, no longer in pursuit of ever-increasing turbine size, but rather to optimise the scale of machine characteristic of where the offshore wind industry's journey of innovation originally began.

Figure 5.18 Multi-rotor prototype installed by Vestas Wind Systems A/S at the DTU test site in Roskilde, Denmark (reproduced with permission of Vestas Wind Systems A/S)

Acknowledgements

The majority of this chapter was written while the author was working for DNV GL and the support and advice from colleagues there during its preparation is gratefully acknowledged.

References

[1] www.telegraph.co.uk/news/earth/energy/10154424/Wind-farms-couldnt-pull-the-skin-off-a-rice-pudding-says-Boris-Johnson.html (accessed 5th October 2018).

[2] UK Government Department for Business, Energy and Industrial Strategy (BEIS) report. *Digest of UK Energy Statistics (DUKES): Renewable sources of energy, Chapter 6: Statistics on energy from renewable sources*. July 2018.

[3] Wind Europe report. *Offshore Wind in Europe – Key trends and statistics 2017*. February 2018.

[4] Global Wind Energy Council report. *Global Wind Report 2017*. April 2018.

[5] Financial Times news article. *Subsidy-free renewable projects on 'cusp of breakthrough'*. March 2018.

[6] Jamieson P. *Innovation in Wind Turbine Design*. Chichester: Wiley, 2011, p. 229.

[7] International Renewable Energy Agency (IRENA) report. *Renewable Power Generation Costs in 2017*. 2018.

[8] DNV GL report. *Energy Transition Outlook 2018*. September 2018.

[9] Landberg L. *Meteorology for Wind Energy – An Introduction*. Chichester: Wiley, 2015

[10] Johnson D., and Erhardt R. Projected impacts of climate change on wind energy density in the United States. *Renewable Energy*. 2016;**85**:66–73.

[11] Van der Hoven I. Power spectrum of horizontal wind speed in the frequency range from 0.0007 to 900 cycles per hour. *J. Met.* 1957;**14**:160–164.

[12] UK Crown Estate Commissioners report (prepared by DNV GL). *Study of UK Offshore Wind Variability*. August 2016. www.thecrownestate.co.uk/media/1772/uk-offshore-wind-variability.pdf (accessed 17th November 2018).

[13] Monahan A. The probability distribution of sea surface wind speeds. Part I: Theory and seawinds observations. *Journal of Climate*. 2006;**19**:497–520.

[14] DNV GL report. Project FORCE – *Offshore wind cost reduction through integrated design*. July 2014.

[15] Fischer T. Study on control concepts suitable for mitigation of loads from misaligned wind and waves on offshore wind turbines supported on monopiles. *Journal of Wind Engineering*. 2011;**35**:561–573.

[16] International Electrotechnical Committee (IEC) design standard. Wind turbine generator systems – Part 1: Safety requirements. *International standard 61400-1 edition 3*. 2005.

[17] International Electrotechnical Committee (IEC) design standard. Offshore wind turbine generator systems – Part 1: Safety requirements. *International standard 61400-3 edition 1.* 2009
[18] Larsen T., Aagaard Madsen H., Larsen G., and Hansen K. Validation of the dynamic wake meander model for loads and power production in the Egmond aan Zee wind farm. *Wind Energy.* 2013;**16**(4):605–624.
[19] Quarton D., and Ainslie J. Turbulence in wind turbine wakes. *Proceedings of the European Wind Energy Conference.* Bristol, UK. 1989.
[20] Vermeulen P. An experimental analysis of wind turbine wakes. *Third International Symposium on Wind Energy Systems.* BHRA. 1980.
[21] Frandsen S., and Thogersen M. Integrated fatigue loading for wind turbines in wind farms by combining ambient turbulence and wakes. *Journal of Wind Engineering.* 1999;**23**(6):327–339.
[22] Larsen G.C., Madsen H.A., Thomsen K., and Larsen T.J. Wake meandering: A pragmatic Approach. *Wind Energy.* 2008;**11**:377–395.
[23] Han T., McCann G., Mucke T, and Freudenreich K. How can a wind turbine survive in tropical cyclone?. *Renewable Energy.* 2014;**70**:3–10.
[24] Hau E. *Wind Turbines: Fundamentals, Technologies, Application, Economics.* Berlin Heidelberg: Springer, 2006.
[25] https://www.senvion.com/global/en/newsroom/press-releases/detail/repower-5m-successful-offshore-premiere/ (accessed 28 November 2018).
[26] Glauert H. The analysis of experimental results in the windmill brake and vortex ring states of an airscrew. *ARCR R&M No. 1026.* 1926.
[27] Sant T. *Improving BEM-based aerodynamic models in wind turbine design codes.* PhD thesis. Delft University Wind Energy Research Institute, 2007.
[28] Burton T., Sharpe D., Jenkins N., and Bossanyi E. *Wind Energy Handbook.* Chichester: Wiley. 2011.
[29] Snel H. Sectional prediction of three-dimensional effects for stalled flow on rotating blades and comparison with measurements. *Proceedings of the EWEC Conference*, 1993.
[30] Bossanyi E. The design of closed-loop controllers for wind turbines. *Wind Energy.* 2000;**3**:149–163.
[31] Bossanyi E. Developments in individual blade pitch control. *Proc. EWEA The Science of Making Torque from Wind Conference*, Delft, the Netherlands, 2004.
[32] Bossanyi E.A. Un-freezing the turbulence: Improved wind field modelling for investigating Lidar-assisted wind turbine control. *Proceedings of the European Wind Energy Copenhagen*, Denmark, 2012.
[33] Dobbin J., Cordle A., Blonk L., *et al.* Fully Integrated Design: Lifetime Cost of Energy Reduction for Offshore Wind. Proceedings of the Twenty-fourth, International Ocean and Polar Engineering Conference, 2014, vol. 3, pp. 415–423.
[34] Kumar A.A., Bossanyi E.A., Scholbrock A.K., Fleming P.A., Boquet M., and Krishnamurthy R. Field testing of LIDAR assisted feedforward control algorithms for improved speed control and fatigue load reduction on a 600kw wind turbine. *EWEA Paris*, France, November 2015.

Chapter 6
Marine solar energy

Luciano Mule' Stagno[a]

In many countries, photovoltaic (PV) installations have become ubiquitous. One can see systems on rooftops, industrial buildings, on facades and in solar farms. In some underdeveloped regions, off-grid systems are providing the only source of electricity for mobile charging, water pumping and other essential uses. By the end of 2017, the installed capacity worldwide was around 400 GWp, and is projected to hit 1 TWp by 2023 [1]. Several countries have exceeded the 100 Wp/capita mark. In southern, sunny climates, this equates to over 160 kWh/year, which amounts to a significant portion of the energy consumed [2,3]. It is therefore easy to forget how recently this technology has developed.

The photovoltaic effect was discovered by Edmond Becquerel in 1839. By the late 1800s the first rudimentary solar devices were being developed by Charles Fritts (1883) using selenium. Key developments in crystal growth, particularly the development of a crystal growth process by Jan Czochralski in 1918, which now bears his name, enabled the growth of large crystals of silicon, germanium and other materials which made the subsequent developments possible. However it took another 36 years for the first practical, high-power solar cell based on silicon to be developed. This had an efficiency of only 6% and was presented by Chapin, Pearson and Fuller, all of Bell Labs, in 1954. By 1958 the technology, though prohibitively expensive for most uses, had found several uses including in space technologies, on the Vanguard 1 satellite.

The first real drive for mass commercialization of solar cells happened during the 1970s oil crisis, primarily in the United States. Several research efforts by private and public entities lead to costs falling from around \$100/W to \$20–40/W, still astronomical by today's standards but a huge improvement nonetheless. In 1982 the first solar power plant was created by Arco Solar in California. It produced 1 MWh annually. The real growth in photovoltaic system installations started happening around the year 2000 but especially after 2004. This is primarily attributed to the German Renewable Energy Sources Act enacted in 2000, which was given even more momentum by a 2004 amendment [4]. This law ensured an advantageous feed-in tariff for renewable energy sources. Its success was quickly

[a]Institute for Sustainable Energy, University of Malta, Marsaxlokk, Malta

replicated in other countries around Europe and the world resulting in the phenomenal growth in photovoltaics we have seen since.

This phenomenal growth especially in the early years had some unintended consequences. The first was a shortage of polysilicon, the raw material used to make solar cells. Until 2004–2005, the majority of polysilicon was used in the semiconductor industry. However, the meteoric growth of the solar industry resulted in the majority of polysilicon being consumed by 2005 to be used in the solar industry and an ensuing major shortage in the raw material. The shortage was so severe that the price of polysilicon jumped from $25–35/kg to over $400/kg by 2008 [5]. This resulted in an increase in the price of solar cells in the following 2 years but also in a rush to build new polysilicon capacity by both the existing and new suppliers. This new capacity started coming online around 2008 and soon resulted in an overcapacity which drove down prices.

Coincidentally, China made a big push into the solar industry at about the same time. This perfect storm, of oversupply of raw materials and new, higher volume, low-cost manufacturing, resulted in a phenomenal drop in the price of PV modules. This price decrease has resulted in a cost for electricity obtained from solar farms to fall, in some cases, below that of fossil fuels – and the situation is only getting better as the price of solar continues to decline and that for power from fossil fuels stays the same or even rises [6]. This has recently resulted in record low prices for electricity from solar farms, below 2c/kWh, in sunny places with plenty of land like the Middle East and South America and a resulting flood of investments [7].

6.1 Solar cell technology

6.1.1 Semiconductor properties and growth

A semiconductor material, as its name implies, is not a good conductor of electricity like metals. Neither is it an insulator. In its intrinsic (natural) form, there are no excess holes or electrons. However, a small amount of doping can be used to add these excess holes or electrons. Take the case of silicon, the most widely used semiconductor. Silicon, like all group four atoms, has four electrons in its outer shell. So introducing a group three atom like boron, which has three electrons in its outer shell will result in a deficit of one electron, or a hole, for every boron atom added. On the other hand, introducing a group five atom such as phosphorus results in an excess electron. For these reasons, these two types of semiconductors are known as P- (positive) and N- (negative) type, respectively. The two situations are shown schematically in Figure 6.1.

The vast majority of solar cells in use today are made of silicon – single crystal or multi-crystalline. Single crystal silicon is typically grown by a process called the Czochralski process. Before large single crystals could be grown however, another ingenious invention was needed – the neck, or the growth of a thin cylinder of silicon before the main body of the crystals are grown (Figure 6.2). This development was made by Dash in 1958 and opened the door to the growth of large diameter single

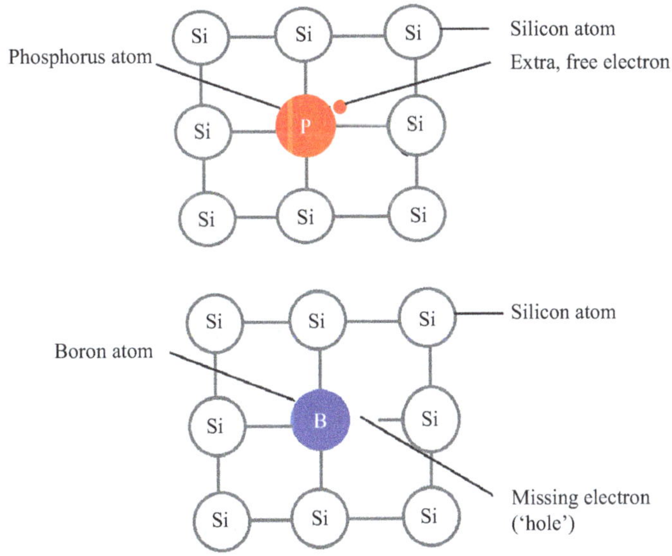

Figure 6.1 A two-dimensional representation of phosphorus and boron doping in a silicon lattice

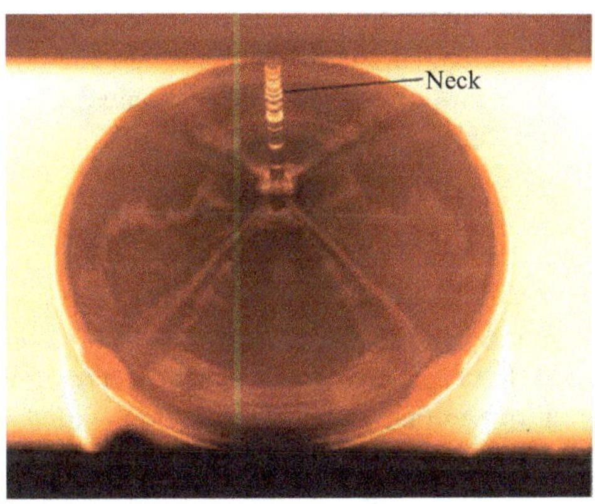

*Figure 6.2 Top view of a silicon crystal during growth showing the neck. The four lines on the taper are the crystal facets. Below the growing crystal is the bright molten silicon. (Courtesy MEMC)**

crystal silicon, thus enabling the development of not only high-efficiency solar cells but also the whole of modern electronics [8].

The starting material for both solar and semiconductor applications is ultra-pure polysilicon. This is made from silica (SiO_2) which is essentially sand.

*Note that the company MEMC no longer exists (bought out)

Given that what is ultimately required is ultra-pure silica, this tends to be mined from locations offering high purity rather than collected from exposed areas such as beaches. The first step is to reduce the silica (SiO_2) to silicon (Si). This is done by mixing with carbon (C) and heating to 1,500–2,000°C in an arc furnace which results in silicon and carbon monoxide.

$$SiO_2 + 2C \rightarrow Si + 2CO$$

The silicon produced in this way is called metallurgical-grade silicon (MG-Si) and has a purity of 98%–99%, with a variety of metal impurities. This is used as is in the chemical industry to produce silicones, silica, silanes and as a component of alloys. Yet this is still not nearly pure enough to make solar cells. To produce photovoltaic-grade silicon, further purification is required. The MG-Si is crushed into a powder form and mixed with anhydrous hydrochloric acid (HCl) to form an intermediate compound that can be further purified. A popular choice in the industry is trichlorosilane ($SiHCl_3$) because it requires relatively low temperatures and it is a liquid at room temperature and can therefore be distilled. The MG-Si and HCl are heated at around 300°C in a fluidized bed reactor which results in trichlorosilane and hydrogen.

$$Si + 3HCl \rightarrow SiHCl_3 + H_2$$

This material is then distilled to remove impurities.

The next step is the actual deposition of photovoltaic-grade silicon. The trichlorosilane is introduced together with hydrogen into the deposition reactor. In this reactor slim rods of (ultra-pure) silicon are thickened when the trichlorosilane decomposes to silicon on their surface.

$$SiHCl_3 + H_2 \rightarrow Si + 3HCl$$

The slim rods are thickened to 150–200 mm by heating for 2–300 h at a temperature of around 1,100°C. This step is one of the most energy intensive steps in solar-cell production and this process is known as the Siemens process and still dominates the industry. However, another process that has gained ground with the growth of the solar industry in the last 10 years or so is one that uses silane and a fluidized bed reactor. The silane and hydrogen are similarly decomposed but into 100–1,500 μm silicon balls known as granular silicon. While cheaper to run due to the lower temperatures used and the higher yield, it is more difficult to control and has some undesirable properties (like captured hydrogen pockets in the granules) so while popular with some producers, it has not been universally adopted.

For the highest efficiency solar cells, mono-crystalline silicon is required. The first step is the melting of this pure polysilicon in a quartz crucible. This is done in a special furnace known as a crystal puller (Figure 6.3). The lower part of the crystal puller contains a graphite heater inside which sits the quartz crucible (usually also held by a graphite susceptor, or outer shell, since quartz becomes soft at the high temperatures of operation). The polysilicon is carefully placed inside the crucible with a small amount of dopant (normally boron or phosphorus containing compounds for P- and N-type doping, respectively) and then the meltdown happens

Figure 6.3 *Typical crystal puller, or growth furnace. The bottom part is the furnace and the column is where the crystal grows. Below the furnace are other components such as pumps and motors (courtesy Linton Crystal Technologies)*

over the course of several hours in an inert environment (usually argon). Once the silicon has completely melted (melting point about 1,420°C), a single crystal thin rod of silicon known as the seed is lowered into the melt.

The seed is carefully pulled out of the melt at a faster speed initially to create the neck described above, and then at a slow speed (of 1–2 mm/min) while rotating in a direction opposite to that which the crucible is rotating. Careful control of the pull-rate and thermal gradients are maintained, and following a tapering to the desired diameter (usually 160–185 mm) steady conditions are maintained during the growth of the main part – the body. When little melt is left, the pull-rate is once again increased resulting in another tapered cone (the tail). As soon as the tail leaves the melt, thermal shock will result in dislocations – but luckily these will be isolated in the tail section which is not used to make solar cells (Figure 6.4).

Once the crystal is complete, the puller is turned off and the crystal carefully cooled. It is then removed from the puller and the tail and taper sections are cut off and recycled. The remaining body is ground to a perfect cylinder or squared off and a wire saw is then used to cut out the wafers. Each wafer is cut to about 200 μm of thickness. The wire itself is a minimum of 100-μm thick, so this kerf loss is the major yield loss in the whole process. The wire is one continuous spool running over wafer guides about 300 μm apart. Each cut thus produces a wafer and a whole rod; 1–2-m long can be cut at once [8].

In the case of multi-crystalline wafers, the process is somewhat simpler. The material is typically produced by casting whereby a very large (1 m × 1 m × 350 mm)

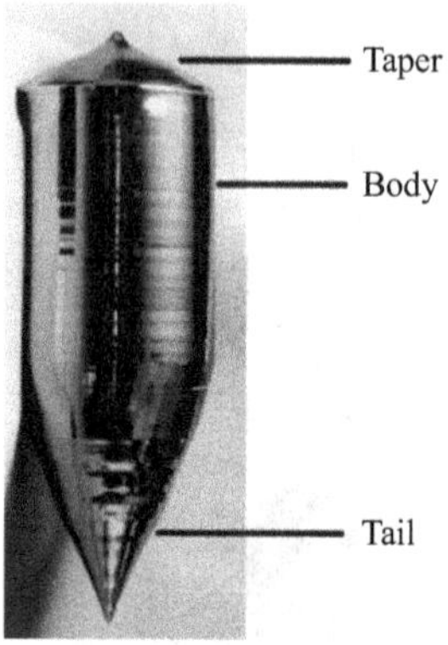

Figure 6.4 Silicon crystal showing the taper, body and tail. (Courtesy MEMC)

crucible is filled with polycrystalline silicon and the material melted and then carefully cooled. The careful cooling is designed to produce the largest crystals in the multi-crystalline cluster.

The primary reason why multi-crystalline silicon is less efficient than single crystal silicon is the losses at the crystal interfaces between individual crystals. Therefore, logically, the larger the individual crystals are, the better. There have even been several attempts to make quasi-single-crystal material by introducing crystal seeds in the melt, but this method has lost steam recently, probably because it is not financially advantageous. The whole block is then cut into $6' \times 6'$ 'bricks' – 36 in the case of the crucible mentioned above. Several of these bricks are glued end-to-end and cut in a wire saw as explained above for a single crystal ingot [9].

Other technologies for making solar cells do exist, most notably thin-film technologies. When the price of silicon was still high, amorphous silicon panels were gaining popularity due to their cheaper manufacturing cost at the time. However, since the price of silicon panels plummeted, this advantage has been reduced and their main disadvantage – the fact that their efficiency is 10%–12% as compared to 16%–22% for crystalline silicon solar cells – implies that almost twice the area is required to generate the same power, apart from the larger number of connections required.

Cadmium telluride (CdTe) panels have been available commercially on the market for over 10 years. Their manufacturers have made remarkable advances in efficiency and worked hard to allay concerns about supplies and recycling. The efficiencies at the time of writing have reached levels of 16%–18%, comparable to silicon. However, the market for CdTe panels has for years been dominated by one manufacturer (First Solar), probably due to the high technological barrier to entry, and that manufacturer has so far focused on large solar farms.

The last viable thin film technology is copper indium (gallium) selenide (CIS/CIGS). This too has been available commercially for several years but the number of major producers is also limited. Its efficiency has also been improved in recent years so modules being sold are also approaching the efficiency of silicon modules. All thin film modules are basically made the same way – thin film layers are deposited in a process such as chemical vapour deposition on glass which has been coated with a transparent conductive layer (such as titanium oxide). The P–N junction is formed by depositing different layers and then a back contact layer is deposited and appropriate front-to-back contacts made [10].

These technologies do offer certain advantages over silicon modules in addition to a potentially lower cost of manufacture (though this has receded in recent years). Thin films tend to offer better performance in hot temperatures and in partially shaded conditions. They are also able to be manufactured in more flexible formats such as in rolls or semi-transparent layers. The reasons they have not gained wider acceptance are probably many but primarily related to the high-technological barrier to entry into the market as compared to traditional modules and possibly raw material supply issues. And as already mentioned, the economic drive to find alternatives to silicon has receded as prices keep falling. For this reason, the total market share of thin film solar panels has remained at around 5%.

Other emerging technologies do exist but none seem promising in replacing silicon in the short or even medium term. If a technology emerged that was cheap to make and with an efficiency that is comparable to silicon (or lower efficiency but very cheap) it could happen, but none have appeared so far. The typical problems in the latest crop of candidates such as dye-sensitized cells or perovskites have been low efficiency and more importantly, stability. It is more likely to see these substances incorporated with silicon solar cells in multi-layered structures to boost efficiency than as a replacement [11].

6.1.2 Semiconductor properties: the P–N junction

The majority of silicon today is grown as P-type by incorporating a tiny amount of boron into the crystal lattice during crystal growth. So naturally, to form a P–N junction a thin layer (a few microns thick) on the surface needs to be N-doped. This is achieved by heat-treating the wafers in a diffusion furnace in the presence of a phosphorus containing gas. At high temperatures (900–1,100°C) enough phosphorus gets incorporated into the top few microns to turn this layer N-type. For simplicity imagine a crystal with all the atoms fixed in a lattice. The nuclei are vibrating in place. The inner electrons are also held mostly in place by electrostatic forces; however, the outermost valence-electrons in metals can be considered to be delocalized or belonging to the crystal as a whole rather than a particular atom (a 'sea' of electrons). Since the charge of the electron is exactly balanced by the nuclear charge, it still takes a certain amount of energy to rip one of these electrons from the atom. According to quantum mechanics, electrons can only occupy certain discrete energy levels or bands. There are more bands than electrons and therefore electrons tend to occupy the lowest level. At absolute zero, the electrons would be still and would fill the band up to the Fermi level and the probability of finding an electron at any location is one-half.

Giving the electrons more energy allows them to jump to higher, empty levels, producing an electric current, in the case of a metal. Every electron jumping to a higher level leaves behind an empty level or 'hole'. So an electron moving around from one energy level to another leaves behind it a hole which in turn appears to move. So while holes are really the absence of an electron they are thought of as an opposing particle to an electron with an equal, positive electric charge.

In the case of an insulator, the number of available energy levels exactly matches the number of electrons. Thus, the valence band is completely filled and while there is a possibility of conduction if the energy applied is high enough, this conduction band is separated from the valence band by a *bandgap*.

If the bandgap is narrow, which is the case in semiconductors (*c.* below 2.5 eV), then energy from a photon or thermal energy might excite an electron enough to jump over this gap to the conduction band. Once there, it is free to move about and conduct electricity as can the hole left behind in the valence band (Figure 6.5). So far we have described pure or *intrinsic* semiconductors. The dopants we mentioned earlier, in this model, create intermediate acceptor and donor levels and make the material an *extrinsic* semiconductor. Thus, the conductivity of doped semiconductors is increased. If the majority dopant is P-type, it will add more acceptor levels and the majority carrier is holes, in the case of N-type, more donor levels are added and the majority carriers are electrons [10].

For completeness sake, vacuum zero is the energy required to move an electron to empty space above the conductor. Work function is the energy difference between the Fermi level and vacuum zero and electron affinity is the energy difference between the bottom of the conduction band and vacuum zero. In the case of a metal, the last two are the same.

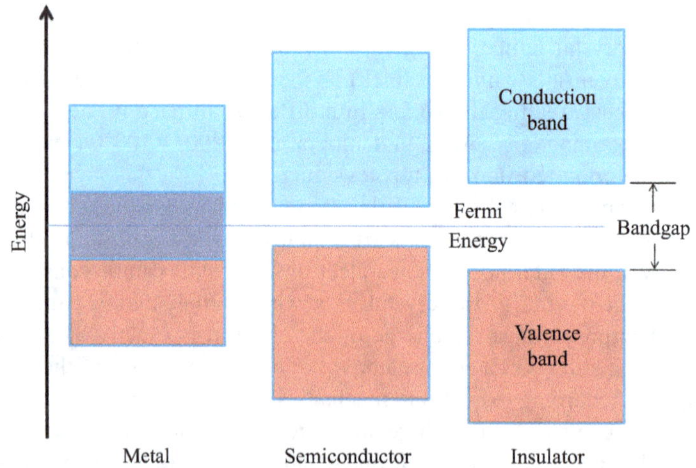

Figure 6.5 Graphical depiction of energy levels and bandgap for metals, semiconductors and insulators

6.1.3 Crystalline solar cells

So now that we understand the basics of the physics and have explained how the material is made, we proceed to understand how a solar cell is made. Various designs have evolved over the past 30–40 years but in its simplest form a solar cell can be made in a few simple steps.

Typically, the wafers, whether single or multi-crystalline, are cut using a wire saw. These large machines consist of a very long spool of very thin wire (<100 μm) which is running in parallel lines and is then slowly pressed against the rod or brick which a slurry of silicon carbide is circulated. While it takes a long time to cut through a 6″ rod or brick, the advantage is that all wafers in a section are cut at once. This technology has replaced the earlier saw technology (such as internal diameter or ID saw) both because of the time required per cut but more importantly because it results in the lowest waste since a wire can be made thinner than a blade. Constant research aims at reducing this waste further by reducing the wire size or increasing the speed of cutting. The drawback is that if the wire breaks while cutting, the section being cut is ruined and can only be used as feedstock for re-melting. So while pushing the boundaries, operators are careful to minimize wire breakage. Some operators also use diamond-embedded wires instead of steel wires with a slurry.

Once the wafers are cut and cleaned, they are usually etched in potassium hydroxide to remove saw damage and to give them texture. Potassium hydroxide etches selectively along the <100> planes so the surface contains a multitude of small pyramidal structures which minimize reflection. And the wafers are then ready to be made into solar cells.

The next step of the process is the formation of the P–N junction. Most modern solar cells are made on P-type crystals, which as explained above means that a tiny amount of boron (grams, as compared to the weight of the crystal which is tens of kilograms) was introduced into the molten silicon during crystal growth. So an N-type layer needs to be created. This is typically done by introducing a stack of wafers in a diffusion furnace at high temperature in the presence of a phosphorus containing gas. At these high temperatures some of the phosphorus atoms will diffuse into the top few microns of silicon creating an N-type layer – and thus a P–N junction. Unless bifacial cells are being made, the back layer is protected either by placing two wafers back-to-back or by some coating which is later removed. Sometimes the doping material is in the form of a paste or liquid applied to the wafer surface or a solid placed in the furnace. If the wafer is N-type, the dopant is of course boron containing [9].

The next step is the formation of front and back contacts. The contact at the back can be a single reflective coating – since it does not matter if it blocks light but might actually help by reflecting photons that reach the bottom back into the bulk silicon. The front contacts are usually made by vacuum evaporated or screen-printed palladium-silver or similarly conductive materials. A lot of research goes on in the choice of material and configuration. The material needs to provide the best conductivity, adherence to silicon and solder or other processes it will endure later. The configuration needs to ensure that the majority of electrons are captured (i.e. minimum distance between contacts) while providing the least coverage or shading of light. The final step is then the addition of an anti-reflective coating on the cell surface. This process is identical to that adopted in lens coatings where a material with an appropriate refractive index ensures that minimal reflection occurs (Figure 6.6).

A solar cell generates about 0.5 V, and a maximum power of output of about 5 W. Typically, 60 or 72 cells are connected in series in a module. This results in panels with a maximum power output of 260–380 W (depending on the efficiency of the cells used). Since the cells are strung in series, the maximum current is still equal to that of an individual cell or about 10 Amperes. If a higher power is needed, strings of solar cells or panels would need to be connected in parallel.

Virtually all panels sold today are guaranteed for 20 or more years. In fact some are routinely offering 25-year guarantees. A lot of emphasis is therefore placed on reliability and durability. The thick busbars (contact strips) of the cells are soldered end-to-end face down on a tempered glass pane. The aim of the sheet is to protect against the elements (including hail) while providing maximum light transmissivity. The backside is then sealed by means of an ethylene vinyl acetate (EVA) sheet. Any joints are sealed with silicone. Finally, the edge is enclosed in a rugged aluminium frame to protect the glass edges and provide an easy way to attach the panels to frames. In some cases, the back layer is also glass. In those cases, sometimes the panels are left 'frameless' and a sealer is simply placed in between the two glass panes.

Various, more complex cell designs exist. Each design improvement over the years resulted in ever-increasing efficiency of solar cells which increased from 14%–16% in the late 1970s/1980s to over 21% for multi-crystalline and over 25% for single crystal solar cells (Figure 6.7) [12].

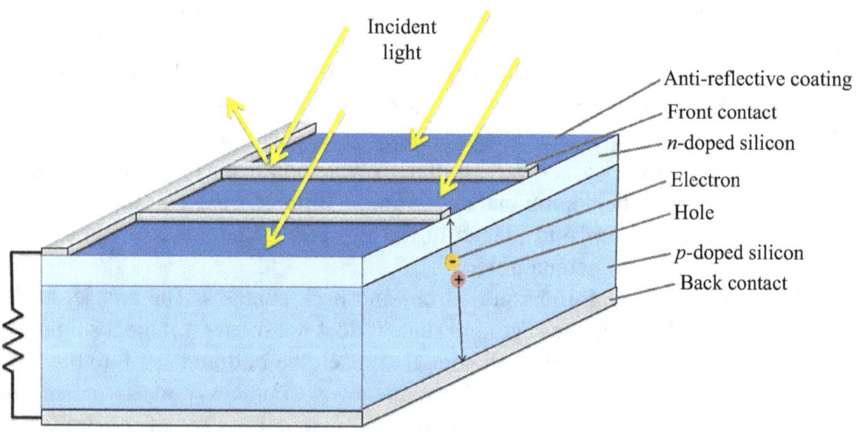

Figure 6.6 Structure of a silicon solar cell showing the various doped layers, contacts and coatings

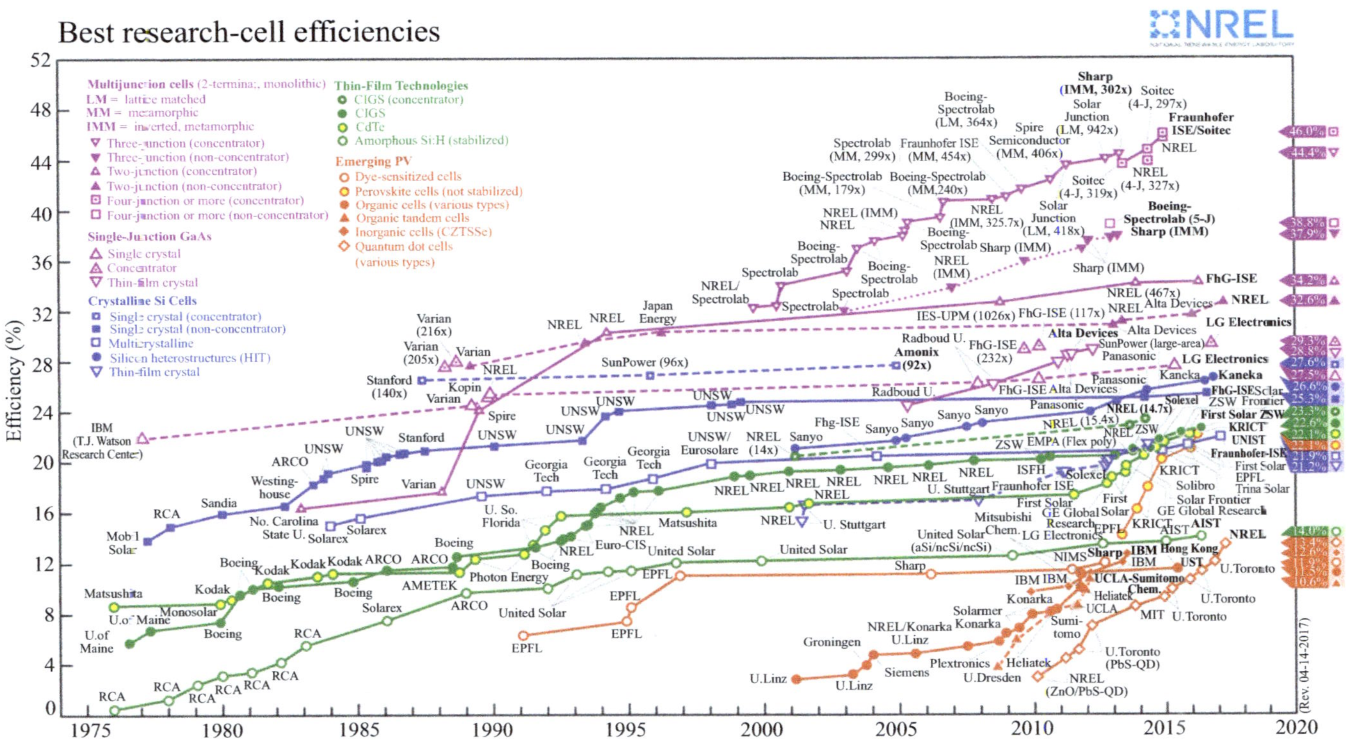

Figure 6.7 Best research cell efficiencies verified over the years. (Courtesy NREL)

6.1.4 Thin film solar cells

Before we proceed further and explain how (crystalline) solar cells are made into panels we should mention again the other type of solar cells, i.e. thin film systems. Having read the description of the process of making silicon, one will appreciate it is not a simple process. Moreover, once a crystal or a block is produced, it has to be sawed into wafers. A substantial percentage of the material is wasted in the process. So almost since the making of the first solar cell, scientists have been devising ways to make thin film solar cells by deposition. The allure of the idea is obvious – if a thin film semiconductor material is deposited, instead of growing it, no waste is generated – so theoretically at least, it should be cheaper, and definitely a more efficient use of materials is achieved. Moreover, if the right materials and processes are used, it could be much less energy intensive, and faster. For it to be successful though, it has to meet the other criteria of a silicon solar cell, i.e. durability, efficiency and simplicity. As mentioned, the only solar thin-film materials that have had any commercial success are amorphous silicon, cadmium telluride and CI(G)S (copper indium (gallium) selenide) [10].

Amorphous silicon was the material that showed the greatest promise initially. A lot of effort was expended in the early to mid-2000s, into raising its efficiency (which at the time hovered in the 5%–7%). By the second half of that decade there were several companies making amorphous silicon panels on glass or in flexible formats. These offered the advantages of flexibility, light weight and cost. Efficiency reached about 10%–11%. Since the price at the time was significantly lower than crystalline panels, they were being used in solar farms where land was cheap and plentiful. For this purpose, extra-large panels were also being manufactured. The flexible panels were being incorporated onto roofs – especially roofs that cannot support weight or where normal panels would ruin the aesthetics. Unfortunately, the efficiency improvements stalled and when the price of silicon solar cells plummeted, their use was no longer economically viable except in some special circumstances – so all but a handful of producers disappeared.

The next major development was cadmium telluride solar cells. The major player in this technology was the company First Solar, which still dominates the CdTe market today. They adopted the technology in the early 2000s and started production by the mid-2000s. At the time the efficiency was in the 7%–9% range but showed promise, since higher efficiencies on small-scale cells was routinely obtained. Today, panels with 17% efficiency are commercially available which is why this technology has survived in spite of the collapse in prices. Their price advantage over silicon has become minimal and might be vanishing altogether but they have managed to survive especially for use in solar farms. Moreover, all thin film technologies have on average a better tolerance of heat (silicon cell efficiency deteriorates considerably at high temperatures) and shading – whereby thin film cell performance drops in direct proportion to the part being shaded while the effect on silicon solar cells is much higher.

CIGS technology emerged about the same time as CdTe. Several processes have been developed over the years for its deposition from vacuum based, chemical vapour deposition (CVD) to a liquid-based method. The latter has not been successfully commercialized so far, though several attempts were made. There are currently several companies making CIGS or CIS modules around the world on traditional glass or in a flexible format; however, the market penetration has remained low. Highest commercially available module efficiencies are in the 13%–14% range though 19% has been reached on cells [12].

What all thin film technologies have in common is simplicity of production and lower material use (and therefore eco-friendlier from an energy and material use point of view). Research into the field continues both in industry and in academia. For the foreseeable future their market share will probably remain the same or shrink unless a major breakthrough or advance is made, e.g. successful implementation of a high-efficiency liquid-based method for CIGs, or continued improvement in efficiency (which makes them more economically competitive). They do have some advantages as highlighted above – better performance in partially shaded conditions, at high temperature and in low/diffused light. They are also easier to make in flexible or semi-transparent formats – which will probably ensure that there will be a market for them in this niche.

Other emerging materials exist. The ones most talked about at the time of this writing are perovskites, (typically) organic tin, or lead halides. The material has been known for a long time but their application in solar cells was only first reported in 2009. The excitement grew as ever-higher efficiencies (over 22%) have been reported in test devices. The allure is even greater since the process to make them is theoretically very simple (deposited from liquid form) and extremely cheap. The snag is that they suffer terrible instability when exposed to air, moisture and sunlight. Robust encapsulation eliminates much of the cost advantage. A lot of research effort is being placed into solving these problems. Over the next 5–10 years, if these are unsuccessful, perovskites will probably remain in the research realm with the possible exception of their use as tandem structures with silicon whereby a perovskite cell transparent in the visible spectrum but active at other wavelengths is used to enhance the efficiency of the silicon solar cell [11].

6.2 Solar systems

6.2.1 Solar panels

Silicon comprises some 95% of the worldwide production market in 2018, which is some 100 GWp of production – or almost 400 million panels. Sixty or 72 cell panels have a maximum power voltage of around 30 or 40 V. Since each cell is around 155 mm^2, 60 cell panels equate to an area of about 1.6 m × 1 m and 72 cell panels an area of 1.9 m × 1 m. The completed panels weigh 18–22 kg.

Once cells are wired (with soldered thin contact strips) in 60/72 groups in the desired pattern (6 × 10 or 6 × 12) they are placed on a sheet of high transmission, anti-reflection coated tempered glass (about 3-mm thick). In panels with an aluminium frame the back is sealed with an EVA sheet which is stretched under vacuum/heat across the back and the edges protected by an aluminium frame which is in turn sealed by silicone rubber. A junction box is placed on the back which is in turn sealed and from which the external leads protrude. Other panels are made with glass on both front and back and in some cases panels are sold frameless.

Over the last decade, solar cell and panel manufacturing has become highly automated. This has improved quality, repeatability and reliability. Very few panels from reputable firms fail on site since there is also a rigorous quality control. Panels are designed to withstand the elements including high wind and hail for their lifetime of 20–25 years (which is guaranteed). They are also guaranteed to perform at 80% or better after 20 years of operation. These factors and the remarkable reliability of production and repeatability from one year to the next have made solar energy such a resounding success on land – and now floating on bodies of water. For example a system in a southern Mediterranean climate will produce 1,550–1,650 kWh/kWp reliably year after year – barring a cataclysmic event such as a major volcanic eruption that would darken skies for an extended period of time [3].

6.3 Floating solar systems

6.3.1 Motivation

One might ask, what is the motivation for installing solar systems on water when there is so much empty space in most countries? Surely systems are cheaper to install on land and less complicated?

Let us start with freshwater bodies. As we shall see in the chronological section below there are many different motivations but they can be summarized as follows:

1. Lack of available land. Yes, land may be available somewhere else, but not in the confines of the business who wishes to install the system. Countries and cities with high population densities also have a problem identifying enough suitable (and affordable) land for solar farms but have plenty of freshwater bodies whose surface is idle and unused.
2. Systems installed on bodies of water can be closer to consumption. This, for example, is the case in many large cities where you would be hard-pressed to find large parcels of available (and affordable) land where to build a farm within the city, where the majority of energy demand is usually originating. Again, reservoirs, ponds and lakes are available within the city confines and are therefore a prime target for floating PV farms.
3. Floating PV systems reduce evaporation of the body of water on which they sit. This provides an added benefit in addition to using this idle space. Evaporation from bodies of water, especially in hot climates can be very significant, and in places where water is scarce this is an added bonus. This is especially the case for bodies of water used for drinking or agriculture.
4. Algal growth on the water is usually decreased, especially if a system is covering a significant portion of the surface area. This improves the quality of the water.
5. Systems installed on water might cost the same or be even cheaper than systems on rooftops. Depending on the design some of the systems that have

been developed are very simple to install and the panels do not require much more than the floats they are mounted on.

6. Systems on bodies of water may perform better due to light reflection on the water surface.
7. Systems on bodies of water will perform better due to the cooling effect of the water below them and in some cases, the lower level of dust. This has been reported over and over again in the literature as we will see in the following.
8. Systems on a body of water can be actively cooled. Active cooling, especially in the hot summer months can improve performance by 10% or more [13]. Even passive environmental cooling can result in 1%–3% better performance [14].
9. A floating system is easy to rotate. This approach has been adopted by several developers who have built rotating single- and dual-axis tracking systems. Except that typically, instead of the individual panels rotating, the whole system rotates about a vertical axis to follow the sun's path.
10. The likelihood or shade on water is lower. Especially if the body of water is large it is easy to place the system in a position which will be completely unshaded year round.

6.3.2 *Components of a floating system*

A floating PV system does not require that many components [15]. Typically:

1. A pontoon or other floating structure. In the case of the leading company installing systems on freshwater bodies, Ciel et Terre and similar systems this is specially designed to hold the panel on its surface with no (or minimal) need for additional structures. Materials are typically high-density, poly-ethylene (HDPE), fibreglass, steel, carbon reinforced fibre or even wood. A few systems do away with floaters by using self-floating panels [16].
2. A mooring system. The system needs to be anchored. While most bodies of water experience limited waves, they may be subject to high winds on occasion and therefore the system needs to be secured. Most large systems are moored at multiple points. However, in the case of rotating (tracking) systems they are sometimes only moored through a single point mooring system.
3. The PV panels. The majority of floating PV systems use normal panels. Panels are designed to withstand storms and heavy rains. A system installed on a lake or reservoir is not likely to get wetter than say a system installed in a northern climate where it rains often. Some installers do apply coatings and special precautions to protect the panels from added corrosion.
4. Inverters, cables and connectors. This is not very different from a land-based system except that an underwater cable is required to connect the system to transmit the power generated to land.

6.3.3 *Chronology*

As already mentioned above, photovoltaic system installations really took off in the last decade. For the first few years it was driven entirely by government subsidies in the form of grants and feed in tariffs but as the market expanded and prices fell it has, in the last few years, become self-sustaining in parts of the world where the conditions are right – plenty of sun year round and cheap available land. Not all countries and regions are as fortunate. Which is why some are considering floating offshore systems. But before we go there, let us take a look at the chronology and development of floating systems on inland bodies of water – lakes, ponds and reservoirs.

The first recorded PV system installed on water was developed by the National Institute of Advanced Industrial Science and Technology in Japan in 2007. This was primarily a research project aimed at comparing the performance characteristics of PV systems installed in a water reservoir with those on land. The land-based system was installed at a 30° tilt. The systems on the water were installed on polystyrene boards and the panels were inclined at 1.3°. The systems were anchored and cables connected to the inverters on the ground. One of the reservoir systems incorporated cooling. As was to be expected the cooled system resulted in higher output [17].

The next two systems, installed in California, USA, in 2008 were practical systems designed to produce electricity and lower the carbon footprint of businesses. Both were installed on wineries – the Far Niente and Gundlach Bundschu wineries. Both were built by the same company which was trying to carve out a niche market. The Far Niente installation received a lot of publicity and is credited with publicizing installations on water [18]. It was also an installation designed for generation with 1,000 solar panels floating on pontoons in an irrigation pond and another 1,300 panels on adjacent land (Figure 6.8). This 478-kWp system is still in operation and was installed primarily to lower the carbon footprint of the winery. Of course by installing 1,000 panels on the reservoir it also avoided having to install them on land – prime real-estate which in this case is used for cultivation of grapes. Moreover, the panels on the water reduced evaporation which in hot climates can be significant. The Gundlach Bundschu system was also composed of a large (450 panel) system on land and 162 panels on pontoons on an adjacent

Figure 6.8 The Far Niente Floatovoltaic™ solar installation in the Martin Stelling Vineyard. (Courtesy Far Niente Wineries)

irrigation pond. The system was reputed to produce up to 60% of the winery's energy needs [19]. While not the first systems on water, these two deserve the distinction of being the first systems on water installed for actual electricity production.

The next developments happened in Italy in 2009. A large 500-kWp system was installed on a lake in Bubano. An environmental impact assessment was carried out before the installation since it was postulated that such a large system on the lake could impact its ecosystem. The construction was a simple one made out of polyethylene hollow cubes similar to the ones used in floating docks. The latter were connected by stainless steel struts on which the panels were installed at a shallow angle. The whole system measured 200 m by 60 m. Site conditions here were challenging given that being in the North, the system is exposed to below freezing conditions and snow in the winter time [20].

A second Italian system was installed that year in Solarolo. It was a 20-kWp system on top of an irrigation pond. The panels were installed in such a way as to allow for air-cooling from underneath. The buoyancy is provided by an outer ring and in fact this system, which provided enough power to meet the needs of the land reclamation authorities and farms in the area, was dubbed the Lotus project because of its circular shape [21].

The other recorded floating system that year happened in Spain, near Alicante, on a reservoir. The primary aim of that installation was to reduce water evaporation in addition to providing power. So special floating pontoons made of plastic were designed each holding two panels on top inclined at 10°. In this case a key target was covering the surface of the water as much as possible.

Even in the first two years of development of the technology one can already appreciate certain trends in the motivation of the installers. These included power generation at source, close to the user, reduction of water evaporation on fresh bodies of water and reduced use of valuable land that could be used for other purposes [22].

The year 2010 saw another installation in a winery, this time at the Petra Winery in Suvereto, Italy. This installation had the panels installed at 40° and included another key innovation – the system could rotate to track the sun. The developers of this system designed it to rotate around a central anchoring point taking advantage of the low amount of energy required to rotate a floating object about a single vertical axis. Such a system can increase the power output by 20%–30% [23]. The same company that designed this system (SCINTEC) designed a second single axis tracking system in 2011, this time on a lake near Pisa. Another innovation was introduced in this system. The panels were installed flat while reflectors were installed to either side of each panel at 60°. The system was rated at 20 kWp. A replica of this system was also installed in the town of Avetrana that same year [24].

The year 2011 saw the first installation in France – a 15-kWp system on a lake outside the village of Piolenc. The system covers a very small fraction of the lake surface [25]. It was later expanded to include a second system as is apparent from Google Earth images. This installation was performed by Ciel et Terre which developed the Hydrelio® floating system. Since then, Ciel et Terre has installed over 120 systems around the world. The patented Hydrelio system is elegant in its simplicity and practicality. The main components are specially designed HDPE blow-moulded floats connected together. Each float has aluminium rails that allow panels to be installed on top. Due to its simplicity and cost-effectiveness the system has gained wide acceptance. It is of course designed for inland bodies of water. Its target market is fresh bodies of water such as lakes, reservoirs and ponds. In fact, in addition to guaranteeing 20+ years of lifetime they certified the system as safe for use on reservoirs used for drinking water [16].

That year (2011) also saw another installation in California by SPG solar. The system having an installed capacity of 350 kWp is located at an irrigation pond in Petaluma. A small (4 kWp) system was also installed in France by another company which introduced the innovation of a complete PVC system. Their aim was primarily to reduce cost and utilize bodies of water with no particular economic, environmental or touristic value. Another innovative system was installed that year in New Jersey on a reservoir at the Canoe Brook Water Treatment plant. The 400 panel system (tilted at 14°) has a unique mooring system in that it was designed to rise and lower itself with the water level [26].

The first commercial floatovoltaic system (a term that has gained popularity in describing floating PV systems) in South Korea (and Asia) was also implemented in 2011 by K-Solar at the Hapcheon Dam. The motivation was similar – the harvesting of energy from otherwise unused water surfaces. The 100-kWp system had panels included at 33°. Special attention in the design was taken due to the harsh conditions the system was expected to endure in its lifetime. It should be mentioned that while this was the first commercial installation in Korea, two research projects had been implemented in 2009–2010 [27].

A system in Israel was also deployed in 2011. The system in Kibbutz Ketura, designed and built by Solaris Synergy used concentrated PV technology instead of the more traditional panels. Solaris Synergy has continued installing systems in Israel and elsewhere, though in later systems they seem to have opted for traditional panel use [28].

In 2012 three photovoltaic systems were launched on reservoirs in South Korea. This was primarily due to government intervention encouraging utilities and electricity providers to move to clean sources of energy to reduce CO_2 emissions and pollution. The three systems were 500 kWp, 20 kWp and 30 kWp in size. The largest system was deployed by K-Water at the same reservoir it had launched a system in the previous year. The second system was done in collaboration with Scintec, and included FTCC (floating tracking cooling concentrator) technology similar to the Italian system made by the same company and the last one, by KRC Corporation, introduced the use of fibre-reinforced polymer for panel support. Another system was also launched in France that year by Osesol – a 100-kWp system on a reservoir in La Pommeraie sur-Sevre.

Three more systems were constructed in Korea in 2012. The system on the Hapcheon Dam (the third one installed there in 2 years) was innovative in that it incorporated four 24.8-kWp circular installations that were able to rotate and therefore track the sun. The system also had tilt control for the panels. The second system, at Dangjin-si, was the first 1-MWp system installed on water. The last system, also by KRC Corporation, was small (30 kWp) and used their fibre-reinforced floats. The design of these floats appears to consist of individual fibre-reinforced floats with a pre-moulded place for the panel to rest at the correct angle of tilt. This company is also still active and busily installing other floating systems around Korea [27].

Southeast Asia also saw the first installation in 2013 in Bishan Park, Singapore. The 5-kWp array was implemented to study the suitability of PV installations on freshwater bodies and their effect on flora and fauna. While this was a modest start to floating PVs, Singapore is another country taking a keen research interest in this technology. The Solar Energy Research Institute of Singapore (SERIS) later installed a test bed comparing several floating technologies and PV systems for their suitability and performance. They have also started organizing an International Floating Solar Symposium in 2017 [29].

The year 2013 also saw the first experiment by a research group in Canada – MIRARCO [30]. The installation consisted of two flexible solar panels floating directly on the surface of the water without the need of a floatation device. A second system was installed next to the body of water on the ground to serve as a control. The elegant idea that MIRARCO was trying to implement is the reduction of components to reduce overall system costs. They approached the problem by using thin-film flexible PV panels which could easily float on their own or with minimum help. At that time thin-film PV panels were still relatively popular due to their lower cost. These types of panels have since lost the cost advantage as explained previously and are, because of their lower efficiency (therefore requiring a larger area per unit of power), now only used in certain specialized uses. In reality, floating on a body of water could be one of the uses where space is not a major issue. Therefore, thin film PV-systems could still offer potential for floating systems if the economics prove that such systems are feasible.

Following the Fukushima Daiichi nuclear accident, Japan changed policy and opted for a shift away from nuclear energy. This necessitated a push on every front, including a major push in renewables. Being a relatively small, mountainous and heavily populated country, large open fields are not always easily available for PV farms. So it was logical that Japan too would start looking at floating installations. In July 2013, the first mega solar farm (1.18 MWp; 4,536 panels) was installed by Ciel et Terre in Saitama Prefecture.

The first installation in Britain was developed in 2014. The first one was a 200-kWp system at the Sheeplands farm in Berkshire and was also installed by Ciel et Terre. The idea was again to use the unutilized surface of calm water for energy generation therefore making farms more sustainable. Kyocera TCL Solar LLC also commenced the construction of several new systems that year: six mega plants ranging in size from the smallest at 696 kWp in Kawagoe, Saitama prefecture, to the largest which was planned to be 13.7 MWp in Yamakura, Ichihara City, Chiba Prefecture. The other four plants were between 1.1 and 2.3 MWp. The Yamakura plant became operational in early 2018. At the time of its commissioning, at 13.4 MWp and 51,000 panels, this was the largest floating solar plant in the world but has since been overtaken by even larger systems [31,32].

Another tracking type system was also installed that year in South Korea by KRC. This 465-kWp system tracked along a fixed central post which would presumably make the tracking more accurate. The year 2014 also saw the second installation by Solaris Synergy in Israel. The new design featured a floatation mechanism that allowed panels to move independently of each other and would therefore provide better tolerance of wind load. This year also saw the first installation of a small (5.39 kWp) system on an aquaculture pond in Samut Songkhram, Thailand. The Ciel-et-Terre system is unusual in that it sits under an area enclosed by high netting which could have some negative impact on the performance, though likely to be minimal.

A flurry of new systems were implemented in Japan in 2014. Eight systems were inaugurated. Seven of the systems were 1–1.5 MWp in size whereas one system (that in Kawajima-machi, Saitama Prefecture) was 7.5 MWp. The last system was implemented by Nikkei BP as were some of the other systems. The rest of the systems were implemented by Ciel et Terre. All these systems are situated in agricultural reservoirs, or irrigation ponds. South Korea also saw the installation of two new 3-MWp systems on reservoirs that year.

The year 2015 saw the installation of the first floating solar plants in Brazil. That country is heavily reliant on hydro-electric power for its electricity. While the energy is renewable it has not been without controversy. Some of the dams were constructed in the rainforest, resulting in large loss of habitat. Second, the building of large dams means flooding of valleys which typically also results in the displacement of people and possibly the loss of cultural heritage. In the case of Brazil, it was also found that the accumulation and decomposition of forest debris at the bottom of the reservoir was resulting in large amounts of methane being emitted. This provided added motivation for the country to make use of the dam-water surface [33]. The construction of two systems, a 5-MWp structure at Balbina Dam and another on the Sobradinho Dam were initiated that year. The latter area is characterized by high winds which could benefit the PV system for cooling but could also prove hazardous [34].

Systems in several new countries were also implemented in 2015: a 108-kWp on a water treatment pond in Sungai Labu, Sepang, Malaysia, and a 157-kWp system in Australia. The 157-kWp system was installed for the usual reasons with the additional motivation of providing another source of energy for remote communities in the large expanses of Australia. The latter is installed on a wastewater treatment plant pond in Jamestown [35]. Its design is very similar to the SCINTEC design whereby the panels are flat and reflectors are placed at a high angle on either side. In India three systems were implemented. First a 10-kWp array in Kolkata, Rajarhat, which was an R&D project by Arka Renewable Energy College. A 100-kWp system at Loktak Lake in Manipur was also implemented by the Manipur Renewable Energy Development Agency. And in Kholpali a 13.5-kWp system was implemented on a hydroelectric power plant pond.

The UK also saw four new installations ranging from 100 to 471 kWp that year. Another new entrant, the Netherlands started a project in Rotterdam area (Slufter). The plan is to develop floating rotating solar systems on a contaminated sludge pond. While the current test contains 120 panels, the pond is very large and there is a potential for a very large installation of up to 540,000 panels [36]. Another small installation was also installed in the Westport Industrial Estate in Groningen City, the Netherlands, while Ciel et Terre installed a 22-kWp system on the Nofar reservoir in Israel.

In 2016 Japan installed a total of 17 MWp in 12 plants across the country. The plants ranged in size from 300 kWp to 2.45 MWp. All sites were irrigation ponds or reservoirs. China also entered this market in 2016. They installed a 20-MWp system on top of a fishpond in Huainan City, Anhui province [37]. China has also expressed its desire to move aggressively with other floating PV projects. A second system was also launched in Malaysia (270 kWp).

The largest system in Europe to date was installed in 2016 – in Godley (Greater Manchester, UK) by Ciel et Terre. The system has a capacity of almost 3 MWp and covers 50% of the reservoir on which it is installed. The reservoir is a drinking water reservoir so the plant is also helping reduce evaporation. The panels have been installed at 12° to optimize performance and do not have excessive wind loading [38]. Other Ciel et Terre systems in 2016 were installed in Savona province, Italy (343 kWp), and the first system in Indonesia – installed by REC. The latter was a 5-kWp pilot project. Two other small systems were also installed in the USA: one of 5 kWp at Florida University and a 9-kWp one at Kenwood, another winery. That year also saw the beginning of a mega project in the USA. A 4.1-MWp project is being implemented in New Jersey (Sayreville Borough) to supply power to the Bordentown Water Treatment plant.

Several projects were also launched in 2017. These include a 572-kWp in Gojiga-ike in Hyogo prefecture, Japan, installed on a reservoir. A 32-kWp system installed in Orlando, Florida, USA, and a 24-kWp system installed in yet another country – in a water retention pond on the Panama Canal, Panama. India continued its march with two more projects (500 and 100 kWp). The 500-kWp project is the largest to date in India and was installed on the Banasura Sagar reservoir in Kerala. And a new system in China became briefly the world's largest – a 40-MWp project in Anhui province at the site of a flooded coal mine. It has since been expanded to 150 MWp (Figure 6.9). This system and many others in China are the work of Sungrow corporation which has also developed their own HDPE-based floating structures. Many other plants are in the works and will undoubtedly continue to

Figure 6.9 Sungrow's 150-MWp installation in Huinan, China. (Courtesy Sungrow Power)

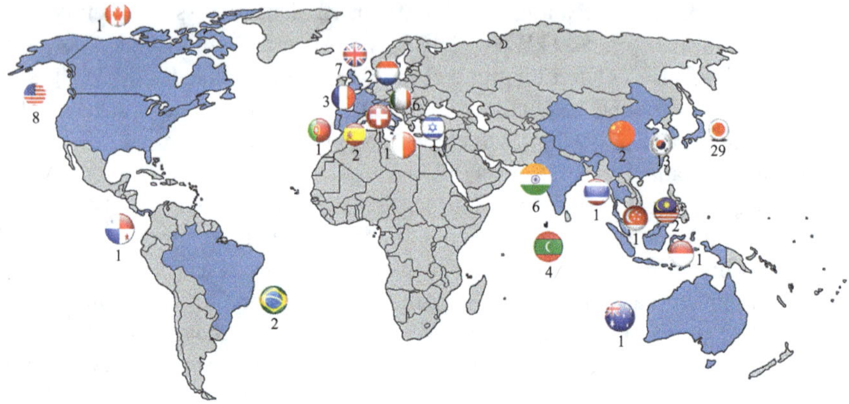

Figure 6.10 World map showing the location of installed systems around the world [45]

break the record for the largest plant worldwide. A map of the systems implemented around the world until 2017 is shown in Figure 6.10.

6.3.4 *Advantages and disadvantages*

The general advantages of floating PV systems have already been highlighted in Section 6.3.1. Over 1 GWp have now been installed around the world so there is significant experience and it has mostly been positive.

Some disadvantages might be as follows:

1. A concern with the higher moisture content – though as already mentioned this is not necessarily much different from rooftop systems in wet climates.
2. The environment might be more corrosive due to contaminants in the water. Durability and survivability of all materials used can therefore be a concern.
3. Access and installation requires more planning and different equipment.
4. Reservoir, lake or pond levels can change considerably over time so the system needs to be designed to cater for this.
5. Environmental impact can be significant and might not be acceptable on certain bodies of water.

While relatively little is published from the commercial systems, it seems that most of the experiences are positive. The fact that the same companies and countries continue installing more and larger systems seems to reinforce this.

However, there have been a few problems. In July 2018 a landslide destroyed a system in South Korea prompting the government to issue new rules on the citing of floating PV systems [39]. In 2016 about 150 kWp of panels in Japan were damaged when wind gusts of over 20 m/s turned over the edge of the system [40]. While it was not mentioned in the report it is also possible that cracks in the silicon developed in additional panels which experienced heavy flexing due to the wind and waves. Not enough long-term information is available on most systems – since most were installed recently. Some of the more unusual designs, e.g. the trackers, will have to prove themselves over time as do the choice of materials in every system. Some systems might encounter specific problems – e.g. systems in large bodies of water might encounter heavier waves, and systems where the panels are in contact with water might experience biological growth on the panels.

In balance though the technology is not only here to stay but is growing exponentially. The mishaps mentioned above will be treated as learning experiences for future systems.

6.3.5 Systems at sea: motivation and special challenges

All the systems mentioned above and virtually all the systems installed so far are located on fresh bodies of water. The reasons being that the challenges are easier than in open sea. However, many of the advantages that apply to fresh water also apply to sea water – namely that you do not use valuable land, that systems can be closer to consumers than otherwise. Cost may not be cheaper – for the installation as such but in small or heavily populated countries or regions land is always at a premium. So in small countries with a large population density like Malta or Singapore and large coastal cities, looking at the sea is an obvious choice.

For example, in the case of Malta, its marine territorial waters are 10 times larger than the entire country. Singapore similarly has very little usable land for PV farms. But the same situation also exists in large coastal cities. While land for large PV farms would typically exist once you leave the city boundaries this could easily mean that the production (PV farm) is 50 or even 100 km from the bulk of the consumption. On the other hand, if you place the PV farm in the sea it can be within hundreds of meters of the city.

The cooling effect and potential gains from reflection also hold for the sea. An additional benefit that has been reported is the gain due to the lower dust accumulation which is especially a problem in hot dusty climates. Active cooling is also possible. A rotating system on the sea is theoretically possible but definitely more challenging.

One potential long-term advantage could also be that if a system becomes redundant, or is needed elsewhere, it can be moved. Moreover, the location can be chosen not only to be close to consumption but also to minimize connection costs to the grid – which is not always the case for land-based systems or systems floating on small bodies of water.

There are also some challenges specific to systems at sea. These include the following:

1. The ability to withstand the elements – waves in addition to the wind. Of course the situation is similar to a lake or reservoir in a sheltered bay – but is much more pronounced in open sea.
2. The corrosive nature of sea water. While not an impassable problem, it is an issue that needs to be tackled.

3. Environmental concerns. Systems in shallow water will shade the sea-bed drastically modifying the ecological environment. Systems in water deeper than about 50 m pose much less of a concern.
4. Higher costs. Installation and mooring are more complex: materials have to be tougher.

All that said, need is the mother of invention and since, as explained, there are legitimate reasons for people to consider in sea and offshore systems, solutions can be sought and found.

6.3.6 Systems at sea: current situation

The first record of any PV system at sea is from 2014 when two systems were launched, one in Malta and one in the Maldives. The one in Malta was part of a University of Malta research project called Solaqua which the author leads. The one in the Maldives was launched by an Austrian company called Swimsol. The only other known groups working on offshore PVs is another Maltese-Canadian group (lead by The Malta Council for Science and Technology, Malta) which announced they are launching an 8-kWp system moored in a fish farm in 2018, and a consortium in the Netherlands lead by Oceans of Energy.

6.3.6.1 The Solaqua Project (Malta)

The Solaqua Project was initiated by the University of Malta in collaboration with two local companies (General Membrane and Pandia Energy) through a funding award by the Malta Council for Science and Technology in November 2012. The aims of the project were to assess the feasibility of offshore solar, to design and test several prototypes at sea for sea-worthiness and evaluate their costs and practicality. Four different configurations were tested.

The first prototype (prototype 1) launched in December 2014 was made out of two connected rafts. The rafts were 7.5-m long and about 2-m wide and 1-m high at the highest point. The design included a gentle slope on the top where the panels would be laid to allow for natural cleaning of the panels by rain, dew or sea-spray. The rafts were constructed using steel reinforced polystyrene and covered with an impermeable canvas membrane to protect the polystyrene. The flexible, amorphous silicon panels were attached on top using Velcro by a process patented by one of the partners [41]. The panels were connected to resistors in a sealed stainless steel box placed inside the raft. The box also contained the data logger which transmitted the data remotely via a mobile communications network. The advantage of putting the steel box in this configuration is that the bottom part of it was immersed in water helping cool the resistors inside. Additional small (2×20 W) multi-crystalline silicon panels were also placed on top of the steel box to power a warning signal and the GPS transmitter (Figure 6.11).

The prototype was launched in an open bay (Qalet Marku) in December 2014 and remained in the water until October 2017. The bay was chosen due to the ease of access. It is in the North of the island and well exposed to the strong Grigale winds which are typically associated with very stormy conditions in the Central

Figure 6.11 The launching of the first Solaqua prototype 1 (Malta 2014). The steel boxes which housed the equipment are seen at the top of the rafts close to the warning light

Figure 6.12 Solaqua prototype 1 during a major storm (2015)

Mediterranean (Figure 6.12). A similar system was installed on the roof of a waste disposal facility (Wasteserv Malta, Maghtab) to provide a reference in the performance analysis of the sea-based system. This site is only about a kilometre from the test site in the water.

The second and third prototypes were launched in March 2016. Both were based on the idea of self-floating PV panels. Prototype 2 consisted of two metal backed panels which were floated on polystyrene sheets and rendered impermeable by coating. Prototype 3 was more original. At concept stage we theorized about the possibility of having panels specifically built to render them buoyant. We envisioned something in the shape of a flying saucer or sphere and many of these panels strung together. Of course nothing like this existed so we constructed a prototype by using two plastic hemispheres (the types used for skylights) and rigging solar panels inside. Since these prototypes were small their data-logging and resistive loads were installed in prototype 1's steel box. A land-based system equivalent to prototype 2 was also installed at the nearby waste management facility of Wasteserv.

The fourth and last prototype 4 was a result of our financial analysis which is explained below. Two rafts were built using 40 mm galvanized steel pipes and floated by polystyrene steel drums which were reinforced by filling by expanded polystyrene. The two rafts each had two panels – aluminium framed in one case and glass–glass in the other. The rafts also included steel boxes containing the resistive loads and data loggers all transmitting via GPS. These prototypes were launched in April 2016 (Figures 6.13–6.15).

Results

Survivability. The first test was that of survivability. All prototypes endured major Mediterranean storms in the time they were deployed. All moorings held but the glass–glass modules were damaged in one such storm which was attributed to the fact that they were installed at too high an angle (30°) and therefore too exposed to wind and waves. All other modules survived without suffering any damage (Figure 6.11).

Output. Most data were obtained from prototype 1 which over the course of 18 months produced about 4% more energy than the equivalent land-based system (Table 6.1). Data gathered from prototype 2 (panels on polystyrene floats) showed an increase in the energy yield of 11% compared to the system on land. The output from prototype 3 was not expected to be high given the non-optimal design which

Figure 6.13 Solaqua project, prototype 2, which consisted of metal backed marine panels made for marine use on polystyrene floats. The panels, box and polystyrene were further coated

Figure 6.14 *Solaqua project, prototype 3, which consisted of a flat solar panel suspended in the centre of a transparent dome. This prototype was intended to test the survivability of a self-floating PV-panel built specifically for floatation*

Figure 6.15 *Solaqua project, prototype 4, which was built as a hollow metal pipe raft with reinforced drums for floatation*

Table 6.1 Power output on W/Wp for the system at sea compared to the system on land

Quarter	Raft PV	Land PV	% difference
2015 Q2	409	408	0.2%
2015 Q3	378	332	13.9%
2015 Q4	134	137	−1.8%
2016 Q1	199	203	−2.0%
2016 Q2	321	304	5.7%
Total	**1,441**	**1,384**	**4.2%**

was more intended to display the concept of a self-floating PV panel and test its durability in the sea. The 11% gain by the panels directly in contact with the water is easily explained via direct cooling; however, biological growth is expected to be a problem with these panels, unless they are treated regularly. It was postulated that the panels in the other raft, which were sitting about 1 m above sea level showed a better output due to a cooling effect of the marine environment and a lower accumulation of dust which was also observed in the summer months when comparing the sea to the land-based systems [14].

Corrosion, salt and ageing. Prototype 1 spent almost 3 years in the sea and the other prototypes one and a half years. Normal wear and tear was observed during the course of regular inspections although physical degradation was not dramatic. An anchoring rope on prototype 1 was replaced mid-way. The structures all survived fairly unscathed. Anywhere that was directly in contact with water saw significant and fast biological growth. One should note that no anti-fouling agents were applied. The affected parts were mainly those under water, but in the case of prototype 2, growth was also starting to appear on the surface. This implies that for panels which are in constant contact with the water they would have to be constantly treated or an alternative solution (such as a coating that inhibits biological growth) has to be found.

Regular inspection showed salt build-up to be surprisingly limited. Even on the prototype 1 rafts which were almost flat. Modest salt build-up was observed and did probably result in a slight drop in performance but this was clearly more than offset by the lower dust build-up and the cooling effect which resulted in an overall gain of almost 4%. A recent thesis by a student under the supervision of the author confirmed that moderate salt build-up has little effect on the overall output [42].

All components and panels were carefully inspected when they were removed from the sea. All panels, including the ones from prototype 1, which included both flexible thin-film panels and 20 W glass–aluminium panels, showed little corrosion, the conclusion being that since they were sufficiently above sea-level they were only getting sprayed or wet during windy and stormy days. Such conditions are not that frequent in the Mediterranean. Prototype 2 did fail after several months of operation. In spite of our best effort to water-proof the system, water did manage to infiltrate probably as a result of the constant motion which resulted in some of the coatings breaking due to material fatigue.

Design, cost and materials

Simulations and cost-modelling was performed to determine both the optimal size and the best materials choices for a low-cost raft. The project cost-target from inception was EUR50–100/m^2. Several materials including wood, steel pipes, concrete and synthetic materials were considered. The simulations focused on the relative forces on the moorings as the size of the platform increased. This, together with cost–benefit analysis confirmed that the best direction for commercially viable systems that could withstand all weather conditions would be systems at least 50 m in diameter [43,44].

The idea of panels very close to the water (prototype 2) or self-floating panels (prototype 3) were investigated because they could offer benefits such as lower cost and higher output but these are likely outweighed by other disadvantages. More mooring would likely be required as well as longer cable lengths resulting in higher overall costs. The waterproofing would have to be robust enough to survive years in constant contact with sea water and biological growth on the surface of the panels would have to somehow be inhibited. Panel prices continue to fall and therefore adding any coating or regular treatment will make any system uncompetitive [45].

The conclusions reached from the Solaqua project were therefore that it is best to go for large (tens to hundreds of meters) rafts – possibly made of smaller hinged sections (making manufacturing easier and cheaper) for a system to work for many years and be cost-competitive with land-based systems. Work continues on this project including studies on the effect of salt and cooling on the output of panels [13,42,46].

6.3.6.2 Swimsol GmbH (Austria)

Swimsol launched their first system in 2014. They have since installed several systems in the 15–100 kWp range and reports a product pipeline of 1.5 MWp [47]. They applied for an Austrian patent for a floating platform design in 2016 (Figure 6.16) [48].

Figure 6.16 One of the Swimsol systems installed in the South Pacific. (Courtesy Swimsol)

Their focus is in Southeast Asian islands and their cost model is based on competition with diesel generation. Many of these islands, especially the very small ones rely completely on diesel generators for their electricity which can be 3–5 times more expensive than electricity obtained from a grid. Therefore, Swimsol have identified a niche whereby they install systems to provide power to resorts and other communities on these small, remote islands [49].

According to their presentation they use glass–glass panels with edge sealing and protected junction boxes for extra durability. The panels sit flat on the platforms and their platforms appear to be very robust and in shallow water. In the presentation available online, they were proposing (in 2015) to build a 10-MWp system for Male, Republic of Maldives. The cost was estimated to be 30 million USD and the proposed FiT was USD 19c fixed, or a variable one tied to oil prices. No information seems available that this project has been launched, but they have installed and seem in the process of installing several other smaller systems in the region and have recently opened an office in Indonesia to target this market [47,49].

6.3.6.3 The Offshore Passive Photovoltaic Project (Malta)

This project was also financed by the Malta Council for Science and Technology starting in 2014. In 2018 they announced that they were nearing the final stage and would launch an 8-kWp system inside a fish farm.

The project lead by The Malta Council for Science and Technology is designed around the concept developed at MIRARCO (who is a partner in the project together with Azzopardi Fisheries) and Laurentian University (Canada). The concept makes use of flexible thin film panels coated to be completely waterproof which are then floated on the water with minimal support. Tests in freshwater (brackish) water were performed by the MIRARCO group and were then to be repeated in this project at sea [30,50].

In the initial results it was found, as expected, that the panels in the water perform better than the ones on land. They reported a 5% increase but the ambient temperature was not very hot so presumably this number could increase further in hotter climates. They also reported a 15% drop in output when the panels were slightly submerged (emphasizing the need to keep them above the water level), and a slight drop initially due to soiling. The costs they reported were very high in the \$7–11 per Wp for such a floating system which they attribute to the small size, customization, and relatively expensive types of panels which are not industry mainstream [30].

6.3.6.4 Solar-at-Sea (The Netherlands)

This was announced in 2018 via a government grant of \$1.48 million from the Dutch government. The stated objective is to have a 2,500 m^2 system on the sea by 2021 and a study is currently underway in the North Sea farm with a 30 m^2 system. The nine-member consortium is led by Oceans of Energy [51,52]. No information is available at the time of writing on the actual setup.

6.3.7 *Offshore solar: the future*

There are currently few of us working on systems at sea, but the number is growing. It is very probable that systems will continue to be installed by companies at sea in niche markets where the cost of electricity is high and the sea conditions are benign. It would also not be a huge leap to install a large system in a large, sheltered bay or port where the only difference from current systems floating on lakes is the sea water instead of fresh water. These systems could even be based on Ciel et Terre or similar technology which is designed for ponds and reservoirs but there is no reason it would not work in a sheltered bay or port where waves would never reach significant wave heights, say higher than 1 m. While it is not entirely clear where the Solar-at-Sea farm will be launched, the Netherlands certainly has that potential of utilizing sheltered areas of the sea which would not be in competition with other activities such as shipping, leisure and fishing.

Open sea is another challenge altogether. For some locations this is the only choice which is the motivation for continued research and development. It will ultimately all boil down to cost. If systems can be made that are in the same ballpark as ground-based systems, and they can withstand the elements, then they will flourish. However, this still has to be proven. The mentioned teams and others are working towards this goal. In the next year or two it is very likely we will see larger funded projects to achieve this goal or even commercial ventures involving researchers and private companies. My prediction is that by 2022 we would have seen at least one sizable system launched in open sea. If we succeed in developing and maturing this technology, it will go a long way towards helping achieve our CO_2 reduction goals since it will, in some cases, increase by orders of magnitude the surface available for PV farm installation.

There would also be considerable synergy with other offshore uses. An obvious one would be with offshore wind farms, which would already have grid connection and possibly mooring. Another one is fish farms or oil-rigs, which might not have a grid connection and make use of the power generated. There could also be synergy with freshwater generation such as reverse-osmosis plants which consume significant amounts of energy and are obviously situated close to shore. The same holds for any other consumer that is located close to shore.

References

[1] *Solar Boom: Installed Capacity Expected to Double by 2023*, Optics.org, 2018, Available from: http://optics.org/news/9/5/35 [Accessed 14 January 2019].

[2] Werner Ch., Gerlach A., Breyer Ch., and Masson G., 'Growth regions in Photovoltaics 2016 – update on latest global solar market development', 33rd EU PVSEC, Amsterdam, 2017, pp. 2794–2807.

[3] Mule'Stagno L., Yousif C., and Vaquero Palacios E.R., 'Solar photovoltaic systems performance in Malta: potential versus real contribution to the 2020 RE target', 26th European Photovoltaic Energy Conference, Hamburg, 2011, pp. 4381–4384.

[4] *A Success Story: Twenty Years of Support for Electricity from Renewable Energies in Germany*, Renews Special, Issue 41, Renewable Energies

Agency (Germany), September 2010, Available from: http://www.planet-biogas.co.uk/wp-content/mediathek/41_Renews_Special_ENGL.pdf [Accessed 14 January 2019].

[5] Mufson S., *Prices Flat in Polysilicon Market*, Washington Post, July 23rd 2013, Available from: https://www.washingtonpost.com/business/economy/prices-flat-in-polysilicon-market/2013/07/23/914479d0-f3e4-11e2-9434-60440856fadf_graphic.html?utm_term=.0ef31f7b8263 [Accessed 14 January 2019].

[6] *Current and Future Costs of Photovoltaics, A Study*, Agora Energiewnede and Fraunhofer ISE, February 2015. Available from: https://www.ise.fraunhofer.de/content/dam/ise/de/documents/publications/studies/AgoraEnergiewende_Current_and_Future_Cost_of_PV_Feb2015_web.pdf [Accessed 14 January 2019].

[7] *Middle East Blazes a Trail with Solar*, Bloomberg Businessweek (Middle-East), Energy, 4 July 2018, Available from: http://businessweekme.com/middle-east-blazes-a-trail-with-solar/ [Accessed 14 January 2019].

[8] Shimura F., *Semiconductor Silicon Crystal Technology*, San Diego, CA: Academic Press, 2012, pp. 114–146.

[9] Green M.A., *Silicon Solar Cells: State of the Art*, Phil. Transactions of the Royal Society, 2013, A371:20110413.

[10] Mertens K., *Photovoltaics – Fundamentals, Technology and Practice*, 2nd edition, Wiley, 2019, pp. 47–123.

[11] Ciammaruchi, L., Oliveira, R., Charas, A., *et al.*, 'Stability of organic solar cells with PCDTBT donor polymer: an interlaboratory study', *Journal of Materials Research*, vol. 33, no. 13, 2018, pp. 1909–1924.

[12] NREL, *Research Cell Efficiency Records*, Available from: https://www.nrel.gov/pv/assets/pdfs/pv-efficiency-chart.20190103.pdf [Accessed 14 January 2019].

[13] Farrugia A., *Design and Analysis of Different Cooling Effects on Photovoltaic Panels*, M.Sc. Dissertation, University of Malta, 2014.

[14] Grech M., Mule Stagno L., Aquilina M., Cadamuro M., and Witzke U. 'Floating PV installation in Maltese sea waters', 32nd European Solar and Photovoltaic Conference, Munich, 2016, pp. 1964–1968.

[15] Choi Y., 'A case study on suitable area and resource for development of floating photovoltaic system', *International Journal of Electrical and Computer Engineering*, vol. 8, no. 5, 2014, pp. 816–820.

[16] Ciel et Terre, *Hydrelio Technology*, 2015. Available from: https://www.ciel-et-terre.net/ [Accessed 14 January 2019].

[17] Ueda Y., Sakurai T., Tatebe S., Itoh A., and Kurokawa K., 'Performance analysis of PV systems on the water', 23rd European Photovoltaic Solar Energy Conference and Exhibition, Valencia, Spain, 2008, pp. 2670–2673.

[18] Sharp Corporation, 'US vineyard uses space saving floatovoltaics', *Ren. Energy Focus*, vol. 9, no. 4, 2008, pp. 64–65.

[19] Napavalley register, *Gundlach Bundschu Winery*, 2008, Available from: http://napavalleyregister.com/wine/gundlach-bundschu-winery-wraps-up-th-year-by-going-solar/article_f0f8640b-e2cb-5d41-b651-d20d9eef0e1e.html [Accessed 27 November 2018].

[20] Ipannellifotovoltaici, *Photovoltaic Floating Platforms*, 2014. Available from: http://www.ipannellifotovoltaici.com/floating_photovoltaic_pv_platforms.html [Accessed 14 January 2019].

[21] *The First Floating Photovoltaic Plant*, 2011. Available from: http://www.luceonline.it/web/en/ricerca/costruzione-del-primo-impianto-fotovoltaico-galleggiante-in-puglia/ [Accessed 14 January 2019].

[22] Ferrer-Gisbert C., Ferrán-Gozálvez J.J., Redón-Santafé M., Ferrer-Gisbert P., Sánchez-Romero F.J., and Torregrosa-Soler J.B., 'A new photovoltaic floating cover system for water reservoirs,' *Renew. Energy*, vol. 60, 2013, pp. 63–70.

[23] Harney B.M., *Improvement in PV Output using a Simple Single-Axis Tracker*, M.Sc. Dissertation, University of Malta, 2014.

[24] Pronk A.D.C., Dizdar D., and Schuurmans W., 'Sun-spotter: gravity point displacement as solar-tracking principle,' Proc. Of the IASS Annual Symp., Amsterdam, 2015, pp. 1–12.

[25] Ciel et Terre, *Floating Solar System – Proto Piolenc – 15 kWp*, 2011. Available from: http://www.ciel-et-terre.net/essential_grid/floating-solar-system-proto-piolenc/ [Accessed 14 January 2019].

[26] S. Chiavari, *Floating on Water*, 2012, Available from: http://www.wwdmag.com/sites/wwdmag.com/files/40_WWD0412.pdf. [Accessed 14 January 2019].

[27] Kim S.H., Yoon S.J., Choi W., and Choi K.B., 'Application of floating photovoltaic energy generation systems in South Korea,' *Sustainability*, vol. 8, no. 12, 2016, p. 1333.

[28] Synergy S., *Solaris Installs First Live Floating Concentrated PV System Connecting to Israel Electric Corp. Grid*, 2011, Available from: http://www.solaris-synergy.com/News.html# [Accessed 14 January 2019].

[29] SERIS, *Floating PV R&D and Testbed*, 2018 Available from: http://www.seris.nus.edu.sg/activities/Floating-PV-R&D-and-Testbed.html [Accessed 14 January 2019].

[30] Trapani K., and Millar D., 'The thin film flexible floating PV (T3F-PV) array: the concept and development of the prototype', *Renewable Energy*, vol. 71, 2014, pp. 43–50.

[31] C. Technica, *Kyocera TCL Solar Completes Two Floating Solar Power Plants in Japan*, 2015, Available from: https://cleantechnica.com/2015/04/27/kyocera-tcl-solar-completes-two-floating-solar-power-plants-japan/ [Accessed 14 January 2019].

[32] Kyocera, *KYOCERA TCL Solar Begins Operation of Japan's Largest 13.7MW Floating Solar Power Plant, 2018*, Available from: https://global.kyocera.com/news/2018/0301_wvfh.html [Accessed 14 January 2019].

[33] Phys.org, *Brazil Uses Dammed Lake Surface for Floating Solar Panels*, 2016. Available from: https://phys.org/news/2016-03-brazil-lake-surface-solar-panels.html [Accessed 14 January 2019].

[34] Esteves Galdino M.A., and de Almeida Olivieri M.M., 'Some remarks about the deployment of floating PV systems in Brazil,' *J. Electr. Eng.*, vol. 5, no. 1, 2017, pp. 10–19.

[35] Rohan Hamden & Associates, *Floating Solar Array – A Case Study, Published for the Local Government Association of Southern Australia and Northern Area Council*, 2016, Available from: *http://www.rohanhamden.com.au/assets/1613-Floating-Solar-brochure-WEB.pdf* [Accessed 14 January 2019].
[36] P. of Rotterdam, *540,000 Solar Panels in the Slufter on the Maasvlakte*, 2015, Available from: https://www.portofrotterdam.com/en/news-and-press-releases/540000-solar-panels-in-the-slufter-on-the-maasvlakte [Accessed 14 January 2019].
[37] Re. Vincent Shaw, '*China Connects 20 MW Solar Plant in Coal Mine Subsidence Area*,' 2016, Available from: http://reneweconomy.com.au/china-connects-worlds-largest-fishpond-solar-plant-to-grid-read-more-httpwww-pv-magazine-comnewsdetailsbeitragchina-connects-worlds-largest-fishpond-solar-plant-to-grid_100024091ixzz45zulk-80007/ [Accessed 14 January 2019].
[38] Scrivener, D., *Europe's Largest Floating Solar Farm Sets Sail*, 2016, Available from: https://www.pagerpower.com/news/europes-largest-floating-solar-farm-sets-sail/ [Accessed 14 January 2019].
[39] Publicover Brian, *South Korea to Set New PV Construction Rules after Landslide Destroys Array*, 2018. Available from: https://www.pv-magazine.com/2018/07/12/south-korea-to-set-new-pv-construction-rules-after-landslide-destroys-array/ [Accessed 14 January 2019].
[40] Kaneko K., and Nikkei BP, CleanTech Institute, *7.55MW Floating Solar Plant Damaged by Typhoo*, 2016. Available from: https://tech.nikkeibp.co.jp/dm/atclen/news_en/15mk/090600817/?ST=msbe [Accessed 14 January 2019].
[41] *General Membrane*, 2018, Available from: http://www.generalmembrane.it/products/en-general-fix-sh-on-synthetic-roofs [Accessed 14 January 2019].
[42] Gretkowska L., *A Study on the Performance of Solar Photovoltaic Modules Exposed to Salt Water*, M.Sc. Dissertation, University of Malta, 2018.
[43] Aquilina M., Mule'Stagno L., Grech M., and Sant T., 'Determining the optimum shape and size for an offshore floating photovoltaic system', OSES 2016, Malta, 2016.
[44] Vassallo E., and Mule'Stagno L., 'Assessing the design options for the optimisation of a 2MWp ground mounted system in Malta', Sustainable Energy 2014 – The ISE Annual Conference, Malta, 2014.
[45] PV Magazine, *Module Price Index, October 2018: Everything must go?!*, 2018, Available from: https://www.pv-magazine.com/features/investors/module-price-index/ [Accessed 14 January 2019].
[46] Muscat M., *A Study of Floating PV Module Efficiency*, M.Sc. Dissertation, University of Malta, 2014.
[47] Swimsol, *Solar Energy for Tropical Islands*, 2015. Available from: http://www.irena.org/EventDocs/Maldives/3FloatingsolarPVinvestmentcaseSwimsol_15.1110.pdf [Accessed 14 January 2019].
[48] Praschl A., Mikats T., and Putschek M., *Floating Platform*, Austrian Patent Application, PCT/AT2015/050325, Swimsol GmbH, 2015.
[49] Energy Monitor Worldwide, *The European Way to Green Energy in Southeast Asia*, Amman, October 2016.

[50] Malta Independent, *OPPV Research Project Approaching Final Stage*, 2018, Available from: http://www.independent.com.mt/articles/2018-03-20/education/OPPV-research-project-approaching-final-stage-6736186591 [Accessed 14 January 2019].

[51] Rolle R., *Insight: Turning the Tide – Towards Offshore Solar Energy, 2018*, Available from: http://www.tribune242.com/news/2018/sep/17/insight-turning-tide-towards-offshore-solar-energy/ [Accessed 14 January 2019].

[52] Probert Andy, *Largest Floating Solar Energy Farm at Sea Planned for the Netherlands*, 2018, Available from: https://greenerideal.com/news/energy/largest-floating-solar-energy-farm-at-sea-planned-for-netherlands/ [Accessed 14 January 2019].

Chapter 7
Offshore support structure design

Erin E. Bachynski[a] *and Maurizio Collu*[b]

7.1 Offshore support structures

The definition of the "support structure" for an ocean renewable energy (ORE) device may vary somewhat depending on the device in question, but we can generally consider the support structure to include those structural components which are not directly considered to be part of the energy extraction system but serve to provide the necessary stability and fixity in order for the device to be able to function. For a bottom-fixed offshore wind turbine or current turbine, this may include the tower, dry or submerged column or truss structures, and the foundation, including any piles which extend through the seabed [1]. For a floating wind turbine (FWT), the tower, hull, mooring lines, and anchors are all part of the support structure.

Support structures for renewable energy devices have been developed with inspiration from the offshore oil and gas (O&G) industry. The most common concepts are described in Sections 7.1.1 and 7.1.2. There are, however, some important differences in the design requirements for support structures for renewable energy devices compared to offshore O&G platforms:

1. The payload for ORE is typically much lighter than for O&G, and its center of mass may be located at a different relative location: wind turbines have a large mass high above the still water level, while marine current energy devices are further below.
2. The requirements for ORE stationkeeping are dictated by the allowed deformations of the power cable, while O&G platforms have requirements related to drilling, extraction, and exploration activities.
3. The dynamic loads induced by the power extraction device for ORE may occur at significantly different frequency ranges compared to O&G. Wind or current turbines induce loads at their rotational frequency (and multiples thereof) in addition to slowly varying loads corresponding to the natural variations in wind and current. Wave energy devices are often required to operate in resonance

[a]Department of Marine Technology, Norwegian University of Science and Technology, Trondheim, Norway
[b]Naval Architecture, Ocean and Marine Engineering, University of Strathclyde, Glasgow, UK

within the primary wave frequency range to maximize energy extraction. This range is normally avoided by designers of O&G platforms.

4. The consequences of failure may differ between ORE and O&G: most O&G platforms are manned, and significant environmental damages (such as oil spills) may result due to a failure of the support structure. For ORE devices, the risks of loss of life or health are reduced since these platforms are typically unmanned and the environmental consequences are relatively minor.
5. Most ORE devices are intended to be mass-produced and deployed in array formations, while O&G platforms are often unique. ORE substructures must therefore be designed for efficient mass construction, transportation, and installation.

Existing concepts for support structure design are summarized in Sections 7.1.1 and 7.1.2. Design criteria and analysis of support structures for ORE are discussed in greater detail in Section 7.2.

7.1.1 Bottom-fixed support structures

Bottom-fixed support structures are characterized by a high degree of fixity at the seabed. Bottom-fixed support structures are typically used in relatively shallow water (less than 40–60 m water depth for offshore wind turbines). They are able to maintain stability through the transfer of loads to the seabed, either through deep piles that transfer the loads to the soil or rock (monopiles, jackets) or through large weight and gravitational forces (gravity-based structures (GBS)). The characteristics and applications of several types of bottom-fixed support structures are discussed separately below. Note that coastal breakwaters and land-based structures for wave and tidal energy are not considered here, as these cannot be considered to be operating in a truly offshore environment. Choosing among different types of bottom-fixed support structures depends strongly on the water depth and soil conditions.

7.1.1.1 Gravity-based structures

GBSs rely on their weight to provide stability and restrain the ORE device against wind, wave, or current loads. In the case of wind turbines—which induce a large overturning moment due to the thrust force acting far above the seabed—a large mass and low center of gravity are required. The most cost-effective way to obtain the required mass has typically been through the use of concrete and water ballast. Many GBS structures therefore take the form of a large-diameter caisson near the sea bed and a more slender cylinder which extends above the caisson (Figure 7.1). In this way, the center of gravity is retained as low as possible while wave loads are reduced. An important design advantage for GBSs is the fact that one can avoid pile-driving, although there may still be work required to prepare the seabed such that it is flat and can support the GBS weight evenly.

The required self-weight of a GBS increases as the water depth increases due to the increased moment arm. As a result, GBSs for offshore wind turbines are so far mostly applied in fairly shallow waters. There are additional challenges related

Figure 7.1 Left: GBS foundation for an offshore wind turbine modeled in SIMA. Middle: Seabased wave energy converter components [5,6]. Right: Wavestar jack-up platform [7]. (Middle) Reproduced, with permission, from ASME; (Right) Reproduced, with permission, from Wavestar

to the construction and transport of GBSs. Significant storage space at the yard is needed if several GBSs are to be installed in a season [2]. Furthermore, due to the large weight of the structures, this storage area must be located on solid and flat ground. The large weight also implies the use of large cranes during installation, although some self-installing GBS designs have been developed to eliminate this disadvantage [3].

The large caisson of a GBS disturbs the hydrodynamic flow. For wind turbines, this implies a need for scour protection around the GBS. For current energy devices, this effect will generally make GBSs less attractive. Some wave energy devices (such as Seabased or Oyster [4]), which require a bottom reference point also use concrete or steel gravity-based housings for the machinery. In this case, the caisson is sufficiently small to avoid disturbing the wave-induced water particle motions.

In addition to traditional GBSs, alternative methods of using gravity to one's advantage have also been considered for wave power devices. The WaveStar platform, for example, employs a jack-up platform as its support structure. A jack-up platform typically has the shape of a ship or barge, with three or four legs which can be extended to the sea floor. When the legs are extended, the platform is lifted out of the water. The stability of the platform comes from the combination of its weight (which presses the legs down on the sea floor) and the arrangement of legs, which helps to resist overturning moments. Because only the legs are wet during operation, a jack-up structure is quite transparent to wave forces and also provides a large, dry, working area. This type of platform is not necessarily well-suited to support wind turbines or marine current energy converters which encounter larger horizontal forces. However, this type of platform may be promising for wave energy converters (WECs).

7.1.1.2 Monopiles, jackets, and tripiles

The overwhelming majority of offshore wind turbines installed as of 2018 are supported by monopile support structures. A monopile is a single, typically cylindrical, and hollow support, which is driven into the seabed like a fencepost. Wind, wave, and current loads are transferred from the pile to the surrounding soil. Monopiles are relatively simple to construct and transport. Although their weight may become large, the structure is simple and can be loaded efficiently onto a barge. For wind turbines, the challenges related to monopiles include their increasing diameter as wind turbines increase in size (requiring more capacity for rolling the metal, heavier lifts, more difficult hammering, and larger wave loads), as well as the noise and damage to the pile head during hammering. The effects of the hammer on the top of pile make it impossible to connect the tower directly to the monopile using bolts or other fasteners. A grouted transition piece is therefore used to connect the foundation to the tower.

Monopiles (Figure 7.2(a)) have also been used successfully as support structures for marine current turbines (MCTs) [8]. The vertical location of the thrust force in a MCT is much lower than in the case of a wind turbine. This reduces the overturning moment that the substructure must support. The monopile solution has an additional advantage for MCTs: it can easily be designed to extend above the sea surface with the same diameter, allowing for a lift system which can raise the turbine out of the water for inspection or maintenance (see Figure 7.2(b)). Installation of a monopile support structure can, however, prove to be challenging in areas with strong currents. A jack-up vessel is typically preferred for performing the piling operations, but the jacking process may be limited by the water depth and current speeds of interest.

Jacket structures (Figure 7.2(c)) derive their name from the way that the legs of the jacket are fitted around the piles which are driven into the ground: like jacket sleeves on a person's arms. The traditional truss structure is very structurally efficient: large forces and overturning moments can be taken up by a much smaller

(a) (b) (c) (d)

Figure 7.2 (a) Monopile wind turbine modeled in SIMA, (b) SeaGen monopile marine current turbine with turbine lifted above the water for maintenance [8]. (c) OC4 jacket wind turbine in SIMA without pile model, and (d) tripile wind turbine in SIMA. (b) Reproduced, with permission, from SeaGen

steel mass than that which is needed for a monopile, and the thin members attract smaller wave and current loads than a monopile. The flow disturbance around small members is also less significant than around a monopile. Despite these potential advantages, jacket foundations are not presently widespread within the offshore wind industry: this may be due to the complexity of construction, also linked to the relatively complex joints and welding required, required storage space during construction, relative lack of experience and supply chain, and the development of ever larger monopiles.

A tripile (or quadpile), which stands on several smaller spread piled supports, represents a compromise between monopile and jacket designs. Tripiles (Figure 7.2(d)) take advantage of the spread piles to take up the overturning moment more efficiently, while the single vertical pile maintains some of the simplicity of a monopile. While this combination is promising, it also maintains some of the disadvantages of both jackets and monopiles for large offshore wind turbines: like jackets, tripiles also require large storage space during construction, and, like monopiles, large crane capacity is needed for lifting operations. Tripiles are seen as a promising solution for marine current energy [9]; the installation of several smaller piles may be easier than installing one large pile in areas with strong currents.

A common disadvantage of monopiles, jackets, and tripiles is the requirement to drive one or more piles into the seabed. Figure 7.3 illustrates the hammering process. This process is time-consuming and noisy, and underwater noise emissions may have adverse effects on marine life [10]. Suction caisson foundations are a recently developed alternative: a caisson resembling an overturned bucket can be drawn down and held in place by pumping water out of the caisson to create a pressure differential. This type of foundation—which can be applied to monopile-, jacket-, or tripile-type designs—can be installed more efficiently than piling and without the harmful noise level [11]. These solutions have been applied to offshore O&G platforms where the loads are primarily vertical, but the dynamic overturning moment from a wind turbine requires further testing and development of suction caissons.

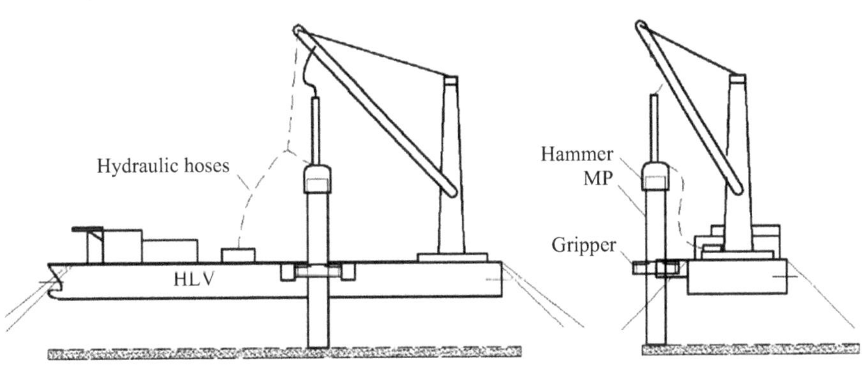

Figure 7.3 System setup for hammering of a monopile (MP) foundation from a floating heavy lift vessel (HLV) [12]. Image courtesy of Lin Li

7.1.2 Floating support structures

Unlike bottom-fixed support structures, floating support structures are characterized by a low degree of fixity to the seabed. They must obtain their stability and resistance to overturning moments through a combination of hydrostatic stiffness and mooring system contributions to stiffness. Hydrostatic stiffness arises from buoyancy and gravitational forces and can provide stiffness in the vertical modes heave, roll, and pitch (see Figure 7.4 for a typical definition of degrees of freedom (DOFs) for a rigid floating body). The mooring system is primarily needed in order to maintain the horizontal position (within certain limits) with respect to the sea bed but can also provide some stiffness in the vertical modes.

Considering the pitch degree of freedom, with the z-axis defined pointing upward and zero at the waterline, the linearized stiffness can be estimated as

$$C_{55} + K_{55} = \rho g I_{wp} - Mgz_g + \rho g V z_B + K_{55} \tag{7.1}$$

where C_{55} is the hydrostatic stiffness and K_{55} is the mooring system stiffness, ρ is the density of water, and g is the acceleration due to gravity. The hydrostatic stiffness arises from the first three terms on the right hand side: the first due to the water plane moment of inertia (I_{wp}), the second due to the mass (M) and vertical center of gravity z_g, and the third due to the displaced volume V and center of buoyancy z_B. A common classification of offshore structures is based on the main mechanism adopted to obtain a sufficiently large, positive $C_{55} + K_{55}$: using a large waterplane moment of inertia (semisubmersibles and barges), using a center of gravity far below the still water level (spars), or using the mooring system (tension leg platforms (TLPs)).

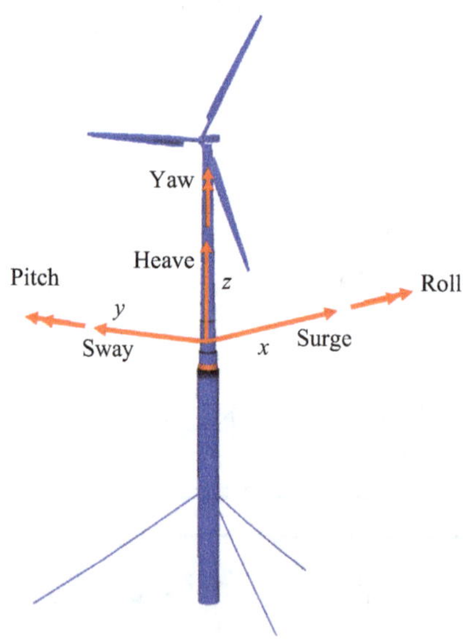

Figure 7.4 Rigid body motions of a floating body

The different types of floating platforms are described separately below; catenary and semi-taut mooring systems are described in greater detail in Section 7.1.2.4.

7.1.2.1 Spar platforms

In order to lower the center of gravity as much as possible, a spar platform is typically long and slender, with heavy ballast placed near the bottom. Equinor's Hywind wind turbine (Figure 7.5) is a well-known example of such a platform [13]. The platform is moored with a catenary mooring system and all of the natural frequencies for rigid body motions are thus placed far below the primary wave frequency range. The slender shape also makes a spar relatively transparent to the wave forces, and the simplicity of the shape is an advantage for construction. On the other hand, the applicability of spar platforms is limited to areas where the water is sufficiently deep: a deep area with calm water (such as a protected fjord) is needed for upending the spar and mounting the ORE device, the route for transporting the spar to its final location must be deep enough to avoid grounding, and the final location for installation must of course be sufficiently deep.

Spar platforms are also susceptible to several interesting dynamic phenomena, which have to be considered during the design. The linear hydrostatic stiffness in heave C_{33} for a floating platform is found from the change in displaced volume due to a heave displacement:

$$C_{33} = \rho g A_{wp} \tag{7.2}$$

Figure 7.5 Hywind Scotland wind park, with spar-type foundations. Image courtesy of Equinor

where A_{wp} is the waterplane area. As spar platforms are designed with a small waterplane area in order to avoid wave excitation, the stiffness in heave is quite low. As a result, the heave and pitch natural periods may also be similar. The linearized pitch stiffness C_{55} in (7.1) should then be modified to account for the change in the volume and center of buoyancy due to the heave motions. The time-varying pitch stiffness leads to a possible Mathieu instability, especially if the pitch and heave natural periods are similar [14].

Spar platforms are also particularly susceptible to time-varying excitation due to vortex shedding [15]. A cylinder in a steady current of a sufficient velocity will shed vortices, alternating from each side. For a spar, this effect may cause a strong oscillatory force (both perpendicular to and in-line with the current). Using the Strouhal number, the frequency of the vortex shedding can be estimated based on the velocity of the flow and the diameter of the cylinder [16]. This effect is particularly problematic if the shedding frequency is similar to the natural frequency of the platform. Small changes to the geometry, such as helical strakes, which prevent a long correlation length in the shedding, can help to reduce vortex-induced motions.

7.1.2.2 Semisubmersible platforms and barges

Semisubmersible platforms and barges primarily use the waterplane moment of inertia I_{wp} in order to achieve stability. For pitch motions, I_{wp} is given by the following equation:

$$I_{wp} = \iint_{A_{wp}} x^2 dxdy \tag{7.3}$$

To obtain the necessary stiffness for stability, it is therefore necessary to increase the waterplane area which is far from the neutral axis. For a barge, this is achieved by a large waterplane area, while a semisubmersible typically uses smaller components placed further from the neutral axis and connected by a system of braces or pontoons. Both semisubmersibles and barges use catenary mooring systems and are designed to have their natural periods for rigid body motions above the wave period. Obtaining a sufficient pitch stiffness for stability without having too much stiffness in heave (resulting in a heave resonance which can be excited by first-order wave forces) can, however, be a challenge during design [17].

Compared to spar platforms, the shallower draft of semisubmersibles and barges represents a significant advantage due to the ease of installation, such as the possibility to install the wind turbine quayside/nearshore and transport the complete platform to the operational site, therefore avoiding the use of expensive, specialized vessels. This whole platform can also be transported back to the maintenance shipyard for major maintenance events. Nonetheless, the increased volume near the waterline can be a disadvantage due to the consequently larger wave forces. Several semisubmersible and barge-like floating platforms for offshore wind are in current development (such as WindFloat, Nautilus, OO Star, Ideol) [18,19]. Different strategies for addressing the challenge of the heave period are apparent: large heave plates or wide pontoons can be used to increase the heave-added mass, or a moonpool can be tuned to aid in damping the motions at certain frequencies.

Semisubmersibles have also been seen as promising for combined wind- and wave-energy devices: the open area in between columns can also house wave energy conversion devices (e.g., W2-Power, Poseidon and 2Wave1wind). Semisubmersible platforms for marine current energy have also been proposed and tested in recent years, such as the Evopod device.[1] Examples of ship-shaped barges used for current turbines are the SR250 and SR2000 tidal energy devices by Scotrenewables,[2] with a rated power of, respectively, 250 kW and 2 MW, tested at the European Marine Energy Centre (EMEC[3]) site in the Orkney Islands. For marine current energy, careful consideration of the semisubmersible form is needed in order to reduce the wave-induced motions and prevent the platform from disturbing the flow to the turbine. Furthermore, the platform should ensure that the maximum angle between the current velocity and the current turbine axis is below a maximum threshold (~10 degrees), as for wind turbines, but in addition should also ensure that a minimum distance is kept between the tip of the blade and the free surface, to avoid cavitation. Examples of semisubmersible support structures for ORE are shown in Figure 7.6.

Semisubmersibles built from concrete, rather than steel, have also received significant research interest [19,20]. Potential cost reductions using concrete can only be realized with reliable design tools which can predict the internal loads in the hull. Splitting and jacking forces which can occur due to wave forces in opposing directions on different columns must be considered—regardless of the material from which the hull is to be built [21].

7.1.2.3 Tension leg platforms

Unlike spars and semisubmersibles, TLPs employ a taut mooring system which provides stiffness and stability in the vertical direction as well as in the horizontal modes of motion. The pre-tension of the tendon system arises due to the excess

Figure 7.6 Semisubmersible support structures for ORE devices. From left to right: WindFloat wind turbine (image courtesy of Principle Power), Nautilus wind turbine (image courtesy of Nautilus Floating Solutions), W2 Power combined wind-wave energy, Evopod current turbine (image courtesy of Ocean Flow Energy Ltd.)

[1]http://www.oceanflowenergy.com/Evopod-Technology.html
[2]http://www.scotrenewables.com/
[3]http://www.emec.org.uk/facilities/tidal-test-site/

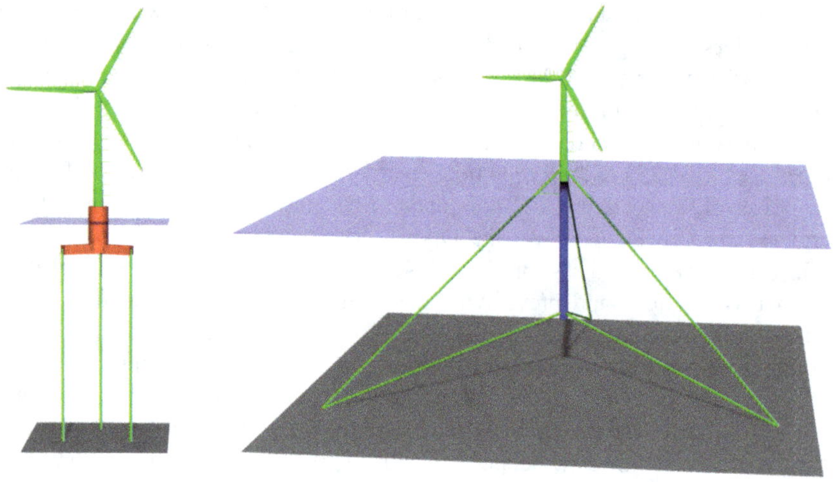

Figure 7.7 TLP and TLB wind turbines in SIMA, designs based on [22,23]

buoyancy of the platform. While a TLP is typically characterized by vertical tendons (left image in Figure 7.7), a tension leg buoy (TLB) may have angled tendons which also provide significant stiffness in surge and sway (right image in Figure 7.7).

For a TLP with n vertical tendons, the linearized mooring system stiffness in pitch depends on the location of the tendon attachment to the hull (at radius r, angle θ in the horizontal plane, and vertical position z), on the pre-tension (F) and stiffness (Young's modulus E and area A) of the tendons, and on the unstretched tendon length (l_0):

$$K_{55} = \sum_{j=1}^{n} \left(\frac{F}{l_0} z^2 + \frac{EA}{l_0} r^2 \right) \cos^2 \theta_j. \tag{7.4}$$

The main contribution to the pitch stiffness generally comes from the EA term, which is much larger than the hydrostatic stiffness. As a result, the heave, pitch, and roll natural periods are quite short, well below the primary wave excitation range. The main advantages of TLP and TLB concepts are the small platform motions in pitch, small footprint on the seabed (which could allow for multiple uses of the same area), and potentially lower steel mass than spar or semisubmersible concepts [22]. On the other hand, the tendon system (including anchors which can handle the vertical load) is expensive, and installation may be both challenging and costly. TLP and TLB concepts may also be more sensitive to variations in water level due to tides or storm surge. During operation, loss of tension in one line can potentially lead to the tendon disconnecting from the anchor, which may result in capsize. The commercial development of TLP and TLB wind turbines has not reached the same level of maturity as spar or semisubmersible platforms, although their development remains an active area of research (also for combined wind-wave platforms) [24–26].

Submerged TLB systems have also been proposed for tidal devices, such as the PLAT-O by Sustainable Marine Energy Ltd.[4] The main advantage of this configuration is that since its operational position is at mid-water depth, its tidal turbine can extract more energy with respect to seabed-fixed current turbines (the closer to the seabed, the lower the current velocity is), and at the same time be less exposed to the wave loads than a semisubmersible or barge configuration.

TLPs and TLBs are also restrained in the horizontal modes of motion as a result of the pretension in the tendons. The surge stiffness of a TLP, for example, is a function of the number of lines and the pretension:

$$K_{11} = \sum_{j=1}^{n} \frac{F}{l_0}. \tag{7.5}$$

The stiffness in surge should not be too large: the natural periods of the TLP in surge, sway, and yaw should be designed to be longer than the typical wave excitation periods. In contrast, TLBs are designed to have horizontal periods shorter than the wave periods.

7.1.2.4 Mooring systems

Mooring systems are used to restrain the motions of floating ORE devices which are intended to produce electricity near a fixed point. The allowable offset is typically dictated by the loads in the power cable or by the needed distance between devices. For floating platforms other than TLPs, typical mooring systems are based on the catenary principle, or employ semi-taut lines.

A catenary mooring system relies on the weight of the mooring line to provide the desired stiffness. The principle can be understood based on a static analysis. Following Faltinsen [16], consider a single element of the mooring line, with an unstretched length ds, cross-sectional area A, Young's modulus E, weight (per length) in water w, and mean hydrodynamic forces F and D (Figure 7.8). The tension in the line is T. The equations for static equilibrium become:

$$dT - \rho g A dz = [w \sin\phi - F(1 + T/(AE))]ds \tag{7.6}$$

$$Td\phi - \rho g A z d\phi = [w \cos\phi - D(1 + T/(AE))]ds. \tag{7.7}$$

These equations are nonlinear and cannot generally be solved explicitly. Iterative methods can be used to find the shape of the line and the tension along the line, and the static equations can be solved for different positions of the fairlead. Based on the resulting tension and angle at the fairleads, the quasi-static stiffness due to the mooring system can be obtained. The quasi-static stiffness is nonlinear, which can lead to effectively different stiffness at different offset positions.

While the quasi-static solution gives useful information about the mooring line stiffness for slowly varying motions, dynamic effects from the mooring lines

[4] https://sustainablemarine.com/plat-o

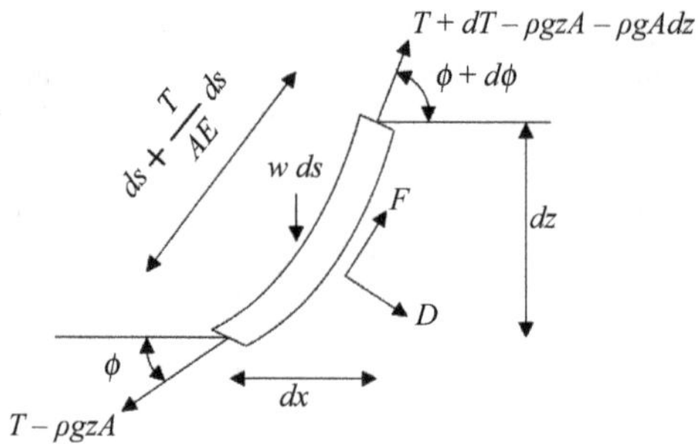

Figure 7.8 Element of a mooring line, based on [16]

should also be considered. Finite element (FE) models using, for example, bar elements with lumped mass models can capture the inertia loads due to the mass and added mass along the mooring lines, the drag forces acting on the lines, and the stiffness effects due to both geometry and axial extension. Dynamic effects of mooring lines have been shown to be quite important for floating wind and WECs [27,28].

7.2 Support structure design

Designing the substructure of a marine renewable energy device is an iterative process, which begins with an initial selection of feasible concepts and preliminary sizing. This is followed by a more refined load and response analysis in order to check that the structural strength is sufficient to withstand the load effects, and that the support structure can provide the necessary conditions for good power production (serviceability). An ideal design would minimize the levelized cost of energy while meeting all safety requirements. As the design is refined, the cycle of analysis is repeated iteratively with updated information. The following subsections examine the initial design process and the relevant load and load effects analysis.

7.2.1 Initial design

As a first step in the initial design stage, one must establish the relevant criteria for a successful support structure design. These requirements might include

- *Stability in intact and damaged conditions.*
  This very simple requirement—that the structure should have a tendency to return to its initial position—is an important condition for initial sizing, which in turn drives the initial cost estimates. For example, a floating platform must have enough buoyancy to support the weight of the ORE device (in addition to the floating platform's own weight), and enough static restoring force to prevent capsizing (as discussed in Section 7.1.2). Stability during both installation and operation should be considered.

- *Restricted ranges of eigenfrequencies to avoid excitation of resonance by, i.e., rotor frequencies, first-order wave forces, or vortex shedding.*
For MCTs or wind turbines, rotor rotation should not excite other resonances. This implies that there should not be any eigenfrequencies of the support structure which coincide with f_{1p} (the frequency of rotation) or $f_{np} - nf_{1p}$, where n is the number of blades. For a wind turbine supported by a monopile substructure, this requirement may be an important driver in the sizing of the pile, see Example 7.1.
While most offshore structures are designed to avoid first-order wave forces, this requirement may differ for certain wave energy devices which are designed for resonant-type response in waves.
- *Maximum offsets or displacements, and limits on dynamic motions.*
The substructure must provide a platform from which the ORE device can effectively operate, which may set limits on the platform motions or on the displacement of certain components. Most ORE devices export power through a flexible cable. The allowable stresses in the power cable due to stretching or bending may set the limits for the allowable displacements of floating platforms. As previously mentioned, this limitation is particularly important for mooring system design. The operation of the ORE device itself may also set limits, such as maximal angular displacements related to the efficiency of MCTs or wind turbines, or maximum relative motions related to power takeoff system limits for WECs. Certain onboard equipment may also set limitations on dynamic responses such as accelerations. Wind turbine manufacturers may set limitations on nacelle accelerations. These limitations can be important design drivers for FWTs.
- *Maximum weight and height of offshore crane lifts during installation.*
The construction phase is a critical consideration for initial design of ORE devices. The feasibility of constructing and installing the support structure and device using existing vessels which are available within the project budget may be a design-driving factor.
- *Human or drone access for inspection or repair.*
Although ORE devices are typically expected to operate autonomously or with some land-based supervision, one must be able to access the devices from time to time for inspection and maintenance. For example, modern turbines are visited approximately four times per year [29]. Although other technological solutions (such as drone-based inspection) may reduce the number of times per year when a human inspector needs to access the device, the ease and safety of human access remains an important consideration. For example, a platform which can support a MCT fully lifted out of the water may have significant operational advantages over a smaller platform which is cheaper to construct but requires underwater inspection.
- *Maximum footprint during construction or in the installed configuration.*
Ocean space is a limited resource, as is storage space in shipyards and onshore. In some cases, considerations related to parallel activities at sea (such as fishing operations near ORE devices) may give an incentive to reduce the

device's footprint. Such considerations affect the design of the mooring system or electrical cable array system in particular, but can also be important in the choice of different types of bottom-fixed structures.

- *Survival of the support structure and/or ORE device in specified severe environmental conditions, or under defined accidental situations.*
  In most design standards, considerations related to structural integrity in severe environmental conditions are discussed with respect to the ultimate limit state (ULS), while accidental situations (such as ship collision or loss of a mooring line) are part of the accidental limit state (ALS). The design standards give some guidelines as to which situations designers must consider in order for the design to be approved, and the evaluation of the structural responses in these conditions is an important part of the design process (see Section 7.2.4). During initial design, these considerations may be significantly simplified and previous experience is used in order to make rough sizing decisions.
- *Sufficient fatigue strength to operate for 20–30 years.*
  The fatigue limit state (FLS) is encountered when a structure fails due to cyclic loading. As with the ULS and ALS considerations, significant engineering analysis is required to evaluate the structure's resistance to such failure. At an early design stage, FLS might only be considered in terms of natural frequency placement, the identification of critical parts of the structure, and a review of similar concepts and previous experience.

One method of obtaining an initial concept design is to consider similar concepts and examine how they might be scaled to meet the present requirements, as, for example, when a larger ORE is to be installed. Some knowledge of how different loads typically scale (see Sections 7.2.1.1 and 7.2.1.2) can therefore be useful for support structure design. Examples 7.1 and 7.2 demonstrate simplified calculations which might be carried out during the initial design procedure. For more detailed design, a better understanding of the loads and load effects is needed, as discussed in Section 7.2.2.

7.2.1.1 Scaling of wind or current turbines

Wind turbines and MCTs follow many of the same operational and scaling principles: power extraction depends on the flow velocity and rotor area. If the main scaling parameter λ represents the scaling of the length of the blades, and we assume the same geometry, materials, tip speed ratio, and incoming flow velocity, both the power and thrust force scale with λ^2. A turbine with blades which are twice as long can produce four times as much power and also produces a thrust force which is four times larger. On the other hand, the weight of the rotor depends on λ^3. Assuming the same density, the volume increases cubically. This implies that the weight of the new and enlarged rotor would be eight times larger. The rotor torque will also increase by λ^3. Given our assumption of constant tip speed ratio, the rotor speed will decrease as $1/\lambda$. A summary of the scaling factors is given in Table 7.1.

The rates of change of different loads—as well as the changes in the excitation frequencies from 1p and 3p—have important consequences when considering

Table 7.1 Turbine scaling, assuming constant tip speed ratio and fluid inflow, geometric scaling, and identical material parameters

Parameter	Scale factor
Length	λ
Power	λ^2
Mass	λ^3
Thrust	λ^2
Generator torque	λ^3
Rotor speed	λ^{-1}

upscaling the support structure for such turbines. For a bottom-fixed structure, the increase in mass compared to the decrease in rotor frequency will tend to work out well for avoiding 1p and 3p excitation. A larger turbine will, however, push the natural frequencies of the structure closer and closer to the region of primary wave excitation. For floating structures, the increase in weight (as well as the typical increased in the distance from the free surface) is an important consideration for stability. Although the thrust force scales slower than the weight, the overturning moment due to the thrust force increases due to the change in arm as well. Geometrically scaling the platform according to the needed additional volume may be feasible, but scaling of various components individually is likely to give a more optimal solution in terms of the resulting natural frequencies. For example, for a semisubmersible wind turbine, the upper columns and lower columns could be scaled differently [17].

Furthermore, scaling of the rotor does not strictly follow the scaling laws given above. Changes in technology (e.g., the shape of the blades, or the material used for construction, or the type of gearbox applied), changes in the resource (such as the wind speed higher above the sea level or further from the coast), or non-scaling aspects (such as access doors for humans, or standard sizes of plates or bolts, or the size of turbulent structures in the flow) can result in deviations from the "typical" scaling laws [30]. Other types of wind or marine current energy devices may also follow different scaling laws, for example, based on the frequency of vortex shedding [31].

7.2.1.2 Scaling of wave energy converters

Scaling of WECs is somewhat more complex than scaling of wind or current turbines. While energy transported by ocean waves for a given crest length is known, the performance of a wave energy device with modified dimensions depends on how the upscaling affects the dynamics of the device, which may vary significantly for different types of extraction principles. A different type of upscaling of the support structure may therefore occur: a larger number of small devices may be attached to the same support structure or moored in an array [32]. Depending on the wave energy extraction principle, special attention should be paid to changes in reactive forces or end-stop forces. Both the local load levels and the phasing among different devices in an array may be important for designing the support structure.

Example 7.1 (Eigenfrequency placement for a 10-MW monopile wind turbine): *Consider a simplified representation of a monopile wind turbine supporting the DTU 10 MW reference wind turbine [33]: a uniform hollow cylinder which is completely fixed at the seabed, with a mass at the top which represents the blades, nacelle, and hub (Figure 7.9). Main particulars of the DTU reference turbine are given in Table 7.2.*

Following [34], the natural frequency of a cantilevered beam with a mass at the end can be estimated as

$$\omega = \sqrt{\frac{k}{(0.23m_{beam} + M)}} \tag{7.8}$$

where m_{beam} is the mass per unit length of the beam and M is the end mass. The stiffness $k = 3EI/L^3$ depends on the Young's modulus (E), area moment of inertia of the cross section (I), and length (L).

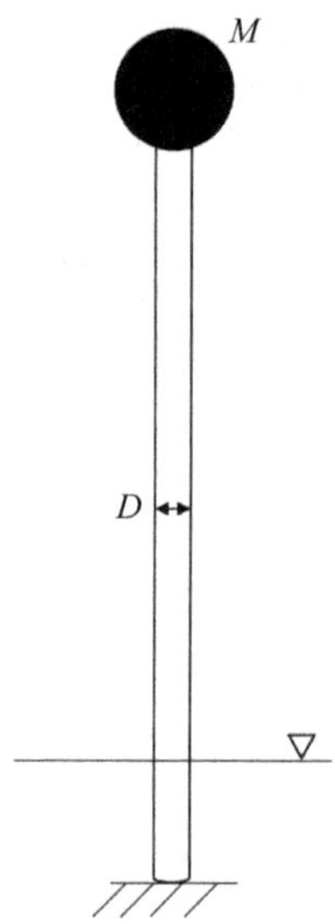

Figure 7.9 Simplified model of a monopile wind turbine

Table 7.2 Main particulars of the DTU 10 MW turbine

RNA mass	674 t
Cut-in rotor speed	6.0 rpm
Rated rotor speed	9.6 rpm
Number of blades	3
Hub height	119 m

Assume that the pile will be made of standard steel and that the diameter-to-thickness ratio is approximately 80. Consider first a location with 20 m water depth: What diameter of the pile/tower could give a reasonable first natural frequency? How much does the required diameter change for 40 m water depth? Discuss the simplifications in the analysis and their consequences.

First, we establish the parameters needed in order to apply (7.8). The density of steel is approximately $\rho = 7{,}850\ \mathrm{kg/m^3}$, and the Young's modulus is approximately 210 GPa. Denoting the diameter as D, thickness as t, and water depth as H, we can find:

$$m_{beam} = \frac{\rho\pi}{4}\left(D^2 - (D-2t)^2\right) \tag{7.9}$$

$$L = 119 + H \tag{7.10}$$

$$I = \frac{\pi}{64}\left(D^4 - (D-2t)^4\right) \tag{7.11}$$

The 1p and 3p ranges can also be established using the given information in Table 7.2. At the cut-in rotor speed, the 1p frequency is 0.6283 rad/s, and at the rated wind speed, the 1p frequency is 1.005 rad/s. (The 3p frequency range is found by simply multiplying these values by 3.) We can then test several values of the diameter (with $t = D/80$) to see where the natural frequencies lie with respect to the 1p and 3p ranges. The results are shown in Figure 7.10. Increasing the diameter moves the natural frequency to the right. For the 20 m water depth, a diameter of

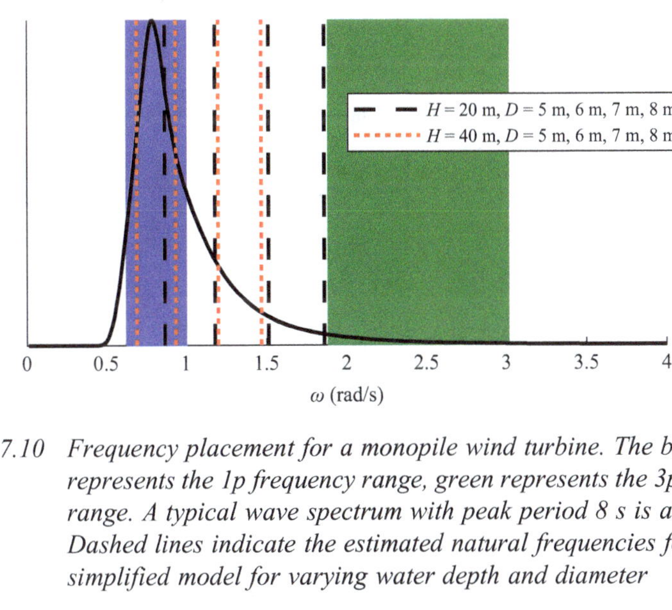

Figure 7.10 Frequency placement for a monopile wind turbine. The blue area represents the 1p frequency range, green represents the 3p frequency range. A typical wave spectrum with peak period 8 s is also shown. Dashed lines indicate the estimated natural frequencies for the simplified model for varying water depth and diameter

approximately 6–8 m gives a natural frequency which lies in between the 1p and 3p ranges. In order to avoid wave excitation, the higher end of this range is preferred. For 40 m water depth, a diameter of at least 7 m is needed in order to avoid 1p excitation.

The analysis considered here gives a reasonable estimate of the size of such monopiles, but the simplifications are significant. One would typically consider a variable diameter (and thickness) over the length of the tower, and the stiffness of the soil must be included in the analysis in order to estimate the needed penetration depth. The stability of the design—considering the maximum allowable deflections and rotations at the mudline, for example—should be considered in addition to the eigenfrequencies. The simplification of considering the blades as a rigid mass with no inertia may not be valid, especially for larger turbines. The rotor inertia as well as the flexibility in the blades can also influence the first natural frequency of the structure as a whole.

Example 7.2 (Case study: preliminary sizing of a 10-MW spar-type wind turbine): *Find a preliminary design for a spar to support the DTU 10 MW wind turbine. The characteristics of the NREL 5 MW wind turbine [35] and the OC3-Hywind spar support structure [36] are given together with some characteristics of the 10 MW turbine in Table 7.3. The static pitch angle at rated wind speed (θ) should not exceed 5 degrees, and the natural period in heave (T_3) should be at least 30 s.*

Table 7.3 Inputs for sizing of a 10-MW spar wind turbine

Parameter	5 MW	10 MW
Spar draft	120 m	–
Spar diameter	6.5 m for 10 m>z>−4 m 9.4 m for z<−12 m Linear taper for −4 m >z>−12 m	–
Spar steel mass[a]	1,053 t	–
Spar ballast mass[a]	6,413 t	–
Spar vertical center of gravity	−89.9 m	–
Turbine and tower mass	599.7 t	1,220 t
Turbine and tower vertical center of gravity	70.3 m	88.4 m
Turbine and tower pitch moment of inertia about still water level[b]	3.4×10^6 t m^2	1.1×10^7 t m^2
Mean thrust at rated wind speed (F_{thrust})	750 kN	1,500 kN
Hub height (h_{hub})	90 m	119 m

[a]Estimated as in [37].
[b]Approximate.

Several approaches to the given task are possible [38,39]; one alternative is given here. The given data provide some suggestions in terms of the floater steel mass per displaced water mass (0.13) and an admissible maximum draft (120 m). A similar parametric design can be defined as in Figure 7.11. In order to simplify the problem, we might assume that:

$$L_1 = 14 \text{ m} \tag{7.12}$$

$$L_2 = 8 \text{ m} \tag{7.13}$$

$$L_3 = 108 \text{ m} \tag{7.14}$$

$$D_1 = \frac{6.5D_2}{9.4} \tag{7.15}$$

Then, we are left with a single design parameter (D_2). From this single design parameter, we can easily estimate the displacement (∇, based on the geometry of the submerged volume), center of buoyancy (c_B), steel mass (m_{steel}, based on the steel mass per displaced water mass), and required ballast mass ($m_{ballast}$, based on equilibrium considerations). Complete equations for the center of buoyancy are not given, but can be found from geometric considerations:

$$\nabla = \rho\pi\left(\frac{D_1^2(L_1 - 10)}{4} + \frac{D_2^2 L_3}{4} + \frac{(D_1^2 + D_2^2 + D_1 D_2)L_2}{12}\right) \tag{7.16}$$

$$m_{steel} = 0.13\nabla \tag{7.17}$$

$$m_{ballast} = \nabla - m_{steel} - m_{towerandturbine} \tag{7.18}$$

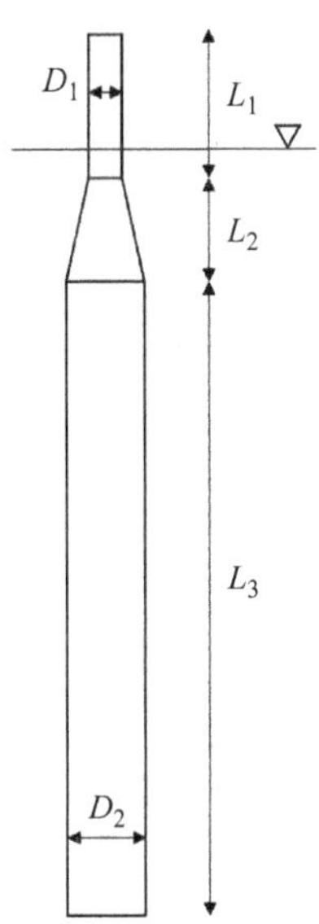

Figure 7.11 Parametric model of a spar wind turbine

Knowing the ballast mass, a lower limit on the vertical center of gravity can be obtained by assuming that the ballast is entirely composed of concrete (with a high density of 2,300 t/m^3). An assumption about the center of gravity of the steel mass needs to be made; a conservative estimate would be to place this halfway in between the top and bottom of the spar. (This is conservative because it will give a higher VCG than what would be expected, given that plate thicknesses near the bottom must be larger in order to withstand the higher hydrostatic pressure.) Using the resulting VCG and (7.1), we can check that an acceptable hydrostatic pitch stiffness—that is, a pitch stiffness which results in $\theta<5$ degrees at rated wind speed—can be obtained. The natural period in heave (T_3) can also be estimated (Equation (7.21)) using approximations of the added mass effects (Equation (7.20)) and the linearized heave stiffness (Equation (7.2)):

$$\theta = \frac{F_{thrust} h_{hub}}{C_{55}} \tag{7.19}$$

$$A_{33} \approx \frac{1}{12}\pi\rho D_2^3 \tag{7.20}$$

$$T_3 \approx 2\pi\sqrt{\frac{\nabla + A_{33}}{C_{33}}} \tag{7.21}$$

Using these assumptions, the results for several designs with varying diameter D_2 are shown in Figure 7.12. The heave natural period does not vary much for different designs—this result makes sense because the added mass, mass, and stiffness all increase at nearly the same rate. The heave natural period is acceptable for all of the designs. Satisfying the criterion for the mean pitch angle at rated wind speed requires a diameter of at least 12 m. Based on the increasing steel mass (and cost) for increasing diameter, a smaller diameter would be preferred. In order to better evaluate the cost trade-offs; however, a more detailed study of the balance between concrete and water ballast is needed.

One could also use simple approximations to check the pitch natural period and ensure that it does not match the heave natural period; however, such a calculation should account for the coupled surge-pitch dynamics and is left as an exercise for the interested reader.

7.2.2 Loads and load effects

ORE support structures are subjected to a combination of different *loads*, including loads acting on the ORE device, which are transferred to the support structure through couplings such as the power takeoff system, as well as loads which act directly on the support structure itself. The design loads are typically divided into permanent loads (Section 7.2.2.1) and variable loads (Section 7.2.2.2) [40]. *Load effects* include motions, displacements and internal forces, and stresses in the structure [40]. In order to evaluate whether or not a structure has sufficient strength and meets the operational requirements of the ORE device, the load effects must be computed and compared to the structural resistance. For ORE devices, the calculation of the load effects can be a time-consuming and complex process, which is described in greater detail in Section 7.2.3.

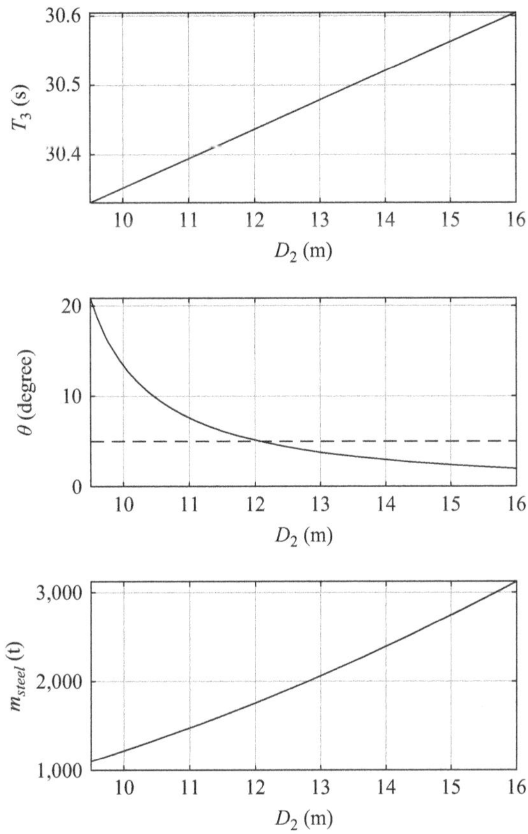

Figure 7.12 Heave natural period, mean pitch angle at rated speed, and steel mass for different values of diameter D_2. The dashed line in subplot 2 indicates the 5-degree limit

7.2.2.1 Permanent loads

Permanent loads are those which do not vary in magnitude, position, or direction [40]. For an ORE support structure in the installed condition, these loads may include the support structure's own weight, the (effective) weight of the ORE device and any permanent ballast, as well as internal and external hydrostatic pressure loads. The reactions due to these loads are also considered among the permanent loads. Such reactions could include the pressure reaction at the base of a GBS, or hinge forces in an articulated WEC. The permanent loads are typically estimated based on the best available data regarding mass and geometry. In some cases, hydrostatic pressure loads can be especially important for designing the plate thickness of ORE support structures.

7.2.2.2 Variable loads

Variable loads include all of the time-dependent loads which act on ORE support structures: both environmental loads and functional loads. ORE devices are subjected to complex and challenging environmental conditions; some of the most important sources of loading are discussed below. Compared to other offshore structures, ORE devices are intentionally located where the environment is particularly challenging: a wind turbine will of course be installed where the wind is strong—usually implying large waves as well—and wave or current energy devices will (following the same principle) be located where the waves or currents are significant. In the present discussion, environmental loads on the ORE device and the support structure are considered together.

Wave loads

Waves (with periods in the range of 5–20 s) are an important source of nonlinear, dynamic loading. A thorough discussion of wave mechanics and wave loads is left to other textbooks [16,41–43], but some basic principles are reviewed here, with an eye towards design and current research topics related to ORE substructures.

Wave loads act as dynamic pressure forces on submerged structures. In general, the pressure includes components from the incoming undisturbed wave (this results in what are called the Froude–Krylov forces), the effect of the stationary structure on the incoming waves (diffraction forces), and the effects of the structural motion on the surrounding fluid (radiation forces). For a slender cylindrical structure with diameter D, where the wavelength to diameter ratio satisfies $\lambda/D > 5$, the diffraction and radiation effects are approximately frequency-independent, and the forces per unit length normal to the slender element (dF) can be computed with reasonable accuracy using Morison's equation (Equation (7.22)). Morison's equation can be applied using only knowledge of the undisturbed wave particle velocity (u) and acceleration ($\dot{u}$) normal to the structure and the structural velocity ($\dot{r}$) and acceleration ($\ddot{r}$) normal to the structure:

$$dF = dz\left(\frac{1}{2}\rho C_D D(u-\dot{r})|(u-\dot{r})| + \rho(C_a+1)\frac{\pi D^2}{4}\dot{u} - \rho C_a \frac{\pi D^2}{4}\ddot{r}\right) \quad (7.22)$$

where C_D is a drag coefficient and C_a is an added mass coefficient, both of which vary as a function of the Reynolds number and Keulegan–Carpenter number.

As mentioned, Morison's equation can only be applied for sufficiently slender structures. Jackets, as well as some spar and monopile designs, and slender components (braces or mooring systems) can often be assessed using Morison's equation. For small sea states, a linear approximation of the horizontal wave particle velocity and acceleration may be sufficient. Considering a single regular wave component with frequency ω, wave number k, amplitude ζ_a, in water depth h, the wave particle velocity and acceleration are

$$u = \omega\zeta_a \frac{\cosh k(z+h)}{\sinh kh}\sin(\omega t) \quad (7.23)$$

$$\dot{u} = \omega^2\zeta_a \frac{\cosh k(z+h)}{\sinh kh}\cos(\omega t). \quad (7.24)$$

The wave particle velocity and acceleration in (7.23) and (7.24) are obtained from the velocity potential, which is a solution of Laplace's equation subjected to the linearized boundary conditions at the free surface, flat seabed, and the radiation condition in the horizontal direction. More details of the fluid dynamics equations can be found in textbooks [16]. Morison's equation and the linear model for the wave particle velocity and acceleration highlight important considerations for ORE support structure design:

- Larger diameters encounter larger wave loads.
- Wave particle velocities and accelerations vary with depth, as illustrated in Figure 7.13. In deep water, there is a rapid decay in the wave particle kinematics, such that large components far from the free surface are subjected to much smaller wave loads. In shallower waters, there is little decay in the wave particle kinematics, and fewer advantages to placing large components far from the free surface.
- The inertia terms are relatively more important for small wave amplitude to diameter ratios (ζ_a/D). Drag effects gain importance for larger ζ_a/D.
- The inertia loads have the same frequency as the waves.
- The drag loads act at the frequency of the waves but also have a higher frequency contribution (at three times the wave frequency) due to the velocity squared term.

As shown in Figure 7.14, for large-volume structures, diffraction effects become more important, and the potential flow equations in frequency domain (FD) are usually solved in order to obtain the frequency-dependent wave excitation and radiation coefficients. The equations are typically solved numerically, and the

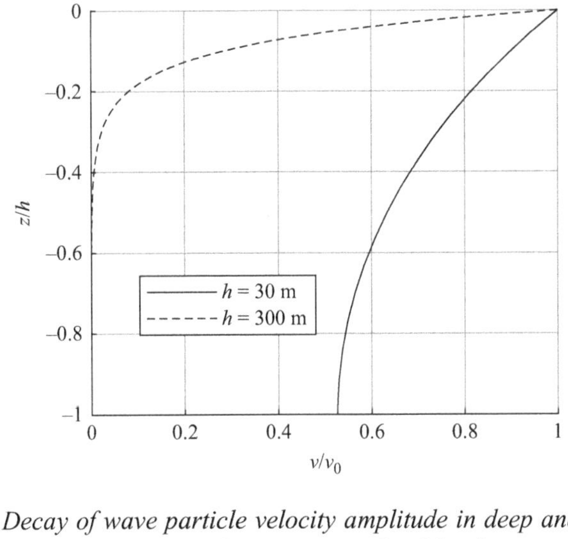

Figure 7.13 Decay of wave particle velocity amplitude in deep and finite water. Note that the vertical axis is normalized by the water depth

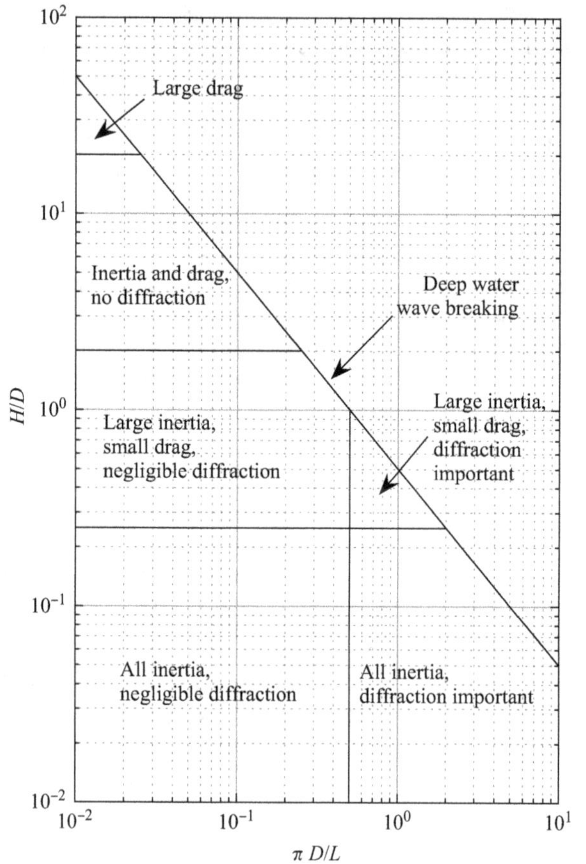

Figure 7.14 Important wave load contributions, adapted from [21]. D is a representative structural diameter, L is the wave length, and H is the wave height

pressure is integrated over the complete submerged structure in order to obtain global coefficients. Beyond the linearized problem, higher order forces may also become important, as these forces can excite the natural periods for rigid body motions. For catenary moored floating platforms, the second-order difference-frequency forces may be especially important [44], while the sum-frequency effects can have greater consequences for TLPs [45].

ORE devices present new challenges with respect to wave load calculations compared to traditional O&G structures. The growth of monopile wind turbine diameters (together with the lengthening of natural periods resulting from the increase in turbine size) requires new developments in order to model second and third-order forces in finite water depths [46–48]. Due to the large diameter, Morison's equation does not capture all of the important diffraction effects, while the finite water depth requires modification of the traditional Faltinsen, Newman, and Vinje [49] methods.

For FWTs, recent results suggest that the second-order difference frequency load models are not sufficient for capturing the low-frequency responses of semi-submersible structures at model scale [44,50]. Whether this discrepancy also occurs at full scale needs further investigation, but some similar differences between simulations and measurements have been observed for spar-type platforms installed at sea [51,52].

The motions of WECs are often large compared to other structures, and better methods for computing the nonlinear buoyancy forces and time-dependent loads have been the topic of several studies [53,54]. Since WECs may be installed in arrays, the interactions among bodies with large motions are also important topics of hydrodynamic research [55]. The influence of waves on MCTs has also been suggested to be a major contributor to fatigue in the blades, and methods for estimating these loads are needed [56,57]. Furthermore, higher fidelity simulation tools such as CFD are increasingly being applied toward more accurate modeling of wave loads on both fixed and floating structures [58,59]. For breaking wave loads, which are especially likely to occur in limited water depth, semiempirical formulations or CFD is applied [60,61].

Wind loads

Dynamic loads due to wind act on all surfaces of a marine structure which are exposed to the air. These loads act across a broad range of frequencies.

For a wind turbine, the largest wind-induced forces in operational conditions are associated with the thrust and torque loads from the turbine itself. It is, however, important to include the wind loads on the tower and other exposed parts of the support structure in more severe wind conditions. Simplified models of the loads on components such as the tower assume that there is a quadratic drag force acting in-line with the relative wind velocity. In reality, the forces on exposed elements may also include components in the cross-wind direction, especially due to vortex shedding effects. Figure 7.15 illustrates the contribution of the tower loads to the total in-line horizontal wind force on a 10-MW wind turbine. For high

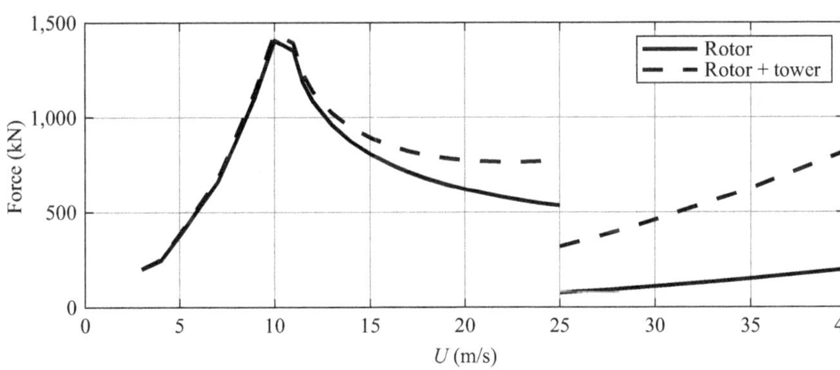

Figure 7.15 Horizontal force due to the rotor alone, and including both rotor and tower, for a 10-MW wind turbine

wind speeds, the effect of tower drag is clearly non-negligible. The overturning moment due to the tower has a smaller relative contribution (due to the lower moment arm compared to the rotor) but may not be negligible.

The frequency content in the wind-induced loads on the rotor of a wind turbine includes the low-frequency variations in the wind speed, as well as components at the rotor frequency and multiples of the rotor frequency (1p, 3p, 6p, etc.). For a FWT, the motions of the nacelle affect the relative wind speed, so the thrust force also includes frequency components at the wave frequency and the platform natural frequencies in surge and pitch, exemplified in Figure 7.16. As shown, the low-frequency components of the thrust dominate, but important contributions can also be seen at the platform frequencies, wave frequencies, and 3p frequencies.

Current loads

Ocean currents—including geostrophic, tidal, or other types of currents—also contribute to the total loading on ORE devices. Marine current energy devices must apply some type of thrust force to the flow in order to extract kinetic energy from the current, but other structures are also subjected to drag and lift loads on submerged components. Drag loads may be especially important for the station-keeping system of a floating platforms. The vortex shedding loads and frequencies should also be checked to avoid resonant motions and vibrations. A major challenge in predicting the current loads is the need for better information and measurements of local current conditions—in terms of the magnitude, direction, and dynamic variations in the flow.

Snow and ice loads

In cold regions, the loads from snow and ice must be considered during the design. Several different types of loads can be expected: snow or ice which settles or freezes to a structure will contribute additional mass—possibly affecting the platform stability and dynamics, platform draft, and wind loads—while drifting ice can impart significant forces on the submerged structure. For offshore wind turbines,

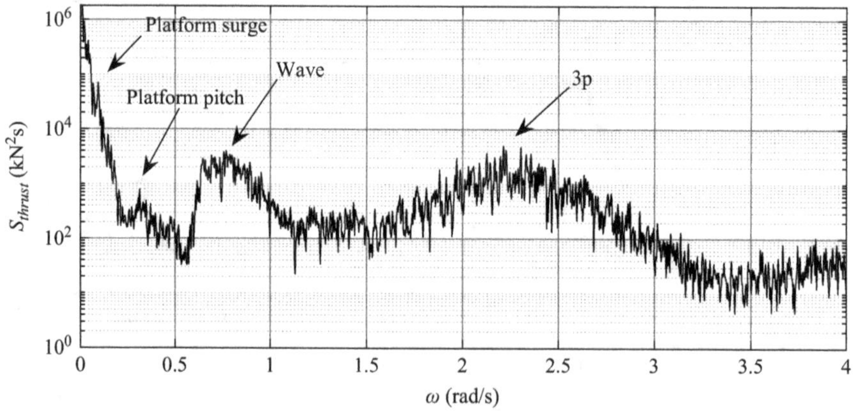

Figure 7.16 Power spectral density of the thrust force on a 10-MW floating wind turbine in below-rated conditions

severe icing on the blades could result in up to 35% loss in power and up to 40% increase in the thrust force [62]. Although the increase in mass due to blade icing for a 5-MW wind turbine is expected to be small compared to the total mass of the rotor, the changes in natural frequency of the rotor could have consequences for the controller and drive train [62]. Drifting ice loads in sea and fresh water have also been studied for offshore wind turbines [63–66]. The magnitude of the loads induced by the crushing of ice depends on the ice thickness as well as the structural geometry: an ice cone can significantly reduce the horizontal loads [66]. Examples of 55 degree ice cones are shown in Figure 7.17. Ice is weaker in bending than in pure compression, so sloping surfaces such as cones can redirect the leading edge of the ice sheet such that it bends and breaks. The frequencies of the ice-breaking loads can also be a concern due to their proximity to typical bending frequencies of monopiles and wind turbine components [63,66].

Figure 7.17 Offshore wind turbines equipped with ice cones. Reproduced, with permission, from Reference [64], Elsevier

Seismic loads

In earthquake-prone regions, the consequences of seismic activity on ORE devices must also be considered, especially for bottom-fixed support structures and TLPs. For bottom-fixed wind turbines, there is extensive numerical and experimental research regarding earthquake loading, as summarized in [67,68]. The natural frequencies of bending modes of modern wind turbines are low enough that the horizontal seismic excitation for design earthquakes (475-year return period) is not likely to cause structural failure or excessive accelerations. There may, however, be cumulative offsets which exceed the strict limits set by wind turbine manufacturers (around 0.75 degrees). On the other hand, vertical earthquake motions have the potential to cause significant accelerations, overstress in the tower, or local buckling [67]. Design standards for wave energy and marine tidal energy also require consideration of seismic loads [69], although the research in these areas is more limited.

Marine growth

ISO-19901-1 defines marine growth (or biofouling) as "living organisms attached to an offshore structure" [70]. The density and thickness of plant or animal life on marine structures depends on the location, depth, and type of structure [71]. Marine growth effectively changes the mass, buoyancy, and hydrodynamic loads on submerged components and also influences the local environment and the rate of corrosion. For WECs, the effects of marine growth can be particularly important for several reasons. WECs are typically smaller structures, such that the effects of marine growth can be relatively large. Furthermore, the performance of a WEC is very sensitive to the natural period, hydrodynamic excitation, and hydrodynamic damping, all of which may be modified by marine growth. For example, a study of marine growth on WECs in Irish waters suggests that there could be an 8% increase in mass within 5 years of deployment [72]. Figure 7.18 shows an example of marine growth on the Oyster 800 WEC.

The effects of marine growth on the mooring systems of FWTs [74] and on MCTs [75] have also been the subject of recent investigations. Experimental results suggest that roughness on an MCT can adversely affect power output (19% reduction), but the extent to which real organisms would stay on the rotating blades was not possible to test [75].

Water level variations

The mean water level at a given location varies over time due to a combination of different factors, including tidal variations and storm surge effects. For some substructures, these variations may be more important than for others: for a TLP, for example, the change in buoyancy due to tidal variations requires more careful consideration than the small change in mooring layout for a catenary-moored floater [76]. For bottom-fixed structures, changes in the water level result in a change in the moment arm where wave and current loads are applied (in addition to a change in the loads themselves) as well as a change in the hydrostatic pressure acting on components near the seabed.

Figure 7.18 Biofouling on the Oyster 800 WEC after 9 months prior to operation [73]. Reproduced, with permission, from ASME

Actuation loads from ORE devices

In addition to the environmental loads on the substructure and those transferred from the ORE device, the substructure is also subjected to actuation loads related to the operation of the ORE device. For example, the substructure must counter the moment imparted by the yaw actuator of a wind turbine. WEC latches and actuators may also contribute large local loads on the substructure.

Other variable functional loads

The category "other variable functional loads" comprises a wide range of loads, many of which are related to maintenance operations. Considering global loads, the loads on access platforms or ladders due to personnel are likely to be small, but these loads are important for the detailed design of such components. The impact loads due to docking of service vessels or accidental collisions may have some more important global load effects. In the case of accidental collision of a 230-t service vessel, Dai *et al.* suggest that even fairly low vessel speeds can lead to significant damage to monopile wind turbines [77]. Operational loads due to cranes should also be considered in the design of ORE substructures [40,78].

Example 7.3 (Wind-wave misalignment effects for offshore wind turbines): *The complexity of the loads acting on ORE devices also leads to complex load combinations and nonlinear effects in load combinations. As an illustration of this complexity, the effects of wind-wave misalignment on offshore wind turbines (as reported by several researchers) are examined here. In general, one can expect that there will be a difference (misalignment) in the wind direction and wave* *direction; in the North Sea, misalignment of up to 30 degrees is quite common [79]. Based on experience from the O&G industry, one might expect that it would be conservative to only consider the aligned wind and wave conditions; however, this cannot necessarily be assumed to be true for OWTs. In operational cases, the fore–aft (aligned with the wind) aerodynamic damping is much larger than the side–side damping: if wave loads result in resonant motions, then the lack of damping could result in larger fatigue or extreme loads during misaligned wind-wave conditions.*

For monopile wind turbines, the relative importance of wind and wave loads depends on the turbine size, monopile diameter, water depth, and environmental conditions. For a 10-MW monopile wind turbine in 30 m water depth, Smilden et al. suggest that the fatigue damage at the most critical point on the cross section at the mudline for similar conditions with 30-degree misalignment could be roughly 1.5 times greater than the corresponding condition without misalignment [80]. Control algorithms which dampen the side–side vibrations through variation of the generator torque are therefore needed [80,81].

On the other hand, studies of FWTs subjected to misaligned wind and wave conditions suggest that the maximum stresses due to misaligned wind and wave conditions are not likely to be larger than those due to aligned wind and waves [82,83]. Nonetheless, interesting results for the platform motions could be observed: while the wave-frequency motions clearly followed the direction of the waves, the low-frequency responses in the wind direction could increase when the waves acted in a different direction. This result was attributed to the fact that the aerodynamic loads at low frequency are also affected by the wave-frequency velocity of the rotor. As a result, the damping of low-frequency wind-induced motions could be reduced when the wave-frequency motions were in a perpendicular direction. An example of a similar result is shown in Figure 7.19 for a

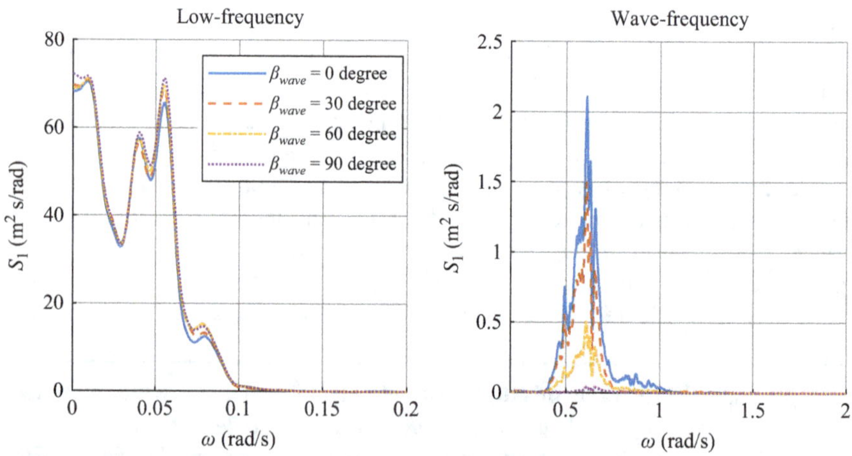

Figure 7.19 Surge response of a 10-MW semisubmersible platform (CSC platform in [84]) in 8 m/s wind and large waves (6 m H_s, 10 s T_p). Surge motion is defined in the same direction as the wind (0 degree), while the wave direction is varied

10-MW semisubmersible floating platform. The low-frequency surge response is seen to increase as the misalignment increases, while the wave-frequency response decreases. In this case, the increase in low-frequency motions is, however, relatively small, compared to the 5 MW platforms in [83].

7.2.3 Short-term and long-term design analysis

Numerical analysis of ORE devices and substructures is often used to check that the structural resistance is sufficient to withstand the load effects the device is likely to encounter during its lifetime. The discussion here focuses on global analysis of the complete device: this type of analysis can be used both for power prediction and for estimating the global responses. For detailed analysis of local loads, the global analysis results may be used as input to local detailed analysis; this type of analysis is not addressed here.

7.2.3.1 Short-term analysis

A short-term analysis is used to evaluate the structural responses during a period of time over which the environmental conditions can be considered stationary. A strict definition of a stationary process $x(t)$, where t is time, requires that all the statistics of the ensemble are invariant under translation of time [85]. More commonly, we refer to a weakly stationary process, which requires that

1. the mean value $E[x(t)]$ is independent of the time t and
2. the auto-covariance $\mathrm{Cov}[x(t), x(t+\tau)]$ depends only on τ for all t.

Over a duration where the wave elevation, current speed, or wind speed can be considered stationary, the environmental conditions can be described by parameters such as those that define the wave spectrum (H_s, T_p, or other parameters), current ($U(z)$), and wind spectrum and coherence (U_w, I, and others). The assumption of stationary conditions is important for being able to relate the wave elevation or time history of wind speed to their frequency-domain representation; however, a challenge (especially for OWTs) is that waves are typically considered stationary over 1–6 h, while wind has historically been considered to have a stationary period of approximately 10 min [86]. There is some consensus toward the application of common 1 h reference periods for wind and waves for analysis of FWTs [86,87].

The description "short-term analysis" may also be used when transient events are considered. Among others, transient events could include a fault event in the control system, loss of a mooring line, wind gusts, yawing operations, loss of connection to the electrical grid, or ship-structure collision. These events are often simulated in a deterministic manner, combined with some stochastically varying environmental conditions, and can be design-driving for some structural components. Blade pitch actuator faults are generally seen to be not only particularly important for the blades and tower top but may also be of particular importance for floating substructures which have low pitch stiffness and ballast compensation for the mean trim angle [88,89].

Time domain and frequency domain analysis

Disregarding transient events, short-term analysis can be carried out in either the time domain (TD) or in the FD. The two approaches are illustrated for a very simple model of a heaving WEC in the following example.

Example 7.4 (Single degree of freedom analysis in time and frequency domain): *A simple model of a WEC which is restricted to only heave motion is shown in Figure 7.20.*

The equation of motion (in the TD) for this system can be written as

$$(m + A)\ddot{\eta}_3 + (c_{PTO} + B)\dot{\eta}_3 + (C_{33} + EA/L)\eta_3 = F_{wave} \tag{7.25}$$

where m is the mass of the buoy, A is the hydrodynamic added mass in heave, c_{PTO} *is the WEC's power takeoff damping, B is the hydrodynamic linear damping (due to, i.e., wave radiation),* $C_{33} = \rho g \pi D^2/4$ *is the hydrostatic stiffness,* EA/L *is the stiffness due to the tether,* F_{wave} *is the excitation force due to the waves, and* η_3 *represents the heave motion.*

To solve this problem in the FD, we would need to assume that this is a linear system: that is, that the hydrodynamic excitation force at a given wave frequency is linearly proportional to the wave amplitude at that frequency, and that the response of the structure to a harmonic load at a given frequency is a harmonic response at the same frequency with an amplitude linearly proportional to that of the load. These assumptions mean that we cannot consider Morison-type quadratic drag loads, but we can consider frequency-dependent excitation, added mass, and damping due to potential flow. Under the assumption of a linear system, we could establish several transfer functions: from the wave amplitude to the excitation force, and from the excitation force to the heave response of the buoy.

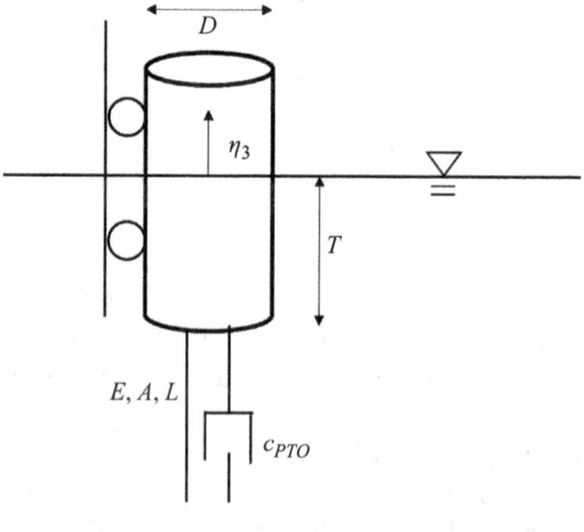

Figure 7.20 Simple heaving WEC

If we only consider the Froude–Krylov forces on the buoy, the excitation force can be written as the integration of the dynamic pressure due to the incident wave on the bottom of the buoy. If we further assume that the buoy is small, a reasonable approximation is to assume the pressure is spatially constant at the center of the buoy. For a regular wave in deep water with elevation $\zeta = \zeta_u \sin(\omega t)$, this gives

$$F_{wave} = \frac{\pi D^2}{4} \rho g \zeta_a e^{-kT} \sin(\omega t), \tag{7.26}$$

where $k = \omega^2/g$ is the wave number. The transfer function H_{hydro} then relates the wave force to the wave elevation as

$$|H_{hydro}| = |F_{wave}|/|\zeta| = \frac{\pi D^2}{4} \rho g e^{-kT}. \tag{7.27}$$

We can further use well-known solutions for second-order differential equations (i.e., [90]) to solve (7.25) for a harmonic F_{wave}. Equation (7.28) shows the amplitude response in heave:

$$|\eta_3| = \frac{|F_{wave}|}{(C_{33} + EA/L)\sqrt{(1-\beta^2)^2 + (2\xi\beta)^2}}, \tag{7.28}$$

where $\beta = \omega/\left(\sqrt{(C_{33} + EA/L)/(m + A)}\right)$ is the frequency ratio, and $\xi = (c_{PTO} + B)/\left(2\sqrt{(C_{33} + EA/L)(m + A)}\right)$ is the critical damping ratio.

The mechanical transfer function relates the response to the forcing, that is,

$$|H_{mech}| = |\eta_3|/|F_{wave}| = \frac{1}{(C_{33} + EA/L)\sqrt{(1-\beta^2)^2 + (2\xi\beta)^2}}. \tag{7.29}$$

When the transfer functions are known, the spectrum of the response (S_{η_3}) can be found immediately from the known wave spectrum S_w:

$$S_{\eta_3} = S_w |H_{hydro} H_{mech}|^2. \tag{7.30}$$

To solve the same problem in the TD, a realization of the time series of the waves based on the spectrum is needed. For a linear realization, the wave elevation is the sum of harmonic components:

$$\zeta(t) = \sum \zeta_{aj} \sin \omega_j t + \phi_j \tag{7.31}$$

where ω_j is the wave component frequency, ϕ_j is a randomly drawn phase angle, and ζ_{aj} is the wave component amplitude, which is related to the wave spectrum S_w with frequency spacing $\Delta\omega$:

$$\zeta_{aj} = \sqrt{2 S_w \Delta\omega}. \tag{7.32}$$

Having obtained the time series of the wave elevation, the wave excitation force time series could be also be found. In the simple case of frequency-independent added *mass and damping,* (7.25) *could then be solved in a discrete step-by-step manner given initial conditions for the heave motion and velocity at time* $t = 0$*, using Euler's method, Newmark-β, or Wilson's method [91,92]. Frequency-dependent added mass and damping coefficients could be included through Cummin's integral or state-space approximation of the retardation function [93,94].*

The two approaches are shown in Figure 7.21. *For the FD approach, the total transfer function, wave spectrum, and response spectrum are shown. The WEC is designed such that its resonance period is tuned to the peak period, such that the*

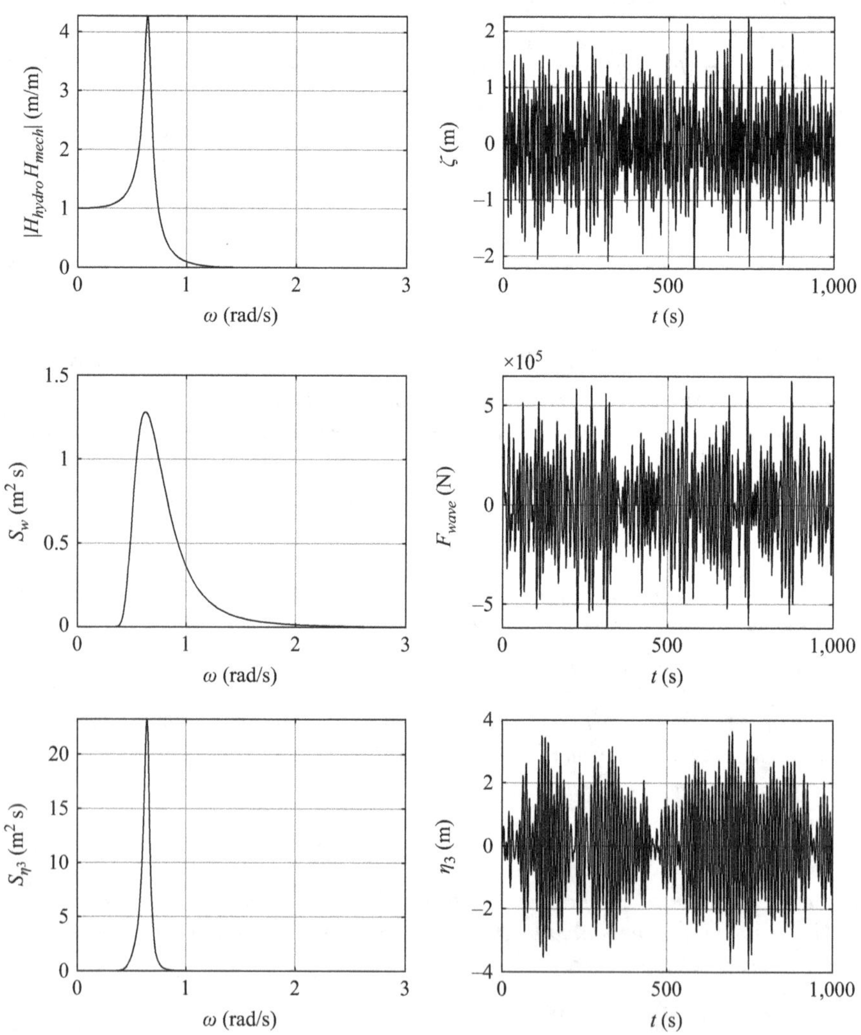

Figure 7.21 Heaving WEC problem solved in the frequency domain (left) and in the time domain (right)

response spectrum is very narrow. The same effect can be seen in the TD results (where a realization of the wave elevation, heave force, and heave response are shown): the visible "beating" effects in the response are related to this narrow bandedness.

As shown, the FD method is very efficient for the analysis of linear systems. While some nonlinear effects (in the environmental loads, control, or structure—such as end stops or mooring system) can be linearized for certain conditions, TD analysis is required when the nonlinearities become important. More complete discussions of the statistical analysis of the results can be found in [93]. TD analysis is also required when considering transient events.

Structural models

In addition to the choice between TD and FD analysis, another important consideration is the representation of the physical structure which is to be applied in the analysis. The desired output from the analysis, as well as the available input, provides some criteria for this choice. A dynamic analysis which considers the heaving WEC from Example 7.4 as a rigid body may be useful for estimating the global motions and the loads in the elastic tether but cannot provide any information regarding deflections within the buoy.

Rigid body models can be combined with models of the couplings between the bodies in a multi-body approach. This type of approach has been widely applied to many types of ORE devices, as evidenced by the WEC-Sim [95] and HAWC2 software programs [96], for WECs and horizontal axis wind turbines, respectively. In order to model the elasticity of elements such as the tower and blades, a hybrid approach is applied in HAWC2, which also incorporates some FE modeling approaches.

FE models discretize the structure into a number of FEs, each of which may have multiple DOFs. A beam element, for example, has DOFs related to axial elongation, torsion, and bending about several axes. By combining the mass, damping, and stiffness matrices corresponding to each degree of freedom in a consistent manner, the dynamic equation of motion becomes [91]:

$$\mathbf{M}_g\ddot{\vec{D}} + \mathbf{B}_g\dot{\vec{D}} + \vec{R}^{int} = \vec{R}^{ext} \tag{7.33}$$

where $\vec{D}$ is a vector of nodal displacements, $\mathbf{M}_g$ is the global mass matrix (which may include hydrodynamic mass), $\mathbf{B}_g$ is the global linear damping matrix, $\vec{R}^{int}$ is a vector of the internal reaction forces (and moments), and $\vec{R}^{ext}$ are the external loads. The internal reaction forces and moments may in some cases be replaced by a linear stiffness matrix multiplied by the nodal displacements. Using an FE model, a piecewise approximation of field quantities such as stress and strain can be obtained.

In global FE analysis of OWTs, beam elements are commonly used to represent the tower and blades, while bar elements are often used to represent chain or rope catenary mooring lines. The hull has often been modeled using a rigid body, but researchers are examining beam and multibody formulations in order to extract the internal loads in the hull within a global analysis [39,97,98]. Through the external load vector and internal reaction vector, FE models can represent nonlinearities in the structural dynamics with straightforward modifications. Changes in the geometric stiffness due to large deflections, nonlinear hydrodynamic loads, material nonlinearities, and even contact between components can be accounted for. On the other hand, the computational effort for nonlinear FE analysis can be significant.

An efficient extension of FE modeling is to apply modal modeling. In cases where the internal reaction vector is relatively linear, the number of DOFs in the structural model can be greatly reduced by selecting a limited number of eigenmodes and treating deflections in those modes as the DOFs. A modal approach is applied in the FAST software for wind turbines [99]. With a proper selection of modes, this type of approach is both accurate and computationally efficient and can capture nonlinearities in the external loads.

Timescales of interest

When setting up a short-term TD analysis, the timescales of interest, or, equivalently, the frequencies which can and must be resolved, should be carefully considered. Resolving these frequencies not only requires a sufficiently small time step but also accurate models of the relevant physical processes and a sufficient duration of the simulation to capture enough cycles after any transient effects. For most ORE devices, accurate calculation of the loads and responses at the primary wave frequencies is a clear requirement. Similarly, the lowest resonance frequencies of the device (such as the first bending mode for a bottom-fixed platform, or the rigid body modes of motion for a floating platform) must be included. As the model fidelity increases, the temporal characteristics of added fidelity in the structural model and load model should be considered together. For example, consideration of flexible blades in the wind turbine model also requires a reasonable discretization of the loads, mass, stiffness, and damping. It also requires that aerodynamic effects such as the tower shadow should be considered. For a WEC, modeling end-stops with high stiffness and low damping can result in poor numerical stability or accuracy unless the time step and integration scheme are correctly chosen. Furthermore, the inclusion of a detailed model of the hydraulic power takeoff system requires a time step and model which can capture those dynamics. Of course, the required model fidelity should also be a primary consideration when planning such analysis.

Example 7.5 (Integrated aero-hydro-servo-elastic analysis for OWTs): *Global short-term analysis of OWTs has received significant attention, resulting in the development of numerous numerical tools which are referred to as "aero-hydro-servo-elastic" software* [100]. *These codes attempt to simultaneously model the wind turbine aerodynamics (aero), the hydrodynamic loads on the substructure (hydro), the blade pitch and generator torque control system (servo), and the structural response (elastic). In some cases, "geo" might be added to this list, recognizing the importance of soil modeling (especially for bottom-fixed OWTs).*

Figure 7.22 *shows an example of the structural and load models which might be considered in such an analysis. In addition to the external loads, the generator*

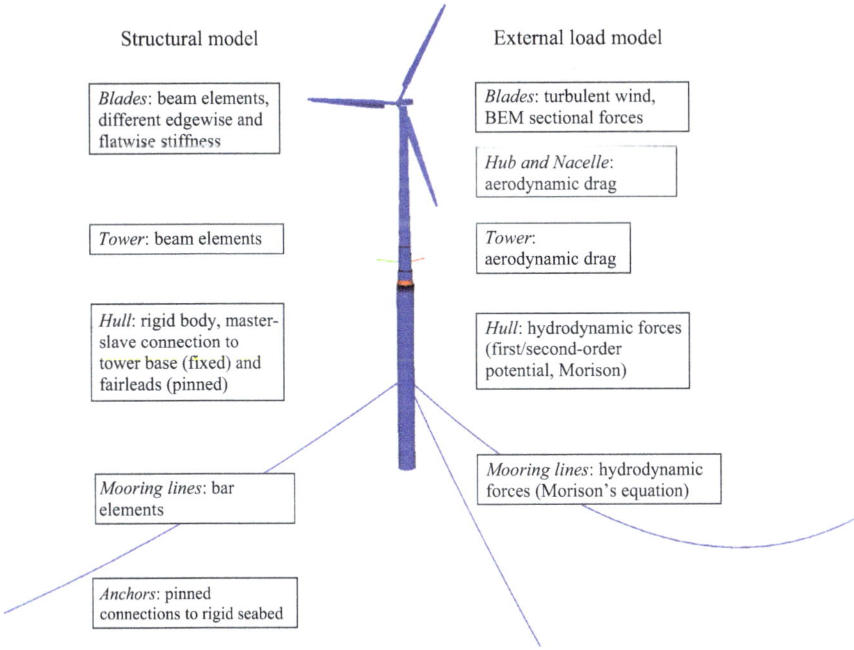

Figure 7.22 Example of aero-hydro-servo-elastic modeling of a spar wind turbine (adapted from [101])

torque must be applied to the shaft based on the wind turbine controller logic, and the blade pitch angle must also be set according to controller commands. The model shown in Figure 7.22 *applies a TD FE formulation, with a time step of 0.005 s in order to capture the high velocities encountered by the blade tip. The Newmark-β approach is used, together with Newton–Raphson iterations on each step such that the updated stiffness matrix (due to large deflections of the blades or significant mooring line nonlinearities) is applied. The external loads which depend on the instantaneous position or velocity of the body—such as Morison-type forces or the aerodynamic loads on the blades—are updated at each time step as well.*

This type of approach is sometimes referred to as "fully coupled" because it simultaneously solves the problem in several domains. This is in contrast to an approach where the aerodynamic problem would be solved separately from the hydrodynamic problem, and the loads would be added together.

7.2.3.2 Long-term analysis

The results from many short-term analyses must furthermore be combined correctly in order to predict the long-term (full lifetime) behavior of the ORE device and substructure. The goal of a long-term analysis might be to evaluate the extreme load effects (ULS design check) or the cyclical load effects (FLS design check).

A full long-term analysis (FLTA) considers the product of the short-term probability distribution of the response and the probability of encountering the environmental conditions in that short-term analysis. For example, considering only three environmental parameters (U, H_s, T_p), the probability distribution of the maximum 1-h response X_{1hmax} might then be written as

$$F_{X_{1h\ \max}}(x) = \int_u \int_h \int_t F_{X_{1h\ \max}|U,H_s,T_p}(x|u,h,t) f_{UH_sT_p}(u,h,t) dt dh du \tag{7.34}$$

where the joint environmental parameter probability density function $f_{UH_sT_p}(u,h,t)$ might be seen as the product of a marginal distribution and several conditional distributions [102]:

$$f_{UH_sT_p}(u,h,t) = f_U(u) f_{H_s|u}(h|u) f_{T_p|U,H_s}(t|u,h). \tag{7.35}$$

for each parameter. $F_{X_{1h\ \max}}$ is the cumulative distribution of the maximum 1-h response and $F_{X_{1h\ \max}|U,H_s,T_p}$ is the conditional distribution of the maximum given the environmental conditions (based on short-term analyses). Equation (7.35) represents the general trend that H_s depends on wind speed, and that T_p is strongly dependent on H_s. In some cases, additional parameters (such as wave and wind direction, or current velocity) may be needed for an accurate representation of the long-term distribution.

The short-term analyses which are needed to find the long-term distribution are typically carried out as independent simulations, and the order in which they occur is not taken into account. This assumption means that memory effects—such as changes in the soil parameters due to previous large loading—are not taken into account. The importance of these effects remains an active research topic.

Fatigue loads

The evaluation of fatigue damage—the effects of cyclic loading on the structure—generally follows either a fracture mechanics approach or an SN-curve approach. In a fracture mechanics approach, crack propagation is estimated based on the stress time history and empirical geometry functions. Once the estimated crack length reaches the structural thickness, the component is considered to have failed. In an SN-curve approach, empirical data relating the number of cycles to failure (SN curve) are compared to the history of stress cycles (most often using the Palmgren–Miner damage accumulation law). For either approach, in principle, an FLTA is needed to provide the lifetime stress history.

Carrying out an FLTA is generally time-consuming. Many of the environmental conditions which must be considered in the FLS check are relatively benign, and a linear FD short-term analysis approach may be justified in order to reduce the computational effort. Further computational savings may be possible through the choices made in "binning" or collecting several environmental conditions into one representative simulation. Figure 7.23 shows an example of a selection of environmental conditions for short-term simulation from Kvittem and Moan [87]. The "base cases" in Figure 7.23 include 155 conditions which

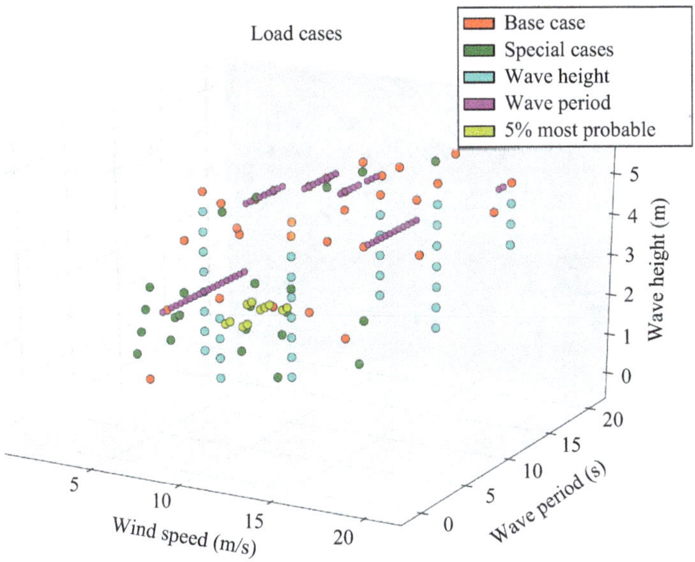

Figure 7.23 Selection of environmental conditions for fatigue analysis [87]. Reproduced, with permission, from Elsevier

were downselected from the 2,201 cases which result from applying criteria from the design standards [103,104]:

- Bin size in mean wind speed: 2 m/s
- Bin size in significant wave height: 0.5 m
- Bin size in peak period: 0.5 s
- Keep bins down to probability level 10^{-4}.

Reducing the computational effort by rational downselection of the most important load cases or by improving the efficiency of the analysis itself continues to be an area of research and would permit more thorough investigation of the design space—possibly reducing the cost of ORE substructures.

Extreme loads

FLTA is considered to be the most accurate method for estimating the extreme load effects during the structure's lifetime. As in the fatigue analysis, however, more computationally efficient methods are desired. In an FLTA for extreme value response, significant computational time may be used for conditions which have little or no effect on the extremes [93].

One simplified approach is the environmental contour method (ECM) [105]. In the ECM, a set of environmental conditions which lie on the contour (a curve in the case of two parameters, or a surface in the case of three parameters) of a particular return period (such as 50 years). The contour is found by the inverse first-order reliability method. The long-term maximum is determined based on the short-term maxima for the conditions along the contour, including an additional factor to

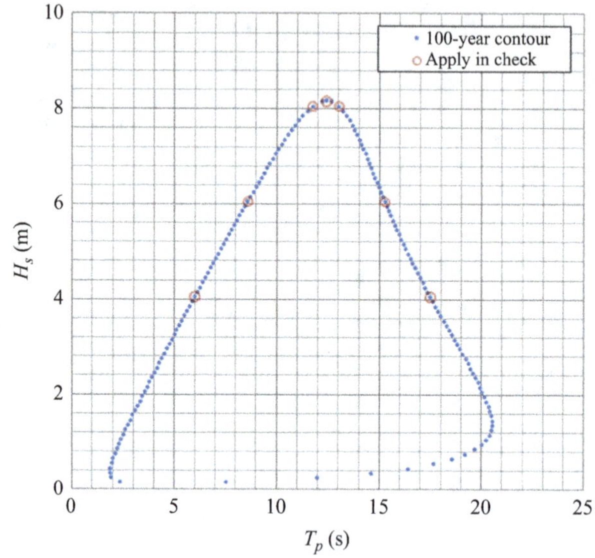

Figure 7.24 Environmental contour for WEC design based on wave conditions at Yeu [107]. Reproduced, with permission, from ASME

account for the variability in the short-term response. This "compensation factor," which is equivalent to selecting a higher target percentile rather than the mean from the short-term analyses, implies that the method should only be applied when the desired response contour is near the environmental contour, and that the response should be a monotonic function of the environmental parameters [106,107]. Modifications to the ECM are therefore needed for ORE devices which have different operational states [108].

Figure 7.24 shows an example of a selection of 100-year sea states for evaluating the mooring system response of a two-body WEC in a shutdown condition [109]. For a two-body system, the 80th percentile results from the ECM gave similar results to FLTA, while the 98th percentile was needed for a one-body configuration of the WEC.

Another approach, termed simplified FLTA, includes only the environmental conditions that contribute to long-term extremes. For example, Li *et al.* showed that only 10% of all environmental conditions are needed in order to reach an acceptable estimate compared to the FLTA for a bottom-fixed wind turbine [110].

7.2.4 Design standards, guidelines, and other considerations

Design standards for ORE devices have been developed in recent years and provide additional guidance toward substructure design. The most common standards follow a limit state approach, where the calculated loads and load effects (described in Section 7.2.2) in short-term conditions must be combined with prescribed safety factors and compared to the calculated *resistance* of the structure. The limit state approach is consistent with the previous discussions of loads and load effects and will not be described further here. Instead, this brief overview includes some comments regarding the available standards, the design philosophy included in the standards, and some important considerations which have not been previously mentioned.

7.2.4.1 ORE design standards

The most widely recognized design standards for OWTs include both bottom-fixed and floating design standards from DNVGL [40,86,111], bottom-fixed and floating design standards from the American Bureau of Shipping [112,113], the FWT design standard from Class NK [114] and the OWT standard from the International Electrotechnical Commission (IEC) [103]. Most of these design standards consider a "load and resistance factor design" approach. This approach aims to obtain a design within the targeted safety level by considering safety factors to account for uncertainties in both the structural loads and the structural resistance [115].

The environmental conditions considered in the standards are largely synchronized, but differences in safety factors and requirements for damage stability, marine operations, and cable design may have more important differences [116]. Furthermore, DNVGL requires scale model testing of FWTs, a requirement which is not present in other guidelines, and the considerations of special conditions such as tsunami loading, wave-in-deck, or ice loads differ among the standards [116].

For WECs and MCTs, an early design guideline was proposed by EMEC [117]; more recent contributions from DNVGL and the IEC have since been published [69,78,118]. The EMEC guidelines provide many suggestions towards generating an appropriate design basis, while DNVGL provides more generic steps toward certification. The IEC provides more specific design load cases and distinguishes between devices which "exploit a resonant response" in the required load combinations.

In addition to the ORE-specific design guidelines, existing standards from the offshore industry may be useful, especially due to the limited experience in the ORE industry today. The development of design standards depends on offshore experience, and revision of the guidelines is to be expected.

7.2.4.2 Other design considerations

The load and response analysis discussed in this chapter largely focuses on the assessment of structural *responses* of *installed* ORE devices. There are also important design considerations which are related to marine operations (such as installation and maintenance) and to the calculation of the structural resistance.

Installation methods and costs (see Chapter 11) can have significant consequences on the design of ORE substructures. For example, optimization of the weight of components and how high they need to be lifted needed storage area in a shipyard, ability to fit within available dry docks, and possibility of using available vessels for installation work may be more important than substructure optimization with respect to steel weight. Over the lifetime of a substructure, the costs of access and maintenance may be significant, and designs which allow for easier inspection or require less maintenance may be favored over designs which are less expensive to construct but more difficult to maintain.

With respect to structural resistance, it is important to note the particular challenges related to corrosion in the marine environment. Designers must account for possible reductions in steel thickness due to corrosion through structural design (cathodic protection or coatings). ORE devices with significant dynamic motions near the free surface—implying surfaces which are at times submerged, at times dry, and also subjected to sea spray—may experience different corrosion rates compared to more static offshore structures.

References

[1] Twidell J, and Gaudiosi G, editors. Offshore Wind Power. Multi-Science Publishing Co. Ltd; 2009.

[2] Thomsen K. Offshore Wind: A Comprehensive Guide to Successful Offshore Wind Farm Installation. Waltham, MA: Academic Press; 2014.

[3] Esteban MD, Couago B, Lpez-Gutirrez JS, *et al.* Gravity based support structures for offshore wind turbine generators: Review of the installation process. Ocean Engineering. 2015;110:281–291.

[4] Babarit A, Hals J, Muliawan MJ, *et al.* Numerical benchmarking study of a selection of wave energy converters. Renewable Energy. 2012;41:44–63.

[5] Seabased Technology. Seabased Technology; 2018. Available from: https://www.seabased.com/the-technology [Accessed July 2018].

[6] Babarit A, Hals J, Kurniawan A, *et al.* Power Absorption Measures and Comparisons of Selected Wave Energy Converters. In: The ASME 2011 30th International Conference on Ocean, Offshore and Arctic Engineering; 2011.

[7] Wavestar. Wavestar – How It Works; 2018. Available from: http://wavestarenergy.com/concept [Accessed August 2018].

[8] Simec Atlantis Energy. Turbines & Engineering Services – Technology History – Seagen 1.2 MW; 2018. Available from: https://simecatlantis.com/services/turbines/history/ [Accessed September 2018].

[9] Rourke FO, Boyle F, and Reynolds A. Marine current energy devices: Current status and possible future applications in Ireland. Renewable and Sustainable Energy Reviews. 2010;14(3):1026–1036. Available from: http://www.sciencedirect.com/science/article/pii/S1364032109002743 [Accessed September 2018].

[10] Bergström L, Kautsky L, Malm T, *et al.* Effects of offshore wind farms on marine wildlife – A generalized impact assessment. Environmental Research Letters. 2014;9(3):034012.

[11] Byrne B, Houlsby G, Martin C, *et al.* Suction caisson foundations for offshore wind turbines. Wind Engineering. 2002;26(3):145–155.

[12] Li L. Dynamic Analysis of the Installation of Monopiles for Offshore Wind Turbines; 2016.

[13] Equinor. Hywind – The World's Leading Floating Offshore Wind Solution; 2018. Available from: https://www.equinor.com/en/what-we-do/hywind-where-the-wind-takes-us.html [Accessed July 2018].

[14] Hooft JP. Advanced Dynamics of Marine Structures. New York: Wiley; 1982.

[15] Li Y, Liu L, Zhu Q, *et al*. Influence of vortex-induced loads on the motion of SPAR-type wind turbine: A coupled aero-hydro-vortex-mooring investigation. Journal of Offshore Mechanics and Arctic Engineering. 2018;140:051903-1–051903-13.

[16] Faltinsen OM. Sea Loads on Ships and Offshore Structures. Cambridge: Cambridge University Press; 1990.

[17] Leimeister M, Bachynski EE, Muskulus M, *et al*. Rational upscaling of a semi-submersible floating platform supporting a wind turbine. Energy Procedia. 2016;94:434–442.

[18] Roddier D, Cermelli C, and Weinstein A. Windfloat: A Floating Foundation for Offshore Wind Turbines; Part I: Design Basis and Qualification Process. In: 28th International Conference on Ocean, Offshore, and Arctic Engineering. OMAE2009-79229; 2009.

[19] LIFES50+. Innovative Floating Offshore Wind Energy; 2018. Available from: http://lifes50plus.eu [Accessed July 2018].

[20] Viselli AM, Goupee AJ, and Dagher HJ. Model Test of a 1:8 Scale Floating Wind Turbine Offshore in the Gulf of Maine. In: ASME 2014 33rd International Conference on Ocean, Offshore and Arctic Engineering. OMAE2014-23639. San Francisco, California, USA; 2014.

[21] Chakrabarti S, editor. Handbook of Offshore Engineering. London: Elsevier Science; 2005.

[22] Bachynski EE, and Moan T. Design considerations for tension leg platform wind turbines. Marine Structures. 2012;29:89–114.

[23] Myhr A, Maus KJ, and Nygaard TA. Experimental and Computational Comparisons of the OC3-HYWIND and Tension-Leg-Buoy (TLB) Floating Wind Turbine Conceptual Designs. In: Proceedings of the 21st (2011) International Offshore and Polar Engineering Conference; 2011. p. 353–361.

[24] Bae YH, Kim MH, and Shin YS. Rotor-Floater-Mooring Coupled Dynamic Analysis of Mini TLP-Type Offshore Floating Wind Turbines. In: 29th International Conference on Ocean, Offshore, and Arctic Engineering. OMAE2010-20555; 2010.

[25] Myhr A, and Nygaard TA. Load Reductions and Optimizations on Tension-Leg-Buoy Offshore Wind Turbine Platforms. In: 22nd International Offshore and Polar Engineering Conference. vol. 1; 2012. p. 232–239.

[26] Bachynski EE, and Moan T. Point Absorber Design for a Combined Wind and Wave Energy Converter on a Tension-Leg Support Structure. In: 32nd International Conference on Ocean, Offshore and Arctic Engineering. OMAE2013-10429; 2013.

[27] Robertson A, Jonkman J, Musial W, *et al*. Offshore Code Comparison Collaboration, Continuation: Phase II Results of a Floating Semisubmersible Wind System. In: EWEA Offshore. NREL/CP-5000-60600; 2013.

[28] Johanning L, Smith GH, and Wolfram J. Measurements of static and dynamic mooring line damping and their importance for floating WEC devices. Ocean Engineering. 2007;34:1918–1934.
[29] van Bussel G, and Zaaijer M. Reliability, Availability and Maintenance Aspects of Large-Scale Offshore Wind Farms: A Concept Study. In: Proceedings of MAREC 2001; 2001.
[30] Jamieson P. Innovation in Wind Turbine Design. Chichester: John Wiley & Sons, Ltd.; 2011.
[31] Bernitsas MM, Raghavan K, Ben-Simon Y, *et al.* VIVACE (Vortex Induced Vibration Aquatic Clean Energy): A new concept in generation of clean and renewable energy from fluid flow. Journal of Offshore Mechanics and Arctic Engineering. 2008;130:15.
[32] Falnes J. Small is Beautiful: How to make Wave Energy Economic. In: 1st European symposium, Wave Energy; 1993. p. 367–372.
[33] Bak C, Zahle F, Bitsche R, *et al.* Design and performance of a 10 MW wind turbine. Wind Energy. 2015. To be accepted.
[34] Stokey WF. In: Harris CM, and Piersol AG, editors. Vibration of Systems Having Distributed Mass and Elasticity. New York: McGraw-Hill; 1961.
[35] Jonkman J, Butterfield S, Musial W, *et al.* Definition of a 5-MW Reference Wind Turbine for Offshore System Development. National Renewable Energy Laboratory; 2009. NREL/TP-500-38060.
[36] Jonkman J. Definition of the Floating System for Phase IV of OC3; 2010. NREL/TLP-500-47535.
[37] Ormberg H, and Bachynski E. Sensitivity of Estimated Tower Fatigue to Wind Modeling for a Spar Floating Wind Turbine. In: The 25th (2015) International Offshore and Polar Engineering Conference. Kona, Big Island, Hawaii, USA; 2015.
[38] Xue W. Design, Numerical Modelling and Analysis of a Spar Floater Supporting the DTU 10 MW Wind Turbine; 2016.
[39] Borg M, Hansen AM, and Bredmose H. Floating substructure flexibility of large-volume 10 MW offshore wind turbine platforms in dynamic calculations. Journal of Physics: Conference Series. 2016;753(8):082024.
[40] Det Norske Veritas. Design of Offshore Wind Turbine Structures; 2007. DNV-OS-J101.
[41] Newman JN. Marine Hydrodynamics. Cambridge, MA: The MIT Press; 1977.
[42] Dean RG, and Dalrymple RA. Water Wave Mechanics for Engineers and Scientists. World Scientific; 1991.
[43] Journee JMJ, and Massie WW. Offshore Hydromechanics; 2001.
[44] Robertson AN, Wendt F, Jonkman JM, *et al.* OC5 Project Phase II: Validation of global loads of the DeepCwind floating semisubmersible wind turbine. Energy Procedia. 2017;137:38–57. 14th Deep Sea Offshore Wind R&D Conference, EERA DeepWind'2017.
[45] Bachynski EE, and Moan T. Ringing loads on tension leg platform wind turbines. Ocean Engineering. 2014;84:237–248.

[46] Kristiansen T, and Faltinsen OM. Higher harmonic wave loads on a vertical cylinder in finite water depth. Journal of Fluid Mechanics. 2017;833:773–805.

[47] Suja-Thauvin L, Krokstad JR, Bachynski EE, *et al*. Experimental results of a multimode monopile offshore wind turbine foundation subjected to steep and breaking irregular waves. Ocean Engineering. 2017;146:339–351.

[48] Bredmose H, Dixen M, Ghadirian A, *et al*. DeRisk – Accurate prediction of ULS wave loads. Outlook and first results. Energy Procedia. 2016; 94:379–387. 13th Deep Sea Offshore Wind R&D Conference, EERA DeepWind 2016.

[49] Faltinsen OM, Newman JN, and Vinje T. Nonlinear wave loads on a slender vertical cylinder. Journal of Fluid Mechanics. 1995;289:179–198.

[50] Berthelsen PA, Bachynski EE, Karimirad M, *et al*. Real-time Hybrid Model testing of a Braceless Semi-submersible Wind Turbine. Part III: Calibration of a Numerical Model. In: 35th International Conference on Ocean, Offshore and Arctic Engineering. OMAE2016-54640; 2016.

[51] Skaare B, Nielsen FG, Hanson TD, *et al*. Analysis of measurements and simulations from the Hywind Demo floating wind turbine. Wind Energy. 2015;18:1105–1122.

[52] Utsunomiya T, Sato I, Yoshida S, *et al*. Dynamic Response Analysis of a Floating Offshore Wind Turbine During Severe Typhoon Event. In: 32nd International Conference on Ocean, Offshore and Arctic Engineering. OMAE2013-10618; 2013.

[53] Merigaud A, Gilloteaux JC, and Ringwood JV. A Nonlinear Extension for Linear Boundary Element Methods in Wave Energy Device Modelling. In: ASME 2012 31st International Conference on Ocean, Offshore and Arctic Engineering. OMAE2012-83581; 2012.

[54] Lawson M, Yu YH, Nelessen A, *et al*. Implementing Nonlinear Buoyancy and Excitation Forces in the WEC-Sim Wave Energy Converter Modeling Tool. In: 33rd International Conference on Ocean, Offshore and Arctic Engineering. NREL/CP-5000-61529; 2014.

[55] Babarit A. On the park effect in arrays of oscillating wave energy converters. Renewable Energy. 2013;58:68–78.

[56] Faudot C, and Dahlhaug OG. Prediction of wave loads on tidal turbine blades. Energy Procedia. 2012;20:116–133. Technoport 2012 – Sharing Possibilities and 2nd Renewable Energy Research Conference (RERC2012).

[57] Lust EE, Luznik L, Flack KA, *et al*. The influence of surface gravity waves on marine current turbine performance. International Journal of Marine Energy. 2013;3–4:27–40. Special Issue Selected Papers – EWTEC2013.

[58] Paulsen BT, Bredmose H, and Bingham HB. An efficient domain decomposition strategy for wave loads on surface piercing circular cylinders. Coastal Engineering. 2014;86:57–76.

[59] Benitz MA, Schmidt DP, Lackner MA, *et al*. Validation of Hydrodynamic Load Models Using CFD for the OC4-DeepCwind Semisubmersible. In: ASME 2015 34th International Conference on Ocean, Offshore and Arctic Engineering. OMAE2015-41045; 2015.

[60] Wienke J, and Oumeraci H. Breaking wave impact force on a vertical and inclined slender pile – Theoretical and large-scale model investigations. Coastal Engineering. 2005;52:435–462.

[61] Bredmose H, Slabiak P, Sahlberg-Nielsen L, *et al.* Dynamic Excitation of Monopiles by Steep and Breaking Waves. Experimental and Numerical Study. In: 32nd International Conference on Ocean, Offshore and Arctic Engineering. OMAE2013-10948; 2013.

[62] Etemaddar M, Hansen MOL, and Moan T. Wind turbine aerodynamic response under atmospheric icing conditions. Wind Energy. 2014;17(2):241–265.

[63] Gravesen H, Sørensen SL, Vølund P, *et al.* Ice loading on Danish wind turbines: Part 2. Analyses of dynamic model test results. Cold Regions Science and Technology. 2005;41:25–47.

[64] Barker A, Timco G, Gravesen H, *et al.* Ice loading on Danish wind turbines. Part I: Dynamic model tests. Cold Regions Science and Technology. 2005;41:1–23.

[65] Yu B, Karr DG, Song H, *et al.* A Surface Ice Module for Wind Turbine Dynamic Response Simulation Using FAST. In: 32nd International Conference on Ocean, Offshore and Arctic Engineering. OMAE2013-11179; 2013.

[66] Shi W, Tan X, Gao Z, *et al.* Numerical study of ice-induced loads and responses of a monopile-type offshore wind turbine in parked and operating conditions. Cold Regions Science and Technology. 2016;123:121–139.

[67] Kaynia AM. Seismic considerations in design of offshore wind turbines. Soil Dynamics and Earthquake Engineering. 2018. DOI: 10.1016/j.soildyn.2018.04.038.

[68] Katsanos EI, Thöns S, and Georgakis C. Wind turbines and seismic hazard: a state-of-the-art review. Wind Energy. 2016;19(11):2113–2133.

[69] Det Norske Veritas. Certification of Tidal and Wave Energy Converters; 2008. DNV-OSS-312.

[70] ISO. Petroleum and Natural Gas Industries – Specific Requirements for Offshore Structures – Part 1: Metocean Design and Operating Considerations. International Organization for Standardization; 2015. ISO-19901-1.

[71] Macleod AK, Stanley MS, Day JG, *et al.* Biofouling community composition across a range of environmental conditions and geographical locations suitable for floating marine renewable energy generation. Biofouling. 2016;32(3):261–276. PMID: 26900732.

[72] Tiron R, Pinck C, Reynaud EG, *et al.* Is Biofouling a Critical Issue for Wave Energy Converters?. In: Proceedings of the 22nd (2012) International Offshore and Polar Engineering Conference. vol. 1; 2012. p. 669–680.

[73] Tiron R, Gallagher S, Doherty K, *et al.* An Experimental Study of the Hydrodynamic Effects of Marine Growth of Wave Energy Converters. In: ASME 2013 32nd International Conference on Ocean, Offshore and Arctic Engineering. OMAE2013-10698; 2013.

[74] Spraul C, Pham HD, Arnal V, *et al.* Effect of Marine Growth on Floating Wind Turbines Mooring Lines Responses. In: 23ème Congrès Franais de Mécanique; 2017.

[75] Walker JM, Flack KA, Lust EE, *et al.* Experimental and numerical studies of blade roughness and fouling on marine current turbine performance. *Renewable Energy*. 2014;66:257–267.

[76] Henderson AR, Argyriadis K, Nichos J, *et al.* Offshore Wind Turbines on TLPs – Assessment of Floating Support Structures for Offshore Wind Farms in German Waters. In: 10th German Wind Energy Conference, Bremen, Germany; 2010.

[77] Dai L, Ehlers S, Rausand M, *et al.* Risk of collision between service vessels and offshore wind turbines. Reliability Engineering & System Safety. 2013;109:18–31.

[78] International Electrotechnical Commission. Marine Energy – Wave, Tidal and Other Water Current Converters – Part 2: Design Requirements for Marine Energy Systems; 2016. IEC TS 62600-2.

[79] Kühn M. Dynamics and Design Optimisation of Offshore Wind Energy Conversion Systems. Delft University of Technology; 2001.

[80] Smilden E, Bachynski EE, and Sørensen AJ. Key Contributors to Lifetime Accumulated Fatigue Damage in an Offshore Wind Turbine Support Structure. In: Proceedings of the ASME 2017 36th International Conference on Ocean, Offshore and Arctic Engineering; 2017.

[81] Fischer T, Rainey P, Bossanyi E, *et al.* Study on control concepts suitable for mitigation of loads from misaligned wind and waves on offshore wind turbines supported on monopiles. Wind Engineering. 2011;35(5):561–574.

[82] Barj L, Stewart S, Stewart G, *et al.* Impact of Wind/Wave Misalignment in the Loads Analysis of a Floating Wind Turbine. In: WINDPOWER 2013; 2013. (poster).

[83] Bachynski EE, Kvittem MI, Luan C, *et al.* Wind-wave misalignment effects on floating wind turbines: motions and tower load effects. Journal of Offshore Mechanics and Arctic Engineering. 2014;136:041902-1–041902-12.

[84] Wang Q. Design and Dynamic Analysis of a Steel Pontoon-Type Semi-Submersible Floater Supporting the DTU 10 MW Reference Turbine. NTNU/TU Delft; 2014.

[85] Ochi MK. Applied Probability and Stochastic Processes in Engineering and Physical Sciences. New York: Wiley; 1990.

[86] Det Norske Veritas. Design of Floating Wind Turbine Structures; 2013. DNV-OS-J103.

[87] Kvittem MI, and Moan T. Time domain analysis procedures for fatigue assessment of a semi-submersible wind turbine. Marine Structures. 2015; 40:38–59.

[88] Bachynski EE, Etemaddar M, Kvittem MI, *et al.* Dynamic analysis of floating wind turbines during pitch actuator fault, grid loss, and shutdown. Energy Procedia. 2013;35:210–222.

[89] Jiang Z, Karimirad M, and Moan T. Dynamic response analysis of wind turbines under blade pitch system fault, grid loss and shut down events. Wind Energy. 2014;17:1385–1409. (first published online June 13, 2013).

[90] Inman DJ. Engineering Vibration. Upper Saddle River, NJ: Pearson Education; 2008.

[91] Cook RD, Malkus DS, Plesha ME, *et al.* Concepts and Applications of Finite Element Analysis, 4th ed. Hoboken, NJ: John Wiley & Sons; 2002.
[92] Chapra SC. Applied Numerical Methods with MATLAB for Engineers and Scientists. New York: McGraw-Hill; 2017.
[93] Naess A, and Moan T. Stochastic Dynamics of Marine Structures. New York: Cambridge University Press; 2013.
[94] Taghipour R, Perez T, and Moan T. Hybrid frequency-time domain models for dynamic response analysis of marine structures. Ocean Engineering. 2008;35:685–705.
[95] Yu Y, Lawson M, Ruehl K, *et al.* Development and Demonstration of the WEC-Sim Wave Energy Converter Simulation Tool. In: Proceedings of the 2nd Marine Energy Technology Symposium; 2014.
[96] DTU Wind Energy. Pre-processing Tools – HAWC2; 2014. Available from: http://www.hawc2.dk/Download/Pre-processing-tools [Accessed July 2018].
[97] Luan C, Chabaud V, Bachynski EE, *et al.* Experimental validation of a time-domain approach for determining sectional loads in a floating wind turbine hull subjected to moderate waves. Energy Procedia. 2017;137:366–381.
[98] Hegseth JM, Bachynski EE, and Karimirad M. Comparison and Validation of Hydrodynamic Load Models for a Semi-submersible floating wind turbine. In: ASME 2018 37th International Conference on Ocean, Offshore and Arctic Engineering. OMAE2018-77676; 2018.
[99] Jonkman JM, and Buhl, Jr ML. FAST User's Guide. National Renewable Energy Laboratory; 2005. NREL/EL-500-38230.
[100] Cordle A, and Jonkman J. State of the Art in Floating Wind Turbine Design Tools. In: 21st (2011) International Offshore and Polar Engineering Conference. vol. 1; 2011. p. 367–374.
[101] Bachynski EE. Design and Dynamic Analysis of Tension Leg Platform Wind Turbines. Norwegian University of Science and Technology; 2014.
[102] Johannessen K, Meling TS, and Haver S. Joint Distribution for Wind and Waves in the Northern North Sea. In: The 11th International Offshore and Polar Engineering Conference & Exhibition; 2001.
[103] International Electrotechnical Commission (IEC). Wind Turbines: Part 3: Design Requirements for Offshore Wind Turbines; 2009. IEC61400-3.
[104] Det Norske Veritas. Column-Stabilised Units; 2015. DNV-RP-C103.
[105] Haver S, and Winterstein SR. Environmental contour lines: A method for estimating long-term extremes by a short term analysis. Transactions of the Society of Naval Architects and Marine Engineers. 2009;116:116–127.
[106] Winterstein SR, Ude TC, Cornell CA, *et al.* Environmental Parameters for Extreme Response: Inverse FORM with Omission Factors. In: 6th International Conference on Structural Safety and Reliability; 1993.
[107] Muliawan MJ, Gao Z, and Moan T. Application of the contour line method for estimating extreme responses in the mooring lines of two-body floating wave energy converter. Journal of Offshore Mechanics and Arctic Engineering. 2013;135:031301-1–031301-10.

[108] Saranyasoontorn K, and Manuel L. On Assessing the Accuracy of Offshore Wind Turbine Reliability-Based Design Loads from the Environmental Contour Method. In: Fourteenth International Offshore and Polar Engineering Conference; 2004. p. 128–135.

[109] Muliawan MJ, Karimirad M, and Moan T. Dynamic response and power performance of a combined Spar-type floating wind turbine and coaxial floating wave energy converter. Renewable Energy. 2013;50:47–57.

[110] Li Q, Gao Z, and Moan T. Modified environmental contour method for predicting long-term extreme responses of bottom-fixed offshore wind turbines. Marine Structures. 2016;48:15–32.

[111] DNVGL. Loads and Site Conditions for Wind Turbines. DNVGL; 2016. DNVGL-ST-0437.

[112] American Bureau of Shipping. Guide to Building and Classing Offshore Wind Turbine Installations; 2010.

[113] American Bureau of Shipping. Guide for Building and Classing Floating Offshore Wind Turbine Installations; 2013.

[114] Nippon Kaiji Kyokai. Guidelines for Offshore Floating Wind Turbine Structures; 2012.

[115] Det Norske Veritas. Structural Design of TLPs (LRFD Method); 2008. DNV-OS-C105.

[116] Gujer P, and Matthias K. Review of FOWT Guidelines and Design Practice. LIFES50+; 2015. D7.1. EU Grant 640741.

[117] Davies P. Guidelines for Design Basis of Marine Energy Conversion Systems. The European Marine Energy Centre Ltd; 2009.

[118] Det Norske Veritas. Certification of Tidal Turbines and Arrays; 2015. DNVGL-SE-0163.

Chapter 8

Electrical power transmission and grid integration

Elisabetta Tedeschi[a] *and Abel A. Taffese*[a]

8.1 Introduction

This chapter provides an overview on the main challenges encountered during the interconnection of marine energy farms to the onshore electric power system. It explains how less technologically mature marine energy converters (MECs), such as wave and tidal ones, are normally integrated into distribution systems, and potentially cause power quality problems. Then it shows how system level issues arise when larger offshore installations, such as offshore wind farms, are interconnected to power transmission systems. The presentation is complemented by illustrative test cases.

8.2 Implications of the grid-side converter topology on the grid integration of MECs

The grid connection of MECs[1] is, in most cases, a necessary step to enable the use of the produced electricity and ensure the economic viability of marine energy projects. If the primary goal of grid integration is to maximize the renewable power injected into the main network, the prospective growth in the installed capacity of offshore energy plants, coupled to the resource-dependent variability of the power production, requires a careful conditioning of the output power prior to the grid interconnection. This is done to avoid detrimental effects on the power system where the interconnection is performed, and concurrently allow the possible provision of grid support services.

The degree of controllability of grid connected MECs strongly depends on the type of grid-side power electronic (PE) interface they employ. As for wind

[a]Department of Electric Power Engineering, Norwegian University of Science and Technology, Trondheim, Norway

[1]In the absence of consensus, but in agreement to several sources, in this chapter the expression *marine energy converters* includes offshore wind turbines, marine solar energy converters and *ocean energy converters*, where the latter expression indicates all those devices that harvest energy stemming from ocean resources (e.g. tides, waves, currents and temperature gradients).

turbines (WTs), the first prototypes of ocean energy converters (OECs) were equipped with squirrel-cage induction generators operating at fixed speed; they could be directly connected to the grid without any PE interface, but this prevents any type of grid support/regulation. However, the increasing penetration of offshore wind and ocean energy made the power conditioning provided by PEs necessary to comply with the increasingly demanding grid code requirements.

Most MECs are currently using back-to-back (B2B) PE interfaces to perform their AC–DC–AC conversion and enable grid integration [1,2]. B2Bs are normally based on voltage source converters (VSCs), either two-level or multilevel, depending on the power range of the application. They can be either partial-scale (i.e. sized for processing a fraction of the nominal power of the generator) or full-scale, if they process the entire power converted by the electrical generator. Partial-scale B2B converters have been widely used with doubly fed induction generators (DFIG) (configuration in Table 8.1 (a)) in the wind energy industry [3,4].

These schemes offer a reduction in the cost of the PEs, which is typically rated for processing 30% of the generator power. Such solutions can provide some grid services, such as reactive power exchange/voltage support; however, they are limited by the converter current capability. Moreover, fault conditions resulting in voltage dips are more critical for this type of drive-train configuration, due to the very high current circulation and increased reactive power absorption [22]. This explains the increasing trend to use full-scale PE converters (configuration in Table 8.1 (c)) in the wind energy sector [23], especially for offshore high-power WTs [5,24]. On the contrary, solutions using DFIG and partially rated PE interfaces found only limited application in wave [8] and tidal [13,25] energy systems, due to the required operation in the harsh and corrosive sea environment, which discourages the use of brushed machines in favour of low- or zero-maintenance solutions [26,27].

Advantages of the fully rated PE interface [9,15,28] are the decoupling between the MEC-side and grid-side dynamics [10] and full controllability of the electrical quantities (e.g. independent control of active and reactive power) at the point of grid connection [11,16]. Furthermore, the presence of the DC link enables the integration of energy storage devices [12,29], to smooth the power and mitigate the grid impact of the MEC interconnection. It is worth noting that, in the offshore wind sector, multicell converters (configuration in Table 8.1 (b)) have been also introduced, due to the increased power capacity of the single WT [30]. In such configurations, each (fully controlled) grid-side inverter is rated for a fraction of the total power processed by the WT [6,7]. This reduces the current and voltage ratings of the converters and, hence, their cost and complexity. However, the entire power flow is conditioned before the grid injection, and the same advantages as the full-scale B2B interfaces are achieved in terms of grid controllability. In some cases, the traditional B2B configuration is modified so that multiple MECs, each equipped with its own power rectifier, share the same inverter (configuration in Table 8.1 (d)) [18]. In such cases, the centralized inverter has a power rating higher than each of the MECs, possibly achieving higher efficiencies and ensuring the same grid flexibility as a full-scale B2B. Marine solar power plants differ from

Table 8.1 *Example of grid-side converters (shaded) used for different types of MECs*

	Partial-scale grid-side converter		Full-scale grid-side converter	Centralized grid-side converter	
	Part of a B2B configuration in a partially rated PE interface	**Part of a B2B configuration in a fully rated PE interface**	**Part of a B2B configuration in a fully rated PE interface**	**Part of a two-stage (AC–DC–AC) PE interface**	**Part of a single-stage (DC–AC) PE interface**
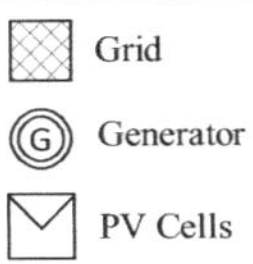	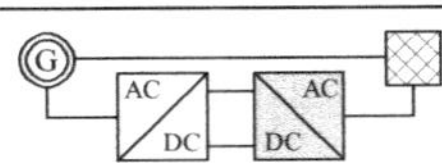 (a)	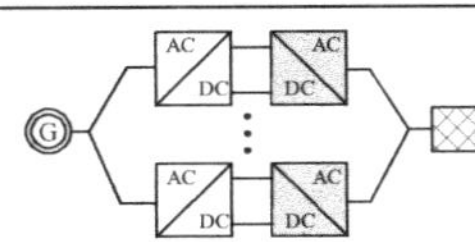 (b)	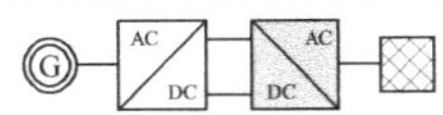 (c)	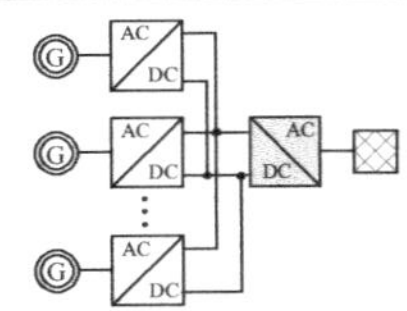 (d)	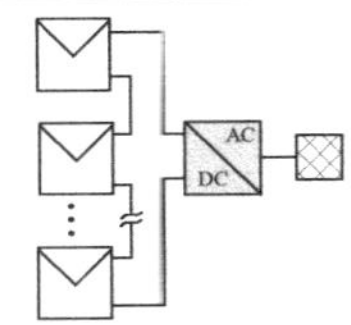 (e)
Offshore wind	[3,4]	[5–7]			
Wave	[8]		[9–11]	[12]	
Tidal	[13,14]		[15–17]	[18]	
Marine solar					[19–21]

other MECs as the power is generated using direct current and only an inverter is technically required for the grid connection, possibly preceded by a DC–DC converter, to increase the power extraction efficiency. As for the onshore counterpart, a central inverter is often used to connect multiple photovoltaic panels (configuration in Table 8.1 (e)) [19–21]. Since the grid-side converter is fully controllable, performances analogous to those of full rated power converters are achievable.

The main PE interfaces used in marine applications are exemplified in Table 8.1.

The effects of the grid interconnection of MECs to onshore electric power systems strongly depend on the installed capacity, which can vary from a few kilowatts in the case of a single OEC to hundreds of megawatts for large wind farms, as well as on the type of network, ranging from coastal AC distribution grids to large transmission systems, possibly using high-voltage direct current (HVDC) technology. Grid effects relevant to MECs are summarized in Figure 8.1.

8.3 Impact of MECs' integration into power distribution systems

8.3.1 Power quality issues in marine energy installations

The least developed MECs, such as wave energy converters (WECs) and tidal energy converters (TECs)[2], are often connected to coastal distribution systems with limited degree of interconnection. From an electrical standpoint, they can be

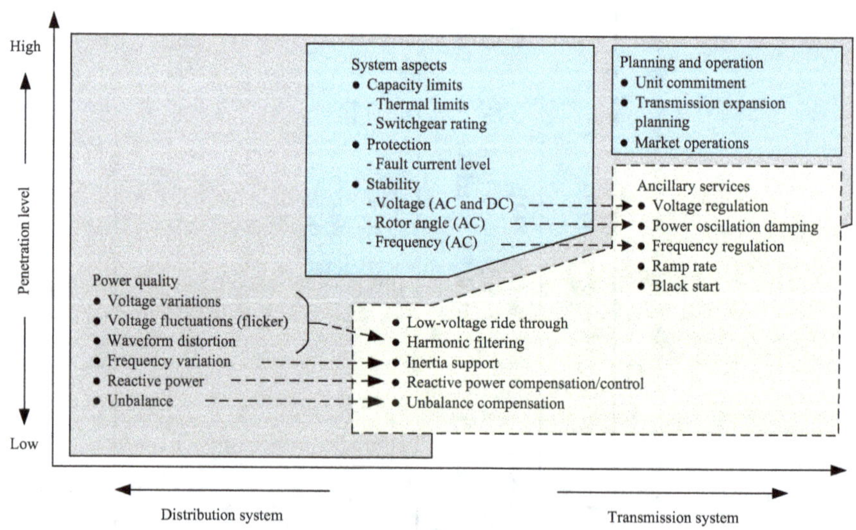

Figure 8.1 Grid impact of marine energy conversion systems and possible grid services (dashed box) provided

[2]Ocean thermal energy converters and osmotic energy converters are not considered hereafter, due to their much lower level of technological maturity.

considered as 'weak grids', characterized by high grid impedances/low short circuit powers. This tends to amplify the power quality issues that are typically produced by MEC devices depending on their operating conditions, power train configuration and applied control strategies [31]. Power quality issues may arise during normal operation as a consequence either of the highly intermittent power generation due to the variability of the natural resource (e.g. voltage variations and flicker), or of the inherent principle of operation of electrical machines and converters (e.g. current and voltage harmonics, producing waveform distortion). However, they can also be a consequence of abnormal operations due to equipment/grid faults (e.g. voltage interruptions and unbalances). The most common power quality issues and possible causes are briefly discussed as follows.

Voltage variations

Variations in the root-mean-square (RMS) value of the voltage at the point of connection of an MEC/MEC farm with the main grid can be classified based on their short- (e.g. less than 1 minute) or long-duration (e.g. more than 1 minute). Specifically, a *voltage sag* (also called *voltage dip*) is a decrease in the RMS voltage between 10% and 90% of its nominal value for a short duration. An RMS voltage decrease between 80% and 90% of the nominal value having a long duration is defined as an *undervoltage*. Conversely, a *voltage swell* is defined as the increase in the RMS voltage above 10% of its nominal value for a short duration. An *overvoltage* is a long-duration increase of the RMS voltage, typically between 10% and 20% over the nominal value[3].

A *voltage interruption* occurs when the voltage decreases to less than 10% of its rated value. The interruption is classified as short duration or long duration if it is sustained for less or more than 1 minute, respectively. Voltage swells or sags can be caused by switching of large loads, capacitor banks connection/disconnection or faults in the main power system; however, in OEC systems, they can also be induced by the inherent power fluctuation of WECs, or the tidal flow turbulence in TECs. Interruptions are normally caused by faults or equipment outages.

Voltage fluctuations (flicker)

They are systematic variations of the voltage waveform envelope, or a series of random voltage changes [32], typically with a magnitude variation between 0.1% and 7% of the nominal voltage and with frequencies less than 25 Hz. Such phenomenon provokes the rapid visible variation in the light output of various lighting equipment, termed as *flicker*. This gives an impression of visual instability that can have adverse consequences on the health of sensitive people. The low-frequency oscillation of wave and tidal resources makes OECs more prone to causing flicker problems compared to other renewable energy systems. Specific studies have been undertaken to assess flicker levels induced at the points of common coupling by both wave [29,33,34] and tidal [35,36] energy farms.

[3]Note that the specific numerical values characterizing power quality phenomena may vary depending on the reference standard considered (e.g. International Electrotechnical Commission (IEC) and Institute of Electrical and Electronics Engineers (IEEE)).

Unbalance

Unbalances in the voltage (or currents) of a three-phase system refer to the presence of any phase and/or amplitude deviations from the ideal conditions, in which the voltage (or currents) in the different phases have the same amplitude and a relative phase displacement of 120 degrees. Unbalances in three-phase power systems mainly affect low voltage grids, due to the higher penetration of single-phase loads. Although MECs are normally three-phase devices, they can be affected by unbalances in the event of asymmetrical grid faults. Their capability to operate properly under unbalances strongly depends on the type of (electrical machine) and PE interface, where the presence of full-scale converters ensures better performance under unbalanced conditions [37,38]. Specific equipment can be installed to provide unbalance compensation when such a condition appears critical for the local power grid.

Waveform distortion

Although electrical quantities such as voltage and currents are expected to have a perfectly sinusoidal waveform with fixed frequency, f_0, (normally 50 or 60 Hz) in AC grids, the increasing presence of non-linear and time-variant loads and undesired effects such as transformers' magnetic saturations may alter the sinusoidal shape of the fundamental waveforms with superimposed, repetitive, disturbances. Specifically, (*integer*) *harmonics* are defined as sinusoidal disturbances superimposed to the voltage and/or current fundamental and whose frequency, f_i, is an integer multiple of the fundamental grid frequency, f_0 (i.e. $f_i = mf_0$, for integer m). *Subharmonics* identify those superimposed components whose frequency, f_s, is below the fundamental component (i.e. $f_s = mf_0$ for $0 < m < 1$). Finally, *interharmonics* are those superimposed components whose frequency is higher than, but not an integer multiple of, the fundamental frequency (i.e. $m_i = mf_0$ for non-integer $m > 1$).

PE converters, which are based on non-linear and switching semiconductor devices, can cause both current and voltage distortion, and MECs equipped with full-scale power converters produce higher harmonic content than partial-scale ones. Such harmonic generation, largely due to the application of pulse width modulation, is concentrated in relatively high-frequency ranges (in the kHz range, depending on the converter switching frequency). However, the need for grid synchronization with consequent use of non-ideal phase locked loops, and the possible interactions between control systems and passive components in the electrical network, can determine the arise of relatively low-frequency components and their potential amplification in case of grid resonances [39].

Frequency variation

This refers to any deviation of the fundamental frequency in the considered AC network from its nominal value, f_0. In traditional power systems, dominated by large synchronous generators, frequency deviations are an immediate consequence of the presence of an (active) power unbalance between generation and consumption, due to the induced change in generators' rotational speed: if the power generation exceeds the total load consumption, the frequency will increase. Conversely, the frequency will decrease if there is a power deficit. The inherent variability of marine renewable sources and limited dispatchability of MEC power production may contribute to this problem.

As frequency is a global variable in AC systems, any deviation from the prescribed range can have serious consequences for the system operation, depending on the characteristics of the load and response of the generation control system to load variations.

Reactive power

Although not strictly classified as a power quality issue, any reactive power circulation, Q, in an electric power system corresponds to a reduction in the overall efficiency. Reactive power is consumed by inductive components (e.g. transformers, motors and reactors) and generated by capacitive components (e.g. capacitor banks and cables). While the use of partial-scale or full-scale PE interfaces allows (a partial or total) control of the reactive power exchanged at the MEC point of connection, the presence of other system components, such as submarine cables and substation transformers in offshore farms, may determine the circulation of significant amounts of reactive power. This causes undesired losses and efficiency drops in the plant operation. Moreover, it affects the voltage profile at the system feeders and requires a tight regulation. An indirect quantification of the reactive power exchange is obtained by the *power factor, PF*, defined, in sinusoidal regimes, as the ratio between the active power, P, and the apparent power, S, at the considered point:

$$PF = \frac{P}{S} = \frac{P}{\sqrt{P^2 + Q^2}} \tag{3.25}$$

8.3.2 System impact of marine energy installations

As MECs are potential sources of power quality problems, when a MEC farm is connected to the onshore power system, checking that such grid integration does not compromise the stability of the entire network is paramount. *Power system stability* is defined as the ability of the power system 'to remain in a state of operating equilibrium under normal operating conditions and regain an acceptable state of equilibrium after being subjected to a disturbance' [40].

The stability of an AC power system can broadly encompass three different aspects, partly interrelated:

1. ***Voltage stability*** is defined as the system capability to maintain an acceptable voltage level at all system busses under normal operations as well as in post-disturbance conditions [40]. Voltage stability can be further divided into *large-disturbance* voltage stability, related to the system reaction to large disturbances, such as faults, contingencies and loss of load/generation, and *small-disturbance* voltage stability, concerning system response to common perturbations of small amplitude. Maintaining voltage stability is related to the capability of the system to meet the demand for reactive power. Hence, MECs will also need to contribute to enhance voltage stability by implementing an

appropriate ***voltage regulation*** method. This can be done using one of the three mutually exclusive control modes: (a) *reactive power control*, setting a constant reactive power exchange reference, independently on the active power generation; (b) *power factor control*, setting the requirements in terms of fixed power factor; (c) *voltage control mode*, based on a V/Q droop action that adjusts the reactive power exchange proportional to the measured local voltage signal [41].

2. ***Frequency stability*** defines the system ability to maintain steady frequency following an event provoking significant imbalance between generation and load, by restoring the load/generation balance with minimal loss of load. The stabilization of the grid frequency can be obtained by means of ***frequency response*** (also called frequency regulation) [42]. It includes three main functions, i.e. (a) primary control, acting on a timescale 15–30 s to limit sudden frequency variations resulting from power unbalances; (b) secondary control, restoring frequency and power references to their scheduled values and acting within a 15-minute timescale and (c) tertiary control, acting to support the secondary control or restore the power reserve used by it on a timescale up to 60 minutes. Prior to primary control action, the frequency change is dictated by the amount of inertia in the system. Since MECs generally have lower inertia compared to traditional synchronous generators, they need to provide *synthetic* (virtual) *inertia* in order to participate in the inertial response phase of frequency regulation.

   Secondary control may entail several types of regulation (e.g. delta control, balance control and ramp rate/gradient control [42]).
3. ***Rotor angle stability*** concerns the ability of synchronous machines of an interconnected power system to remain in synchronism after being subjected to a disturbance. It relies on the capability to maintain/restore equilibrium between mechanical torque and electromagnetic torque of each of the interconnected synchronous machines. Rotor angle stability can also be sub-divided in *large-disturbance* stability (also known as transient stability), related to the system reaction to large disturbances and *small-disturbance* stability, concerning the ability of the machines to keep synchronism when subjected to small disturbances [40].

In order to ensure the stability and safe operation of any power system integrating MECs, the above-mentioned aspects need to be regulated. This is normally performed through specific grid codes, which set requirements for the subsystem operation under both normal operating conditions and fault situations.

Specific grid codes for offshore energy installations are still under development, because offshore energy is relatively new development, and wave and tidal energy conversion systems have not reached the commercial stage, yet.

In Europe, at the time of writing, only UK and Germany have included specific provisions for 'Offshore Power Generating units powered by intermittent sources' (National Grid [43]) and offshore wind farms (TenneT [44]), respectively. ENTSO-E has also made a substantial effort to harmonize national grid codes [45]. A review of grid codes relevant to MECs can be found in [46,47].

MEC farms are required to comply with such grid codes in order to be allowed to interconnect with the main power system. This is normally validated by numerical models and analyses. In particular, by ***load flow analyses*** the network performance is assessed based on parameters such as (active and reactive) power circulation, voltage and frequency levels and component loading conditions. Simulation studies are normally required also to evaluate the system behaviour under ***fault conditions***, aimed at ensuring that no grid components exceed their transient ratings (e.g. compliance with thermal limits), protection systems are properly sized, and that grid subsystems (including MECs) comply with prescriptions for operation under fault conditions (e.g. requirements for low voltage ride through or voltage regulation, Section 8.4.1).

8.3.3 Case study

The following study considers the interconnection of a 20-MW-wave farm into a distribution system [48]. The wave farm is composed of 20 point-absorbers, each equipped with a permanent magnet synchronous generator and an electric drive rated 500 kW and a 0.69 kV/13.2 kV transformer. Groups of 10 point-absorbers are connected to the same DC link and they have a common grid-side inverter controlling the active and reactive power injection into the point of common coupling (PCC) (configuration in Table 8.1(d)). The local electric infrastructure is modelled resembling the Biscay Marine Energy Platform (*bimep*) test site [49], where the four offshore benches are composed of subsea cables, length between 3 and 6 km. Once onshore, subsea cables are replaced by 1 km of land cables. This ensures the interconnection to the local power substation that hosts two 25 MW, 13.2 kV/132 kV transformers operating in parallel. The PCC is upstream of the onshore substation and a fixed load, represented by a 50-Ω resistance, is connected to such point to represent the local power consumption. *Bimep* is connected to a strong electric grid, which can be modelled by a Thevenin equivalent with a purely inductive series impedance ($Z_{\text{grid}} = j4.21\ \Omega$). The study comprises system analyses on the effects of the wave farm grid integration, with a particular focus on voltage regulation and efficiency of the power transfer. The recommendation of the local distribution system operator (DSO) [50], allowing a maximum ±7% voltage deviation at the PCC, is considered, and the effect of the grid strength on voltage support is analysed by considering the additional case of $Z_{\text{grid}} = 50.52\ \Omega$ to emulate the *bimep* interconnection to a hypothetical significantly weaker grid. The analysis is based on load flow numerical simulations performed with DIgSILENT Power Factory©.

As a first test, the voltage level, V, active power injection, P, and reactive power exchange, Q, are measured at the PCC, when all the centralized grid-side inverters are controlled to inject power into the PCC with unity power factor ($PF = 1$, i.e. no reactive power injection at WEC point of connection).

The corresponding variability is shown through their time evolutions in Figure 8.2. Moreover, the corresponding maximum, minimum and average values of V, P and Q are calculated and shown by the two leftmost histograms of Figure 8.3(a), (b) and (c), respectively. It can be noticed that the efficiency of the power transfer is higher and the effect of the active power variability on the PCC voltage is negligible in case of a strong grid. In a weak grid scenario, however, the

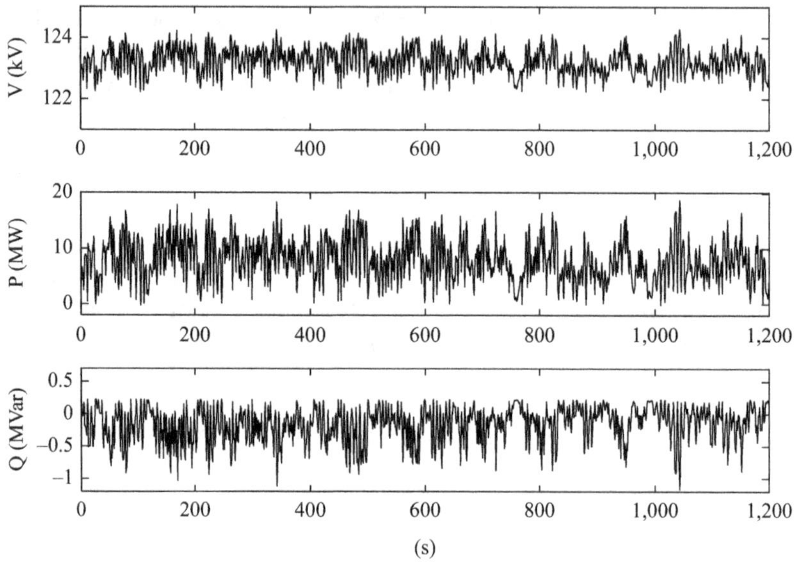

Figure 8.2 Voltage, active power and reactive power at the PCC with grid-side inverters

minimum voltage, occurring at no active power injection, may violate the set voltage limit ($V_{\text{lim}} = 122.76$ kV), although the average voltage will be still above such value.

Furthermore, it can be noted that the reactive power measured at the PCC is due to the generation/absorption of the passive elements of the grid (cables, transformers), as all the four inverters are injecting pure active power at their respective points of connection.

As a second test, the study verifies if power factor control (as a means of voltage regulation) can avoid the undervoltage condition detected in the case of weak grid. In this case, all inverters are controlled to inject power with a constant power factor, $PF = 0.9$ *(leading)*. Although this strategy improves the efficiency of the power transmissions (+3.5% at PCC), this is obtained at the cost of a very high (average) reactive power injection, and power factor control fails to avoid the voltage limit violation (as $V_{\text{min}} < V_{\text{lim}}$). This is because the maximum voltage drop at the PCC occurs when there is a zero active power injection from the WEC farm and the local load needs to be fully supplied by the main grid. However, with a constant power factor control, when no active power is injected from the farm, there is no reactive power injection either, although these are the instants when voltage support is most needed.

As a possible solution, voltage control mode at PCC is performed by using a real-time decentralized strategy based on a *V/Q* droop constant, which allows higher reactive power injection when the active power production from the WECs is low, while inherently respecting the capacity rating of the inverters. Details on

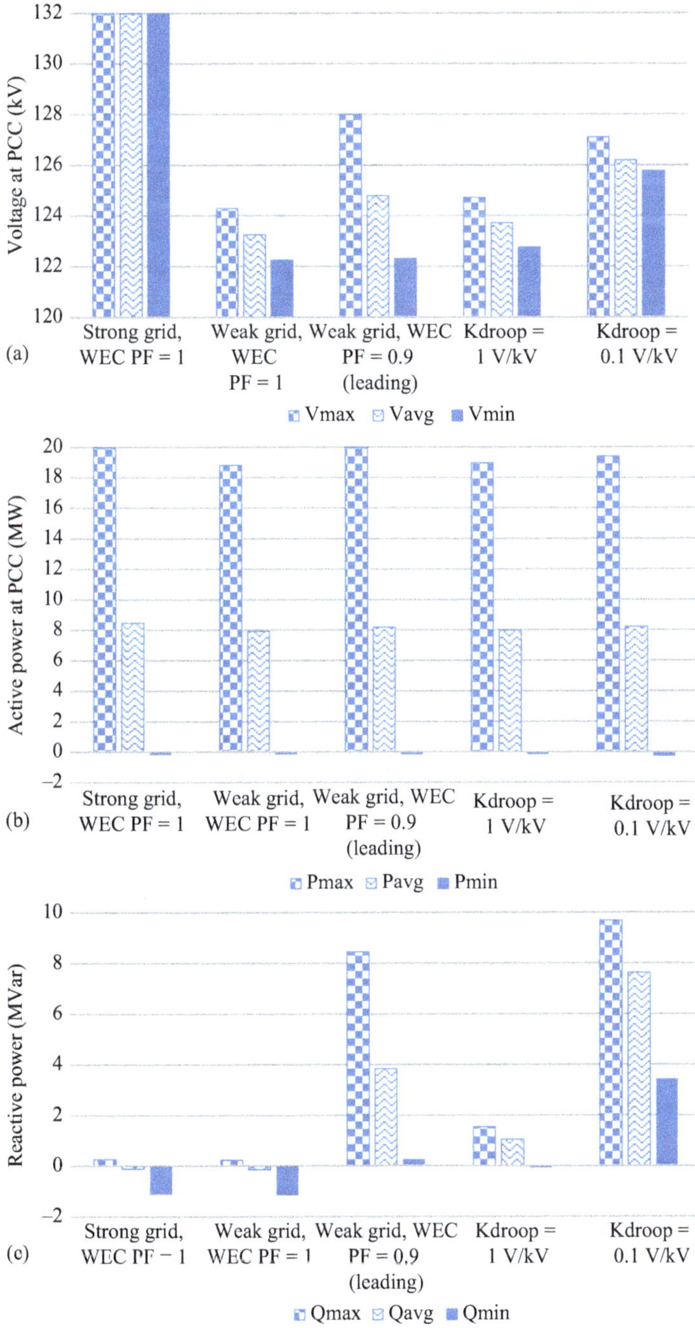

Figure 8.3 Maximum, minimum and average value of the voltage (a), active power (b) and reactive power (c) at the PCC under different grid conditions and WEC grid side

the implementation of such decentralized droop control strategy can be found in [48]. It can be observed that the use of a droop constant of 1 V/kV reduces the voltage variability at the PCC and ensures that any voltage limit violation is avoided. Such a benefit is obtained at the cost of an (average) reactive power injection that is significantly lower than the one required by the above-presented power factor control. It can also be noted that any decrease in the value of the droop constant (see the rightmost histogram in Figure 8.3(a)–(c)) ensures higher voltage at the PCC and a slightly better efficiency in the average power extraction. However, this is obtained at the cost of a significantly higher reactive power injection at the PCC (and hence a potentially critical decrease of the power factor).

8.4 Impact of MECs' integration into power transmission systems

The offshore wind energy sector underwent an exponential growth in the last years, and it is expected that the offshore installed capacity can reach 70 GW by 2030 just in Europe [51]. As the installed capacity and distance from shore of new wind farms are also increasing, they need to be connected to the power transmission grid. Moreover, the increasing power penetration implies that it is not sufficient anymore to ensure that newly connected farms do not negatively affect the stability and efficient operation of the pre-existing power system. Large wind farms need to become active players in the energy system. They must be capable of supporting the grid, e.g. by helping avoid stability and congestion problems, reducing power quality issues, and making the operation of the entire power system more efficient and reliable. Thus, they are increasingly required to provide ancillary services, i.e. grid support functions that are additional to their primary scope of injecting renewable energy into the grid. Although some of these grid services are also beneficial in power distribution systems (as for voltage and frequency regulation, explained in Section 8.3.2), their importance becomes critical in the highly interconnected transmission systems.

8.4.1 Additional ancillary services that can be provided by marine energy installations

The above-mentioned ancillary services can be grouped into two broad categories: services provided during normal operation (e.g. harmonic compensation), and those required during contingencies (e.g. low-voltage ride through (LVRT), black start, and power oscillation damping). These services are briefly discussed in the following.

Harmonic compensation

This refers to the capability of passive and active components to mitigate the effect of current or voltage harmonics, and/or prevent them from propagating through the power system. MECs equipped with a fully controllable grid-side converter, if this is properly controlled, can provide harmonic compensation for frequencies within their control bandwidth [52]. Alternatively, dedicated passive or active filters may be required.

Low-voltage ride through

This refers to the capability of a generator to stay connected to the grid and remain operational, possibly providing reactive power injection, in the event of a voltage dip. LVRT has become a requirement of significant importance with the increase in the wind energy penetration, because the additional loss of generations capacity could further contribute to the unbalance if large wind farms are allowed to disconnect during voltage dips. This brings the grid closer to instability. Conversely, the possibly required injection of reactive power during voltage dips is meant to help the voltage recovery.

Black start

This indicates a sequence of actions (e.g. re-energization, frequency management and resynchronization) performed to restore the normal operation of an electrical system after a disturbance that provoked a blackout or state of emergency [53]. MECs using VSCs may be able to perform black start, sometimes supplemented with energy storage or small diesel generators.

Power oscillation damping

This is an ancillary service aimed at improving small-signal (small-disturbance) stability in an AC system by providing damping of electromechanical oscillations (typically in the frequency range of 0.2–2 Hz) by injecting a damping power. It can be provided by MECs using appropriate modulation of either active or reactive power [54].

The capability of different types of MEC to provide ancillary services is shown in Table 8.2.

8.4.2 Transmission technologies

HVAC versus HVDC technologies

Power transmission based on high-voltage alternating current (HVAC) is the most widely used technology for offshore marine renewable applications [55]. The configuration of an offshore HVAC transmission system typically comprises the following:

- The *collection platform*: It is the point of connection between the offshore power collection grid, collecting the power from all the MECs, and the HVAC cable. The offshore platform hosts the power transformers required to increase the voltage of the collection grid (typically operated at 33 or 66 kV), from medium to high voltage level.
- *HVAC cables*: They are needed for the long distance power transmission to shore. They are composed of a copper or aluminium conductor with an cross-linked polyethylene (XLPE) insulation, which ensures high current capabilities, ease of installation and limited maintenance requirement, and metallic screens or shields ensuring mechanical robustness and confinement of the electrical field.
- *Reactive power compensators*: The high, distributed capacitance of submarine cables causes the circulation of high charging currents, which increases with the voltage level and the length of the line. Being the current capacity of the

Table 8.2 Capability to provide ancillary services of offshore energy systems and traditional power systems

Ancillary service		MEC with partial-scale grid-side converter as part of a B2B configuration	MEC with full-scale grid-side converter as part of a B2B configuration	Traditional power systems based on synchronous generators
Frequency regulation	Fast freq. regulation	−	++	−−
	Inertia support	−	−*	++
	Primary control	−	−*	++
	Secondary and tertiary control	−−*	−*	++
Voltage regulation	Fast reactive current injection	−	++	+
	Slow V/Q support	−	++	++
Harmonic compensation		+	++	−−
LVRT		−	+	NA
Black start		−−	+	+
Power oscillation damping		−	+*	++

(++) very good capability, (+) sufficient capability, (−) limited, (−−) not feasible and (NA) not applicable.
*This capability will improve if energy storage is integrated.

cable is fixed, the circulation of high reactive currents reduces the cable capability to concurrently transfer active currents, hence limiting the active power transfer. In marine energy farms using HVAC, reactive power compensators, e.g. stationary, quasi-stationary (static var compensators (SVCs)) or dynamic (static synchronous compensators (STATCOMs)) units, are generally connected at both ends of the HVAC cable to limit the circulation of such capacitive currents.

HVAC transmission relies on power transformers and AC breakers, which are mature technologies, and the operation of a multi-terminal/meshed grid is fully consolidated. However, the need for reactive power compensation constrains HVAC application to offshore farms of limited capacity and relatively short distance from shore, whereas HVDC is preferred for further and higher power applications. The power transmission efficiency of HVDC cables is higher compared to that of corresponding HVAC cables, and HVDC cables provide more compact and cheaper solutions for the same power transfer. However, an HVDC system requires the deployment of high power/high voltage AC/DC converters, which cause additional losses, costs and, potentially, harmonic pollution in the interconnected system. Overall, the choice between HVAC and HVDC technologies is always based on a detailed techno-economic assessment, which takes into account the availability, technological maturity and cost of all the required components, in addition to the installed capacity and distance of the installation. For each installed capacity a break-even distance can be identified, above which the cost of the HVDC solution is lower than that of the HVAC one.

Although the first subsea applications of the HVDC technology, dating back to the 1950s (Gotland island), were based on current source converters (i.e. line-commutated converters, LCC, using first mercury arc valves, and then thyristors), the breakthrough paving the way to offshore HVDC applications was the development of HVDC systems based on VSCs (VSC-HVDC). Such systems use self-commutated converters, have lower footprint, and reduced harmonic filtering requirements compared to thyristor-based solutions. Moreover, they are capable of a full and independent control of active and reactive power at their terminals. The first HVDC offshore wind farm, Bard Offshore 1, which was commissioned in 2013 in the North Sea, uses VSC-based transmissions to transfer 400 MW to the German mainland, over a cable distance of 200 km [56].

The main components of HVDC transmission systems are as follows:

- The *HVDC converters:* They perform the conversion from AC to DC at the rectifier station, and from DC to AC at the inverter station. They are operated at selected DC voltage levels (e.g. $\pm$525 kV).
- *AC and DC harmonic filters:* These may be required to limit the amount of harmonics produced by the converter operation, depending on the topology of the selected HVDC converters. Adding passive filters, however, causes increased footprint and weight, which are critical factors in offshore systems. Especially for very high power applications, converters with minimum filtering requirements (e.g. modular multi-level converters (MMCs) [57]) are preferred.
- The *HVDC cables:* Two different technologies are currently in use for HVDC submarine cables: paper lapped cables, which are a consolidated technology available for voltages up to $\pm$700 kV and power ratings of 3,400 MW (bi-pole rating), and polymeric cables (e.g. XLPE), which are used for HVDC-VSC applications that do not require polarity reversal to reverse the power flow. XLPE cables have been applied so far for voltage levels up to 640 kV and a power capacity of 3,000 MW, and have many advantages compared to paper lapped ones. They allow compacter solutions for the same power level due to the higher conductor temperature, and have lighter moisture barriers, reducing the overall cable weight. Furthermore, jointing extruded cables is easier and the environmental hazard of oil leaks associated to paper lapped cables is avoided. It is worth noting that VSC-HVDC technology, unlike the LCC-HVDC one, is also suitable for (parallel) multi-terminal/meshed developments [58], enabling the development of a future, highly interconnected, offshore DC grid [59].

Stability in DC power systems

DC voltage stability is one of the main concerns in DC grids because DC voltage is used as a global stability indicator in DC systems. It is often compared with frequency in AC systems, since both quantities reflect the balance between active power generation and consumption in their respective systems [60]. There are a number of properties unique to DC systems that give rise to new stability challenges [61]. The first one is that the controllers are significantly faster than the conventional controllers are in AC systems. Therefore, they can be affected by high-frequency (a few hundred of hertz) resonances, which are common in cable-connected offshore MECs. This has an implication on the level of detail necessary in modelling of components for studying stability of such systems. Most of the simplifications applied in AC systems are no longer applicable. A second challenge is that each node in a DC system is actively controlled by a power converter using proprietary control schemes. This makes it difficult to guarantee stability when making interconnections due to the lack of critical information necessary to perform compliance studies. This has introduced the need for new stability studies focusing on the interoperability of components in a power system [62]. Another challenge in DC systems is that they have a very limited energy storage, mostly obtained from DC link capacitors and cables, compared to AC systems [61]. Thus, unless specific precautions are taken, they cannot sustain large disturbances and faults for extended durations (hundreds of milliseconds) on their own, like AC systems with large synchronous generators do. This problem can be mitigated by exchanging ancillary services with the AC grid. For example, the AC grid can provide DC voltage support to improve transient stability of the DC grid, which, in return, provides power oscillation damping service to improve small signal stability of AC grids [63].

8.4.3 Case study

One of the challenges of integrating large-scale remote offshore renewable sources is that they are often connected via HVDC lines to weak points in the onshore side, which implies that the point of common coupling is prone to power oscillations. This can be alleviated by implementing power oscillation damping (POD) service using a converter connected to the onshore grid. Another consequence of having a weak connection is that the effect of power quality problems due to the variability of offshore resources, such as wind, is more pronounced. Therefore, it is important to reduce the variations in the active power injected in to the onshore AC grid by applying active power smoothening, which can be implemented on any of the offshore converters. A common feature between the two services discussed in this section, i.e. power oscillation damping and active power smoothening, is that they both manipulate active power to provide grid support. The main challenge with active power manipulation is that it creates a disturbance on all the other connected grids unless energy storage is included in the converters. However, storage deployment may not be feasible due to its cost and the space/weight restrictions often present in offshore projects. With the emergence of MMC, which has inherent energy storage capability distributed across it capacitors, the aforementioned POD and power smoothening services can be provided without propagating the problem or requiring external storage.

A case study demonstrating the provision of such services using MMC [63] will be presented. The case study considers a four terminal meshed system [64] (shown in Figure 8.4) with two 900-MW offshore wind farms (T1 and T2) and two onshore AC grids (T3 rated 900 MW and T4 rated 1,200 MW) interconnected via HVDC cables and MMC converters. The onshore AC grids provide DC voltage support by using droop control strategy [65], injecting active power in to the DC grid proportional to the DC voltage deviations in order to avoid large changes in DC voltage and to improve transient stability. In the case of POD (Figure 8.5(a)),

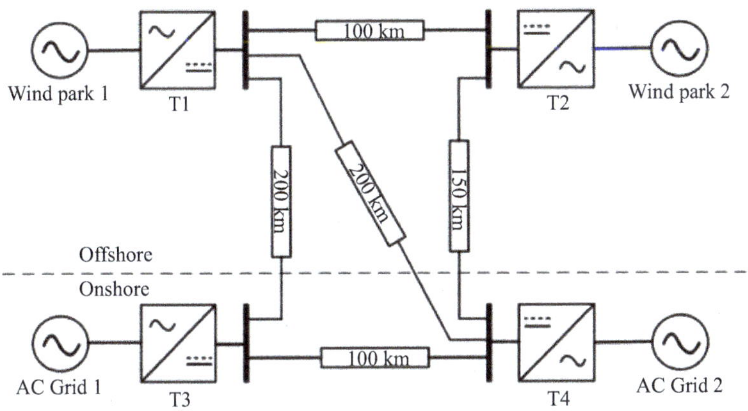

Parameters	Value for T1,2,3	Value for T4
Rated power (MW)	900	1,200
AC grid voltage (kV)	380	380
AC grid reactance (Ω)	17.7	13.4
AC grid resistance (Ω)	1.77	1.34
Arm capacitance (μF)	29.3	39
Arm inductance (mH)	84.8	63.6
Arm resistance (Ω)	0.885	0.67

Figure 8.4 A four-terminal HVDC test grid together with the main parameters

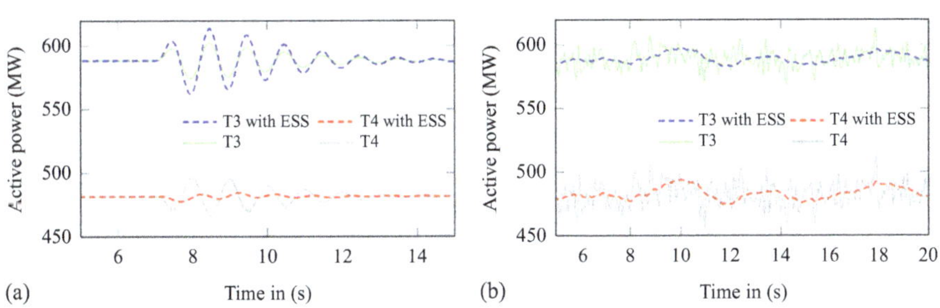

Figure 8.5 The impact of ESS: (a) power oscillation damping and (b) wind variation filtering

the oscillating power injected into the AC grid by T3 causes the DC voltage to oscillate, which in turn leads to a similar oscillation in T4. This implies that the oscillation propagates into the AC grid connected to T4. This is undesired because it can lead to power quality problems in the AC system and trigger controller detrimental interactions. The solution proposed in [63], and here referred to as energy storage support (ESS), is to use the MMC energy storage to provide the oscillating power to counteract the original oscillation at T3, and hence, stop propagation of the disturbance to T4 (see T4 with ESS in Figure 8.5(a)). The same approach can be used to reduce the fast power variations induced by wind energy variability. The simulated case shows one of the wind farms (T1 in Figure 8.4) injecting fluctuating power in to the DC grid, which is absorbed by the two AC grids via T3 and T4. The results show that it is possible to reduce the variability of wind power by using the energy storage capability of the MMC (see T3 and T4 with ESS in Figure 8.4(b)) by absorbing and releasing power when wind power increases and decreases, respectively. This is analogous to connecting a mass (inertia) in AC systems to smoothen the frequency variations. It should be noted that the amount of energy storage in the MMC is limited and is only suitable for compensating rapid (a few milliseconds) and short-term (a few seconds) variations in active power.

8.4.4 Hybrid HVAC/DC systems and expansion planning

The large-scale penetration of offshore renewable energies and, even more, the widespread use of HVDC transmission will require new approaches in the way grid expansions are planned. Transmission expansion planning (TEP) implies defining the plan for installing new electric facilities, identifying their number, type, location and scheduled year of commissioning. The goal is to supply the expected power demand at the minimum possible cost and comply with techno-economic and reliability constraints. The grid integration of intermittent energy sources, such as MECs, will require TEP to refine and adopt approaches capable of taking into account the generation stochasticity. Furthermore, significant HVDC integration into the traditional power system will originate hybrid AC/DC high voltage transmission systems, for which planning techniques should also differ from traditional ones, e.g. by taking into account the additional flexibility ensured by the HVDC converters. A review of the implications that the HVDC penetration has on transmission planning techniques with a specific focus on the offshore scenario can be found in [66].

References

[1] Anaya-Lara O., Tande J.O.G., Uhlen K., and Merz K.O. *Offshore Wind Energy Technology* (Chichester: Wiley, 2018).

[2] Alcorn R., and O'Sullivan D. *Electrical Design for Ocean Wave and Tidal Energy Systems* (London: The Institution of Engineering and Technology, 2013).

[3] Nordex. *N90/2500 HS Offshore Wind Turbine Datasheet* [online]. Available from http://www.nordex-online.com [Accessed September 2018].

[4] Sinovel Wind Group. *SL6000 Series Wind Turbine-Products* [online]. Available from http://www.sinovel.com/english/content/?109.html [Accessed September 2018].

[5] Vestas. *Vestas V164-8.0 - 8,00 MW - Wind Turbine* [online]. Available from http://www.homepages.ucl.ac.uk/~uceseug/Fluids2/Wind_Turbines/Turbines/V164-8MW.pdf [Accessed September 2018].

[6] Enercon. *Grid Technology* [online]. Available from https://www.enercon.de/en/technology/grid-technology/ [Accessed September 2018].

[7] Siemens Gamesa. *Offshore Wind Turbine SWT-6.0-154* [online]. Available from https://www.siemensgamesa.com/en-int/products-and-services/offshore/wind-turbine-swt-6-0-154 [Accessed September 2018].

[8] Garrido I., Garrido A.J., Alberdi M., Amundarain M., and Barambones O. 'Performance of an ocean energy conversion system with DFIG sensorless control'. *Math. Probl. Eng*. 2013.

[9] Gato L.M.C., Carrelhas A.A.D., Correia da Fonsec F.X., and Henriques J.C.C. *Open Sea Operating Experience to Reduce Wave Energy Costs: Turbine-generator Set Laboratory Tests in Variable Unidirectional Flow* (H2020 OPERA project, Deliverable D3.2, 2017).

[10] BioPowerSystems. *ODrive: Self-contained Power Conversion Module for Wave Energy Applications* [online]. Available from http://bps.energy/ [Accessed September 2018].

[11] Mekhiche M., and Edwards K.A. 'Ocean power technologies Powerbuoy®: System-level design, development and validation methodology'. *Proceedings of the 2nd Marine Energy Technology Symposium METS2014*; (2014).

[12] Sjolte J., Sandvik C., Tedeschi E., *et al*. 'Exploring the potential for increased production from the wave energy converter lifesaver by reactive control'. *Energies* 2013: **6** (8): pp. 3706–3733.

[13] Park J.W., Lee K.W., and Lee H.J. 'Wide speed operation of a doubly-fed induction generator for tidal current energy'. *30th Annual Conference of IEEE Industrial Electronics Society, 2004. IECON 2004*; (IEEE, 2004), pp. 1333–1338.

[14] Lee K.S., Yum K.-D., Park J.-S., and Park J.W. 'Tidal current power development in Korea'. *East Asian Sea Congress*; (2009).

[15] Atlantis Resources. *AR1500 Tidal Turbine* [online]. Available from https://www.atlantisresourcesltd.com/wp/wp-content/uploads/2016/08/AR1500-Brochure-Final-1.pdf [Accessed September 2018].

[16] Marine Current Turbines. *SeaGen S 2MW Datasheet* [online]. Available from https://www.atlantisresourcesltd.com/wp/wp-content/uploads/2016/08/SeaGen-Brochure.pdf [Accessed September 2018].

[17] Andritz Hydro Hammerfest. *Renewable Energy from Tidal Currents* [online]. Available from https://andritz.com [Accessed September 2018].

[18] Schottel Hydro. *Schottel Tidal Generator - Cost-Effective Power from Currents* [online]. Available from https://www.schottel.de [Accessed September 2018].

[19] SolarEdge. *One Floating PV Installation — Double the Benefits* [online]. Available from https://www.solaredge.com/sites/default/files/solaredge_case_study_de_krim_floating_installation_netherlands_eng.pdf [Accessed September 2018].

[20] SMA Solar Technology. *Kagoshima Nanatsujima Mega-Solar Power Plant* [online]. Available from https://www.sma.de/ [Accessed September 2018].

[21] SUNGROW Power Supply. *The World's Largest Floating PV Power Plant of 40MW Connected to the Grid Using Sungrow's Inverters* [online]. Available from https://www.prnewswire.com [Accessed September 2018].

[22] Lopez J., Sanchis P., Roboam X., and Marroyo L. 'Dynamic behavior of the doubly fed induction generator during three-phase voltage dips'. *IEEE Trans. Energy Convers.* 2007: **22** (3): pp. 709–717.

[23] Serrano-González J., and Lacal-Arántegui R. 'Technological evolution of onshore wind turbines—a market-based analysis'. *Wind Energy* 2016: **19** (12): pp. 2171–2187.

[24] Blaabjerg F., Ma K., and Yang Y. 'Power electronics for renewable energy systems-status and trends'. *Integrated Power Systems (CIPS), 2014 8th International Conference on*; (VDE, 2014), pp. 1–11.

[25] Park J.-W., Lee K.-W., Kim D.-W., Lee K.-S., and Park J.-S. 'Control method of a doubly-fed induction generator with a grid synchronization against parameter variation and encoder position'. *2007 IEEE Industry Applications Annual Meeting*; (IEEE, 2007), pp. 931–935.

[26] O'Sullivan D.L., and Lewis A.W. 'Generator requirements and functionality for ocean energy converters'. *2010 XIX International Conference on Electrical Machines (ICEM)*; (IEEE, 2010), pp. 1–7.

[27] Benelghali S., Benbouzid M., and Charpentier J.-F. 'DFIG versus PMSG for marine current turbine applications'. *Revue des Energies Renouvelables* 2012: **15** (1): pp. 29–41.

[28] Taaffe D., and Baldock A. 'MeyGen Phase 1A: Lessons learned to date and knowledge sharing plans'. *All Energy*; (2016).

[29] Clemot H., Babarit A., Dupriez-Robin F., and Tran T.Q. 'Development of a wave-to-wire model to calculate flicker caused by wave energy converters and study power quality'. *2017 IEEE Manchester PowerTech*; (IEEE, 2017), pp. 1–6.

[30] Blaabjerg F., Liserre M., and Ma K. 'Power electronics converters for wind turbine systems'. *IEEE Trans. Ind. Appl.* 2012: **48** (2): pp. 708–719.

[31] Wang L., Isberg J., and Tedeschi E. 'Review of control strategies for wave energy conversion systems and their validation: the wave-to-wire approach'. *Renew. Sustain. Energy Rev.* 2018: **81** (P1): pp. 366–379.

[32] IEEE recommended practice for the analysis of fluctuating installations on power systems'. *IEEE Std 1453-2015 Revis. IEEE Std 1453-2011* 2015: pp. 1–74.

[33] O'Sullivan D., Salcedo F., Blavette A., Santos M., and Lewis A.W. 'Case studies on the benefits of energy storage for power quality enhancement: Oscillating water column arrays'. *Proceedings of the 4th International Conference on Ocean Energy (ICOE2012)*, Dublin, Ireland; (2012).

[34] Santos M., Blavette A., Tedeschi E., O'Sullivan D., and Salcedo F. 'Case study on the benefits of energy storage for power quality enhancement: Point absorber arrays'. *Proceedings of the International Conference on Ocean Energy, Dublin, Ireland*; (2012), pp. 17–19.

[35] Blavette A., Multon B., Ahmed H.B., *et al.* 'Wave-induced flicker level emitted by a tidal farm'. '2015 IEEE Power & Energy Society General Meeting'*2015 IEEE Power & Energy Society General Meeting*; (IEEE, 2015), pp. 1–5.

[36] Thiringer T., MacEnri J., and Reed M. 'Flicker evaluation of the SeaGen Tidal Power Plant'. *IEEE Trans. Sustain. Energy* 2011: **2** (4): pp. 414–422.

[37] Fortmann J., Mendonça Â., Göksu Ö., Schaube F., Morales A., and Soerensen P. 'Latest developments of negative sequence extensions for generic RMS models of wind turbines'. *Proc. of the 15th Wind Integration Workshop*; (2016).

[38] Mendonça H., de Castro R.M., Martínez S., and Montalbán D. 'Voltage impact of a wave energy converter on an unbalanced distribution grid and corrective actions'. *Sustainability* 2017: **9**(10): p. 1844.

[39] Kocewiak L.H., Hjerrild J., and Bak C.L. 'Wind turbine converter control interaction with complex wind farm systems'. *IET Renew. Power Gener.* 2013: **7** (4): pp. 380–389.

[40] Kundur P., Paserba J., Ajjarapu V., *et al.* 'Definition and classification of power system stability IEEE/CIGRE joint task force on stability terms and definitions'. *IEEE Trans. Power Syst.* 2004: **19** (3): pp. 1387–1401.

[41] Slootweg J.G., de Haan S.W.H., Polinder H., and Kling W.L. 'Voltage control methods with grid connected wind turbines: A tutorial review'. *Wind Eng.* 2001: **25** (6): pp. 353–365.

[42] de Alegría I.M., Andreu J., Martín J.L., Ibanez P., Villate J.L., and Camblong H. 'Connection requirements for wind farms: A survey on technical requirements and regulation'. *Renew. Sustain. Energy Rev.* 2007: **11** (8): pp. 1858–1872.

[43] National Grid Electricity Transmission. *THE GRID CODE* [online]. Available from https://www.nationalgrid.com [Accessed September 2018].

[44] TenneT TSO GmbH. *Grid Code - High and Extra High Voltage* [online]. Available from https://www.tennet.eu [Accessed September 2018].

[45] European Commission. *Establishing a Network Code on Requirements for Grid Connection of Generators* (Commission Regulation (EU) 2016/631, 2016).

[46] Santos-Mugica M., Robles E., Endegnanew A.G., Tedeschi E., and Giebhardt J. 'Grid integration and power quality testing of marine energy converters: Research activities in the MaRINET Project'. *2014 Ninth International*

Conference on Ecological Vehicles and Renewable Energies (EVER); (IEEE, 2014), pp. 1–9.

[47] Robles E., Haro-Larrode M., Santos-Mugica M., Etxegarai A., and Tedeschi E. 'Comparative analysis of European grid codes relevant to offshore renewable energy installations'. *Renewable and Sustainable Energy Reviews* 2019: **102**: pp. 171–185.

[48] Tedeschi E., and Santos-Mugica M. 'Modeling and Control of a wave energy farm including energy storage for power quality enhancement: The Bimep case study'. *IEEE Trans. Power Syst.* 2014: **29** (3): pp. 1489–1497.

[49] Tethys Database. *Biscay Marine Energy Platform (BiMEP)* [online]. Available from https://tethys.pnnl.gov [Accessed September 2018].

[50] *Iberdrola* [online]. Available from https://www.iberdrola.com/home [Accessed September 2018].

[51] Wind Europe. *Wind Energy in Europe, Scenarios for 2030* [online]. Available from https://windeurope.org [Accessed September 2018].

[52] Kocewiak Ł.H., Kramer B.L.Ø., Holmstrøm O., Jensen K.H., and Shuai L. 'Resonance damping in array cable systems by wind turbine active filtering in large offshore wind power plants'. *IET Renew. Power Gener.* 2017: **11** (7): pp. 1069–1077.

[53] Holttinen H., Cutululis N.A., Gubina A., Keane A., and Van Hulle F. 'Ancillary services: Technical specifications, system needs and costs Deliverable D 22'. 2012. Available from http://orbit.dtu.dk/en/publications/ancillary-services-technical-specifications-system-needs-and-costs-deliverable-d-22(c57219c9-b2d1-4681-9374-311fbb161b64).html.

[54] Harnefors L., Johansson N., Zhang L., and Berggren B. 'Interarea oscillation damping using active-power modulation of multiterminal HVDC transmissions'. *IEEE Trans. Power Syst.* 2014: **29** (5): pp. 2529–2538.

[55] Fosso O.B., Tedeschi E., Amin M., Zadeh M.K., and Suul J.A. *Large-scale Integration of Marine Energy – Offshore Transmission System Characteristics and Dynamic Challenges* (FP7 Marina Platform Report DL8.4, 2014).

[56] Rohn H., Wiesen K., and Teubler J. 'Resource use of wind farms in the German North Sea—The example of Alpha Ventus and Bard Offshore I'. *Resources* 2013: **2** (4): pp. 504–516.

[57] Sharifabadi K., Harnefors L., Nee H.-P., Norrga S., and Teodorescu R. *Design, Control, and Application of Modular Multilevel Converters for HVDC Transmission Systems* (Chichester: John Wiley & Sons, 2016).

[58] Rao H. 'Architecture of Nan'ao multi-terminal VSC-HVDC system and its multi-functional control'. *CSEE J. Power Energy Syst.* 2015: **1** (1): pp. 9–18.

[59] Van Hertem D., Gomis-Bellmunt O., and Liang J. *HVDC Grids: For Offshore and Supergrid of the Future* (Chichester: John Wiley & Sons, 2016).

[60] Rouzbehi K., Candela J.I., Gharehpetian G.B., Harnefors L., Luna A., and Rodriguez P. 'Multiterminal DC grids: Operating analogies to AC power systems'. *Renewable and Sustainable Energy Reviews* 2017: **70** pp. 886–895.

[61] Beerten J., Gomis-Bellmunt O., Guillaud X., Rimez J., van der Meer A., and Van Hertem D. 'Modeling and control of HVDC grids: A key challenge for

the future power system'. *2014 Power Systems Computation Conference (PSCC)*; (IEEE, 2014), pp. 1–21.

[62] Despouys O., Petit A., Rault P., *et al. First Recommendations to Enhance Interoperability in HVDC-VSC Multi-vendor Schemes* (BestPath Project Deliverable D4.3, 2016).

[63] Taffese A.A., and Tedeschi E. 'Coordination of modular multilevel converter based HVDC terminals for ancillary services'. *2018 Power Systems Computation Conference (PSCC)*; (IEEE, 2018), pp. 1–7.

[64] Leterme W., Ahmed N., Beerten J., Angquist L., Hertem D.V., and Norrga S. 'A new HVDC grid test system for HVDC grid dynamics and protection studies in EMT-type software'. *11th IET International Conference on AC and DC Power Transmission*; (2015), pp. 1–7.

[65] Vrana T.K., Beerten J., Belmans R., and Fosso O.B. 'A classification of DC node voltage control methods for HVDC grids'. *Electric Power Systems Research* 2013: **103**: pp. 137–144.

[66] Elahidoost A., and Tedeschi E. 'Expansion of offshore HVDC grids: An overview of contributions, status, challenges and perspectives'. *2017 IEEE 58th International Scientific Conference on Power and Electrical Engineering of Riga Technical University (RTUCON)*; (IEEE, 2017), pp. 1–7.

Chapter 9
Offshore energy storage

Seamus D. Garvey[a] *and Rupp Carriveau*[b]

Renewable energy harvested offshore faces the same issue faced by most renewable energy resources – intermittency. This applies in equal measure to offshore wind, offshore solar power, wave power and power from tidal stream sources. The problem associated with intermittency of supply has two aspects: (a) power may not be available when there is a strong demand for it and (b) power may be available when there is very little demand for it. Many commentators focus only on the first of these two issues, and they propose alternative generation facilities for when renewable energy resources are weak. Solutions of this nature intrinsically result in overcapacity of total generation. The issue of power generation from renewables at times of low demand is no less serious; it reduces the value of the total energy output from the generation facility and thus makes the economics more challenging.

Energy storage has the potential to solve both aspects of the imbalance problem simultaneously – absorbing power from renewables when demand is low and supplementing power from renewables when the demand exceeds what can be supplied from the renewable resources. If energy storage systems were available at negligible cost and with very high performance, then it would be the perfect solution to the intermittency problem. In reality, whether or not energy storage is viable depends on numerous factors including the elasticity of the demand side, the relative costs of backup generation (including the true costs of any emissions produced) and the sensitivity of the cost of transmission between generation and load.

This chapter focuses on energy storage situated offshore. Large amounts have already been written on energy storage generally (e.g. [1–8]) and there would be little value in adding to these outputs. However, there are good justifications in concentrating specifically on storing energy offshore. First, the environment is rather special and it provides resources that may be helpful for energy storage. These resources include (a) hydrostatic head between surface and seabed that may sometimes be large, (b) an effectively infinite amount of thermal ballast enabling a stable reference temperature to be maintained and (c) an unlimited supply of

[a]Faculty of Engineering, University of Nottingham, Nottingham, UK
[b]Turbulence and Energy Laboratory, University of Windsor, Windsor, Canada

saltwater that may be useful for electrolysis to support hydrogen production. Second, energy storage at the site of renewable energy generation potentially makes better use of expensive electricity transmission lines joining the generation to consumption. Finally, there are opportunities for integrating storage with the primary harvesting of energy that can afford substantial effective reductions in cost and increases in effective performance.

Subsequent sections in this chapter are devoted to different categories of technology addressing energy storage offshore. For brevity, only the main categories of storage are dealt with in any depth. The final section lists some of the other emerging technologies. We do not address electrochemistry at all in this chapter – notwithstanding that electrochemical batteries have become almost synonymous with *energy storage* in the minds of most people. This omission is not due to bias. Batteries certainly do serve important roles in storing energy for discharge over relatively periods (mainly less than 2 h). The timescales associated with intermittency of offshore renewable energy are much longer than this – more often measured in tens or hundreds of hours.

In the remaining sections in this chapter, we will make extensive use of the term *exergy*. Most of the material written about energy storage should strictly refer to exergy storage. Exergy is a well-defined concept in thermodynamics describing the ability to do (mechanical) work when a system is allowed to come back into equilibrium with its environment. Most undergraduate textbooks on thermodynamics contain clear sections defining and explaining exergy. Exergy is measured in the same units as energy – joules (J). Electricity is pure exergy; 1 J of electricity is 1 J of exergy. Gravitational potential energy is also pure exergy. However, 1 J of heat energy is not equivalent to 1 J of exergy. The concept of *Carnot efficiency* must be applied to heat energy to assess its exergy content. Conceptually, the reader may find it convenient to think of electrical energy as a suitable interpretation wherever we use the word exergy in the following sections. In much of what follows, we will use the term 'energy storage' where we know it would be more accurate to say 'exergy storage' because the misnomer is already well established and understood.

9.1 Underwater compressed air energy storage

Compressed air energy storage (CAES) is one of the main energy storage technology classes strongly suited to long-duration energy storage (typically 4–20 h). CAES actually represents a wide class of technologies spanning how the pressurised air is stored, how the compression and expansion processes take place and how heat is managed. CAES is undergoing something of a reincarnation at present as attention shifts towards longer-duration storage. An excellent recent review of CAES is given by Budt *et al.* in [9].

Underwater compressed air energy storage (UWCAES) is a specific subset of CAES which takes advantage of the hydrostatic pressure in deep water to provide a means of storing large amounts of pressurised air without expending very large sums of money on pressure containments. UWCAES has received significant attention in recent years [10–25]. A further major advantage of UWCAES is that the air pressure in the store is naturally (almost) constant pressure in most varieties of underwater store irrespective of the degree to which it is charged. This contrasts with the use of salt caverns for storage of natural gas or air where the pressure in the cavern varies with the level of fill. This chapter devotes substantial attention to UWCAES because of its very specific linkage to offshore energy storage. However, it is also a very useful context in which to develop and explain some concepts and definitions that are applicable to other technologies.

Figure 9.1 shows only one of the many possible arrangements for UWCAES. In this figure, all air is compressed electrically on board a floating vessel, and electricity is brought to the floating vessel via a cable from land. Energy recovery

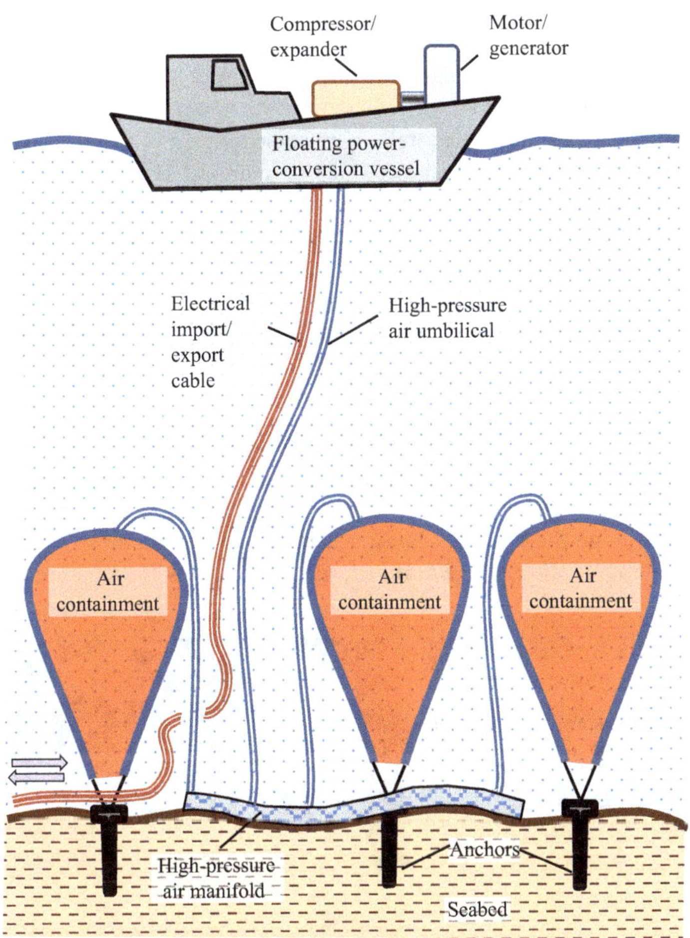

Figure 9.1 One possible configuration of a UWCAES system

from storage would follow the reverse path. The high-pressure (HP) air stores are located underwater and they comprise flexible containments that change shape as the quantity of air stored in them rises and falls. This configuration has advantages with respect to machinery maintenance, but a downside is that the high-pressure air duct to the air stores may be rather long.

The set of possible configurations for UWCAES systems includes all of the following options:

- The power-conversion equipment may be (i) onshore with ducts to convey pressurised air back and forth between the equipment and the stores, (ii) offshore floating at the surface as Figure 9.1 indicates, (iii) offshore at the sea surface but supported on a bottom-fixed structure or (iv) located at the seabed close to the stores. All of these have merits. A later subsection discusses the important matter of the size and cost of ducts for the pressurised air.
- The pressurised air stores themselves may be flexible containments or relatively rigid ones. If they are rigid, they may be operated either as isobaric (constant pressure) or isochoric (constant volume) stores. For UWCAES in particular, isobaric operation is an obvious choice in nearly all cases. The flexible containments are naturally almost isobaric.
- The power-conversion machinery may comprise reciprocating or turbomachines or some combination. For large power-ratings, turbomachines are generally preferred and they tend to be more cost-effective. Turbomachines are normally quite sensitive to changes in the operating pressure ratio and for UWCAES this is generally relatively constant so that turbomachines suit quite well. Against this, reciprocating machines may still be a more natural choice at the high-pressure end (typically between pressures of ~10–20 bar and ~100–250 bar).

Figure 9.2(a)–(c) shows a real flexible containment for storing pressurised air underwater as reported in [16], where it is referred to as an *energy bag*.

Compressing air releases heat. Consequently, the exergy stored in a CAES installation can comprise two parts: (a) exergy stored in the form of pressurised air and (b) exergy stored in the form of heat. For CAES installations where the

(a) (b) (c)

Figure 9.2 (a) 'Energy bag' being tested prior to deployment. (b) 'Energy bag' being deployed. (c) 'Energy bag' installed underwater

pressurised air is stored in tanks without the advantage of large external hydrostatic pressure, the costs of the tanks per unit of exergy stored in the pressurised air are typically much higher than the costs of the thermal stores per unit of exergy stored in the form of heat. Thus, there is a very strong economic motivation to exploit thermal storage of exergy as much as possible for these contexts [10].

With UWCAES, the balance of economics is different. Exergy storage in the form of pressurised air may be relatively inexpensive and the economic advantage of storing heat may be very much lower. For this reason, we focus on isothermal (constant-temperature) compression and expansion processes here. Isothermal compression does release heat but there is no exergy contained in this heat because the heat is at the same temperature as the ambient environment. Similarly, isothermal expansion does draw in heat but no exergy is taken in because this heat is at the same temperature as the environment. For UWCAES, there is obviously an abundance of seawater present to act as sink/source of heat at ambient temperature and this greatly facilitates the operation of isothermal processes.

9.1.1 How much exergy is stored per unit volume of air containment

Consider a volume, V_s, of air stored underwater at pressure p_s. Ambient pressure at sea level averages to 101,325 Pa and we denote this as p_0. Define:

$$r := (p_s/p_0) \tag{9.1}$$

The density of seawater, ρ_w, is typically $\rho_w = 1{,}025$ kg/m^3 and the acceleration due to gravity at the earth's surface is $g = 9.8106$ m/s^2. For UWCAES, storage pressure is usually identical to the local water pressure. Thus at a depth, h, from the sea surface, storage pressure is given by

$$p_s = p_0 + \rho_w\, g\, h \tag{9.2}$$

For each 10 m of seawater, storage pressure rises by 100,560 Pa (1.0056 bar). Assuming that the pressure of the stored air remains constant even as the quantity of air stored reduces to zero, the exergy stored underwater is written as

$$B_u = p_s V_s \log(r) = p_0 V_s\, r \log(r) \tag{9.3}$$

This exergy itself comprises two parts: the exergy contained in the air that is stored underwater, B_{air}, and the exergy that is present as a result of seawater having been pushed out of a volume that it once occupied, B_{water}. It is useful to split the two components – especially since we will shortly consider a different offshore energy storage methodology where only B_{water} is stored:

$$B_{\text{air}} = p_0 V_s \times (r \log(r) - (r - 1)) \tag{9.4}$$

$$B_{\text{water}} = p_0 V_s \times (r - 1) \tag{9.5}$$

Table 9.1 summarises the exergy stored in each 1 m^3 of constant-pressure underwater storage volume at different depths, h – based on (9.3). At water depths

Table 9.1 Exergy density (at the seabed) of underwater cavities

Depth, h (m)	Pressure ratio, r ( )	B_{water} (kWh/m^3)	B_{air} (kWh/m^3)	B_u (kWh/m^3)
10	1.992	0.02793	0.01073	0.03866
20	2.985	0.05587	0.03601	0.09187
50	5.962	0.1397	0.1600	0.2996
100	10.92	0.2793	0.4558	0.7351
150	15.89	0.4190	0.8176	1.237
200	20.85	0.5587	1.224	1.782
300	30.77	0.8380	2.130	2.968
400	40.70	1.117	3.128	4.245
500	50.62	1.397	4.195	5.591
700	70.47	1.955	6.484	8.440
900	90.31	2.514	8.934	11.45
1,200	1,200.9	3.352	12.83	16.18
1,500	1,498.7	4.190	16.94	21.13
2,000	1,994.9	5.587	24.15	29.73

below ~35 m, the exergy content associated with the displacement of water exceeds the exergy content in the pressurised air itself. At higher depths, B_{air} exceeds B_{water} by some way.

9.1.2 Corrections for air density and non-ideal gas behaviour

In the above formulae, we have assumed that air is an ideal gas and that its volume is inversely proportional to pressure. For UWCAES applications, storage temperatures for air remain fairly constant at (deep) sea temperature – around 280 K. At such temperatures, the behaviour of air as a gas remains close to ideal but it does deviate slightly. Ideal gas behaviour would suggest that the density of air increases by a factor of 200 when its pressure increases by this factor from ambient whilst temperature remains constant. In fact, the density increase is about 3.4% more than this due to real gas effects. The assessments of the stored exergy density, B_{air}, in Table 9.1 for high pressures are increased by an amount less than this since a greater mass of air is stored per unit volume but the exergy per unit mass is slightly reduced.

A second note about the calculations behind Table 9.1 is that they calculate exergy density of the air at the seabed – not at the sea surface. The pressure of air in the store is obviously higher than the pressure of air when it emerges from the pipe at sea level – even if there are negligible pressure drops associated with the flow rate. To be accurate, we must account for the work required to lift the air from the seabed back up to the level of the sea surface where it will be expanded. The density of air at high pressures is not insignificant. For example, in a UWCAES system with depth $h = 2{,}000$ m, the air pressure in the store might be 20.213 MPa (202.13 bar) and the density of that air would be approximately 251.5 kg/m^3. The density of dry air at temperature 280 K and pressure p_0 is approximately

1.260 kg/m^3 and we denote this as ρ_{a0}. The relationship between air pressure at the level of the sea surface, p_{surface}, and that in the air store is

$$p_{\text{surface}} = p_s \exp(-\rho_{a0} g h / p_0) \tag{9.6}$$

In the case of a UWCAES system with h = 2,000 m, the air pressure at the level of the sea surface would be 15.834 MPa (158.34 bar) – 43.8 bar lower than storage pressure. The correct total stored exergy expression taking account of the exergy needed to raise the air back up to sea surface level is

$$B_u = p_s V_s (\log(r) - \rho_{a0} g h / p_0) \tag{9.7}$$

For the case of air storage at depth h = 2,000 m, this correction represents a reduction in the assessment of total stored exergy of 4.6%.

9.1.3 Structural capacity and its relevance to energy storage

The concept of *structural capacity* was introduced in [26]. Structural capacity is a measure of how much material is required to carry known sets of loads over known distances. The concept is highly useful in making comparisons between different possible ways of achieving the same structural end. It is especially useful in the context of energy storage. The unit of *structural capacity* is N m – the same as the unit of energy (1 N m = 1 J). A quantity of material that can carry a force of 100 N over a distance of 20 m corresponds to (100 × 20) N m = 2,000 N m. A flywheel comprising a spinning rim composed of 10,000 N m of structural material can store a maximum of 5,000 J of kinetic energy (pure exergy) – i.e. 50% of the value of structural capacity. If this structural material was in the form of a continuous carbon fibre (wound to form a composite rim), one might ask whether the same amount of material could store more or less exergy if it was used to form the reinforcement around a spherical pressure vessel containing compressed air at 100 bar. The spherical vessel in this case would have a volume of 336.8 × 10^{-6} m^3 and the exergy stored would be 12,130 J – much larger than would be stored in a flywheel made from the same material.

The concept of structural capacity enables conclusions to be made that do not depend on the type of material used. In the above case of a comparison between a composite flywheel rim and a pressure vessel storing compressed air, it does not matter whether the fibre material used was steel fibre, glass fibre or carbon fibre. Using structural capacity does not allow perfect comparisons. Specifically, it intrinsically ignores self-weight loads and the fact that using weaker materials may result in the proportions of the loaded structure being changed significantly. It also does not take account of the fact that some materials are naturally capable of bearing stresses in two directions where others (such as carbon fibre in a composite) are intrinsically limited to carrying stress in one direction only. A flywheel composed of steel plate could store larger amounts of kinetic energy than a flywheel formed from the same amount of steel used as a reinforcing fibre wound into a radially thin rim if the maximum allowable stress in the steel was the same in both cases. Similarly, a spherical pressure vessel formed from a continuous steel skin can store substantially higher pressures of air than can be held in a spherical vessel where the same amount of steel is used as a reinforcing fibre and where the maximum allowable stress is the same in both cases. One final comment that should be made about structural capacity is that it does not automatically distinguish between stresses that are compressive and those that are tensile. For offshore applications in particular, concrete is an especially important material and concrete provides an extremely economical way to carry compressive loads but this does not extend to tensile loads.

Provided that the above caveats are borne in mind, the concept of *structural capacity* can be a powerful tool to explore the relative effectiveness of different forms of energy storage – and it is particularly relevant for offshore energy storage.

For the purposes of evaluating the cost-effectiveness of different forms of offshore energy storage, we define the ratio *EC2SC* as the ratio between exergy storage capacity for a given system and the structural capacity of the material required to make that system. For a flywheel composed as a radially thin rim formed by winding a high-strength fibre in a composite, $EC2SC = 0.5$. For a thin spherical shell containing compressed air at 100 bar and formed using high-strength fibre as the main load-carrying material, $EC2SC = 1.213$. We will explore *EC2SC* for flexible containments and subsequently for other offshore energy storage options.

9.1.4 Exergy versus structural capacity for underwater containments

UWCAES may be based on either rigid or flexible containments. Flexible containments were first proposed by Laing and Laing [10] while rigid containments were proposed by Seymour [11]. Rigid containments have the advantage that the structure itself provides most (if not all) of the holding-down force required for the containment to counteract the buoyancy of the stored air. However, they have the disadvantage that stresses in the walls are a combination of bending stresses and membrane stresses [20] and this makes analysis complex. Flexible containments, by contrast, can only carry membrane stresses. Arguably these are more straightforward to analyse – though the geometry does change significantly with fill level. Most UWCAES containments for air achieve equality between the pressure of the stored air and the pressure of the surrounding water at some particular height above seabed. However, this is not the case in all propositions and it is feasible to have over-pressure within the containment – as used in the hydro-pneumatic energy storage concept described in [21] and [22]. Provided that the pressure within the containment is not many times greater than that of the surrounding water, some useful advantage is gained from the hydrostatic pressure.

A first point to note in this subsection applies to both rigid and flexible containments alike: the (variable components of) stresses in containment walls are directly proportional to the difference in density between seawater and the air within the containment. This difference in density causes a gradient in the pressure difference between the air and the water at any one height above seabed.

This gradient in turn is responsible for the buoyancy force. At one particular seabed depth, this gradient reduces to zero. That depth is $h \cong 8{,}210$ m. There are some places in oceans where such depths can be found but these are few. At these depths, a containment for pressurised air requires zero strength since the fluid within the containment has the same density as that outside and the pressure within matches the pressure outside. Denoting the density difference between water and air at storage pressure, p_s, within a storage containment as ρ_{diff}, we have:

$$\rho_{\text{diff}} = \rho_w - r \times \rho_{a,0} \tag{9.8}$$

where $\rho_{a,0}$ represents the density of the air at the surface of the water. Typically $\rho_{a,0} = 1{,}260 \text{ kg/m}^3$.

The total strength requirement for a containment is determined from the density difference, ρ_{diff}. If the cost of manufacturing the containment is considered to be proportional to $(\rho_{\text{diff}} \times V_s)$, we can see how the ratio between exergy storage capacity and cost varies with depth. Using (9.3) and (9.7), we obtain:

$$\text{exergy\_cost\_ratio} \propto (r \times (\log(r) - \rho_{a0} g h / p_0)) / \left(\rho_w - \rho_{a,0} \times r\right) \tag{9.9}$$

Figure 9.3 shows a plot of this value as a function of water depth – normalised for 100 m depth. The message is very striking; the economics of UWCAES depends very strongly on depth. Over the range of depths from 20 to 2,000 m, the graph is quite linear on a log–log plot with slope 1.3. In words, the exergy storage capacity associated with a given amount of structural cost increases with $h^{1.3}$. Increasing depth by a factor of 4 increases the exergy storage capacity value by a factor of almost exactly 6.

At depths above 4,000 m, the exergy storage capacity per unit of manufacturing cost begins to approach exceptional levels. Other costs such as the cost of deployment and the cost of the air ducts usually begin to dominate at greater depths and thus the optimal UWCAES system is generally not so deep.

In fact, the manufacturing cost for a complete underwater flexible air store has three main components: (a) surface area, (b) the requirement for a holding-down

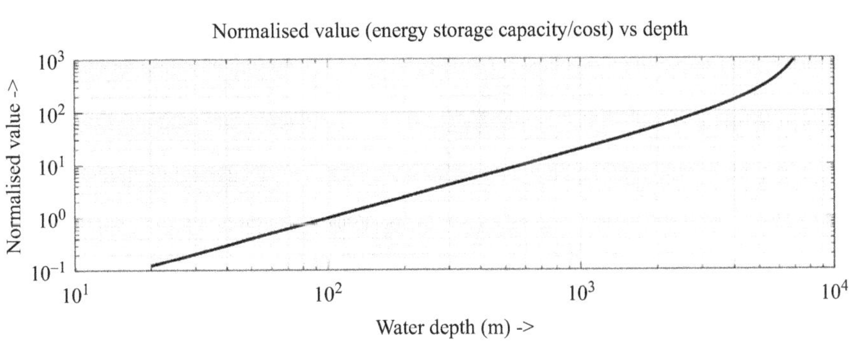

Figure 9.3 Variation of 'value' with depth (normalised to 1 at 100 m)

force and (c) structural material to transmit that holding-down force around the periphery of the containment. Let D be some representative dimension of the flexible containment at complete fill, such as the maximum diameter of containment in any horizontal plane. Then the three cost components above vary in proportion to D^2, D^3 and D^4, respectively. Since the energy storage capacity is itself proportional to D^3, the optimal scale for a flexible underwater store (based on manufacturing cost only) is found to be the value of D for which the costs associated with surface are identical to the costs associated with holding-down force [18]. Based on typical expected values for the cost elements, it was concluded in [18] that the optimal scales for flexible underwater stores would be around 20 m for the mean diameter. However, handling difficulties reported in [16] probably motivate a reduction of unit size towards 10 m.

If the size of an underwater containment is determined, then the proportions of costs (surface costs, holding-down force and structural strength) are also determined. It is already seen that the cost of surface will closely match the cost of structural strength. In [23], deployment methods for flexible containments were discussed and one particular option highlighted there is to envelop a suitably large ballast mass in structural material and connect this to the base of the air containment as Figure 9.4(a) indicates.

In this ballasted arrangement, the cost of holding-down force is dominated by the cost of structural material for the ballast mass envelope. Then, if the average density of the material within the ballast space is around twice the density of water, the volume contained in that space is identical to the volume of the air containment and the requirement for structural material is similar. Thus we may determine *EC2SC* (see previous section) for flexible balloon-shaped containments as follows:

$$\text{Stored\_Exergy} \cong p_0 \times V_s \times r \times (\log(r) - \rho_{a0}gh/p_0) \tag{9.10}$$

Figure 9.4 (a) Schematic of a ballasted flexible containment. (b) Quadrants of a steel shell built to form ballast for the 'energy bag' of [16]

$$\text{Cost}_{\text{SC}} \cong 4 \times (V_s \times \rho_{\text{diff}} \times g) \times (\pi D/2) \tag{9.11}$$

$$EC2SC_{\text{UWCAES}} \approx \frac{p_0 \times r \times (\log(r) - \rho_{a0} g h / p_0)}{2\pi \times g \times D \times (\rho_w - r \times \rho_{a,0})} \tag{9.12}$$

In (9.11), the factor of 4 arises as the product of (1.33 × 3). The 1.33 is present because the total tension force in the meridional straps running from the connection point to the apex of the flexible air containment is greater than the total buoyancy force by approximately this factor. The factor 3 is present because there are three similar components of cost (surface, holding-down force and structural strength for the flexible containment). The first bracketed term in (9.11) represents the buoyancy force and the second bracketed term represents the approximate length of each meridional strap. If we accept that the optimal diameter D is in the order of 10 m, we find that for $h = 1{,}000$ m, $r = 100.2$ and $EC2SC \approx 820$. This number provides a strong illustration of the primary attraction of UWCAES.

9.1.5 The air ducts

A UWCAES must provide for air ducts to convey the air between the storage containments and the compression/expansion facilities. As noted earlier, the compression/expansion facilities are invariably at (or slightly above) the level of the sea surface. The air duct(s) must span the distance between the underwater air stores and the machinery. At low depths, the pressure difference between inside the duct(s) and outside is minimal. However, at levels close to the surface, this pressure difference is large and the duct(s) must have sufficient wall strength to contain the pressure difference. The duct(s) must have sufficient internal cross section to ensure that the additional pressure drop associated with flow through the duct at full power of charging/discharging is not excessive. In general, it will be found that transmitting power in the form of compressed air over a few hundreds of metres is not problematic but when power must be transmitted over several kilometres or more, the cost of the duct/pipeline itself begins to become significant for systems with small power ratings. For larger systems, greater distances are less problematic.

The pressure drop associated with flow in a pipe of length L and diameter d is determined by the Darcy–Weisbach equation:

$$\Delta p = f_D \times \rho v^2 \times (L/2d) \tag{9.13}$$

where ρ represents the density of the fluid, v denotes the mean velocity of the fluid in the pipe and f_D represents the so-called *friction factor*. The friction factor is a function of the internal pipe roughness and the value of Reynolds number:

$$Re = (\rho v d/\mu) \tag{9.14}$$

The friction factor can be calculated using the Colebrook equation:

$$(f_D)^{-0.5} = -2\log\left(\varepsilon/(3.7d) + 2.51(f_D)^{-0.5}/(Re)\right) \tag{9.15}$$

Using (9.13) and (9.15), the total pressure drop associated with a given mass flow rate of air can be determined. This pressure drop is additional to the pressure change associated with the gravitational head in the pressurised air. For relatively small mass flow rates, this additional pressure drop does not depend on whether the air is flowing into or out of storage. When significant pressure changes occur in the air ducts linking the air stores with the power-conversion machinery, it is necessary to calculate the pressure drop associated with flow as the sum of contributions from different lengths of duct. This is straightforward: the duct is divided into a number of discrete sections and an average pressure (hence density) is used for each one. The mass flow rate is common to all of the sections.

Typically, a system designer might accept a pressure drop between 5% and 10% of total storage pressure associated with the flow of air at system rated power. If the same pressure drops occur during charging and discharging, this would mean that the pressure available during expansion is then between 9.75% and 19% lower than the pressure that was achieved during compression. However, the sacrifice in exergy is not nearly so bad. If we assume that pressure drops arising from air flow in the ducts accounts for a reduction in pressure ratio of Δr each way (i.e. the compressors produce pressurised air at the pressure $p_0 \times (r + \Delta r)$ and the expanders receive air back at the pressure $p_0 \times (r - \Delta r)$), then the proportionate loss of exergy associated with flow losses in both directions is given by

$$(\Delta B/B) = 1 - \log(r - \Delta r)/\log(r + \Delta r) \tag{9.16}$$

With a nominal (storage) pressure ratio of $r = 100$ and a proportionate pressure drop $(\Delta r/r) = 0.1$ (i.e. 10%) occurs each way, (9.16) shows that the proportionate round-trip exergy loss connected with flows is only 4.3%. This is very surprising at first sight. In pumped-hydro, e.g., flow losses of 10% of the total head would account for associated round-trip exergy losses of ~20%. The reason that the exergy loss is not higher for the compressed air case is that as pressure drops, volume increases proportionately. Table 9.2 indicates pipe internal diameters required for air flows associated with 10 MW of mean exergy flow for various different levels of flow-associated pressure drop per kilometre of pipe as a function of storage pressure. We assume relative roughness values of $(\epsilon/d) = 1 \times 10^{-4}$ for the pipe bores.

Table 9.3 repeats this calculation set for power ratings of 100 MW. The increases in pipe diameters are very close to 2.5 in all cases – obviously less than 3.13 ($\sqrt{10}$).

Table 9.2 Duct bores required for given flow-loss targets (10 MW rating)

	0.2%/km	**0.4%/km**	**0.7%/km**	**1.2%/km**
$r = 50$	0.436 m	0.380 m	0.339 m	0.316 m
$r = 70$	0.369 m	0.321 m	0.287 m	0.267 m
$r = 100$	0.310 m	0.269 m	0.241 m	0.224 m
$r = 150$	0.254 m	0.221 m	0.198 m	0.184 m
$r = 200$	0.222 m	0.193 m	0.172 m	0.160 m

Table 9.3 Duct bores required for given flow-loss targets (100 MW rating)

	0.2%/km	0.4%/km	0.7%/km	1.2%/km
$r = 50$	1.093 m	0.950 m	0.849 m	0.791 m
$r = 70$	0.993 m	0.803 m	0.718 m	0.669 m
$r = 100$	0.775 m	0.674 m	0.603 m	0.561 m
$r = 150$	0.637 m	0.554 m	0.496 m	0.461 m
$r = 200$	0.555 m	0.483 m	0.432 m	0.402 m

9.1.6 Using thermal storage in conjunction with air storage

Compressing air releases heat. Indeed, a fact that most people find highly surprising is that for every 1 J of energy that is used up in compressing air, exactly 1 J of heat falls out of that compressed air as it cools back down to ambient temperature. The only thing that the system designer can control in relation to this heat is the range of temperature over which it emerges.

If all of the heat of compression emerges at temperatures very close to ambient, then there is no exergy contained in this heat and all of the exergy remains in the compressed air. By contrast, if this heat emerges at a relatively high temperature, then there is substantial exergy associated with that heat and it may be that most of the exergy invested in compressing the air emerges in the form of heat.

A common approach in CAES systems is that air is inducted into the compression stages at ambient temperature and then compressed *adiabatically* (i.e. without allowing any heat to emerge from that air) before being exhausted at a pressure and temperature substantially higher than ambient conditions. For an ideal gas, we have:

$$p = \rho R T \tag{9.17}$$

where T represents absolute temperature, R is the specific gas constant (it is 287.15 J/kg K for dry air), ρ denotes the gas density and p denotes pressure. Additionally, we know that (pV^{γ}) remains constant for any adiabatic process where γ denotes the ratio of specific heats for the gas. For air, $\gamma = 1.4$.

We can deduce a relationship between pressure and temperature for any adiabatic process. This is

$$(T_2/T_1) = (p_2/p_1)^{\chi} \tag{9.18}$$

where

$$\chi = ((\gamma - 1)/\gamma) \tag{9.19}$$

For air, $\chi = 0.2857$. Applying (9.18), we see that if T_1 is ambient temperature (~280 K) and if we do not wish for T_2 to rise above 600 °C (873 K), then the pressure ratio (p_2/p_1) for any one adiabatic stage of compression should not exceed 53.5. There is no definite maximum acceptable value for the temperature of air emerging from one compression stage. However, materials that can carry reasonable stresses at high temperatures are very few and they are expensive when the highest temperature ever sustained exceeds 600 °C (873 K). For this reason it is sensible to restrict the pressure ratio in any one stage of compression to below 50. It is possible to use more than one stage of compression, of course, and correspondingly more than one expansion stage.

When CAES systems are designed with storage of the pressurised air in tanks not benefitting from a high external hydrostatic pressure, the relatively high cost of the tanks motivates the designer to make extensive use of thermal storage to help contain the marginal cost for each unit of energy storage capacity. This motivation is much less strong in the context of UWCAES because the marginal costs of additional units of underwater storage are relatively low. However, there is still some strong motivation for coupling thermal storage with UWCAES. The compression/expansion machinery is generally less expensive with fewer stages and if the heat of compression is not then stored and re-used, the system performance will be affected seriously.

High-pressure air is never stored hot – not in any CAES system and especially not in a UWCAES system. There are three main reasons for this: (i) the mass of air that can be stored in a given containment would be much lower if the air was stored hot, (ii) the temperature of hot air in a containment would tend to degrade the containment more rapidly and (iii) it would be much harder to insulate the surface of the pressurised air containment to prevent rapid heat loss than it is to insulate the surface of a separate thermal store. The third point is especially important for UWCAES systems where the outer surface of the air containment is in direct contact with cool seawater which will convect away any heat extremely effectively. The next section of this chapter discusses the storage of heat offshore and the same methods and reasoning are applicable whether that heat is to be used in conjunction with compressed air or in isolation.

If a UWCAES system uses single-stage compression and expansion and if the heat of compression is stored as in parallel with the compressed air, the total exergy stored by the system is given by

$$B_t = p_s V_s((r^{\chi} - 1)/\chi - \rho_{a0} gh/p_0) \tag{9.20}$$

If a UWCAES system uses several stages of adiabatic compression and expansion, it is normal that each stage uses the same pressure ratio. Then, if the heat of compression is stored as in parallel with the compressed air and if there are n separate stages, the total exergy stored by the system is given by

$$B_t = p_s V_s \left(n \left(r^{(\chi/n)} - 1 \right) \chi - \rho_{a0} gh/p_0 \right) \tag{9.21}$$

It is easily found that as n (the number of stages of compression/expansion) approaches infinity, (9.21) rapidly approaches the same value as (9.7). Figure 9.5 shows the ratio between the exergy stored per unit volume of HP air storage for the cases of multi-stage adiabatic compression/expansion and simple isothermal compression/expansion. Different curves are present for each of ten different water

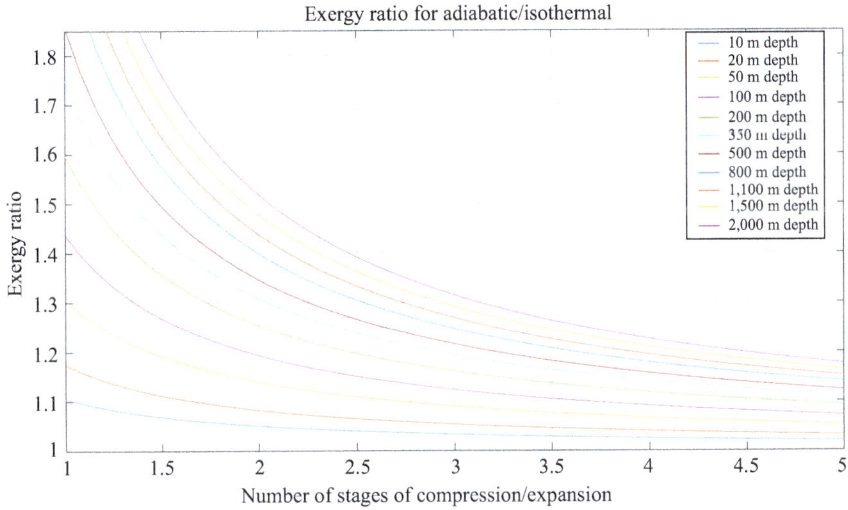

Figure 9.5 Ratio of exergy densities for 'adiabatic' and isothermal storage

depths (corresponding to different total pressure ratios). The vertical axis of the graph is truncated at 1.85 because the curve associated with 500 m (50 bar) reaches 1.85 with single-stage compression and the temperature at which air would emerge from compression would exceed 600 °C at this pressure ratio.

For (9.21) and Figure 9.5, it is assumed that the air entering each stage of compression is at the same temperature as the ambient. This need not be the case. Some developers (e.g. [27]) have proposed CAES systems in which air is preheated prior to each stage of compression in order that the heat of compression emerges at much higher temperatures. In such systems, a very high fraction (e.g. ~60%) of all exergy stored can be retained in the form of high-grade heat and this can potentially reduce the overall cost of the exergy stores. These systems have less relevance for UWCAES than for systems where compressed air is stored in tanks that are not aided by a high external hydrostatic pressure.

9.1.7 An example system design

We illustrate the points made in this section by addressing a hypothetical system design for a UWCAES system where depths of 2,000 m are accessible at a distance of 7 km from shore. The system rated power is 500 MW (for both input and output) and the energy capacity required for the storage system is 10 GWh – equivalent to 20 h of output at full rated power. This energy storage capacity is similar to that of the largest pumped-hydro energy storage plant in the UK. We will assume that compression, expansion and heat transfer processes are ideal. In reality, they would not be, of course, but the calculations give a useful initial impression of scale.

For a 2,000 m depth, storage pressure is 202.1 bar but the air pressure at the surface is 197.2 bar. The pressure difference arises because of the hydrostatic head in the air column itself. Applying (9.7), we find that each cubic metre of underwater air storage volume retains 102.1 MJ (or 28.36 MWh). To store 10 GWh, we require 352,600 m^3 of high-pressure air storage. Individual flexible containments might be ~200 m^3 each – in which case there would be around 1,800 such bags. The bags would be arranged into small groups (of perhaps nine per group) with each group feeding a common small manifold. These groups would themselves feed into large manifolds (with perhaps 20 groups per larger manifold). Then each of the ten large manifolds would be connected to the main air duct.

The maximum diameter of each bag would be ~7.25 m and the height from bottom-fixing to top of any one bag when fully inflated would be ~10 m. The air would have density 251.4 kg/m^3 (a quarter of the density of the surrounding water) and the pressure difference between air at the top of the bag and the seawater immediately outside would be almost exactly 1 bar. Evidently, these bags are indeed pressure-vessels (at the top) but the pressure differences across the containment are relatively small relative to other vessels. For comparison, note that the pressure difference present in a typical car tyre between inside and outside is normally between 2 and 3 bar.

To achieve 500 MW of input power, a volume flow rate of 4.9 m^3/s would occur into the main manifold. This equates to a mass flow rate of 1.23 tons of air per second. To accommodate this mass flow in a single pipe, the pipe internal diameter would be approximately 0.6 m. With a cross-sectional area of 0.287 m^2, the mean air velocity in the duct entering the manifold would be ~17.3 m/s. Reynolds number is 1.49×10^8 and the friction factor is 0.00227 (with $\varepsilon/d = 1 \times 10^{-4}$). The pressure drop in the kilometre closest to the main manifold would be 1.43 bar. The pressure drop in the other 1 km lengths is very similar – but they are slightly larger (1.46 bar/km) close to the surface where pressure has fallen. Total pressure drop in the duct at full-rated power would be 10 bar (~5% of storage pressure). The exergy loss associated with this pressure drop alone is <1%. The volume of air contained in the duct would be 1,979 m^3 – around 0.56% of the total volume in the main air stores.

It is interesting also to calculate how much steel would be used for the main pipe. If maximum stresses of 200 MPa were allowed in the pipe walls, the radial thickness of the pipe walls would have to be at least 30.3 mm. To allow for some small amount of corrosion over time, the pipe would probably be designed with 35-mm thick walls. Then the total mass of steel in the main pipe would be ~3,800 tons. At material costs in the order of ~£5/kg (~$7.5/kg) including processing, this pipe alone would cost upwards of £20M ($30M). At 500 MW rating, the power-conversion equipment would very probably cost in the region of £150M ($225M). The containments (including installation and manifolds and connection tubes) would probably cost around £300M ($450M). Altogether, a total system cost in the order of £500M ($750M) would be expected.

9.1.8 Sites available for UWCAES

There are several considerations that must be made when trying to locate the optimal site for UWCAES. Physical factors will affect both the exergetic efficiency and the costs of operation. It should be intuitive that with increased depth comes increased storage density. It follows that locations with the greatest depth would be very attractive prospective sites. However, there remains a limit to what is practical when cost constraints are applied. The cost implications of ultra-deep deployments and large riser lengths are fairly obvious. The other challenge of course is that great depths often come at great distances from shore (and from the load). Two seminal site surveys were conducted in the wake of the first commercial UWCAES development [24,25]. Pete *et al.* [24] focused on a scan of candidate sites in the Gulf of Maine between the US and Canada, while Pimm *et al.* [25] delivered a more global perspective. In order to offer visualisation of the potential for prospective UWCAES sites off this East coast of the US, Pete *et al.* utilised a storage density function closely related to (9.7) except that it omitted to account for the exergy required to raise the air from the storage containment up to the surface prior to expansion. They then integrated this function into geographic information system (GIS) software to illustrate zones of favourable storage density in and around the Gulf of Maine as shown in Figure 9.6.

The data illustrated in Figure 9.6 revealed that the New England study area boasts over 110 TWh of underwater energy storage potential. This is equivalent to

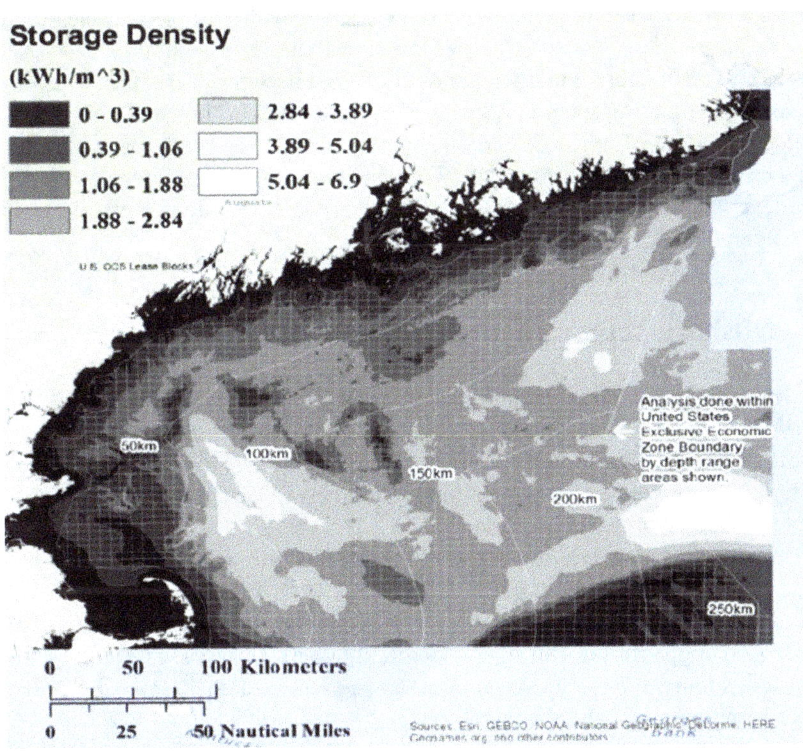

Figure 9.6 Adiabatic UWCAES storage density map for the New England area in the Gulf of Maine (with permission) [24]

supporting the entire US electrical load for several consecutive days. Pimm *et al.*, however, emphasised that systems with onshore turbomachinery should not consider air storage more than 5 km from shore. Beyond this, the entire operation would need to be at sea to be economically feasible [25]. Pimm *et al.* further reveal that air storage depths of 400 m or more could be serviced by familiar turbomachinery schemes similar to those at the historic terrestrial facilities at Huntorf and McIntosh. Norway, Greece, the Azores, the Canary Islands and Italy are identified as European sites of particular promise. In North America, California shows the greatest potential as its deepwater is within 12 nautical miles of the shore. Other favourable American locations include the Hawaiian islands. Similarly, essentially all of the Caribbean islands boast deep water within national territorial limits.

While nearshore deep water is a natural prerequisite, other physical factors include the ready availability of transmission infrastructure near the coastal interface to the offshore system. More ideally, when large load centres and generation sources are co-located, nearshore transmission requirements are minimised. This is a natural advantage for offshore CAES, given that over half the planet's population lives within 50 miles of the coast. Considering all of these factors, densely populated island nations are prime candidates for UWCAES installations. Many of these islands that have been greening their power infrastructure still rely on diesel generators to firm renewable generation. UWCAES provides a potential clean alternative to infuse these systems with necessary enabling grid flexibility.

Finally, as offshore wind generation moves to deeper water, chasing stronger winds, the potential for integration with offshore UWCAES remains significant. In these scenarios, the UWCAES turbomachinery and power conditioning would be hosted on an offshore platform at sea. The development of these integrated systems could leverage electromechanical synergies that could include aspects like shared mooring and electrical transmission infrastructures.

9.2 Offshore pumped hydro

An alternative to compressing air and storing this underwater is to pump water out of a rigid containment deep underwater – or to pump water from sea-level up to an onshore reservoir. The same principles apply in both cases to calculating pressure heads, volume flow rates for a given input/output power and the amounts of energy stored per unit volume of water moved.

Underwater pumped-hydro (UWPH) comprises pumping water out of fixed-geometry underwater containments. UWPH has been progressed in at least two separate projects and the configuration is quite similar in both cases. The *StEnSea* project [28] proposes the use of thick-walled concrete domes as volumes to be evacuated. Figure 9.7(a)–(c) shows a prototype concrete shell from the *StEnSea* project being prepared and then deployed.

A different project [29] proposes vessels that may be formed from a concrete that includes substantial amounts of fly-ash. That project also proposes that the underwater containments may also serve as mooring points for floating wind

(a) (b) (c)

Figure 9.7 (a) Shell preparation for test. (b) Shell lifted into water. (c) The shell being sunk to its final location underwater

turbines. Water is pumped out of the containment and the space above the water is effectively evacuated. The pressure in that space would be the vapour pressure of water at the storage temperature – very small fractions of atmospheric pressure.

9.2.1 Exergy storage density for UWPH

For UWPH to be effective, we require that $p_{a,c}$ should be lower than $(\rho_w gh)$ but this is the case for all depths below about 70 km and so this criterion is satisfied for all water depths accessible on earth. The exergy stored in the evacuated space is given by

$$B_{\text{uwph}} = (p_0 + \rho_w gh) \times V_s \tag{9.22}$$

In the above equation, $\rho_w gh$ dominates over p_0 for all practicable values of h – typically between 250 and 2,500 m.

9.2.2 Key distinctions between UWPH and UWCAES

UWPH is very different from UWCAES in numerous ways:

- The net pressure difference between inside the containment and outside is positive for UWCAES but negative for UWPH.
- With UWPH, the magnitudes of pressure difference encountered between inside the containment and outside the containment are much higher than with UWCAES. For UWCAES, the maximum pressure difference is proportional to the height of the containment and is independent of depth. For UWPH, the maximum pressure difference is proportional to depth.
- With suitable design for the containment in UWPH, the stress is compressive at every point in the wall material at every state of fill. This enables much cheaper material to be used than would be needed to cater for tensile stresses.

- The UWPH containments naturally provide their own holding-down force in the sense that net buoyancy on these structures is lower than the containment weight. By contrast, the UWCAES containments generally require either to be fixed to the seabed or weighted down.
- With UWPH, the power-conversion equipment is a simple pump/turbine set. Because seawater is effectively incompressible, there is no heat management to be done during either the charging or discharging stage. This has both advantages and disadvantages. The advantage is simplicity. The disadvantage is that UWPH offers much more restricted options for integration with other subsystems than UWCAES does.
- The turnaround efficiency of UWPH is likely to be higher than that for UWCAES for a variety of reasons but chief amongst these is the fact that UWPH uses pumps and (liquid) turbines to manage the power conversion whereas UWCAES requires that mechanical power conversion and heat transfer are done simultaneously. The naturally higher density and viscosity of seawater automatically mean that sealing solutions will naturally tend to work better for water than for air so that machine efficiencies are naturally higher for similar levels of expense in the machine.

Perhaps the most significant contrast between UWPH and UWCAES is the amount of exergy stored per unit volume of the containment. This ratio is given by

$$\left(B_{\text{uwph}}/B_{\text{uwcaes}}\right) = \frac{(\rho_w gh - p_0 \times (\exp(\rho_{a0}gh/p_0) - 1))}{(r \times \log(r) - \rho_{a0}gh/p_0)} \tag{9.23}$$

where r is the pressure ratio for the compressed air storage as defined in (9.1) and (9.2). Figure 9.8 shows how this ratio varies with the storage depth. The figure shows that at low depths the exergy stored per unit volume of containment is very similar but at depths in the order of 1,000 m, the UWPH concept stores significantly less exergy per unit volume.

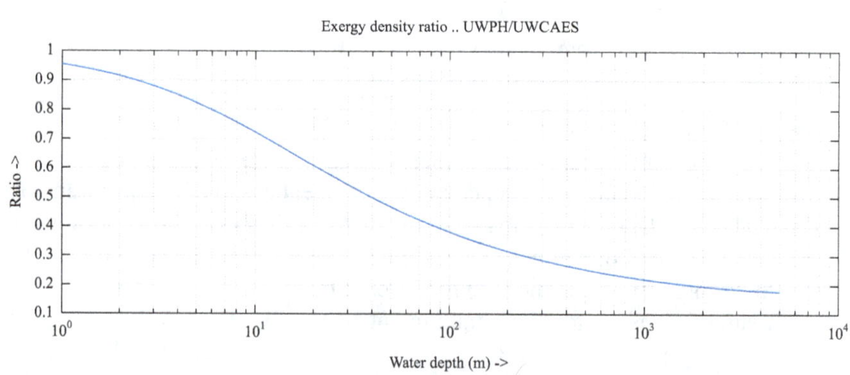

Figure 9.8 Ratio of exergy densities; UWPH/UWCAES

9.2.3 The EC2SC ratio for UWPH

A UWCAES system using flexible containments must provide structural material serving two distinct purposes: (i) maintaining a pressure difference between the air and the water at the upper extremes of the containment and (ii) holding the containment down against the net buoyancy force. The material used in both cases is predominantly in tension. For UWPH, the pressures outside the containment are larger than those inside and the boundary of the containment therefore carries compressive loads. This means that the materials that may be used include concrete, composites comprising resin-bonded discrete pieces of aggregate and other low-cost options. Net buoyancy loads are usually negligible for UWPH containments because the containment walls are normally sufficiently thick that the positive buoyancy developed as a result of the negligible density of the air within the containment when the system is charged is more than counteracted by the negative buoyancy of the containment walls which are much more dense than water. In effect, the ballast mass can be integrated very easily into the containment itself at negligible marginal cost. Then it is legitimate to consider the cost of the containment purely in terms of the cost of the structural material required to sustain the pressure differences between the water outside the containment and the air within. With suitable design of the containment, it is possible to ensure that the stresses everywhere in the containment walls are compressive and that the maximum stresses occur when the system is fully charged (i.e. the containment has been emptied of seawater). The optimum shape is close to spherical but not exactly spherical. Here, we develop the analysis based on a purely spherical shell with internal diameter D and wall thickness t. The minimum value of thickness, t, is determined by equating the load arising from the pressure difference (tending to crush the sphere) with the load arising from compressive stresses in the sphere walls preventing it from collapse:

$$(\rho_w gh) \times \pi(D+2t)^2 - p_0 \times \exp\left(\rho_{a,0} gh/p_0\right) \times \pi D^2 = 4\sigma_{\max} \times \pi(D+t)t \quad (9.24)$$

The volume of containment wall material is given by

$$\begin{aligned} V_{\text{cw}} &= \pi \times \left((D+2t)^3 - D^3\right)/6 \\ &= \pi \times (6D^2 t + 12Dt^2 + 8t^3)/6 \end{aligned} \quad (9.25)$$

and the total amount of structural capacity in the wall is given by $V_{\text{cw}} \times \sigma_{\max}$. Thus, we may calculate $EC2SC_{\text{UWPH}}$ as

$$EC2SC_{\text{UWPH}} \approx \frac{(\rho_w gh - p_0 \times (\exp(\rho_{a0} gh/p_0) - 1))}{V_{\text{cw}} \times \sigma_{\max}} \quad (9.26)$$

where V_{cw} is determined from (9.25) and where t is found from (9.24). Table 9.4 summarises $EC2SC_{\text{UWPH}}$ for different depths, h and for different values of $\sigma_{\max}$. At significant depths, the pressure in the water exceeds the maximum allowable stress in the containment walls and then there is no geometry that can serve. Close to this boundary, the utilisation of material is very poor because the wall-thicknesses are

Table 9.4 EC2SC values for UWPH

	σ_{max} = **10 MPa**	σ_{max} = **15 MPa**	σ_{max} = **20 MPa**	σ_{max} = **30 MPa**	σ_{max} = **45 MPa**	σ_{max} = **70 MPa**
h = 200 m	0.5304	0.5869	0.6154	0.6442	0.6634	0.6772
h = 400 m	0.3557	0.4615	0.5159	0.5712	0.6085	0.6353
h = 700 m	0.1377	0.3008	0.3901	0.4829	0.5464	0.5924
h = 1,100 m	–	0.1160	0.2383	0.3749	0.4713	0.5412
h = 1,600 m	–	–	0.0768	0.2494	0.3823	0.4826
h = 2,200 m	–	–	–	0.1154	0.2811	0.4137

large relative to the internal diameter of the containment. It is found that the EC2SC values approach 0.7 when the maximum allowable compressive stress in the wall material is much higher than the hydrostatic pressure. For structures made from normal concrete, allowable compressive stresses in the order of 30 MPa are typical. Hydrostatic pressure in water reaches 30 MPa (300 bar) at 3,000 m depth. Evidently, the EC2SC values for UWPH are much lower than those for UWCAES (there is a factor of more than 1,000 between 0.7 and 820). In considering this comparison, it is important to remember that the UWPH containment walls carry only compressive stresses whereas the UWCAES containment walls carry tensile stresses.

9.3 Buoyancy energy storage systems

There have been several propositions for energy storage systems exploiting buoyancy. In the simplest of these (e.g. [30,31]), a buoyant element is pulled down in the water against the net buoyancy force and the work done on pulling it down can be recovered when the buoyant element is allowed to rise again back to its original position. Figure 9.9 illustrates this. The work done during charging (like the work recovered during discharging) is transformed by electrical machines.

If the depth is significant relative to the vertical dimensions of the buoyant element, the fully charged state of the system bears a very strong resemblance to the case of UWPH discussed in the previous section. This similarity arises because in both cases, the stored energy could be released by allowing the chamber/containment to flood. Hence, we may immediately use the results from Section 9.2.3 above to reason that the structural capacity (measured in N m) required to keep the water out of the containment at full depth is at least 1.425 (1/0.7) greater than the maximum energy storage (measured in J). More structural capacity is needed, however, for these systems. The cable pulling the buoyant chamber downwards to the seabed must have a structural capacity (measured in N m) equal to the stored energy capacity (in J) *if the generator and cable drum are at the seabed.* It would seem to be more common to have the generator and cable drum above water level (as Figure 9.9 shows). In such cases the total structural capacity required for the

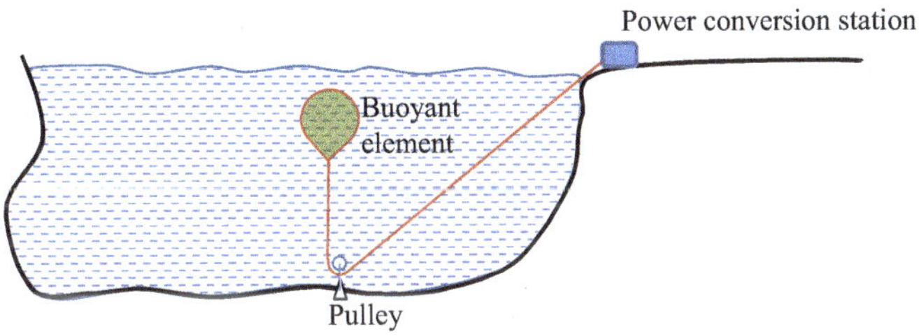

Figure 9.9 Schematic of a simple Buoyant Energy storage scheme

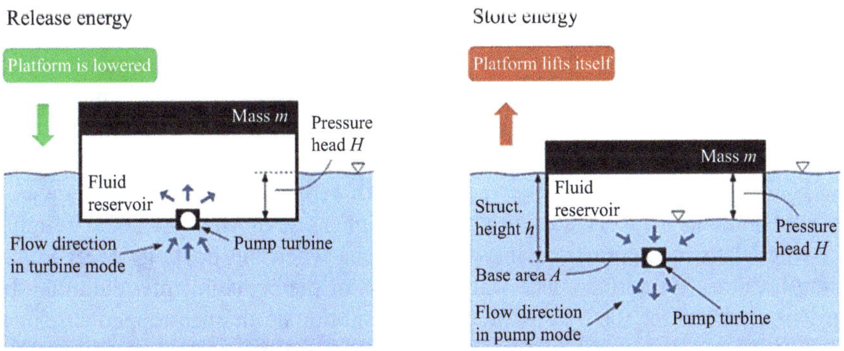

Figure 9.10 The Buoyant Energy concept

cable is double the value of the stored exergy and the requirement for a holding-down force may also be almost doubled.

Then it is necessary also to recognise that the seabed fixing also requires some structural material. If the seabed itself comprises high-integrity rock, the additional structural material required for the seabed fixing may be small but a more general approach might be to adopt the same methodology used above for the holding-down force required for flexible containments for underwater compressed air storage. Accounting for all components of the structural capacity required, it is clear that these systems are unlikely to achieve EC2SC ratios of more than about 0.3 if the generator and cable drum are sited at the seabed. If the cable used is fixed to a winch unit at the water surface, then the vertical lift force acting on the seabed anchor is doubled and the structural capacity required for the cable is also doubled. In these cases, the EC2SC ratio is unlikely to exceed 0.2.

Note also that over 50% of the structural material required for these systems is loaded with tensile loads so that the scope for using very inexpensive material (e.g. concrete) is rather restricted.

In other varieties of buoyancy energy storage, the net buoyancy of a containment is adjusted by allowing water to flow into it through a turbine (during discharging) or by causing water to be pumped out of it (during charging). One example is the Buoyant Energy concept developed by the University of Innsbruck [32,33]; refer to Figure 9.10. Water pumps and water turbines have the potential to be extremely efficient energy-conversion processes so these varieties of buoyancy energy storage have the potential to achieve high energy round trip efficiencies. Indeed, exactly this observation was applied to UWPH earlier to explain why it can normally achieve higher turnaround efficiencies than UWCAES. In these cases, the energy stored can be regarded as comprising mainly of gravitational potential in the mass of the containment being raised and lowered by the charging and discharging processes directly. These systems can be open-topped since there is no advantage to them ever becoming completely submerged – as no further water would be displaced by a fully submerged containment. The energy stored in this class of buoyancy energy storage is precisely *mgh* where *h* is the depth that the containment sinks as a body between the charged and discharged states and where *m* is the mass of the containment. A significant attraction of this variety of buoyancy energy storage is that the only requirement for a seabed fixing is to prevent the containment from drifting – and thus negligible strength is required in the mooring cable. The pressure differences occurring across the walls of the containment are relatively small.

It is slightly more complex in this case to assess the EC2SC ratio for this form of offshore energy storage. We can achieve some estimate of this by supposing that the containment shape is something like a sphere with wall thickness roughly proportional to the depth from the water surface when the floating object is just barely breaking the surface. The wall thickness of the containment would then be (virtually) zero at the top – consistent with the notion of an open-topped structure. The relationship between wall thickness, t, and the distance from the top of the sphere, z, would then be determined by

$$t(z) = R(\rho_w g z)/2\sigma_{\max} \tag{9.27}$$

where R represents the mean radius of the containment. By integrating $(t(z) \times \sigma_{\max})$ over the area of the sphere, we gain an impression of the total structural capacity in this object. Ignoring the weight of the containment itself, we can assess the exergy storage capacity of the unit as

$$B_{\text{buoy},2} = \left(4\pi R^3 \rho_w/3\right) \times g \times R \tag{9.28}$$

It transpires, unsurprisingly, that both the exergy storage capacity and the structural capacity for this sphere are proportional to R^4 so that the EC2SC ratio is independent of R. The EC2SC ratio is also independent of the value of ρ_w, g and $\sigma_{\max}$ and emerges as exactly (2/3).

9.4 Offshore thermal energy storage systems

If the costs of power-conversion equipment are not considered, thermal energy storage is, without doubt, the most cost-effective means of storing exergy at large scale. It is entirely conceivable that thermal pumping systems can be created which realise energy storage system costs below \$5/kWh of energy storage capacity with present-day material and process costs. Nothing else has any prospect of achieving lower-cost exergy storage. To support this assertion, we might consider what is the exergy storage capacity of 1 kg of a common rock such as basalt that has been heated up to 600 °C from ambient. Data provided in [34] indicate that the specific heat of basalt rises from 750 J/kg K at 0 °C, through 1,000 J/kg K at 300 °C, up to about 1,170 J/kg K at 600 °C. If we consider that heat is injected into this rock in increments of 1 °C, it is found that the exergy stored in each 1 kg of basalt is >280 kJ. Present-day lithium-ion batteries achieve exergy densities of ~900 kJ/kg. Evidently, hot rock is rather exergy dense.

One significant downside of thermal energy storage is that the ratio between overall work converted during charging (or discharging) and the net exergy stored (or extracted from storage) is relatively high. This has two consequences for pumped thermal systems. First, the total power rating of machinery required to absorb 1 MW of power in thermal pumping will be significantly higher than 1 MW and hence the costs of power-conversion equipment will be relatively high compared with other forms of storage. Second, the losses during the charging or discharging phases of the cycle are based on the overall mechanical power being handled so turnaround efficiencies will tend to be relatively low.

In [35], the authors outline a system that, although not proposed explicitly for offshore energy storage, has very high merit in the offshore environment in at least some respects. This is the classical *pumped-thermal* energy storage system in which a Brayton cycle is used for discharging and a reverse-Brayton cycle is responsible for charging. Figure 9.11 illustrates this system and Figure 9.12 shows the hot and cold storage tanks whose development was begun by *Isentropic Ltd.* in the UK and is being continued by the University of Newcastle. In this system, there are two thermal stores, one *hot* and one *cool*. The *hot* store resides on the high-pressure side of the system and the *cold* store resides on the low-pressure side. In [35], the pressure ratio between high pressure and low pressure is 12:1. The working fluid is

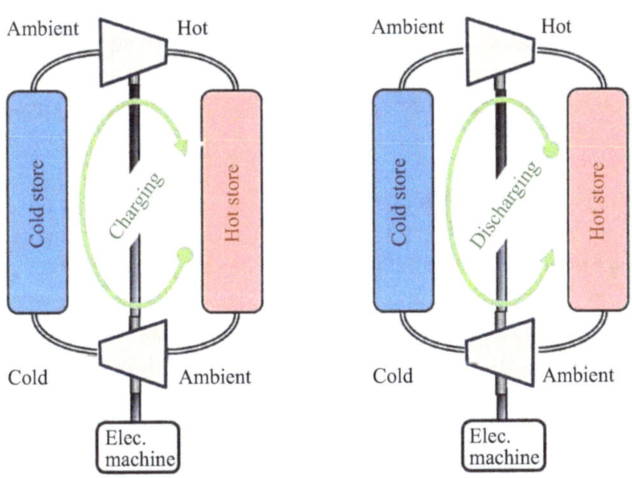

Figure 9.11 Schematic of a pumped-thermal exergy storage system

Figure 9.12 The thermal stores for a prototype thermal pumping system [35]

argon gas which, being monatomic, has a value of $\gamma = 1.6$. Most gases, including air, are (predominantly) diatomic and thus they have a value of $\gamma = 1.4$. If compression and expansion were perfectly isentropic (i.e. reversible), then the temperature ratio would be almost exactly 2.5:1 (or 5:2). For charging processes, this ratio of (5:2) would be identical to the ratio between the work converted in the compressor and the work delivered from the expander. Thus, we can deduce that for every 3 J of net exergy placed into storage, 5 J of exergy is used up in compression and 2 J of exergy is returned from the expander. Thus losses are based on the conversion of 7 J of exergy.

In this present case ([35]), each store retains heat spanning between a common 'ambient' temperature and an extreme temperature. The hot-store retains heat at temperatures up to 600 °C. As noted previously in the section on UWCAES, practical considerations to do with finding materials that can still hold reasonable stresses at high temperatures typically restrict this hot temperature to ~600 °C. The power-conversion machinery coping with the high-pressure air must be able to deal with such high stresses directly. The cold store sees temperatures down to ~−135 °C.

Packed beds of thermal storage material are one of the default choices for thermal storage of exergy at very large scale. Views differ on whether or not the thermal storage medium should be held within the pressurised system or outside of it. This decision is especially important for the high-pressure side where the pressurised system sees both high pressures and high temperatures. Heat transfer is especially good when the packed bed is held within the pressurised system but obviously the costs of the pressure vessel can be substantial. In [35], this choice is made. For very long-duration storage, retaining the packed-beds in ambient-pressure containers provides the possibility of extremely low costs, although this requires an indirect heat-transfer process which causes lower performance. Ambient-pressure packed beds are especially attractive for offshore energy storage.

The mean density of the content of a packed bed is typically quite similar to the density of seawater. For example, a volume that is 45% filled with rock whose solid density is 2,400 kg/m^3 has a mean density of around 1,080 kg/m^3. If a volume of this nature is 20 m tall, there is a lithostatic pressure[1] of approximately 2.12 bar. Although the bed of rocks can and does develop some internal shear stresses initially helping to prevent it from bulging-out the bottom of the containment, these internal shear stresses are lost when there is uneven cooling or heating of the rock particles as the packed bed is either charged or discharged. An external hydrostatic pressure that is approximately the same as the internal lithostatic pressure is an ideal way to ensure that the containment does not bulge out continuously. There is a slight caveat here, of course. If the external pressure is delivered by cold seawater, a thick insulation layer is required to prevent excessive heat loss and the radial pressure must be transmitted through this insulation layer.

There are, of course, other pumped-thermal storage solutions that may be applicable to the offshore context. Laughlin [36] describes a specific solution where the main hot and cold storage units each comprised two-tank systems. The main hot store uses molten salt held in two separate tanks. The temperature of the salt in both tanks must be above the melting temperature for the salt (above 150 °C for most commercial thermal storage salts). The main cold store uses a fluid (like CO_2 or methane) that would be gaseous at ambient temperature and ambient pressure but stores this in two non-pressurised vessels – both colder than the evaporation temperature of the fluid at ambient pressure. At the centre of the system of [36] is a packed-bed recuperator that enables a complete separation of temperature to occur between the cool side of the hot store and the warm side of the cold store.

One further pumped thermal storage arrangement is described in [37] and is especially suited to, and designed for, the offshore environment. Here the sea itself is used as thermal ballast to define the cold side of the reversible thermal pumping. During charging, heat is pumped from the sea up to a high temperature (again ~600 °C) and stored in a packed bed using a reverse Brayton cycle. Then during discharge, a normal Brayton cycle operates to convert the stored heat back into work. Reference [37] recognises and exploits the strong potential usefulness of the offshore environment in two key respects: (a) using the sea itself as thermal ballast and (b) exploiting external hydrostatic pressure in the sea to counter-act internal lithostatic pressure in the packed beds.

9.5 Other concepts

The set of possibilities for offshore energy storage is very wide. In the sections above, we have covered the main categories but naturally there are others. One area that we have not covered in depth above is flywheel energy storage [38]. As noted earlier, it is relatively straightforward to show that the maximum energy that can be stored in the rotor is a simple multiple of the total amount of structural capacity

[1]Lithostatic pressure in soils or granular materials is a perfect analogy to hydrostatic pressure in liquids.

employed in the rotor. For a flywheel which manages the centrifugal forces through hoop stresses in a radially thin rim on the rotor, the EC2SC ratio is a maximum of 0.5. It is noted in [38] that the *Laval disc* achieves EC2SC ratio of 1.0. In the case of offshore storage, it is at least conceivable that a flywheel rotor could be formed from ice and the bearing system could comprise that ice simply floating in the water. The power-conversion system could be, in effect, a linear motor 'wrapped around' the flywheel rim. Obviously, ice is rather poor as a structural material, but it is effectively free. In general, flywheels tend to be cost-effective only for very short durations of energy storage so that the system can get through very high numbers of (partial) cycles in its lifetime and this would not be different for the offshore case.

Another area not covered in depth here is the whole concept of forming synthetic fuels [39] at the offshore location using electricity generated at the site. Such fuels include pure hydrogen, ammonia and organic fuels based on carbon dioxide that may have been captured from combustion or from oil-and-gas exploration. This action extends a little beyond energy storage since the formation of fuels also provides alternative power/energy transmission options.

Finally, we mention liquid air energy storage (LAES) [40]. This is closely related to pumped thermal energy storage and, in one view, it is precisely pumped thermal energy storage where ambient environment (the sea, in the context of offshore energy storage) would be used as thermal ballast for the hot side and liquefied air would be retained as the cold store. LAES has been studied extensively in the UK and commercial activities have formed around it. To the authors' knowledge, it has not been proposed specifically as an option for offshore energy storage but it is clear that it could be. LAES certainly has the potential to achieve extremely low marginal costs per unit of exergy storage capacity.

9.6 Integrating offshore energy storage with generation

It is commonplace to imagine that energy storage necessarily means taking electricity in from a generation source, converting that into some other energy form that is more easily stored and then recovering that energy to convert it back to electricity. This has been the interpretation adopted in all previous sections of this chapter. However, there is another important class of energy storage whereby energy is retained in some compact and dense form prior to conversion into electricity and where the conversion to electricity occurs only when the electricity is in demand. It is noted in [41] and [42] that precisely this format of energy storage has enabled electricity systems around the world to function ever since these systems were introduced. Moreover, there are two strong advantages to such systems in that (i) we can potentially avoid the costs of power-conversion machinery to transform energy from electricity into a storable form and back again and (ii) (for energy that passes through storage) we can also avoid the losses associated with every transformation of energy between forms. Generation-integrated energy storage (GIES) systems are not rare. They include all instances of natural hydro where there is a dam and upper reservoir capable of storing water, all biomass facilities, many concentrated solar power plant (where energy may be stored as heat prior to it being used to raise steam to generate electricity) and some photo-catalysis applications where energy may be held in the form of hydrogen produced by sunlight prior to being converted to electricity in a fuel cell. GIES systems can exist for several of the offshore renewable energy forms. In [21], e.g., Garvey describes a wind turbine that compresses air directly from the harvested wind energy and that air is stored in underwater containments prior to the generation of electricity. The authors of [43] outline a different arrangement for a wind turbine wherein the wind turbine drives a hydraulic pump that may directly drive a compressor, and they propose that compressed air can be stored in the tower of the wind turbines. In [44], a relatively new method of integrating energy storage with offshore wind power is described. This concept blends together the function of a power-transmission system for a wind turbine (taking mechanical power from a very slowly rotating turbine shaft and converting this into electrical power) and blends this with the capability to pump heat from a cold store into a hot store. At a subsequent time when the wind resource is weak but electricity is in high demand, the heat pumping can be reversed to deliver a Brayton cycle in which work is extracted as the thermal stores are emptied. For all cases of GIES systems implemented offshore, there is one further significant advantage: the transmission lines conveying power to shore may be utilised much more effectively than they might otherwise be.

References

[1] T. Letcher (Ed.). *Storing Energy: With Special Relevance to Renewable Energy Generation*. Cambridge, MA: Elsevier. (2016).

[2] H. Chen, T.N. Cong, W. Yang, C. Tan, Y. Li and Y. Ding. Progress in electrical energy storage systems: A critical review. *Progress in Natural Science*, 19(3) (2009): 291–312.

[3] F. Diaz-Gonzalez, A. Sumper, O. Ginus-Bellmunt and R. Villafafila-Robies. A review of energy storage technologies for wind power applications. *Renewable and Sustainable Energy Reviews*, 16(4) (2012): 2154–2171.

[4] M. Aneke and M. Wang. Energy storage technologies and real life applications – A state of the art review. *Applied Energy*, 179 (2016): 350–377.

[5] S. Sabihuddin, A.E. Kiprakis and M. Mueller. A numerical and graphical review of energy storage technologies. *Energies*, 8(1) (2014): 172–216.

[6] V. Jülch. Comparison of electricity storage options using levelized cost of storage (LCOS) method. *Applied Energy*, 183 (2016): 1594–1606.

[7] B. Zakeri and S. Syri. Electrical energy storage systems: A comparative life cycle cost analysis. *Renewable and Sustainable Energy Reviews*, 42 (2015): 569–596.

[8] A.B. Gallo, J.R. Simões-Moreira, H.K.M. Costa, M.M. Santos and E. Moutinho dos Santos. Energy storage in the energy transition context: A technology

review. *Renewable and Sustainable Energy Reviews*, 65 (2016): 800–822, https://doi.org/10.1016/j.rser.2016.07.028.

[9] M. Budt, D. Wolf, R. Span and J. Yan. A review on compressed air energy storage: Basic principles, past milestones and recent developments. *Applied Energy*, 170 (2016): 250–268.

[10] O. Laing and J. Laing. Energy storage for off peak electricity. USA Patent US4873828, 31 March, 1989.

[11] R.J. Seymour. Undersea pumped storage for load levelling. In: *California and the World's oceans '97*, San Diego 1997.

[12] A.J. Pimm and S.D. Garvey. Underwater compressed air energy storage. In: *Storing Energy: With Special Relevance to Renewable Energy generation*. T. Letcher (Ed.). Cambridge, MA: Elsevier. 2016. pp. 135–154.

[13] Z. Wang, W. Xiong, H. Wang and Z. Wang. A review on underwater compressed air energy storage. *Energy Storage Science and Technology*, 4(6) (2015): 585–598.

[14] B.C. Cheung, R. Carriveau and D.S.-K. Ting. Parameters affecting scalable underwater compressed air energy storage. *Applied Energy*, 134 (2014): 239–247.

[15] D. Fiaschi, G. Manfrida, R. Secchi and D. Tempesti. A versatile system for offshore energy conversion including diversified storage. *Energy*, 48(1) (2012): 566–576.

[16] A. J. Pimm, S.D. Garvey and M. de Jong. Design and testing of energy bags for underwater compressed air energy storage. *Energy*, 66 (2014): 496–508.

[17] Z. Wang, W. Xiong, D.S.-K. Ting, R. Carriveau and Z. Wang. Conventional and advanced exergy analyses of an underwater air energy storage system. *Applied Energy*, 180 (2016): 810–822.

[18] A.J. Pimm, S.D. Garvey and R.J. Drew. Shape and cost analysis of pressurised fabric structures for subsea compressed air energy storage. *Proc IMechE, Journal of Mechanical Engineering Science*, 225(5) (2011): 1024–1043. https://doi.org/10.1177/0954406211399506.

[19] Z. Wang, D. S.-K. Ting, R. Carriveau, W. Xiong and Z. Wang. Design and thermodynamic analysis of a multi-level underwater compressed air energy storage system. *Energy Storage*, 5 (2016): 203–211.

[20] B. Kantharaj, A.J. Pimm and S.D. Garvey. Flexible vs. rigid containments for underwater storage of pressurized air. *Proc. OSES2016, Offshore Energy and Storage symposium*, Malta, July 8–10, 2016.

[21] D. Buhagiar and T. Sant. *Modelling of a novel hydro-pneumatic accumulator for large-scale offshore energy storage applications*. Energy Storage 14(2) (2017): 283–294.

[22] D. Buhagiar, T. Sant, R.N. Farrugia, L. Aquilina, D. Farrugia and F.M. Strati. A first embodiment of a marine floating platform with integrated energy storage. *Proc. OSES2018, Offshore Energy and Storage*, Ningbo, China. July 6–8, 2018.

[23] S.D. Garvey, M. Chee and A.J. Pimm. Deployment methods for flexible air containments in deep water far offshore. *Proc. OSES2016, Offshore Energy and Storage symposium*, Malta, July 8–10, 2016.
[24] C. Pete, J. McGowan and W. Jaslanek. Evaluating the underwater compressed air energy storage potential in the Gulf of Maine. *Wind Engineering*, 39(2) (2015): 141–148.
[25] A.J. Pimm and S.D. Garvey. Potential locations for underwater compressed air energy storage in Europe and North America. *Proc. OSES2015, Offshore Energy and Storage*, Edinburgh, July 1–3, 2015.
[26] S.D. Garvey. Structural capacity and the 20MW wind turbine. *Proc. IMechE, Part A, Journal of Power and Energy*, 224(8) (2010): 1083–1115. https://doi.org/10.1243/09576509JPE973.
[27] G.L. Guidati. Patent EP2687702A1 – filed 20/07/2012.
[28] M. Puchta, J. Bard, C. Dick, *et al.* Development and testing of a novel offshore pumped storage concept for storing energy at sea – STENSEA. *Energy Storage*, 14(2) (2017): 271–275. https://doi.org/10.1016/j.est.2017.06.004.
[29] A.H. Slocum, G.E. Fennell, G. Dundar, B.G. Hodder, J.D.C. Meredith and M.A. Sager. Ocean renewable energy storage (ORES) system: Analysis of an undersea energy storage concept. *Proc. IEEE*, 104(4) (2013): 906–924. http://hdl.handle.net/1721.1/78934.
[30] K. Bassett, R. Carriveau and D.S-K. Ting. Experimental analysis of buoyancy battery energy storage system. *IET Journal of Renewable Power Generation*, 10 (2016): 1523–1528.
[31] K. Bassett, R. Carriveau and D.S-K. Ting. Underwater energy storage through application of Archimedes principle. *Energy Storage*, 8 (2016): 185–192.
[32] R. Klar, B. Steidl, T. Sant, M. Aufleger and R.N. Farrugia. Buoyant energy – Balancing wind power and other renewables in Europe's oceans. *Energy Storage*, 14(2) (2017): 246–255.
[33] R. Klar, M. Aufleger, G. Scheide and V. Neisch. PrepareBE (Buoyant Energy) – Exploratory project presentation. *Proceedings of Offshore Energy and Storage (OSES2016)*, Malta, July 12–14, 2016.
[34] P. Hartlieb, M. Toifl, F. Kuchar, R. Meisels and T. Antretter. Thermophysical properties of hard rocks and their relation to microwave-assisted comminution. *Minerals Engineering*, 91 (2016): 34–41.
[35] A.J. White, G. Parks and C.N. Markides. Thermodynamic analysis of pumped thermal electricity storage. *Applied Thermal Engineering*, 53(2) 2013: 291–298.
[36] R.B Laughlin. Pumped thermal grid storage with heat exchange. *Renewable and Sustainable Energy*, 9 (2017): 044103.
[37] J. Ruer. Energy storage for offshore wind power. *Conference Proceedings REM2012*, Ravenna, Feb. 2012.

[38] M.E. Amiryar and K.R. Pullen. A review of flywheel energy storage system technologies and their applications. *Applied Sciences*, 7(3) (2017): 286. https://doi.org/10.3390/app7030286

[39] X. Yan, O.R. Inderwildi and D.A. King. Biofuels and synthetic fuels in the US and China: A review of well-to-wheel energy use and greenhouse gas emissions with the impact of land-use change. *Energy Environmental Science*, 3 (2010): 190–197.

[40] R. Morgan, S. Nelmes, E. Gibson and G. Brett. *Liquid air energy storage – analysis and first results from a pilot scale demonstration plant. Applied Energy*, 137 (2015): 845–853. https://doi.org/10.1016/j.apenergy.2014.07.109

[41] S.D. Garvey. Integrating energy storage with renewable energy generation. *Wind Engineering*, 39(2) (2015): 129–140.

[42] S.D. Garvey, P.C. Eames, J.H. Wang,*et al*. On generation-integrated energy storage. *Energy Policy*, 86 (2015): 544–551.

[43] C. Qin, E. Innes-Wimsatt and E. Loth. Hydraulic-electric hybrid wind turbines: Tower mass saving and energy storage capacity. *Renewable Energy*, 99 (2016): 69–79.

[44] S.D. Garvey, A.J. Pimm, J.A. Buck, *et al*. Analysis of a wind turbine transmission system with intrinsic energy storage capability. *Wind Engineering*, 39(2) (2015): 149–173.

Chapter 10
Multipurpose platforms

Maurizio Collu[a] *and Erin E. Bachynski*[b]

10.1 Introduction

10.1.1 Context

A multipurpose platform is an offshore system designed to serve the purposes of more than one offshore industry. Indeed, a number of industries have expanded, or are expanding, from onshore to offshore locations, adapting to the harsh environment in order to extract energy (conventional or renewable) or useful materials (deep mining), to produce food (aquaculture), nutraceuticals, and pharmaceuticals (blue biotechnology), to expand the urban areas, or to simply develop the local tourism, to name a few. This process starts to appear like, and certainly has the potential to become, a further 'agricultural/industrial revolution', characterised by

- a substantial shift from 'gathering of resources where available' to the systematic, settled use of the resources offered by the ocean,
- a shift from offshore facilities of a specific industry operating independently to an 'ecosystem' of offshore industrial activities, and
- the resulting development of new technologies.

Expanding the above-mentioned industries towards this new frontier can help to respond to two of the main challenges facing mankind: the sustainable, safe, and reliable production of (1) energy and (2) food, driven by an increasing population and rising prosperity [1]. For these reasons, and also due to the huge potential to create jobs and revenues, many countries around the world are proposing and implementing strategies to develop the so-called Blue Economy.

In its communication entitled 'Blue Growth – opportunities for marine and maritime sustainable growth' [2], the European Union (EU) formally acknowledged three important factors:

- Rapid progress is being made in the development of offshore technologies, including those targeting deep waters.

[a]Department of Naval Architecture, Ocean and Marine Engineering, University of Strathclyde, Glasgow, UK

[b]Department of Marine Technology, Norwegian University of Science and Technology, Trondheim, Norway

- The finite nature of land and freshwater resources which are at an increased strain as a result of economic development.
- The opportunity offered by offshore renewable energy resources to reduce the anthropogenic greenhouse gas emissions, which has been identified as the major cause of climate change [3].

The 'Blue Economy' already accounted for 5.4 million jobs across the EU way back in 2011. It was generating a gross added value of 500 billion euro every year, with huge potential for innovation and growth [4,5]. Therefore, a number of initiatives were funded (see Section 10.2.1), furthering the knowledge and understanding in this field to find sustainable, techno-economically feasible solutions.

Outside Europe, China is another example where the importance of the emerging 'Blue Economy' is being recognised. The 'Made in China 2025' strategic plan focuses on 10 priority sectors, highlighting areas of opportunity for economic growth [6]: within the marine resources priority sector, the focus is on ocean renewable energy (ORE) technologies. Furthermore, 'the 13th five year plan of Marine renewable energy development' [7] is promoting the use of a combination of renewable energy technologies to provide at least 60% of the total energy demand of the Shandong, Zhejiang, Fujian, Guangdong island communities. In fact, in China, a number of small/isolated communities do not have a reliable access to basic utility services (fresh water, energy), due to the remote nature of their location, and are being abandoned by the local communities. The sustainable, local production of basic utility services is seen as a promising approach to solve this issue.

The further development of the aquaculture sector is also high in the agenda: the 'Outline of the National Marine Economic Development Plan' [8] specifically targets the development of novel ocean farming methods that are not only more productive but also more socially and environmentally compatible.

10.1.2 Why multipurpose platforms?

Usually, due to the distance from the shore and to the tough metocean conditions, an offshore system needs to be self-sufficient in many aspects for a certain period of time (sometimes extending to several weeks). For example, oil and gas platforms located in far offshore and deep water locations need to be self-sufficient from a basic utility services' point of view (energy, water) and need to have a minimum set of skills covered by their crew: not only the technical staff to operate and maintain the oil and gas-related operations but also catering staff, security staff, fully qualified medics, radio operators, and others.

Based on the Blue Growth strategy and on the current trend of the offshore wind and aquaculture industries, it is likely that a number of offshore facilities will be co-located in the same offshore area. It follows that, rather than having the necessity to be self-sufficient at single platform level, there is an opportunity to create an '*industrial (sustainable) ecosystem of offshore facilities*', where the common services will be shared and provided by specialised systems.

The multipurpose platform concept is a possible implementation of this ecosystem: in this case, the facilities of this ecosystem are based on the same platform or closely co-located, exploiting the specific potential synergies and avoiding by design the arising tensions.

10.1.2.1 Potential synergies

Many of the potential synergies identified for hybrid wind-wave energy devices (e.g. [9]) can also be applied to a generic multipurpose platform concept. These synergies are not only at technological level but also at political, economic, social, technological, and environmental levels (PESTLE approach).

A high-level list of potential synergies, divided by discipline, is provided in Table 10.1. At the moment, the experience with these platforms is quite limited, since these concepts have only been investigated at a numerical level or tested at small scale, in indoor experimental facilities. No fully commercial multipurpose platform is available, with the only small-scale prototype, tested outdoor being the hybrid wind-wave energy device developed by Floating Power Plant [10]. Therefore this table should be considered as non-exhaustive and likely to expand.

In the United Kingdom, the Energy Act 2004 [11] and the Climate Change Act 2008 [12] have created a framework that has allowed the British Offshore Renewable Energy sector to lead at an international level. A similar common regulatory framework (P.1), establishing a common reference for the development of all these systems using the ocean resources, would substantially increase the development of both single-purpose and multipurpose platforms.

A key performance indicator of any ORE system is the capital expenditure (CAPEX) per MW of rated power. By utilising the same platform for multiple uses, the total CAPEX cost should be lower than the sum of the CAPEX where different

Table 10.1 Some of the main synergies that can be enabled by a multipurpose platform, PESTLE approach: P, Political; SE, Socio-Economic; T, Technological; L, Legal; E, Environmental (expanded from [9])

No.	Synergy
P.1	Encouraging a common regulatory framework
SE.1	Lower CAPEX/system due to shared infrastructure
SE.2	Lower OPEX/system due to shared O&M facilities, vessels, and staff
SE.3	Provision of local, sustainable source of energy, food, fresh water, to isolated/remote communities
T.1	Enhanced renewable energy yield/more energy per unit area
T.2	Smoothed and more consistent power output
T.3	Common grid infrastructure
T.4	Shared mooring system/connection to seabed
T.5	Wave energy converter wake effect creating milder wave conditions for other systems and services
L.1	Encouraging a coordinated plan over the marine and coastal regions
L.2	Encouraging a common licensing procedure
E.1	Reduced environmental impact compared to single systems

devices are built as separate systems (SE.1). Furthermore, this consideration can be extended also to the operational expenditure (OPEX) (SE.2): as already mentioned, the costs associated with the daily operation and maintenance can be shared, since the majority of the ports, vessels, and staff could provide this service to multiple offshore systems.

From a social point of view, an important consequence of providing a sustainable, local source of energy, food, fresh water should be considered: the sustainment of isolated/small communities. Although this aspect has not been properly considered in the EU, where the focus has mainly been on 'farms' of large, multi-MW multipurpose platforms, in other countries like Indonesia and China, and continents like Africa, the possibility to provide local, sustainable basic utility services (and the jobs linked to these) is seen as a key factor to enable the development of these isolated communities (SE.3).

As far as the technological aspects are concerned, several potential synergies have been identified ([9,13] and others). Similarly to hybrid wind-wave, wind-tidal, wind-wave-tidal offshore renewable energy devices, the rated power per unit of area is higher when multiple ORE devices are co-located. This, in general, also leads to an enhanced energy yield (T.1). On top of a quantitative enhancement, from a power quality point of view, the energy produced is smoother (short term) and more consistent (longer term) (T.2), thanks to the different physics underpinning the different energy sources. This is beneficial not only for the national/international energy grid, but also to the co-located systems relaying on the provided energy, as, for example, the aquaculture systems present in the multipurpose platform.

Furthermore, the needed infrastructure to export the energy produced can also be shared – sharing therefore also its costs (T.3). The same consideration can be extended to the mooring system (floating platform), or the connection to seabed (fixed platform) (T.4). In addition to these aspects, there is an evident synergy between wave energy converter (WEC) and aquaculture systems. On one side, WEC are designed to absorb wave energy, therefore creating milder wave conditions in their wake. Aquaculture systems, especially those systems used to produce finfish, are negatively impacted by waves, and for this reason, so far they have been installed only in sheltered and shallow waters in the nearshore. Therefore, if properly located, WEC systems can open new areas for aquaculture systems. In addition to this, the milder wave conditions are also beneficial for the maintenance operations of the various systems (T.5).

There have been some initiatives aiming at creating a coordinated planning framework for the utilisation of marine and coastal regions resources, such as the Maritime Spatial Planning and the Integrated Coastal Protection Management EU initiatives [15]. Furthermore, the MUSES project (Multi-Use in European Seas, 2016–18) [14] explored the opportunities for multi-use of the ocean resources across five EU sea basins (Baltic Sea, North Sea, Mediterranean Sea, Black Sea and Eastern Atlantic) and presented a series of case studies, identifying both the main drivers and justifying the multi-use of the ocean space and the perceived barriers to be considered (Figure 10.1). Nonetheless, a well-defined framework able to coordinate all the activities taking place in an offshore environment (L.1) is still

Figure 10.1 Overview of the multifarious nature of systems, technologies, and stakeholders of the ocean space (courtesy of MUSES project [14])

lacking. In general, the necessary licensing procedures tend to be different for each offshore system typology, making the transfer of knowledge from the most mature industries to the new players in the offshore scenario difficult.

The environmental impact assessment is an important step of the overall licensing procedure, as it can determine the feasibility of a project. The environmental impact of a multipurpose platform, where many of the infrastructures and services are shared, is likely to be reduced when compared to the installation of the single-purpose platforms [9] (E.1), since, for example, some of the infrastructure will be shared, leading to less material used and smaller footprint. Furthermore, this effect can be enhanced if the by-product of one system can be used as input by another system: an example of this is the integrated multi-trophic aquaculture approach, where, for example, the waste nutrients from higher trophic-level species are used for the production of lower trophic-level crops of commercial value, enhancing the overall sustainability [16].

10.2 Multipurpose platform projects and concepts

10.2.1 EU projects

Between 2010 and 2013, 'The ocean of tomorrow' calls [17] aimed at funding initiatives that encouraged the marine and maritime research and industrial communities to come together and develop concepts and technologies with a multidisciplinary, cross-cutting approach, facilitating cross-fertilisation, and a holistic attitude. A multidisciplinary approach does not only cover techno-economic aspects but also the social, legal, and environmental aspects from the early phases of the project.

Among the 'Ocean of tomorrow call', one of the topics of the call FP-7-2011 was 'Multi-use offshore platforms', and three projects were funded on this topic: H2Ocean [18], Mermaid [19], and Tropos [20].

The project H2Ocean [18,21] focused on the '*Development of a wind-wave power open-sea platform equipped for hydrogen generation with support for multiple users of energy*'. A series of techno-economic feasibility assessments, complemented by environmental impact analyses, were performed considering far offshore, multipurpose farms in the Mediterranean Sea, North Sea, and North Atlantic Ocean. These farms include offshore floating hybrid-wind-wave energy units, aquaculture units, and a number of service platforms. The latter are used to coordinate all the activities, for the production of hydrogen, considered the energy vector of the future, and as an environment monitoring station. In Figure 10.2, a 3D sketch of a part of a multipurpose farm is represented.

The development of closely co-located offshore wind, wave, and aquaculture infrastructures, assessing also their environmental compatibility and their economic feasibility, was at the core of the Mermaid project [19]. Focusing on four areas (Baltic Sea, North Sea, Atlantic Ocean, and Mediterranean Sea), this project

Figure 10.2 A proposed wind-wave-aquaculture multipurpose concept, from the H2Ocean project [18]. In the final version, a farm of hybrid wind-wave devices, with 5 MW vertical axis wind turbines and 3.6 MW WEC was planned, sheltering a separate but closely co-located aquaculture system (courtesy of Floating Power Plant [10] and Fusion Marine [22])

performed an assessment at technical, environmental, and socio-economic level of the solutions proposed. One of the solutions proposed is the hybrid wind-wave multipurpose platform, composed of a triangular floating platform with three vertical columns at the corners and one in the middle, which supports a 5 MW wind turbine, and an oscillating water column (OWC) WEC installed in each cornet column (Figure 10.3 [23]). As a final result, guidelines for the development, management, and implementation of multipurpose offshore platforms in these European scenarios were provided.

The Tropos [20,24] project, like the previous ones, looked at multipurpose platform systems, but adopting a 'modular' approach: the platform is composed of specialised modules and a central platform. The specific activities covered by the different modules are maritime transport, renewable energy, aquaculture, and leisure. Three main multi-use platform concepts were defined and analysed [25] (Figure 10.4):

- the 'Green & Blue' concept, located north of Crete, integrating offshore wind turbines and aquaculture (fish, algae) systems;
- the 'Leisure Island' concept, off the coast of Gran Canaria, combining leisure facilities and solar energy systems; and
- the 'Sustainable Service Hub', located in the Dogger Bank (North Sea, UK), providing energy and transport services to the offshore wind farms planned in this region.

Figure 10.3 MUP platform developed by IH Cantabria within the Mermaid project (FP7-288710), one of the multipurpose platform concept proposed within the Mermaid project: a four column semisubmersible, with a triangular base plate, hosting a 5 MW wind turbine and three OWCs WEC (one in each corner column). 3D model (left) and experimental test (right) [23] (courtesy of IH Cantabria, Spain)

Figure 10.4 The three TROPOS multipurpose platform concepts: Sustainable Service Hub (top), Green & Blue (bottom left), and Leisure Island (bottom-right) (courtesy of TROPOS Consortium)

In parallel to these, other two projects were funded, focusing on multipurpose platforms capable of utilising more than one offshore renewable energy source, but not including aquaculture systems: MARINA and ORECCA.

The MARINA platform project [26], or 'Marine Renewable INtegrated Application Platform' project, brought together the offshore wind, wave, and tidal research and industrial communities, in order to assess the techno-economic feasibility of new systems, coupling multiple offshore wind and ocean energy converters. A number of concepts have been proposed and analysed, but eventually three main conceptual designs have been proposed, coupling floating offshore wind turbine systems and wave energy converters: the 'Spar Torus combination (STC)', the 'Semisubmersible Flap Combination (SFC)', and the 'OWC array with wind turbines' (Figure 10.5).

The STC concept couples a 5 MW spar floating offshore wind turbine with a torus-shaped, coaxial body oscillating vertically with respect to the main spar body. This oscillation, through the use of a suitable power take-off system, can be used to extract energy from the waves [27]. The numerical analyses conducted suggest that, if compared with the separated systems, the STC combined wind-wave concept can reduce the total capital costs and increase the energy yield. The SFC [28] is a three-column, semi-submersible wind turbine combined with a rotating flap-type wave energy converter, hinged at the submerged pontoons of the semi-submersible. Based on the configurations analysed, comparing the platform with and without the wave energy converters, the dynamics of the floating wind turbine does not seem to be substantially affected, while the total power can be up to 8% higher. The OWC array consists of a large floater having multiple wave energy converters and supporting a single wind turbine.

The ORECCA project [29] focused on bringing together the offshore renewable energy communities (wind, wave, tidal), in order to assess the state of the art in this field. It also prepared a roadmap for the further development of these technologies, trying to identify synergies whenever possible.

More recently, the European Commission funded two projects under the topic 'Multi-use of the oceans marine space, offshore and near-shore: enabling

Figure 10.5 MARINA Platform project [26], two hybrid wind-wave energy devices: (left) the spar–torus combination (STC), (right) the semisubmersible-flap combination (SFC), 3D numerical models (courtesy of NTNU, Norway)

technologies (H2020, BG-04-2017)': Space@Sea [30] and The Blue Growth Farm [31]. These two projects build upon the experience and results accumulated in the previous initiatives, by proposing innovative multipurpose platform concepts and testing them in a relevant environment, therefore bringing the technical maturity to TRL 5 [32].

There have been a number of companies proposing multipurpose offshore platform concepts. A notable example is provided by Floating Power Plant [10], which claims to operate '*the world's only offshore-proven and grid-connected combined floating wind and wave device*'. Their P37 platform (Figure 10.6) is 37 m wide and is grid-connected. It combines flap-type wave energy converters, with a total rated power of 134 kW, to three 11 kW wind turbines [33]. This platform has been operated since August 2013 and equipped with a number of sensors to assess its performance and to support the numerical modelling and analyses activities [34]. The full-scale concept proposed by the company is the P80, an 80 m wide platform capable to combine wave energy converters with a total rated power between 2 and 3.6 MW (depending on location) and a wind turbine with a rated power between 5 and 8 MW.

Figure 10.6 Photos of the Floating Power Plant P37 hybrid wind-wave platform: the turret mooring system (top-right in the picture) allows the platform to align perpendicularly to the main wave direction, enhancing the efficiency of the dynamically ballasted oscillating floaters. Three 11 kW wind turbines are coupled with the WEC system (courtesy of Floating Power Plant [10])

Table 10.2 Overview of multipurpose platform concepts and their multiple functions

Concept	Aquaculture	Wind	Wave	Other ORE	Tourism	Transport
H2Ocean [18]	✓	✓	✓			
Mermaid (IHC) [19,23]		✓	✓			
Tropos [20] Green & Blue	✓	✓				
Tropos [20] Leisure Island				✓	✓	
Tropos [20] Sust. Service Hub		✓				✓
Marina (STC) [27]		✓	✓			
Marina (SFC) [28]		✓	✓			
Marina (OWC) [26]		✓	✓			
Floating Power Plant P37 [10]		✓	✓			
Pelagic Power W2Power [36]		✓	✓			

Another concept, proposed by Pelagic Power, is the W2Power plant [35]. It combines a three-floater semisubmersible floating offshore wind turbine platform, accommodating two wind turbines with a rated power of 4–6 MW each, and a number of hemispherical point absorbers actuating hydraulic pumps, for a total rated power up to 3 MW, for a potential total rated power of $2 \times 6 + 3 = 15$ MW. A number of numerical analyses have been conducted, complemented by small-scale ocean basin tests [36].

Table 10.2 summarises the discussed multipurpose platforms and their main functions.

10.3 Design and analysis of multipurpose platforms

10.3.1 Multidisciplinary design methodology

As reported by Zanuttigh *et al.* [13]: '*The design of these (multipurpose) solutions is a complex interdisciplinary challenge, involving scientists and technical experts with different backgrounds*'. Although the same can be stated for any offshore systems, in the case of multipurpose platforms, the multidisciplinary aspect of the challenge is enhanced by the number of different systems considered, usually designed and analysed by different research and industrial communities. For example, the aquaculture community and the ORE community have been working independently until few years ago.

A detailed multi-criteria design methodology, suitable for closely co-located rather than integrated multipurpose concepts, is presented by Zanuttigh *et al.* [13], based on four main steps:

- pre-screening phase,
- preliminary design of single-use platforms,
- ranking phase, and
- preliminary design of the selected multi-use platform.

In the pre-screening phase, the importance of a combined resource assessment approach, to identify locations suitable from a multi-renewable energy point of view, is highlighted (see Section 10.3.2). In the second step, based on the pre-screening results, single-use platforms are designed at a preliminary level, assessing the potential overall production and combining the single-use platforms in different multi-use layouts. In the third step, the different multi-use concepts proposed are evaluated against a set of multidisciplinary criteria, producing a ranked list of the proposed concepts. It is noteworthy to mention that, among the several criteria considered in the ranking phase, the 'synergy with other uses' is considered. This is highlighted as an important aspect for the final choice. In the fourth step, the selected optimum multi-use concept is further developed, refining some of the approaches adopted in the previous phases.

10.3.2 Resource assessment: combined wind-wave resources

Multipurpose platforms which can exploit multiple forms of renewable energy must be located at sites which fulfil several criteria:

1. All of the resources in question are available, at levels which can be economically exploited.
2. The platform can survive the worst environmental conditions.
3. The water depth and/or soil conditions are conducive to the platform.
4. Power can realistically be delivered to users, also considering important environmental consequences of cabling.

Furthermore, since one advantage of multipurpose platforms is that they may be able to provide more uniform power output – for example, by producing electricity from waves during periods of low wind speeds – it is interesting to examine the temporal correlations among different electricity resources. For wind-generated waves, one can expect there to be a strong correlation between wind and waves. This would imply high variability in the power production, while a location which is exposed to swells or very long fetch would imply a lower correlation and lower variability in the total power production [37].

Several authors have studied the wind and wave resource at European locations, using either measured metocean data (typically at ORE tests sites, e.g. [38]) or hindcast data [37,39,40]. Hindcasts are numerical atmospheric and ocean models which use available historical measurements as inputs or boundary conditions to the simulations. Measurement data are also used to calibrate and check these models. Due to the limited number of co-located and temporally synchronised long-term measurements of wind and wave conditions, hindcasts are often used to obtain the necessary long-term statistics for different locations.

Kalogeri *et al.* studied a range of European sites with promising wind and wave resources and found some locations where the wind and wave power have reasonably low correlation (0.5–0.6) and good seasonal stability [37]. The power at each site was estimated using the power curve of a representative wind turbine and representative wave devices. The most promising locations were all sites which are exposed to Atlantic swells. The correlation C for different time lags τ was defined as

$$C(\tau) = N^{-1} \sum_{i=1}^{N-\tau} \frac{(x_i - \mu_x)(y_{i+\tau} - \mu_y)}{\sigma_x \sigma_y} \tag{10.1}$$

where μ represents the mean value, σ represents the standard deviation, and wind and wave power are represented by x and y. The index i refers to the N available observations.

Higher correlations between the wind and wave conditions were seen in closed or semi-closed basins (such as the Baltic Sea) where the waves are primarily wind driven. Figure 10.7 illustrates some of the results of their study: very high correlations can be seen for the Baltic Sea and English Channel, and those sites also show small variations in the time lag at maximum correlation. The more exposed sites have lower correlation – though the correlation can vary significantly during the year – and larger variations in the time lag which corresponds to maximum correlation. It is important to note that the mean wind and wave power both tend to be the largest during autumn (September, October, November (SON)) and winter (December, January, February (DJF)) for the selected sites.

Qualitatively, measurement data agree well with the hindcasts regarding the temporal correlation between wind and waves: the seasonal correlation between

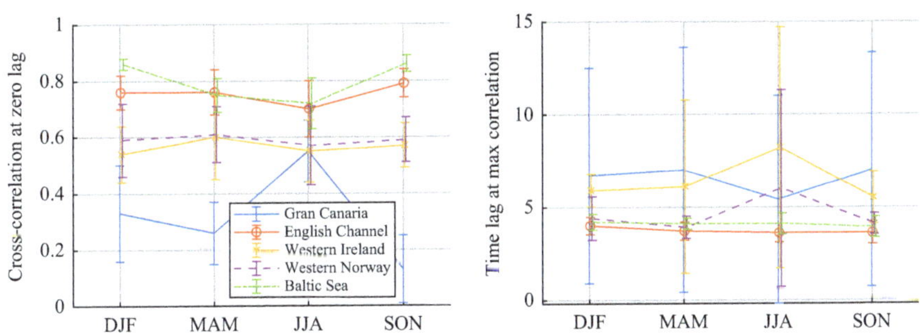

Figure 10.7 Seasonal correlation between wind and wave power at selected sites (left) and time lag between wind and waves at maximum correlation (right). Selected European sites, data from Kalogeri et al. [37]

wind and wave power on the West and South-West coast of Ireland was estimated to be significantly lower (0.2–0.35) compared to the East coast (0.6–0.7) [38].

An evaluation of the joint probability distribution of the wind and wave conditions for similar European sites suggested that the extreme wind speeds do not vary significantly over the region, but the extreme wave conditions differ significantly [39]. The sites with the best wave power resource also tended to have the most severe extreme sea states: Atlantic sites had generally higher wave power potential and larger extreme significant wave heights than the North Sea sites [39].

10.3.3 Modelling and analysis

An overview of the main characteristics of some of the numerical approaches applied for the modelling of multipurpose platforms is given in Table 10.3. Five main aspects have been analysed: aerodynamics, hydrodynamics, structural dynamics, mooring models, and the power take off (PTO) system dynamics.

Two main approaches are adopted for quantifying the aerodynamic forces acting on the platform, and in particular on the wind turbine subsystem: a quasi-static (QS) thrust (QST) approach and the well-known blade element momentum approach [36], which is corrected in some cases to take into account the effects of dynamic stall and dynamic inflow/wake. The QST is a simplified approach whereby the aerodynamic thrust acting on the wind turbine, at hub level, can be calculated at each instant of time (time domain simulation) as the product of the air density ρ, the velocity of the wind relative to the wind turbine, at hub level $V_{R,HUB}$, a thrust coefficient c_T, a function of the tip speed ratio (λ), and of the angle

Table 10.3 Numerical modelling approaches

Platform	Aerodynamics	Hydrodynamics	Structural dynamics	Mooring model	PTO
H2Ocean [41] VAWT-WEC	BEM+DS	PF	All rigid	LM [42]	N/A
Mermaid IHC [23]	QST [43]	P: PF W: PF [44]	All rigid	QS	LD+DP
Marina SFC [46]	BEM [45]	P: PF+ME+NA W: PF+ME	P: rigid T: FE W: rigid/FE	FE [45]	LD [45]
Marina STC [27]	QST	PF+ME	T: rigid W: rigid	QS	LD+LS
FPP P37 [34]	BEM+DS	PF	MB	LM/Dyn	N/A

T, Turbine; P, platform; W, WEC; PTO, power take off; MB, multibody; BEM, blade element momentum; DS, dynamic stall; QST, quasi-static thrust; PF, potential flow; ME, Morison equation; NA, Newman's approx; FE, finite element; LM, lumped mass; QS, quasi-static; Dyn, dynamic; LD, linear damping; LS, linear stiffness; DP, dynamic pressure; N/A, not available.

between the average wind speed vector and the vector perpendicular to the rotor plane ψ:

$$T = 1/2\rho V_{R,hub}^{2} A c_T(\lambda, \psi) \tag{10.2}$$

A number of approaches are adopted to quantify the hydrodynamic forces acting on the multipurpose system. For large volume structures, a linear potential flow (PF) approach is used, complemented by a Morison equation approach to take into account viscous forces, especially for the small volume components of the platform. A Newman's approximation approach has been adopted to model the slowly varying force and moments acting on the support platform in the analysis of the MARINA – SFC system [46]. In general, the codes are limited when evaluating the hydrodynamic interactions between the platform and the elements of the WEC moving relatively to the platform – this is an active area of research. Furthermore, the PF numerical implementations and solvers are subject to numerical instabilities and errors, especially when there are some narrow gaps between the main platform body and the WEC elements, forcing to simplify the overall geometry of the multipurpose platform (e.g. [34,41]).

Several levels of approximations are adopted for representing the structural dynamics of the platform. The support structure is usually represented as a rigid body. The WEC system can, at the simplest level, be considered as a rigid body, or modelling the main body interacting with the waves as a rigid body, while modelling with a finite element (FE) approach the connection of the WEC with the main platform (e.g. [46]). The wind turbine has been modelled as a rigid body or as a flexible body (FE), depending on the main aim of the analyses conducted: when evaluating the global response, the elasticity of the rotor and the tower of the wind turbine can be neglected.

The mooring line dynamics can be represented, at the most basic level, with a constant stiffness matrix, evaluated at a suitable position, ignoring any dynamic effect due to the inertia of the lines and the hydrodynamic forces (mainly viscous drag and added mass), therefore adopting a QS approach. The inertia of the lines can be considered when adopting a lumped mass (LM) approach, and when also the hydrodynamic forces are considered, an LM/Dyn approach is adopted. A more accurate, but more computationally expensive, approach is the fully FE representation of the lines (FE).

In a WEC, a PTO system is, in general, defined as the system which converts the power absorbed by the main element interacting with the waves into useable power (electricity or others). Depending on the type of WEC, different models are adopted to represent this effect. In the case of OWC, as in [23], a linear damping approach to simulate the hydrodynamic viscous forces and a dynamic pressure approach to evaluate the pressure inside the chamber are used. When, instead, the relative movement between the WEC system and the main platform is exploited to produce power, the PTO is modelled as a linear damping force, or a linear damping plus linear stiffness force (e.g. [27]).

10.4 Conclusions

The terms 'Blue Economy' and 'Blue Growth' are now widely accepted terms. The World Bank defines these two terms concurrently as the '*sustainable use of ocean resources for economic growth, improved livelihoods and jobs, while preserving the health of ocean ecosystem*' [47,48], while the World Wildlife Fund (WWF) states in its report 'Principles for a sustainable Blue Economy' [49] that the '*Blue Economy means the use of the sea and its resources for sustainable economic development. For others, it simply refers to any economic activity in the maritime sector, whether sustainable or not*'.

While there are still no commercially mature products, and just a few prototypes proposed by the industry, several research initiatives have been funded in this area, aimed at investigating the feasibility and sustainability of a number of multi-use concepts, adopting a multidisciplinary approach, i.e. considering the Political, Economical, Social, Technological, and Environmental (PESTLE) relevant aspects. Among these innovative concepts, an offshore multipurpose platform can be defined as a platform (or a number of closely co-located platforms) that serves (or serve) the purpose of multiple offshore industries (Aquaculture, Biotechnology, Maritime Transport, Tourism, Renewable Energy, and others).

In the present chapter, first of all, the potential synergies arising when closely co-locating these offshore activities, as identified in the literature, are summarised, highlighting the fact that these synergies emerge at different disciplinary levels (PESTLE). Then, an overview of the main multipurpose platform projects and concepts is given: while the majority focused on hybrid offshore renewable systems, others also proposed concepts integrating aquaculture, tourism, and maritime transport systems.

Focusing on the design and analysis of multipurpose platforms, the main differences and challenges when compared to the design and analysis of single-purpose systems are discussed. The enhanced multidisciplinary nature of the process, due to the number of different systems considered and to the legacy of sectors that have been evolved, so far, independently, is still perceived as one of the main challenges in this field. Furthermore, the necessity to perform a combined resource assessment is highlighted, providing as example an overview of the methodologies adopted to perform a combined wind-wave energy resource assessment.

Although this is a relatively new and rapidly evolving research field, an overview of the so-called aero-hydro-servo-elastic engineering approaches, which model the aerodynamic and hydrodynamic forces acting on the platform, as well as their structural dynamics response and the control strategy adopted, is presented. A number of numerical approaches, ranging from quick, robust, low-accuracy techniques to high-accuracy but computationally expensive methodologies, are discussed, also highlighting the emerging research gaps.

To summarise, the 'Blue Growth' can be a further agricultural/industrial revolution, with its potential to provide (in a sustainable manner) food, energy, jobs, economic prosperity to a world population that is both increasing and demanding higher living standards. Multipurpose platforms could very well be one of the optimum way to unleash this potential, but the focus has to be on identifying the synergies and tensions arising when closely co-locating these offshore activities and find a way to maximise the first while eliminating or minimising the last.

References

[1] BP. BP Energy Outlook 2018. BP; 2018.

[2] European Commission. Blue Growth – Opportunities for Marine and Maritime Sustainable Growth (Brussels, 13.9.2012 COM (2012) 494 final). EC; 2012.

[3] Pachauri RK, Allen MR, Barros VR, *et al.* Climate Change 2014: Synthesis Report. Contribution of Working Groups I, II and III to the Fifth Assessment Report of the Intergovernmental Panel on Climate Change. IPCC; 2014.

[4] European Commission. Scenarios and Drivers for Sustainable Growth from the Oceans. Seas and Coasts, First Interim Report, ECORYS for EC, DG MARE, Rotterdam/Brussels; 2011.

[5] European Commission. Blue Growth. Available from: http://ec.europa.eu/maritimeaffairs/policy/blue_growth_en. Accessed on July 2018.

[6] China-Britain Business Council. Made in China 2025; 2015. Available from: http://www.cbbc.org/mic2025/.

[7] State Oceanic Administration People's Republic of China. China 13th Ocean Energy Development Five Year Plan (2016–2020); 2016.

[8] China State Council. Outline of the National Marine Economic Development Plan; 2003.

[9] Pérez-Collazo C, Greaves D, and Iglesias G. A review of combined wave and offshore wind energy. Renewable and Sustainable Energy Reviews. 2015;42:141–153.

[10] FPP. Floating Power Plant; 2018. Available from: http://www.floatingpowerplant.com/.

[11] Britain G. Energy Act 2004. Stationery Office; 2004.

[12] Britain G. Climate Change Act 2008: Chapter 27: Explanatory Notes. Stationery Office; 2008.

[13] Zanuttigh B, Angelelli E, Kortenhaus A, *et al.* A methodology for multi-criteria design of multi-use offshore platforms for marine renewable energy harvesting. Renewable Energy. 2016;85:1271–1289.

[14] MUSES Consortium. Multi-Use in European Seas (MUSES); 2018. Available from: https://muses-project.eu/.

[15] European Commission. Proposal for a Directive of the European Parliament and of the Council Establishing a Framework for Maritime Spatial Planning and Integrated Coastal Management; 2013.

[16] Troell M, Joyce A, Chopin T, *et al.* Ecological engineering in aquaculture-potential for integrated multi-trophic aquaculture (IMTA) in marine offshore systems. Aquaculture. 2009;297(1–4):1–9.

[17] European Commission Directorate-General for Research and Innovation. The Ocean of Tomorrow Projects (2010–2013); 2014.
[18] H2Ocean consortium. H2Ocean; 2015. Available from: http://www.h2ocean-project.eu/.
[19] DTU. The MERMAID Project; 2014. Available from: http://www.vliz.be/projects/mermaidproject/.
[20] TROPOS Consortium. The Tropos Project: Modular Multi-Use Deep Water Offshore Platform Harnessing and Servicing Mediterranean, Subtropical and Tropical Marine and Maritime Resources; 2012. Available from: http://www.troposplatform.eu/.
[21] AWS TRUEPOWER SL. H2OCEAN Report Summary; 2016. Available from: https://cordis.europa.eu/result/rcn/177009_en.html.
[22] Fusion Marine. Fusion Marine; 2018. Available from: https://fusionmarine.com/.
[23] Armesto JA, Sarmiento J, Ayllón V, *et al.* Numerical and experimental study of a multi-use platform. In: ASME 2016 35th International Conference on Ocean, Offshore and Arctic Engineering. American Society of Mechanical Engineers; 2016. pp. V006T09A018–V006T09A018.
[24] TROPOS Consortium. TROPOS Report Summary; 2015. Available from: https://cordis.europa.eu/result/rcn/171930_en.html.
[25] Quevedo E, Carton M, Delory E, *et al.* Multi-use offshore platform configurations in the scope of the FP7 TROPOS Project. In: 2013 MTS/IEEE OCEANS – Bergen. IEEE; 2013. pp. 1–7. Available from: http://ieeexplore.ieee.org/document/6608061/.
[26] MARINA Consortium. MARINA Platform; 2017.
[27] Muliawan MJ, Karimirad M, and Moan T. Dynamic response and power performance of a combined spar-type floating wind turbine and coaxial floating wave energy converter. Renewable Energy. 2013;50:47–57.
[28] Michailides C, Luan C, Gao Z, *et al.* Effect of flap type wave energy converters on the response of a semi-submersible wind turbine in operational conditions. In: ASME 2014 33rd International Conference on Ocean, Offshore and Arctic Engineering. American Society of Mechanical Engineers; 2014.
[29] ORECCA Consortium. Offshore Renewable Energy Conversion Platforms Coordination Action; 2017. Available from: https://cordis.europa.eu/project/rcn/94058_en.html.
[30] Space@Sea Consortium. Space@Sea; 2018. Available from: https://spaceatsea-project.eu/.
[31] Blue Growth Farm Consortium. The Blue Growth Farm; 2018.
[32] Horizon 2020. G. Technology Readiness Levels (TRL); 2015.
[33] Kallesøe BS, Larsen T, Paulsen U, *et al.* Aero-Hydro-Elastic Simulation Platform for Wave Energy Systems and Floating Wind Turbines. Roskilde: Danmarks Tekniske Universitet, Risø Nationallaboratoriet for Bæredygtig Energi (Denmark Forskningscenter Risoe Risoe-R. 2011); 1767.

[34] Yde A, Larsen TJ, Hansen AM, *et al.* Comparison of simulations and offshore measurement data of a combined floating wind and wave energy demonstration platform. Journal of Ocean and Wind Energy. 2015;2(3): 129–137.

[35] Pelagic Power AS. Pelagic Power: W2Power; 2010. Available from: http://www.pelagicpower.no/.

[36] Hanssen JE, Margheritini L, O'Sullivan K, *et al.* Design and performance validation of a hybrid offshore renewable energy platform. In: Ecological Vehicles and Renewable Energies (EVER), 2015 Tenth International Conference on. IEEE; 2015. pp. 1–8.

[37] Kalogeri C, Galanis G, Spyrou C, *et al.* Assessing the European offshore wind and wave energy resource for combined exploitation. Renewable Energy. 2017;101:244–264.

[38] Fusco F, Nolan G, and Ringwood JV. Variability reduction through optimal combination of wind/wave resources An Irish case study. Energy. 2010; 35(1):314–325.

[39] Li L, Gao Z, and Moan T. Joint environmental data at five European offshore sites for design of combined wind and wave energy concepts. In: 32nd International Conference on Ocean, Offshore, and Arctic Engineering. OMAE2013-10156; 2013.

[40] Cradden L, Kalogeri C, Barrios IM, *et al.* Multi-criteria site selection for offshore renewable energy platforms. Renewable Energy. 2016;87:791–806.

[41] Collu M, Shires A, and Borg M. H2Ocean – D4.1 Report Assessment on Wind Turbine Types. Cranfield University; 2014.

[42] Cevasco D, Collu M, Rizzo C, *et al.* On mooring line tension and fatigue prediction for offshore vertical axis wind turbines: A comparison of lumped mass and quasi-static approaches. Wind Engineering. 2018; 42(2):97–107. Available from: http://journals.sagepub.com/doi/10.1177/0309524X18756962.

[43] Martini M, Guanche R, Armesto JA, *et al.* Met-ocean conditions influence on floating offshore wind farms power production. Wind Energy. 2016;19(3): 399–420. Available from: http://doi.wiley.com/10.1002/we.1840.

[44] Iturrioz A, Guanche R, Armesto JA, *et al.* Time-domain modeling of a fixed detached oscillating water column towards a floating multi-chamber device. Ocean Engineering. 2014;76:65–74. Available from: http://dx.doi.org/10.1016/j.oceaneng.2013.11.023.

[45] Luan C, Michailides C, Gao Z, *et al.* Modeling and analysis of a 5 MW semi-submersible wind turbine combined with three flap-type wave energy converters. In: Volume 9B: Ocean Renewable Energy. ASME; 2014. Available from: http://proceedings.asmedigitalcollection.asme.org/proceeding.aspx?doi=10.1115/OMAE2014-24215.

[46] Michailides C, Gao Z, and Moan T. Experimental and numerical study of the response of the offshore combined wind/wave energy concept SFC in extreme environmental conditions. Marine Structures. 2016;50:35–54.

Available from: https://linkinghub.elsevier.com/retrieve/pii/S0951833916301149.

[47] World Bank. What Is the Blue Economy?; 2017. Available from: http://www.worldbank.org/en/news/infographic/2017/06/06/blue-economy.

[48] World Bank and United Nations Department of Economic and Social Affairs. The Potential of the Blue Economy: Increasing Long-term Benefits of the Sustainable Use of Marine Resources for Small Island Developing States and Coastal Least Developed Countries. Washington, DC: World Bank; 2017. Available from: https://openknowledge.worldbank.org/handle/10986/26843.

[49] World Wildlife Fund. Principles for a Sustainable Blue Economy. WWF. 2015. Available from: http://wwf.panda.org/?247477/Principles-for-a-Sustainable-Blue-Economy.

Chapter 11

Installation, operation and maintenance of offshore renewables

Vincenzo Nava[a,b], Pablo Ruiz-Minguela[a], Germán Pérez-Morán[a], Raúl Rodríguez-Arias[a,c], Joseba Lopez-Mendia[a], and José-Luis Villate-Martínez[a]

The chapter deals with the basic concepts of installation, operation and maintenance of offshore renewable energy systems. Whilst focus is given to the offshore wind industry, the extension to ocean energy (wave and tidal) offers a wider perspective on the major issues concerning the installation and maintenance. A reliability-based approach has been adopted for the analysis of the failures, providing an overview about the most common functional decomposition methodologies as well as logistic requirements for the different operations at the various stages of the lifetime of an offshore renewable project. The economic modelling of the operations, based on strategies for their planning, briefly completes the chapter.

11.1 Introduction

11.1.1 Impact of installation, operation and maintenance activities in offshore renewable systems

Offshore activities during installation, operation and maintenance have a relevant and measurable contribution in the final levelised cost of the energy produced by an offshore renewable project. The importance of estimating the impacts of operations on the overall economy of a project is further increased by the growth of the sector of offshore renewables during the last years. Offshore wind, e.g., has reached 18.5 GW of cumulative power and more than 4,500 turbines connected to the grid across 11 countries in Europe by the end of 2018 [1] while other offshore renewable technologies such as tidal stream of wave power, despite being still in their infancy, can count with a number of projects in the range of megawatts. Offshore wind

[a]Tecnalia Research and Innovation, Derio, Spain
[b]Basque Centre for Applied Mathematics (BCAM), Bilbao, Spain
[c]NAUTILUS Floating Solutions, Derio, Spain

deployment is mainly concentrated in areas with a continental shelf, such as the North Sea in Europe, where shallow waters can extend hundreds of kilometres far offshore. However, new solutions for deeper waters are emerging (mainly based on floating platforms) allowing offshore wind deployment in many places around the world.

Installation costs will impact the capital expenditures (CAPEX) while operation and maintenance activities are part of the operational expenditures (OPEX), the latter representing between a quarter and a third of the total lifetime costs of offshore renewable projects. These costs might highly depend on the specific conditions of the location (e.g. depth, distance to shore and type of seabed).

For an offshore wind farm, the main cost elements to be considered for the calculation of installation, operation and maintenance costs are usually: turbines, foundations, array cables, offshore substation, export cable and onshore facilities. In some cases where the connection point to the grid is offshore, the cost of export cables and onshore facilities are not included in the CAPEX but in the operational expenditures as a transmission fee.

Figure 11.1 shows a typical cost distribution of an offshore wind farm. The shares have been estimated from relevant literature and based on the figures

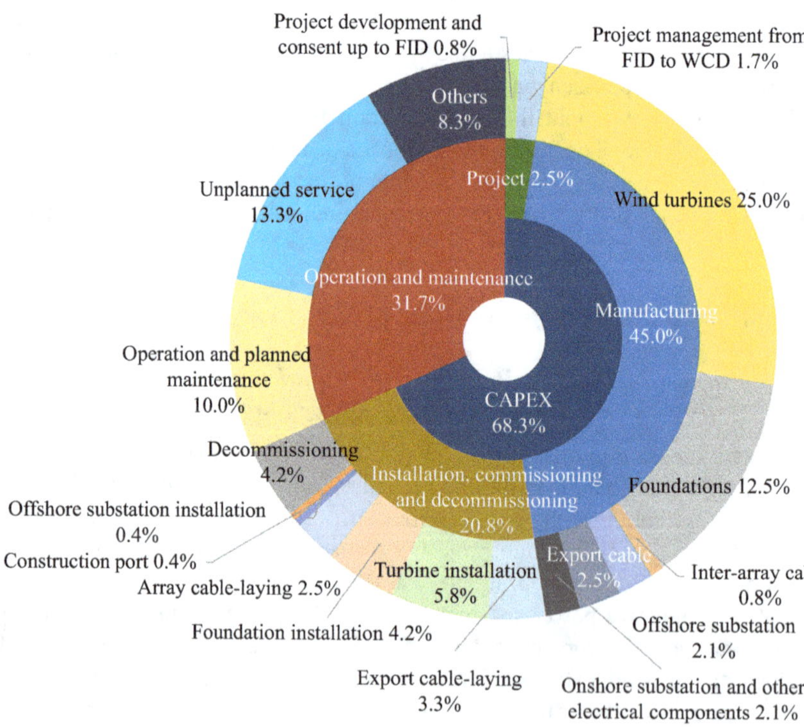

Figure 11.1 Breakdown of costs in the life cycle of an offshore wind farm (FID: final investment decision; WCD: work completion date)

provided in [2]. It can be noticed how the total cost of installation, commissioning and decommissioning activities and operation and maintenance is estimated to amount to more than 50% of the total cost of the farm.

Novel approaches for installation, operation and maintenance tasks, based on digital technologies such as digital twins, big data or deep learning can help in reducing costs. For example, digital tools can be used for vessel management to save fuel, integrated data management to take daily decisions, data analytics for medium-term planning and digital twins for maintenance optimisation.

11.1.2 Functional decomposition of offshore renewable systems

In this section, the concept and main applications of the functional decomposition of complex systems, like offshore renewable systems, are presented. The problem is tackled using a general systems-engineering-based approach focused on the taxonomy of the systems for the analysis of failures, and therefore particularised for some technologies (i.e. fixed wind, floating wind, wave and tidal energy).

A system is a 'set of interrelated components working together as an integrated whole to achieve some common objective' [3]. Systems engineering is an inter-discipline, dealing with all the fields pertaining to complex systems throughout their life cycle; functional decomposition is a systems engineering methodology consisting in de-aggregating a complex system into elementary components according to their function. Thus, a functional decomposition provides the following benefits:

- It is easier to understand the system, its interactions and interfaces and is a useful tool in providing requirements insights for both analysis and design.
- It provides answers to the user's stated and unstated needs by providing direct traceability to capabilities and to requirements.
- It is a powerful tool to prepare the functional specification of the complete system, i.e. the performance that a technology must achieve under a set of environment and operational conditions.

Functional decomposition provides practical applications throughout the entire system life cycle in order to organise and correlate elements for cost, weight, specifications, system function and effectiveness, design, production and maintenance studies. The system taxonomy is widely applied both for qualification of novel technologies and failure data analysis.

Two main techniques or methods are available for developing the functional decomposition diagram of a complex system, as shown in Figure 11.2 (for a full review see [4]).

- SOP (Subtract and Operate Process) method: This is a bottom-up technique applicable when the functions of the main subsystems are known and hence the decomposition into subfunctions can be easily carried out.

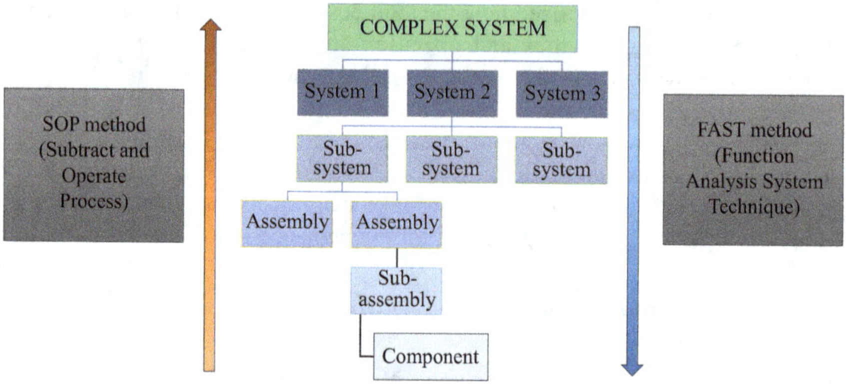

Figure 11.2 Functional decomposition diagram depicting five levels

- FAST (Function Analysis System Technique) method: This is a top-down approach in which the starting points are the smallest or simplest functions and the tree structure is developed by combining them.

A practical implementation of the functional decomposition in the development, deployment, O&M, decommissioning and disposal phases of an offshore renewable project is the work breakdown structure (WBS). These complex projects are made manageable by breaking them down into individual components following a hierarchical structure. WBS defines tasks that can be completed independently of other tasks, facilitating resource allocation, assignment of responsibilities, and measurement and control of the project. A WBS is a product-oriented family tree composed of hardware, software, services, data and facilities. It displays and defines the product, or products, to be developed and/or produced. It relates the elements of work to be accomplished to each other and to the final product. The WBS was originally issued by the US Navy in March 1973 which has been recently updated [5]. It is a structured system (three-digit numbers providing five levels of breakdown) which was intended for use in specification preparation, cost estimating, cost progressing, management, weight control, drawing numbering, shipyard job order coding, and similar purposes. An advanced approach is the extended ship work breakdown structure (ESWBS), providing two additional levels of breakdown of functional systems.

The criteria proposed to define the functional decomposition of an offshore renewable system are as follows:

- Identify the main function of each system in the complete system which is an offshore wind turbine (OWT), a floating offshore wind turbine (FOWT) or a tidal or wave energy converter (TEC or WEC).

- Functions are broken down to the appropriate level, often to subfunctions and functional elements, depending on the range of functions addressed by the entity.
- The functional decomposition has to be the most basic form of the hierarchical decomposition of the functions with their associated performance requirements.
- It is highly recommended to use the nomenclature proposed by relevant international standards for the different offshore renewable systems technologies, namely:
  - IEC 61400-3:2009 (wind turbines – part 3: Design requirements for offshore wind turbines) [6]. This standard considers the following main systems: rotor-nacelle assembly, tower, substructure and foundation.
  - IEC TS 61400-3-2 (Wind energy generation systems – Part 3-2: Design requirements for floating offshore wind turbines) [7]. In the case of FOWT, the systems defined include rotor-nacelle assembly, tower, floating substructure and station-keeping system.
  - IEC TS 62600-2:2016 Ed.1.0 (Marine energy – Wave, tidal and other water current converters – Part 2: Design requirements for marine energy systems) [8]. It defines the primary structure as the collective system comprising the structural elements, foundation, mooring and anchors, piles, and device buoyancy designed to resist global loads.

Recently, some examples of the functional decomposition devoted to the risk analysis of four different concepts of FOWTs and of an oscillating water column (OWC) device have been produced in the framework of two European H2020 funded projects, namely Lifes50+ (qualification of innovative floating substructures for 10 MW wind turbines and water depths greater than 50 m [9], deliverable D6.2 [10]) and OPERA (open sea operating experience to reduce wave energy cost [11]).

Finally, it should be noted that an appropriate functional decomposition of the offshore renewable system is essential for carrying out properly a detailed failure analysis (see also Section 11.1.3).

11.1.3 Concepts of reliability and failure analysis

Functional decomposition is the basis for the assessment of the 'reliability' of complex systems, such as offshore renewable energy systems. Indeed, reliability-based approaches applied to the design of these technologies (either as isolated components or as aggregated in subsystems, devices and arrays) may help in reducing the cost of energy (reducing the OPEX costs and increasing the availability) through the reduction of the frequency of failures, or the correction of the causes of failures, or identifying novel approaches for the correction of failures or identifying new designs. The term reliability commonly refers to the capability of a system to fulfill the functional requirements for which the system itself has been designed under given operational conditions for a given time interval.

Using a probabilistic approach (for a full review, see [12] and [13]), the reliability of a system at time t $R(t)$ is the probability $P[T > t]$ that the system is working with no failures in a time interval larger than t:

$$R(t) = P[T > t]. \tag{11.1}$$

The mean time to failure (MTTF) represents the expected time of failure for a component or system and it is defined as

$$MTTF = E(T) = \int_0^\infty R(t)dt. \tag{11.2}$$

The hazard function $\lambda(t)$, often referred to as the failure rate, for a component/system working at time t is the instantaneous probability of failure per unit time. Reliability can be calculated as

$$R(t) = exp\left[-\int_0^t \lambda(t')dt'\right] \tag{11.3}$$

The modelling of the hazard function, therefore, influences the reliability of the system. The expected typical behaviour of the failure rate, indeed, is the one represented in the blue curve in Figure 11.3, generally named as bathtub curve. Failure rates are very high, but decreasing, at the beginning and they are usually referred to as 'early infant mortality' failures, generally due to defects in the manufacturing or installation. Then, there might be a period of time in which the failure rates could be considered to be constant (operational life of the component/system). Finally, due to wearing out and ageing, the failure rate increases again. One component/system can exhibit all the three pieces of the curve, or just two or only one, depending on the modelling capabilities, the knowledge about the

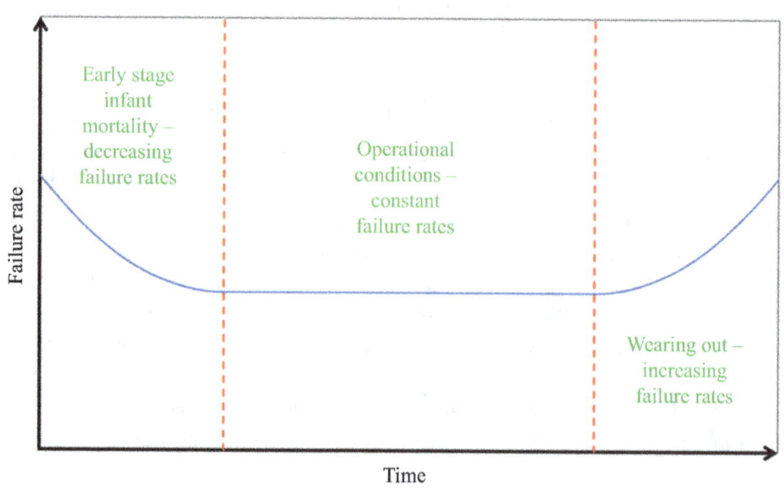

Figure 11.3 Failure rate versus time (bathtub curve)

component/system and the characteristics of the component/system itself. For example, following (11.3), in case of failure rate constant with time, the hazard function follows an exponential distribution.

For a repairable component/system after failure, the maintenance time is the time needed to carry out all the activities needed to restore it to the previous functional state. The maintainability cumulative distribution function $M(t)$ is, therefore, the probability to perform those activities in a time interval shorter than t.

$$M(t) = P[T_R \leq t] \tag{11.4}$$

The related density function is $m(t)$, and mean time to repair (MTTR) is defined as the expected time needed to carry out the repair itself:

$$MTTR = \int_0^\infty tm(t)dt \tag{11.5}$$

It could be defined the mean time between failures (MTBF) as the expected time between two failures for a repairable system. For a repairable system, indeed

$$MTBF = MTTF + MTTR \tag{11.6}$$

MTBF can be approximated to *MTTF* only for highly reliable systems (high *MTTF*) or for components/systems easy to be repaired (low *MTTR*).

The MTTR is related to the downtime of the system, intended as the time in which the plant does not produce energy; the complementary concept is the availability, i.e. the fraction of time in which the system is working properly.

The analysis of the reliability of complex systems, made of several subsystems, would depend on the interconnections among components. For example, if components are connected in series, then the reliability of all the system will rely on the reliability of the less reliable component, as the system will fail when just one of the components fails. Similarly, if components are connected in parallel and some redundancy is built in the system, then the system will fail when all the components fail. In order to describe, therefore, the failure of a complex system, a plethora of approaches have been developed, most of them based on a combination of the failure analysis and the functional decomposition. The fault tree analysis (FTA), for example, is a top-down process: the undesired event (a failure) is decomposed – following the path indicated by a functional decomposition analysis – into the possible causes that may have determined the failure. Therefore, FTA can be considered both quantitative and qualitative, providing not only information on what may have caused the event, but also estimating its probability. Among the bottom-up procedures, the failure modes and effects analysis (FMEA) can provide a cause–effect estimation of the faults in a system. The starting point of the procedure is the analysis of the failure modes of given components, and then the effects on the systems can be extrapolated. Such a procedure is particularly efficient in the case of systems with a reduced level of redundancy. An extension of the FMEA is the failure mode, effects and criticality analysis (FMECA), including an evaluation of the risks (impact versus probability of occurrence) that the failure may cause. Other extensions of these procedures, such as the failure reporting and corrective action system (FRACAS), hazard identification study (HAZID) and the hazard and operability analysis (HAZOP) will deal, respectively, with the corrective actions, the identification of the hazards and the corrective procedures in case of occurrence. A more detailed description of these methods is beyond the scope of this chapter. For a full review, see, e.g., [14].

11.2 Life cycle activities for offshore energy systems

Offshore logistics can be defined as the management of the flow of resources required to perform the various activities during the life cycle of an offshore renewable energy project. Main logistics phases are installation, O&M and decommissioning. Key resources are ports, vessels and associated equipment, as well as personnel.

Offshore logistical costs can be quite significant in comparison with traditional onshore logistics. They usually involve a broad variety of vessels in complex offshore operations such as lifting, towing, positioning and manipulating heavy structures [15].

The next sections describe the logistical operations and requirements necessary to execute the three main phases of an offshore renewable energy project.

11.2.1 Installation phase

This section describes the installation requirements for the most important offshore energy systems, namely the electrical infrastructure, the moorings and foundations and the offshore renewable energy devices (fixed and floating).

The offshore electrical infrastructure comprises subsea power cables (static and inter-array), electrical connectors, the offshore substation and dynamic cables in the case of floating devices.

Station keeping of fixed devices is done through a certain type of foundation, whereas floating devices will require a mooring system comprising anchors, mooring lines and a variety of auxiliary components.

This chapter addresses offshore renewable energy devices such as wind turbines or ocean energy converters, for instance wave and tidal devices. All of them can be fixed or floating, although fixed wind farms are predominant since the other technologies are still in a validation phase.

Figure 11.4 presents the typical installation order for offshore renewable energy systems. Logistic activities that come before or after offshore installation are highlighted in dark blue.

The static cable (also known as the export cable) is installed after completing the onshore infrastructure (i.e. onshore substation and cables). The installation of foundations can start before that of subsea power cables. The top-side module of the surface piercing substation (or the subsea mounted substation) is installed in between the static cable and inter-array cables. Electrical connectors are then

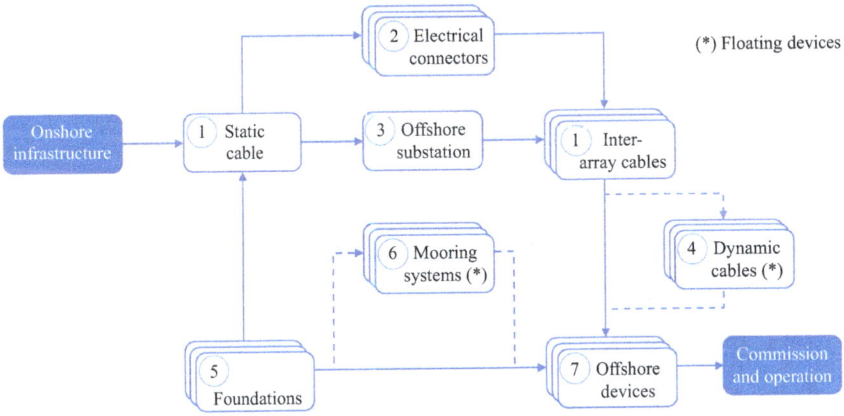

Figure 11.4 Typical order of installation activities

Figure 11.5 Generic sequence of key installation activities

linked to the cable ends, pulled up and connected to the offshore substation. In the case of floating devices, it is also required to install dynamic cables that run from the sea bottom to a position near the water surface. Monopiles, jackets and gravity foundations can be used for station keeping of fixed devices. In the case of floating devices, normally anchors are used in the station-keeping system. Offshore renewable energy devices are then mounted on top of the foundations or hooked up to the mooring system (for floating devices). Installation activities end after the inter-array cables or dynamic cables are finally connected. Multiple installations can be completed simultaneously provided vessels are available.

The generic sequence of the logistical activities during installation is shown in Figure 11.5. It is worth noting that surveying pre- and post-installation is not required for all offshore renewable energy systems.

Surveying operations usually occur before and after the installation of systems interfacing with the seabed such as subsea power cables, anchors or the offshore substation. A survey of the cable routes requires a thorough study of the seafloor morphology and geology.

Loading and preparing the vessel (and transport to site) is an important step that occurs before any specific offshore installation operation. Actually, vessel selection greatly conditions the design of the operation, its duration and underlying risks.

The *specific offshore installation operations* are to be followed by *visual inspection and testing* in accordance with the relevant standards.

11.2.1.1 Subsea power cable installation

The installation of submarine cables is one of the most critical logistical activities over the course of the grid connection of an offshore renewable energy project. Indeed, the cost of installation of submarine cables represents one to three times the price of the cable itself [16].

Subsea power cables comprise export and inter-array cables. Cable installation usually involves (i) trenching of the seafloor, (ii) cable laying and tensioning and (iii) cable protection. Static cables also require landing at the beach manhole, from where the cable is extended on land to connect to the onshore infrastructure. It is highly advisable to avoid vessel anchorages during installation to minimise risks of damage to the cables [17].

11.2.1.2 Cable-to-cable connection

Electrical connection of cables is one of the most challenging activities carried out during the construction of a farm. Actually, this operation is closely related with cable laying and protection tasks and is often performed alongside. There are two main methods depending on the type of electrical connector used, often referred to as 'dry-mate' and 'wet-mate' connectors.

Dry-mate connections are performed out of the water whereas wet-mate connection takes place subsea. Dry-mate connection involves (i) cable protection removal, (ii) connector lifting to the surface, (iii) connection and (iv) re-deployment. Wet-mate simply requires the underwater connection assisted by divers or a remote operated vehicle (ROV).

11.2.1.3 Offshore substation installation

Offshore substations are used to step-up the transmission voltage and reduce electrical losses. Most designs consist of a foundation support structure and topside module which houses the electrical equipment (surface piercing substation). Monopiles and jackets are the two most frequent foundation types for offshore substation platforms. Installation operations are similar to other type of foundations and are therefore described in Section 11.2.1.5.

Due to the size of the structure, the topside module will normally be transported to the offshore site by a barge and moved onto it by self-propelled modular transporter (SPMT) using ramps. The installation of the topside module of a surface piercing substation involves lifting onto the transition piece of the monopile foundation or the top of the jacket foundation before the matting operation.

In the case of subsea mounted substations installation comprises (i) seabed preparation, (ii) subsea unit installation and (iii) subsea unit protection. Seabed preparation consists of the removal of soft, mobile sediments and/or the levelling of an area by addition of gravel or rock dumping [18].

11.2.1.4 Cable-to-device connection

Power cables connect offshore energy devices to the inter-array electrical components in the seabed. For fixed wind turbines, this connection to the static electrical components is normally made from subsea to surface, whereas for floating devices the process can be reverted (surface-to-subsea). When pile foundations are used, a J-tube is attached to the outside to carry the electrical cable. The end of the inter-array cable is fed up through the J-tube via a winch. The process of feeding the cable usually requires divers and/or an ROV.

Subsea-to-surface connection involves (i) deployment of guide wire and pulling line, (ii) cable lay, (iii) cable pull and (iv) cable termination.

Surface-to-subsea is an alternative method to the described above. In this case, the cable end is wet-mate connected instead of pulled. No termination is needed.

11.2.1.5 Foundations

Piles require drilling equipment and seafloor penetration, whereas gravity base foundations are directly lowered to seafloor. In the case of gravity base foundations, ballast tanks can be used to transport the floating structure and be flooded with water when into position. Jacket platforms may require the use of piles and/or pin piles, i.e. first the pin piles are installed using templates and then the jacket will be placed on the pre-installed pin piles.

Depending on the type of foundation, installation activities may involve (i) seafloor preparation, (ii) lowering substructure to the seafloor, (iii) seafloor penetration and (iv) positioning of foundation over substructure.

Foundations of offshore substations are installed in a similar way to the activities described above.

11.2.1.6 Mooring systems

Mooring systems comprising anchors and lines are required for floating energy devices. In more complex configurations buoys and clump weights may also be needed. Although floating devices are still in the validation phase, experience of mooring installation in the oil and gas (O&G) industry can serve as a useful reference.

Usually the mooring line is pre-attached to the anchor as post-installation using an ROV will involve extra costs. Mooring installation involves (i) installation of anchors and mooring lines, (ii) pre-tensioning of lines, (iii) securing the line to a surface pennant buoy and (iv) device hook-up.

11.2.1.7 Offshore renewable energy devices

Given the large range of offshore renewable energy devices, a close cooperation is needed among the marine contractor, site developer and the technology developer to define the particular installation strategy. This section describes the common logistical activities shared by all devices.

Devices will be assembled at port and then loaded out. The most common load out operations, largely used in oil and gas sector, are the trailer, skidded, lifted and float-away load outs [19]. The devices are then towed out or loaded onto vessels to transport them to the site, where they are positioned, lowered and connected to already installed foundations or hooked up to moorings. Then the devices are connected to the power cable and commissioned. In some cases, some device assembly activities can be performed in site prior to electrical grid connection and commissioning. At the end of the commissioning stage, the devices are officially operational and can feed power to the onshore electrical network.

11.2.2 Operation and maintenance phase

This section describes the installation requirements for the O&M phase of offshore renewable energy systems. Maintenance can be defined as the combination of all technical and administrative actions intended to retain an item in, or restore it to, a state in which it can perform its required function [20]. O&M activities comprise three main types of interventions, namely inspection and monitoring to determine health status of the various systems, on-site replacement of components and onshore repairs of major elements.

11.2.2.1 Inspection and monitoring

Farm condition monitoring can be performed either remotely using ad-hoc installed instrumentation or in specific monitoring campaigns (on a temporary basis) to assess a particular area or system. Additionally, a small team of technicians/divers can travel to the site to inspect farm systems and determine their actual state of health. Minor repair actions can be done at the site directly. The generic sequence of inspection and monitoring activities is outlined in Figure 11.6.

There is a large range of floating buoys and bottom-fixed equipment for continuous resource monitoring (e.g. met-ocean buoys, light detection and ranging (LiDAR), masts, acoustic Doppler current profiler). Other instrumentation, such as condition monitoring and fault detection units, can be installed in offshore renewable energy devices. Underwater inspection can be assisted by remote operated vehicles – ROVs equipped with video recording, sensors and manipulators. A supervisory control and data acquisition (SCADA) system is commonly used for data acquisition and processing. This system provides real-time status information, trends, alarms and remote control around-the-clock to facilitate a safe working environment, comprehensive assessment and safe operation of the farm. A request for O&M intervention can be initiated as a result of the analysis of monitored data or direct in-situ inspection.

11.2.2.2 On-site interventions

On-site interventions comprise repair and replacement of components within the electrical infrastructure, mooring systems and devices. The generic sequence of the in-site intervention activities is outlined in Figure 11.7.

Careful evaluation of interactions between the various actors, mobilisation requirements and constraints are the key to planning a successful O&M mission. The maintenance crew is transferred to the site offshore to perform the maintenance works. Accessibility strongly depends on the type of system. Floating devices can often be accessed directly from the water surface; other systems might need underwater operations (diving, ROV). After the crew has returned to port, analysis of the intervention will allow for continuous improvement of methods.

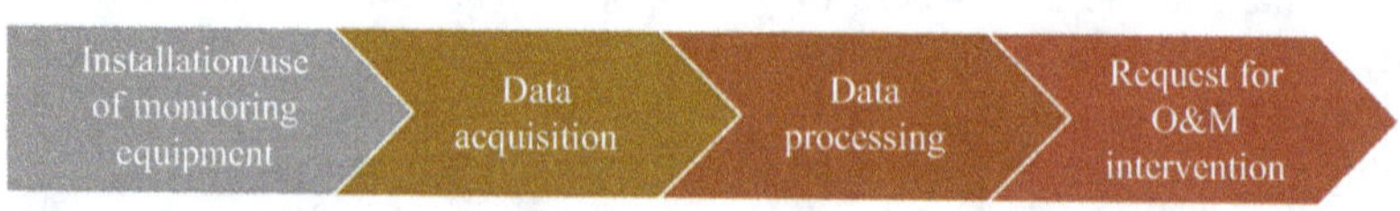

Figure 11.6 Generic sequence of key inspection and monitoring activities

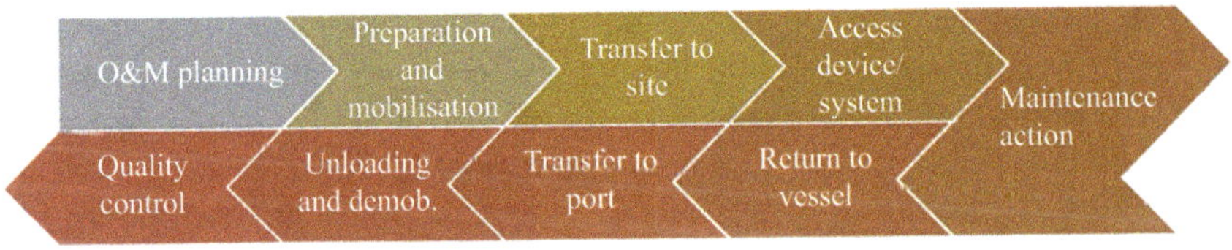

Figure 11.7 Generic sequence of key on-site replacement activities

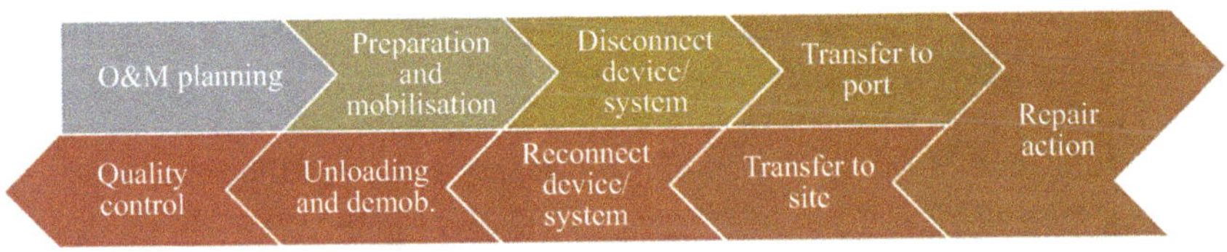

Figure 11.8 Generic sequence of key onshore repair activities

11.2.2.3 Onshore interventions

Offshore renewable energy devices can be designed to be retrieved during operational lifetime for replacement, refitting, upgrading or major repairs which cannot be performed on-site. In case of larger systems, the spare lead time might take a significant share of the overall maintenance intervention. Careful planning is required to minimise costs, risks as well as farm downtime.

The generic sequence of onshore repair activities is shown in Figure 11.8.

These logistics combine activities of both the installation phase and on-site interventions. They can be split into three main stages: (i) disconnect system and transport to port, (ii) onshore repair action and (iii) transport to site and reconnect system. Disconnecting a system usually involves activities similar to installation but reversed. For instance, retrieval of offshore renewable energy devices involves (i) disconnect from power cable, (ii) disconnect from moorings and foundations and (iii) lift device to vessel or de-ballast it for towing.

11.2.3 Decommissioning phase

Upon the operational lifetime of an offshore farm comes to an end (usually after 20–25 years), companies must remove all structures and return the site to the status before the start of the project [21]. Despite this requirement, there is little experience with offshore farm decommissioning. One of the few examples is the Vindeby Offshore Wind Farm [22].

Decommissioning of offshore systems is highly tied with their design. Generally, it can be seen as reversing the installation method. The typical decommissioning order for offshore renewable energy systems is as follows: (1) device removal, (2) umbilical and dynamic lines retrieval, (3) mooring and foundations removal, (4) offshore substation removal, (5) export and inter-array cables removal, (6) site clearance and verification and (7) material disposal. However, the offshore substation, export cable and scour protection may stay after farm decommissioning provided they can be reused in a new project.

Decommissioning may involve specific techniques such as cutting, unfastening, unsoldering, dredging, pumping and lifting operations. Farm components may be reused, scrapped, placed offshore as an artificial reef or disposed in designated landfill. Steel components can be recycled in many situations. However, concrete structures and composites generally require alternative solutions for disposal.

11.2.4 Vessels and equipment

All vessels employed in the offshore sector can be classified as either construction or support vessels. The first category is used in the installation and decommissioning phases of offshore renewable energy farms. The second category belongs to the O&M activities.

The total weight of the system to be installed is one of the most important factors to be considered when selecting an appropriate vessel to perform offshore operations (i.e. payload and storage capabilities). Other important factors for vessel selection are the deck space (to accommodate the equipment/devices), positioning capabilities and the bollard pull.

Table 11.1 presents vessel, equipment and additional considerations for the most common operations described in the previous sections. A more comprehensive

Table 11.1 Vessels and equipment for offshore operations

Operation Figure 11.4.4	**System**	**Vessels and equipment**	**Other considerations**
Surveying	1	Survey vessel	Lifting equipment, eco-sounders, ROV, samplers
Trenching	1	Plough, water jet	
Laying	1	Cable laying vessel (CLV), barge, multi-purpose vessel	Dynamic positioning, carrousel or turntable, tensioners, winches
Protecting	1	Rocky dumping vessels	Concrete mattresses
Landing	1	Trencher, drilling rig	
Connection	2, 4	CLV, barge, tugboat	Deck covered area ([23] dry-mate), lifting equipment, ROV and divers
Positioning	5, 7	Offshore construction vessel, crane or jack-up barge	Dynamic positioning, drilling rig, divers
Mooring	6	Anchor handling vessel	Divers
Load-out	7	Multi-wheel hydraulic trailers, SPMTs, steel rails, cranes	
Hook-up	7	Tug, anchor handling vessel	Divers, ROV
Monitoring	Any	Survey vessel, monitoring equipment	Divers, ROV
Maintenance	Any	Crew transfer vessel, Heavy Lift Vessel	Divers, ROV, spare parts
Repair	Any	Crew transfer vessel, heavy lift vessel, multi-purpose vessel, barge, tugboat	

list and examples of resources employed in offshore operations can be found in [23] and [24].

11.3 Planning the operations

This section is focused on the operations planning. This is a critical activity for the installation and O&M cost reduction of renewable energy, and especially for offshore renewables which are conditioned by the environment. Different operations strategies as well as the concepts of access time and weather window are introduced in this section.

11.3.1 Strategies for planning the operations

This section will introduce the main strategies for the operations planning, focusing on the O&M phase, but the application to the installation phase of offshore renewables, e.g., is straightforward.

The main cost drivers in the installation and O&M of offshore renewables are related to the offshore operations. The following aspects are closely related and must be carefully defined for the deployment area:

- Weather restrictions. Operations are conditioned by the oceanographic and meteorological conditions of the deployment site. A detailed study of the area is key, not only to characterise the available resource in terms of power generation, but also to plan the offshore operations and define the type of access system in the different stages of the project.
- Duration of the operation. The operation timing must be defined for the analysis and selection of the weather window to carry out the operation. It is critical for unplanned corrective operations.
- Access time. Conditioned by the base port in the vicinity of the farm, weather conditions and type of means – logistics – for the access to the asset. The aim of the operation planning is to minimise the access time, in order to reduce the downtime.
- Logistics and means to perform the operation. As described in previous sections, there are different types of vessels and access systems suitable for different operations and different weather conditions. This is one of the major cost drivers for the operations and it is also conditioned by the available means in the deployment area – vessels, access and accommodation means, skilled personnel, etc.

The O&M stage of an offshore renewable farm is conditioned not only by the above-mentioned factors, but also by the type of device. Required means and installation plans are completely different for a floating wind farm or for a tidal farm. Nevertheless, for the installation stage, the operations planning, the access systems and the installation strategies are analogous. Figure 11.9 presents an overview of the key offshore wind O&M activities and logistics involved [25].

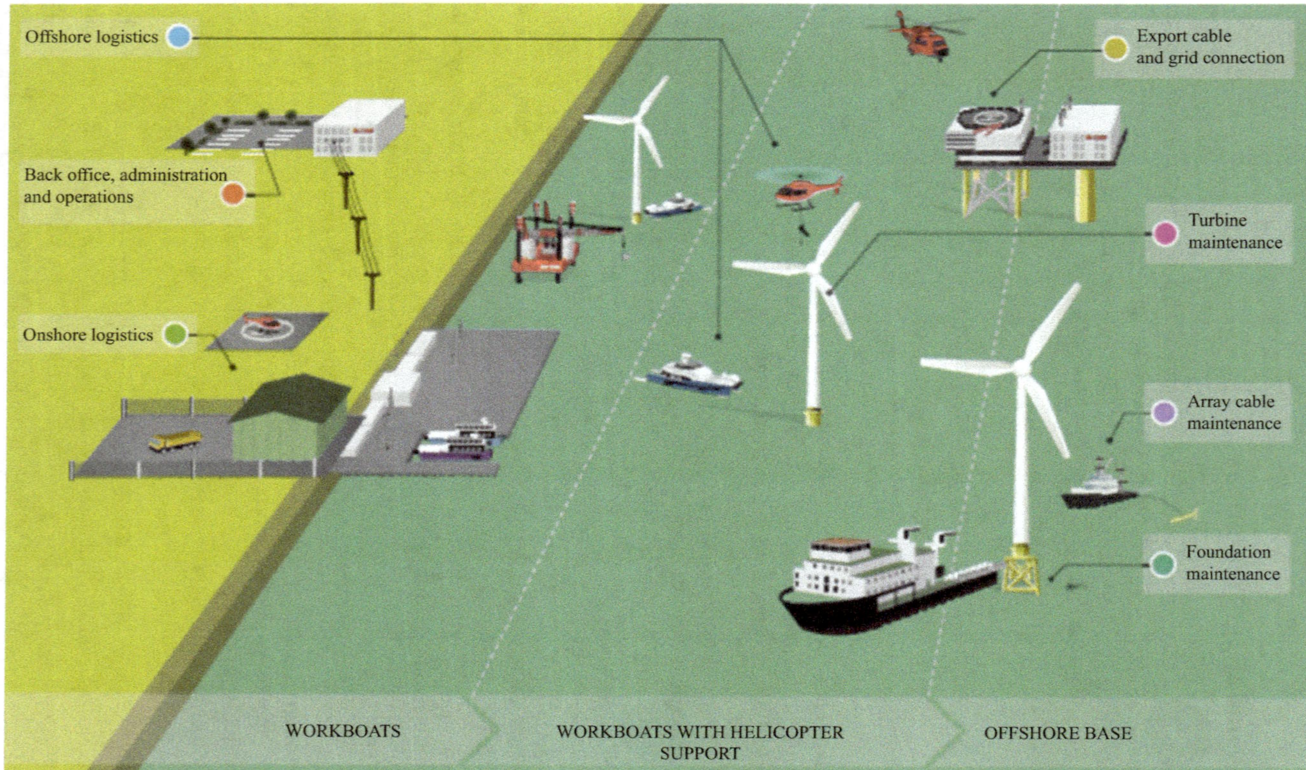

Figure 11.9   Overview of the key offshore wind O&M activity (image courtesy of [2])

Offshore renewable O&M strategies are similar to onshore ones, but are conditioned by the particularities of the offshore environment which also impose specific logistic means. Figure 11.10 summarises the typical O&M strategies [26].

Preventive maintenance, defined as proactive replacements or repairs based on routine inspections or information from condition monitoring systems, is carried out at predetermined intervals or corresponding to prescribed criteria, and intended to reduce the probability of failure or the performance degradation of an item.

- Calendar-based maintenance is carried out in accordance with established intervals of time or number of units of use but without previous condition investigation. It is based on the failure rate of components.
- Condition-based maintenance is preventive maintenance based on performance and/or parameter monitoring. Sensing techniques and data analytics for the interpretation of the measurements are in the core of condition-based maintenance, which is very important in offshore to prevent failures and reduce the unscheduled operations.
- Weather-based maintenance. Many preventive maintenance activities cannot be performed when scheduled and need to be done if the weather and calendar permit.

Corrective maintenance is the reactive replacement or repairs of failed or damaged components and it is carried out after a failure has occurred. It is intended to restore an item to a state in which it can perform its required function.

- Scheduled maintenance based on the observed degradation of a system or component by means of condition monitoring or inspection.
- Unscheduled maintenance necessary after an unexpected failure of a system or component. This is the most common corrective maintenance and has important consequences on cost.

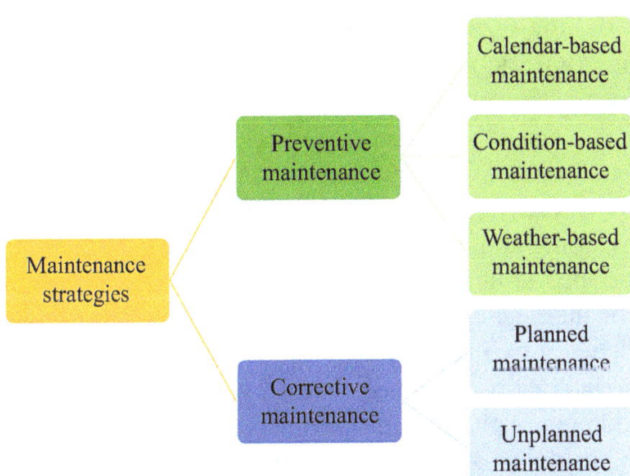

Figure 11.10 Maintenance strategies

- Direct cost which is the component replacement cost including spare parts, personnel and access system.
- Indirect cost caused by the power generation downtime and the subsequent loss of revenue. This cost can be very high as the component affects the whole farm, i.e. export cable, or if the weather conditions lead to a long downtime, i.e. failure in winter with bad wave conditions.

The selection of the strategy and the operations planning, both for the installation and O&M stages, are critical tasks in the deployment of an offshore renewable farm. These activities and the means required to carry them out – logistics, personnel, etc. – are usually planned well in advance to the manufacturing and installation stages. As a reference, O&M strategy and means are defined some 5 years in advance of the farm installation, in order to define the access systems, weather restrictions or onshore logistics for the spare parts.

Several interrelated aspects, like weather windows, access time and logistics, must be carefully defined and taken into account in order to take decisions on the strategy to adopt.

11.3.2 Weather windows

The choice of the strategy for planning the operations is based on economic and logistic criteria, which – at any case – cannot overlook the severity of the environment where the operation should be carried out and the duration of the operations itself. It becomes straightforward, therefore, the definition of weather window as the time interval, long enough to guarantee the fulfilment of given operations or a full phase of it, during which the severity of the weather conditions will be below a set of thresholds, namely the operation limit conditions (OLCs). The operation (or phase of it) is therefore characterised by its duration and the OLCs: in the ocean environment, they usually consist in limiting conditions in terms of significant wave height, as well as peak period, current speed and wind speed. In general, the duration of the operation should include not only the time needed to carry out the operation itself, but also the transportation time to/from the site. In terms of OLCs, they are depending on the specific operation to complete, the presence of a crew and the vessels equipment in order to guarantee safe conditions.

The duration of the weather window, longer than the duration of the operation, can be defined sometimes as persistence. In case of planning several operations, the concepts of overlapping and not overlapping weather windows become important; indeed, the persistence of not overlapping weather windows is an integer multiple of the duration of the operation. The (average) delay time or waiting time is the (average) time between two subsequent weather windows.

Given a time interval of reference τ (e.g. a year, or a month, or a season), then accessibility is defined as the cumulative persistence of all the weather windows occurring in τ expressed as a percentage with respect to τ.

Let us consider Figure 11.11, which plots the simulated significant wave height at a certain location as a function of time during a week ($\tau = 168$ hours total).

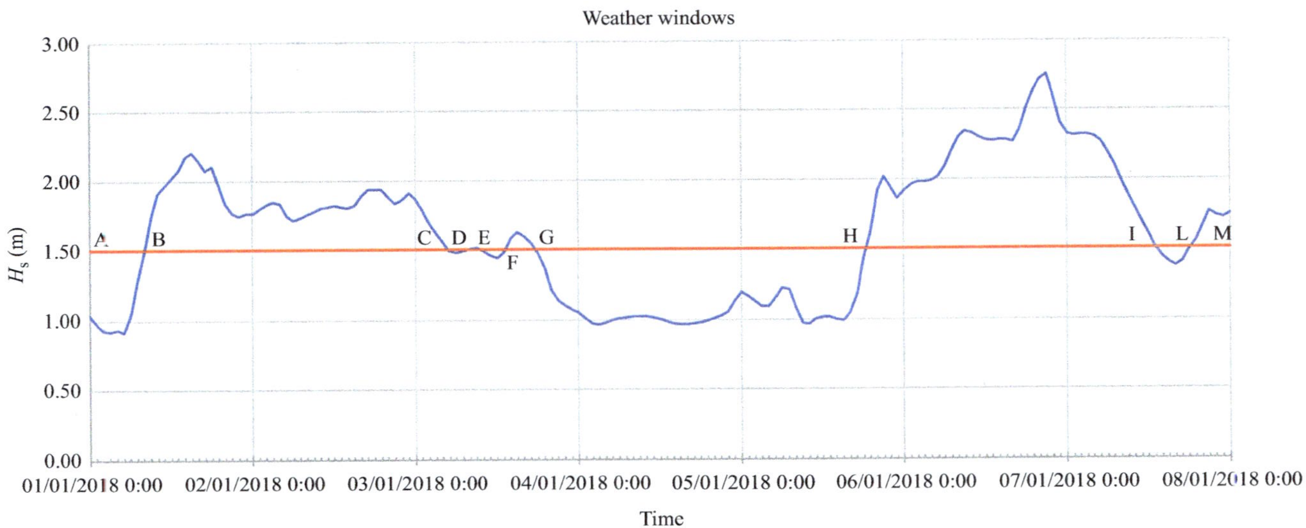

Figure 11.11 Weather windows (simulated data)

Let us consider an operation of duration $d = 6$ hours, and a limit significant wave height $H_s = 1.5$ m, with significant wave height considered as the only OLC. In this case, it results the following:

- Section AB: H_s is below the threshold for 8 hours, longer than the duration of the operation. In case of overlapping weather windows, the persistence is therefore 9 hours, while in case of none-overlapping weather window, the persistence is 6 hours, as just one operation can be carried out.
- Section BC: H_s is higher than the OLC, so it cannot be considered as a weather window.
- Section CD: despite H_s is below the threshold, its duration (3 hours) is too short.
- Section DE: H_s is higher than the OLC, so it cannot be considered as a weather window.
- Section EF: despite H_s is below the threshold, its duration (4 hours) is too short.
- Section FG: H_s is higher than the OLC, so it cannot be considered as a weather window.
- Section GH: H_s is below the threshold for 49 hours, longer than the duration of the operation. In case of overlapping weather windows, the persistence is therefore 49 hours, while in case of none-overlapping weather window, the persistence is 48 hours, i.e. six times the duration of the operation.
- Section HI: H_s is higher than the OLC, so it cannot be considered as a weather window.
- Section IL: despite H_s is below the threshold, its duration (5 hours) is too short.
- Section LM: H_s is higher than the OLC, so it cannot be considered as a weather window.

In this case, therefore, the accessibility of the site is 33.9% in case of overlapping weather windows and 32.1% in case of not overlapping weather windows.

11.3.3 Estimation of the delay time

In literature, several methods have been presented for the estimation of the delay time. An accurate estimation of the weather windows and delay time, indeed, is required for the estimation of the availability of the farm, the accessibility and the waiting time for performing the operations. The delay time must be taken into account in all the techno-economic analyses for evaluating the viability of an offshore renewable energy project.

Generally, the methods can be divided into two major groups: the family of the statistical approaches and the probabilistic approaches.

The statistical-based methods represent the most common approach to the problem. These methods are based on a time history of the meteoceanic climate variables (significant wave height, peak period, current speed and/or wind speed). Therefore, a time domain routine for the identification of the weather windows, given a range of duration of operations and several combinations of OLCs, is produced. The outcome can be post-processed estimating statistics, and the results in terms of average availability, average delay time, etc., can be plotted into graphs or distributed into look-up tables. Of course, the availability of good quality and long term time series data is crucial. The data can come from direct measures, but also hindcast models; in the latter case, the uncertainty in the modelling will also affect the outcome. The statistics can be calculated on yearly basis, seasonal basis, etc. as, in general, variability is expected with the period of the year.

The other big family of approaches, herein named as probability-based methods, requires a calibration of probability distributions from historical series of data, or uses the results of already calibrated models. For example, the NMI method [27] is based on Weibull model of the significant wave height, while the modified NMI method [28] is based on a log-normal distribution. Other methods belonging to this family adopt approaches based on Markov chain models. An extension of these models is proposed by [29], in which also other higher and lower order delay times are taken into account. At any case, the major drawback, as in the case of the statistical models, is that these models require good-quality long series of data to calibrate the distributions; moreover, in any case, statistical tests for checking the goodness of the fit should be carried out. Finally, coupled processes, in case of simultaneous OLCs to be satisfied, would require complex relationships among the distributions, as in general the hypothesis of independence among the variables is not guaranteed.

11.3.4 Offshore standards and technical recommendations for operations

The attempt of providing an exhaustive collection or a set of references of offshore standards and technical recommendations inherent to operations at sea applicable to offshore renewable energy systems is a particularly cumbersome task: from one side, indeed, just a few standards specific to operations for offshore renewable energy devices have been released, given the novelty of the sector; from the other side, for all the fields not covered by specific regulations or recommendations, more generally all the standards released for any kind of operation at sea for the oil and gas industry, e.g., as well as fishing and other maritime industries may be adopted and therefore the literature in this sense is too wide for the scope of the present book. For this reason, in this section the sources have been categorised based on the releasing institution and for themes.

In terms of sources, this kind of documentation is generally provided by

- institutions, entities or bodies managing the open sea field site (e.g. BiMEP (Biscay Marine Energy Platform) [30] and EMEC (European Marine Energy Centre) [31]), or more generally by port or coast authorities in charge of the public domain where the installation and operation of offshore energy devices take place;
- international and national classification and standardisation companies, such as Lloyd's Registers, BSI (British Standard Institution), DNV-GL (Det Norske Veritas and Germanischer Lloyd), RINA (Registro Italiano Navale), Bureau

Veritas, ABS (American Bureau of Shipping), International Maritime Organisation (IMO), among others;
- governmental and non-governmental agencies and bodies dealing with specific themes, such as HSE (Health and Safety Executive).

The thematic categorisation of all these sources cover the areas of major impact of the operation at sea:

- health and safety: this set of regulations deals with the fulfilment of the installation, operations and maintenance, and navigation and in safe conditions as well as the development of models for the risk assessment, plans and strategies on how to control emergencies;
- environmental aspects for the prevention and reduction of marine pollution;
- port operations and load out, dealing with the design and assembly of components and procedures related to the device load out;
- transport to site, dealing with the operation specific requirements and recommendations related to marine device sea transport operations, transit, towing and position of devices and mooring lines;
- installation, covering design and operational aspects of launching, upending, positioning and setting, lifting, foundation and moorings installation;
- operation and maintenance, dealing with several issues, taking into account the influence of technical, operational and weather factors on reliability, maintainability and survivability;
- decommissioning, in order to reduce risks for the marine operations needed during removal of offshore installations.

11.4 Economic modelling of installation, operation and maintenance

The viability of any project of renewable energies at sea relies on an economic model to analyse the offshore farm costs and incomes balance, besides taking into account other factors, such as environmental and social impacts.

In general, any economic model considers the following cost concepts.

CAPEX: CAPital EXpenditures, representing the initial investment in the farm, including procurement, manufacturing, installation and commissioning costs.

OPEX: OPeration EXpenditures, defined as the costs related with the operation and maintenance of the farm.

AEP: Annual Energy Production represents the annual energy generated by a farm:

$$AEP = P_{\mathrm{nom}} * CF * 8{,}760 \tag{11.7}$$

where CF is the capacity factor and P_{nom} is the nominal or power of the device. CF represents the percentage of time in which the device is producing energy at its rated power yearly, and it includes in its definition the availability of the plant.

COE: Cost Of Energy, i.e. the cost (€ per kWh produced) at which the energy produced is sold.

LCOE: the Levelised Cost Of Energy, i.e. the cost (€ per kWh produced) at which the energy produced should be sold in order to cover the expenses during the project lifetime (PL in years):

$$LCOE = \frac{CAPEX + \sum_{n=1}^{PL} \frac{OPEX_n}{(1+r)^n}}{\sum_{n=1}^{PL} \frac{AEP_n}{(1+r)^n}} \tag{11.8}$$

where r is the interest rate, $OPEX_n$ is the OPEX at the year n and AEP_n is the annual power production at the year n. Therefore, in order to be economically viable, the energy produced by the plant should be sold at a cost COE> LCOE, if other regulatory mechanisms – as feed-in tariffs, etc. – are not operating.

NPV: Net Present Value represents the difference between the present value of cash and the present value of cash over a period of time.

$$NPV = COE \sum_{n=1}^{PL} \frac{AEP_n}{(1+r)^n} - \sum_{n=1}^{PL} \frac{OPEX_n}{(1+r)^n} - CAPEX \tag{11.9}$$

In the framework of the economic models for offshore renewables, O&M costs are defined as the annual costs needed to maintain the performance of the converters in the farm. O&M expenditure can be expressed by any of the three following metrics [25]:

- Percentage of initial cost: O&M is calculated as a straight percentage of the initial investment in the farm.
- €/MWh: This metric provides a cost based on the relationship between the annual maintenance costs and the annual energy production.
- Percentage of cost of electricity: O&M costs are represented as a percentage of the total cost of the project.

Different models can be used to estimate the OPEX for commercial and pilot farms. The simplest approach consists in considering OPEX as a percentage of CAPEX.

A more detailed OPEX model aims at identifying the most critical points in order to improve components design, operations requirements and optimisation of resources. In general, more advanced models account for more variables, such as the reliability and the MTTF of components and subsystems in order to estimate the number of operations to carry out during the lifetime of the project, the expected delay time for the proper weather window in case of corrective maintenance, and so on. Based on the functional decomposition on the system and approaches for the identification of failure modes (e.g. FMEA), all the operations must be specified by a series of characteristics, such as equipment, vessels, labour and spares, duration of the operation and MTTR. The total cost of the operational expenditure will be obtained from the summation of different components (see also Figure 11.12):

- Vessel cost
- Personnel cost
- Equipment cost
- Cost of consumables

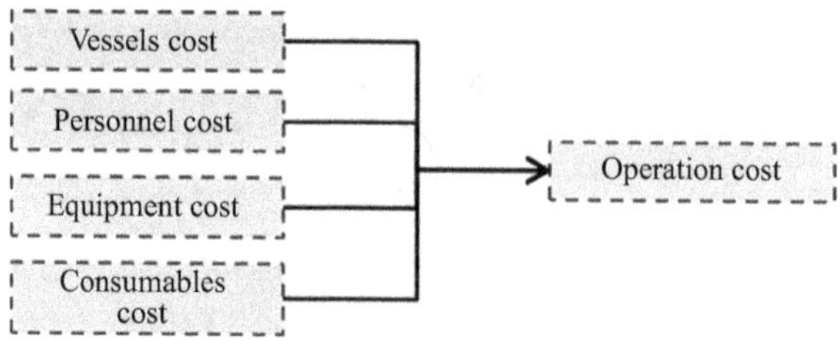

Figure 11.12 Operation cost distribution

The overall OPEX cost, taking into account corrective and preventive operations, can be represented as the summation of each operation cost per year according to the following formula:

$$OPEX = \sum_{n=1}^{PL} (N_{PO_N} * \bar{t}_{PO} + N_{CO_N} \bar{t}_{CO}) * (V_C + P_c + C_c) \tag{11.10}$$

where

N_{pO_N} represents the number of preventive operations at year N;
$\bar{t}_{PO}$ is the average operation time for a preventive operation;
N_{CO_N} is the number of corrective operations at year N;
$\bar{t}_{CO}$ is the average operation time for a corrective operation;
V_C represents the vessels renting cost per unit of time;
P_c represents the personnel cost per unit of time; and
C_c represents the consumables costs per unit of time.

The number of preventive operations N_{pO_N} depends on the frequency of the planned inspections, while the number of corrective operations N_{CO_N} depends on the MTBF of the components. The average operation time for a preventive operation $\bar{t}_{PO}$ will depend on the type of operation to perform, etc., but in general it will be shorter than the average operation time for a corrective operation $\bar{t}_{CO}$. Indeed, in this case the accessibility of the site, i.e. the expected delay time for the occurrence of an appropriate weather window, should be accounted for. It is clear, therefore, that even in a simplified approach as the one described in (11.10), the OPEX can be minimised by playing with the frequency of inspections, or using components more reliable with lower MTTF, or using equipment with less restrictive OLCs (as far as this is possible) given that all the other variables are kept constant. A more complex model, also taking into account if the operation must be taken in situ or in port, was described in deliverable D6.2 [32] within the framework of the already mentioned OPERA project and has been described in the flow chart in Figure 11.13.

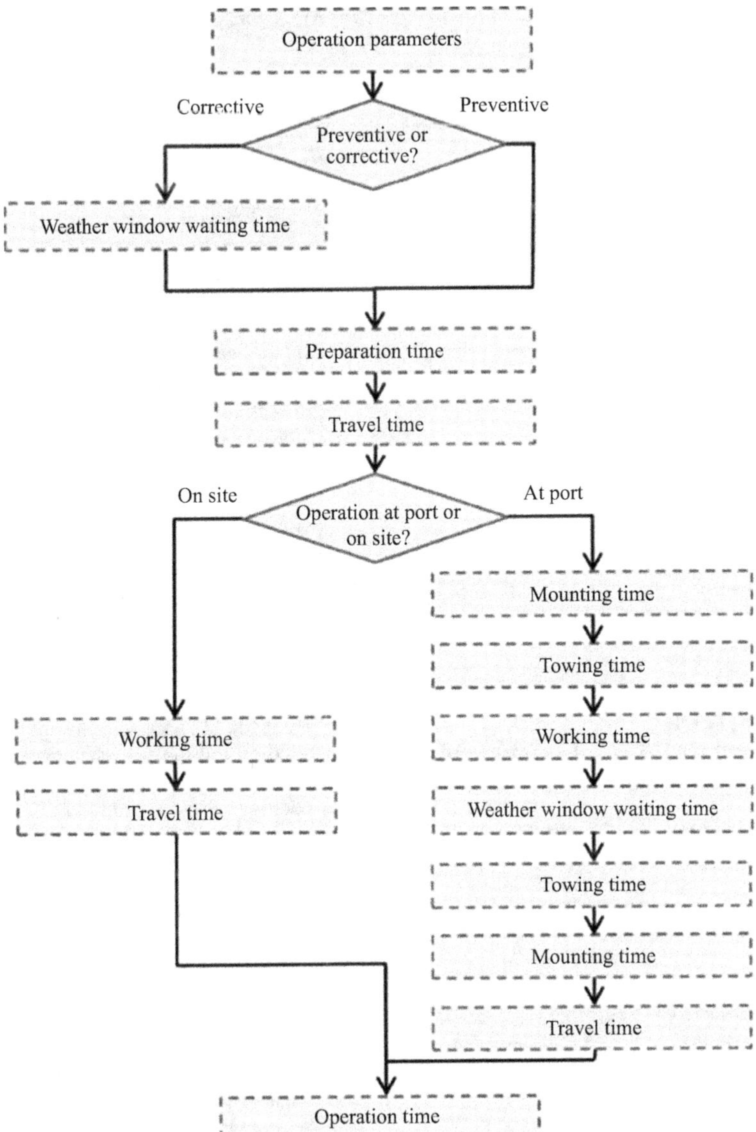

Figure 11.13 Flow chart representing the operation time estimation depending on whether the operation is planned, non-planned, on-site or at port

More advanced tools, based on either statistical or probabilistic approaches for logistics models, have been developed in the last years. As a reference, Table 11.2 collects a non-exhaustive list of some of these tools and their characteristics.

Table 11.2 Summary list of relevant existing tools for supporting offshore logistics management

Tools developer	Name of the tool	Overview of the tool
Fraunhofer IFF and ISL	Offshore logistics diagnostic tool	The toolset consists in several tools to support logistic planning for the wind farm industry (onshore and offshore)
Mojo Maritime Ltd.	Mermaid: marine economic risk management aid [33]	The software estimates the impact of scheduling and metocean conditions on complex marine operations
Energy Research Centre of the Netherlands (ECN)	Offshore wind O&M optimisation software [34]	The tool can compare a large variety of maintenance scenarios and provides the impact on the techno-economic performance
National Renewable Energy Laboratory (NREL)	Balance-of-System Model [35]	Combination of a customised NREL offshore cost model BOS and the ECN O&M tool to optimise installation and O&M strategies for offshore wind
EDF Group	ECUME [36]	Mean cost of operation of an offshore wind farm project and risk measurement of O&M operations (based on ECN's tool)
SINTEF-NOWITECH	NOWIcob- [37]	Life-cycle cost and O&M optimisation tool for offshore wind farms
Overspeed GmbH Co. KG	OutSmart: Offshore wind O&M strategy simulator [38]	Based on the logistic scenario and historical weather data, the tool delivers the downtime and OPEX
University of Strathclyde	StrathSim: offshore wind OPEX model [39]	It is an OPEX model for decision-making, asset modelling and ocean engineering in general
DTOcean	DTOcean toolset [40]	It consists in tools for planning logistics, installation and O&M, within a design toolset for deploying the first generation of wave and tidal energy converter arrays
NORCOWE	Maintenance and marine logistics tool [41]	The model simulates maintenance planning and execution, as well as marine logistics for the operation and maintenance life cycle phase

References

[1] WindEurope, 'Offshore wind in Europe. Key trends and statistics 2018,' February 2019.

[2] The Crown Estate, 'A Guide to an Offshore Wind Farm', 2010.

[3] A. Kossiakoff, W. N. Sweet, S.-J. Seymour and S. Biemer, *Systems Engineering Principles and Practice*, Wiley, Hoboken, NJ, 2011.

[4] K. Otto and K. Wood, *Product Design Techniques in Reverse Engineering and New Product Development*, Upper Saddle River: Prentice Hall, 2001.
[5] Department of Defense, 'Military Standard 881D – Work Breakdown Structures for Defense Matericl Items', 2018.
[6] International Electrotechnical Commission IEC, 'IEC 61400-3: Wind Turbines – Part 3: Design Requirements for Offshore Wind Turbines', 2009.
[7] International Electrotechnical Commission IEC, 'IEC TS 61400-3-2 – Wind Energy Generation Systems – Part 3-2: Design Requirements for Floating Offshore Wind Turbines', 2019.
[8] International Electrotechnical Commission IEC, 'IEC TS 62600-2: Marine energy – Wave, Tidal and Other Water Current Converters – Part 2: Design Requirements for Marine Energy Systems', 2016.
[9] H2020, 'Lifes50+: Qualification of Innovative Floating Substructures for 10MW Wind Turbines and Water Depths Greater than 50m'. [Online]. Available: http://lifes50plus.eu/. [Accessed 2018].
[10] Lifes50+, 'Deliverable D6.1 – Risk Management for Deep Water Substructures', 2016.
[11] H2020, 'OPERA: Open Sea Operating Experience to Reduce Wave Energy Cost', 2018. [Online]. Available: http://opera-h2020.eu/.
[12] A. Papoulis and P. S. U., *Probability, Random Variables and Stochastic Processes with Errata Sheet*, McGraw-Hill Education, New York, 2015.
[13] A. P. Teixera and C. Guedes Soares, 'Fundamentals of Reliability', in *Thermal Power Plant Performance Analysis*, Springer London, 2012, pp. 91–122.
[14] A. Hassanin and P. Jukes, 'Risk Analysis', in *Encyclopedia of Maritime and Offshore Engineering*, John Wiley & Sons, Ltd., Hoboken, NJ, 2018.
[15] DTOcean, 'D5.2 Characterization of Logistic Requirements', 2014.
[16] J. Green, A. Bowen, L. J. Fingersh and Y. Wan, 'Electrical Collection and Transmission Systems for Offshore Wind Power (PrePrint)', Offshore Technology Conference (OTC), Houston, Texas, 2007.
[17] T. Axelsson, 'Submarine Cable Laying and Installation Services for the Offshore Alternative Energy Industry', Ocean Energy, Galveston, Texas, 2008.
[18] D. Bank, P. Environmental, I. Pei and T. Pei, 'Dogger Bank Teesside – Appendix A: Project Description, Preliminary Environmental Information 1 (PEI1)', Forewind, 2012.
[19] S. K. Chakrabarti, *Handbook of Offshore Engineering – Volume 2*, First edn., Elsevier, Oxford, 2005.
[20] Institute of Electricals and Electronics Engineers, *The Authoritative Dictionary of IEEE Standards Terms* (IEEE 100), New York: IEEE Press, 2000.
[21] M. J. Kaiser and S. B., 'Offshore Wind Energy Installation and Decommissioning Cost Estimation in the U.S. Outer Continental Shelf', U.S. Dept. of the Interior, Bureau of Ocean Energy Management, Regulation and Enforcement, Herndon, VA, 2011.
[22] E. Topham and D. McMillan, 'Sustainable Decommissioning of an Offshore Wind Farm', *Renewable Energy*, vol. 102, Part B, pp. 470–480, 2017.

[23] Navigant Consulting, 'Global Evaluation of Offshore Wind Shipping Opportunity – Section 3', The Danish Shipowners' Association, 2013.

[24] 4C Offshore Ltd, '4C Offshore', 09 May 2019. [Online]. Available: https://www.4coffshore.com/.

[25] GL Garrad Hassan, 'A Guide to UK Offshore Wind Operations and Maintenance', Scottish Enterprise and The Crown Estate, 2013.

[26] P. Tavner, *Offshore Wind Turbines Reliability, Availability and Maintenance*, The Institution of Engineering and Technology, London, 2012.

[27] S. Kuwashima and N. Hogben, 'The Estimation of Wave Height and Wind Speed Persistence Statistics from Cumulative Probability Distributions', *Coastal Engineering*, vol. 9, pp. 563–590, 1986.

[28] M. Mathiesen, 'Estimation of Wave Height Duration Statistics', *Coastal Engineering*, vol. 23, pp. 167–181, 1994.

[29] J. Feuchtwang and D. Infield, 'Offshore Wind Turbine Maintenance Access: A Closed-form Probabilistic Method for Calculating Delays Caused by Sea-state: Probabilistic Method for Calculating Access Delays Caused by Sea-state', *Wind Energy*, vol. 16, no. 7, pp. 1049–1066, 2013.

[30] BiMEP, 'Biscay Marine Energy Platform', 2018. [Online]. Available: https://bimep.com/.

[31] EMEC, 'European Marine Energy Centre', 2018. [Online]. Available: http://www.emec.org.uk/.

[32] H2020, OPERA: Open Sea Operating Experience to Reduce Wave Energy Cost, Deliverable 6.2 'Operational model for offshore operation of wave energy converters', 2018. [Online]. Available: http://opera-h2020.eu/wp-content/uploads/2017/09/OPERA_D6.2_Operational-model-for-offshore-operation-of-WECs_TECNALIA_20170727_v1.0.pdf.

[33] Mojo Maritime, [Online]. Available: http://mojomermaid.com/. [Accessed 16 July 2018].

[34] M. Asgarpour and R. van de Pieterman, 'O&M Cost Reduction of Offshore Wind Farms – A Novel Case Study', ECN-E–14-028, 2014.

[35] NREL, [Online]. Available: https://www.nrel.gov/. [Accessed 18 7 2018].

[36] B. Maples, G. Saur, H. R., R. van de Pietermen and T. Obdam, 'Installation, Operation, and Maintenance Strategies to Reduce the Cost of Offshore Wind Energy', 1 National Renewable Energy Laboratory 2 Energy Research Centre of the Netherlands., 2013.

[37] F. Douard, C. Domecq and W. Lair, 'A Probabilistic Approach to Introduce Risk Measurement Indicators to an Offshore Wind Project Evaluation – Improvement to an Existing Tool ECUME', *Energy Procedia*, vol. 24, pp. 255–262, 2012.

[38] OutSmart, [Online]. Available: http://out-smart.eu/. [Accessed 16 July 2018].

[39] D. McMillan, '8-Minute Guide to Cost Reduction', Technology & Innovation Centre, University of Strathclyde.

[40] DTOcean, 'DTOcean', EU, 2017. [Online]. Available: http://www.dtocean.eu/. [Accessed 19 September 2018].

[41] Norwegian Research Cluster for Offshore Wind Energy, [Online]. Available: http://www.norcowe.no/. [Accessed 18 July 2018].

Chapter 12
Challenges and future research

Neil Kermode[a]

12.1 Challenge one – proving it *works*

As the preceding chapters have shown – the ocean energy technologies we are looking at today are all at very different points in their development pathways.

Some technologies – notably offshore wind – are now commercial and evolving rapidly (albeit with some level of public subsidy), whilst others are still making their way out of their land-based laboratories and into the sea.

What they all have in common is that each and every technology must make a journey through the technology readiness levels until we reach the Eureka moment of 'it works'.

For some technologists, this is the ultimate goal – to show simply that it can be done.

At one level this is correct – we need to first prove that something works before we can begin to imagine how it will develop and improve in the years ahead.

In the case of most marine technologies shown earlier in this book, we can see that the 'Eureka' moment has passed and now the journey to competence in the market is underway. However just as you cannot learn to ride a bicycle simply by listening to another cyclist describe the process, nor can any technology learn all its lessons from its predecessors. On the other hand, only a fool would not listen to any of the advice or experiences of others. The challenge is knowing who to listen to about what, and when to commit, to push off and begin your own journey.

However, to develop technologies which will become as ubiquitous and commonplace as onshore wind or solar power – then we need to examine and explore how the technologies will make their journey. How will they progress through their various technology readiness levels, until they reach the point that we have a product we can *sell* and also a product that people want to *buy*.

This journey will involve a continual development and must address the actual practicalities of a myriad of elements including installation, reliability, operability, fatigue and mean time to fail of everything from components to systems to people.

For ocean energy technologies, this will ultimately involve significant test and development in the real sea environment – at specialist facilities such as European

[a]The Marine Energy Centre, Stromness, UK

Marine Energy Centre (EMEC) – where developers can discover the weak points in their design and then resolve them.

As we discussed in Chapter 1, this process has been characterised by Wave Energy Scotland via their own four-stage progression. EMEC provides an overview of the technology development pathway in Figure 12.1.

This pathway, which could be replicated for any of the technologies covered in this book, will take a technology up to the point where we can say – with some confidence – 'it works'.

Wave Energy Scotland (WES) stages 3 and 4 are a major challenge and require developers to take technologies out of the lab and into the sea.

At EMEC, we are aware of over 200 active wave technology developers across the world; the majority of companies are still at WES stage 3, working at Technology Readiness Level (TRL) 5 or below (Figure 12.2) [1]. This indicates that wave energy is at an earlier stage of progression than tidal energy.

For tidal technology – where there are around 80–100 technology companies in existence – we can see a more advanced picture, where some companies have been categorised as having reached TRL 8 (system completed and service qualified through test and demonstration) (Figure 12.3).

Some of the most advanced tidal energy technologies include Andritz Hydro Hammerfest and SIMEC Atlantis Energy's AR1500 turbines which have been deployed at MeyGen's tidal array project off Caithness, and Orbital Marine Power's SR2000 at EMEC, both in Scotland. Each project has devices deployed and operating through much of 2018 and into 2019 with significant gigawatt hours produced.

Sadly, although OpenHydro's technology had also reached this significant milestone, the company was liquidated in 2018 by major shareholder Naval Energies [2]. This underscores the considerable challenge in taking a technology beyond the point where it works reliably, and onto where it works commercially.

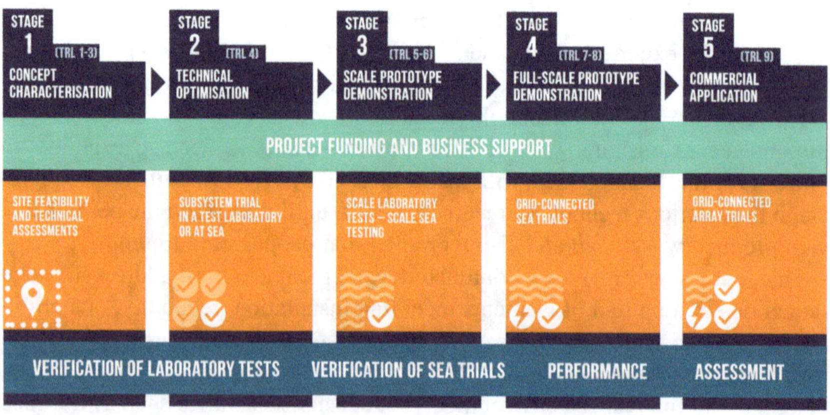

Figure 12.1 EMEC pathway to commercialisation

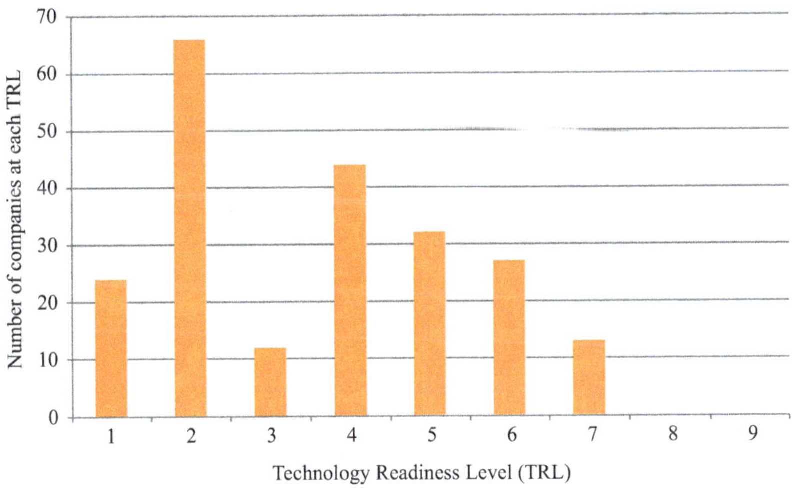

Figure 12.2 Wave energy technology TRLs (EMEC)

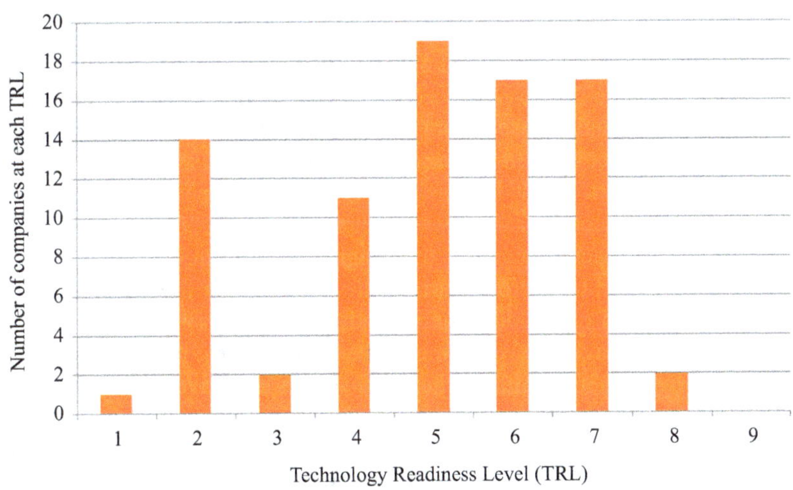

Figure 12.3 Tidal stream energy technology TRLs (EMEC)

12.2 Challenge two – keeping it working

In many of the early deployments of devices, there are failures. These tend to be as a result of an underestimation of the input forces and operational conditions. This is entirely understandable since many of the locations into which these devices are being placed are poorly understood. They are rarely visited by humans and the way in which our new machines behave is predicted by our understanding derived from mental and mathematical models.

All of these assumptions can be wrong and often cumulatively so. Indeed, the history of engineering is littered with examples of underestimates or unexpected loads causing failure.

So, the second challenge is all about making the device last beyond the 'Eureka' moment and demonstrating that it works in the longer term – in other words getting from TRLs 5–6 to TRL 8 and beyond.

Here we encounter a systematic blind-spot; the failure to systematically learn from previous work. This is a problem more of start-ups than large companies, but it is a significant flaw and one often driven by the unrealistic demands of funders overriding good engineering.

Most developers of machines are dependent upon assistance for their first devices, but the temptation to over-promise on its performance has become the norm. In a competitive investment environment each developer wants to attract the money to their idea. An 'arms race' of ever-more ambitious claims has led to some untenable outcomes. Furthermore, many schemes are only funded on the basis of what they could lead to rather than what they can achieve. So money is raised on expectation rather than upon realism. Despite being called 'business cases' they are often less about business and more about hope.

As an example, one wave developer was designing their second wave machine (nearly three times the size of the first one) before operational experience had been gained on the first one. Indeed the second one was built before the first one had been removed, so potential lessons about decommissioning and of the performance of hidden parts was not derived before the next machine was ready and deployed.

So rather than learn all the lessons from the first machine the opportunity was squandered and the second machine was built sub-optimally and failed.

The better approach would be to design, install, run, decommission *and examine* the first machine before starting the build of the next device. A more conservative approach of building the second machine as a better version of the first machine would have allowed some of the lessons learnt to be applied.

This requires 'patient capital', whereas investors are driven by an unrelenting urge to get their money back quickly. Similarly, governments have developed unrealistic expectations as to what they can and should fund, so they look to retreat from supporting projects too early, hoping the market will take over.

Unfortunately, most marine devices run into a fundamental problem at this point. Having proved that it works, the funders then want the next machine to work perfectly and indefinitely. Neither are possible. It is often the case that funders want to see the next machine run for years or they want to see an immediate scale up to commerciality. In my experience, this is unfair and leads to poor outcomes.

It would be unreasonable to ask a new cyclist to take a job as a cycle courier the day after they managed without their training wheels; however, it does seem that this is the model some funders would like to see marine developers adopt.

In fact, we need to be ready to recognise that technical development requires many iterations before problems can be detected, identified and solved and before best practice can be developed. Indeed it is axiomatic that one needs to do some things badly before you can work out what best practice actually is.

One of EMEC's unwritten objectives is to make marine energy dull. We believe we need to get the industry to the point where it is no more exciting to put a machine at sea than it is to see a something roll off a production line. Yes, it is exciting to see the first car or aircraft off the line, or the millionth, but quickly after the media froth has blown away, the solid business of churning out dependable saleable products takes over. That is effectively where we need the technologies in this book to be able to get to.

So, what are the challenges of making it all very dull?

They break into the following challenges.

12.3 Challenge three – technical improvements

12.3.1 Servicing

It has been said that marine energy is not a rocket science. Indeed it is not. In fact, satellites are launched into a nice dry vacuum, whereas our machines are being launched into biologically active electrolyte full of grit. Our machines are under corrosive, erosive, colonial, impact and digestion attacks from the moment they go out. This is hard on materials to endure and it is hard on the engineers to seek to combat them all simultaneously.

I am not for a moment trying to denigrate what our space brethren achieve; I simply want to emphasise that the nature of the engineering challenge marine energy faces is comparable in many respects to that of space.

Take one example: maintenance. We need to understand what we are asking of our devices and put that in the context of other complex machines. In space it is rare to be able to undertake maintenance – with the exception of projects on the International Space Station. Most space equipment is 'fit and forget'.

If we take the area I know best – tidal turbines – the cost of intervention on machines is a direct function of accessibility. In some cases, the machines can be accessed on site and in some cases, they need to be returned to shore for maintenance. Taking some of the bottom-mounted tidal turbines, there is a general target of seeking to intervene no more often than every 5 years. At such an intervention there would be a full service before return to the site. This would have meant that the machine would have been in the sea for approximately 44,000 hours and been working for around 90 per cent of the time (allowing for slack water as the tides change).

If one compares that to a high-end modern internal combustion car (say a Mercedes Benz), then there would be a general requirement to service it after 10,000 miles and then every 20,000 miles thereafter. In effect, this is a service after 1 year and then 2 thereafter based on average mileages. So in our 5-year period for a single service for a tidal turbine, the Mercedes would have had three.

And if we assume the average speed of a car when travelling is a conservative 30 miles per hour then the 10,000 miles would have involved the car being used for approximately 333 h per year. Of the 8,760 h in a year, this means the car was in use for under 4 per cent of the year. Comparing that with our tidal turbine we see we are asking over 20 times the usage.

And we should not forget that Mercedes has been making cars since 1886 when they produced the first gasoline powered car and have since probably produced millions of cars. The point is that after these uncountable numbers of lessons they have still got a car that needs an annual return to a dealership after seeing less than 1/20th of the load we are expecting of our tidal turbines. We are making a big ask of our designers and engineers.

12.3.2 Access

It is therefore unsurprising that the costs of intervention in servicing are becoming a significant factor in marine energy, and this is a difficult sell to politicians and government funders. It is hard to make the fact that you will not need to go near the machine a week longer into a major selling point. Not doing things is always harder to sell than a sexy new kit.

In the case of offshore wind, the means of intervention have had to evolve. The first offshore turbines could be accessed by a walkway from a pier, so making them 'barely offshore', but lessons were learned. The first generation of machines out in the sea was accessed by boat from the nearby port. This has led to new types of vessels being produced to rapidly and safely take operatives out to turbines for maintenance work. In addition, the means of getting them on and off in worse and worse conditions remains an area receiving attention. The issue being that if operatives can only access turbines in flat calm conditions, then the numbers of days per year when this occurs is extremely limited. This in turn would lead to long periods before an intervention can be accomplished and therefore there will be long periods after a machine is shut down for a fault before it could be accessed and re-started.

So, increasing the size of access opportunities by better sea-keeping and better personnel transfer remains an area of great interest and is a major learning area. Technically it has nothing to do with anything on the turbine or the wind, but it is absolutely relevant to the success of the industry.

However, there are also limits on how far one can travel by boat from shore before this becomes uneconomic, so the use of helicopters is now routine for some larger offshore facilities. The work patterns now involve extended stays on some machines and accommodation is provided on platforms in the wind farm [3] and there is even emergency accommodation aboard some machines.

12.3.3 Data

Of course, the rise in the number of sensors on devices and the ability to avoid putting an operative onto devices is also a rich area for expansion. Already we are seeing robots able to undertake blade inspections on wind turbines and we expect to see many more interventions proposed to undertake servicing and repair work remotely in many of the technologies covered in this book.

12.3.4 Materials

There is the entire area of new materials. One only has to look at the inventory of materials used in new wind turbines to see how they have moved away from the engineering stalwarts of steel and wood and are now increasingly using Kevlar composites that are expected to be overtaken by graphene in due course. And of course graphene also has remarkable electrical properties that we are only just exploring, so we can anticipate the copper content of future machines to be lower than now.

But it is not just what goes into making the machines, it is also a matter of what keeps the weather and the environment out. The use of slippery materials to shed bio-fouling rather than poison it, the use of adaptive materials to shed excessive load rather than resist it and the use of smart materials to report on their condition will all expand the envelope of possibilities and reduce the cost per kilowatt hour delivered.

But it is the combination of better monitoring, experience of what works and access to unimagined materials that has permitted the rate of wind turbine scale up seen in recent years.

Indeed the sizes of turbines now becoming available, such as GE's 12-MW Halidade machine, are truly gigantic, with blades over 100-m long. And with higher capacity factors available offshore than onshore and fewer issues with 'neighbours' I firmly believe we will continue to see these giants grow and grow (Figure 12.4). It will be a brave engineer who says 'it will get no bigger'.

There is a rule of thumb of materials that is dangerous in this arena. It is claimed by some that the mass of the device per rated kilowatt is a good measure for comparison of technologies. This is probably true for production models and

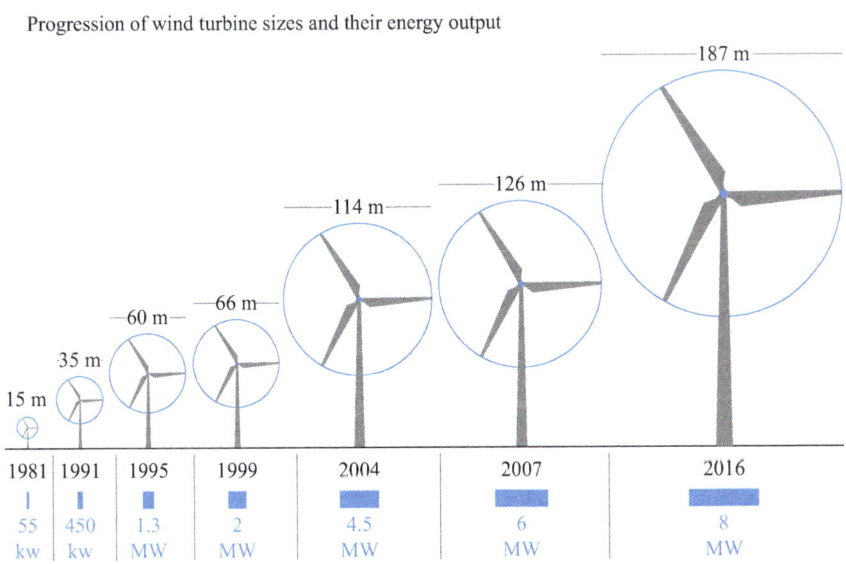

Figure 12.4 Progression of wind turbine size

can compare one with another. However, it is a crude measure as it assumes that the material is the major cost. In fact, it is the work done on the material that is often the most expensive cost. So a 1-tonne lump of steel ballast has very little effort committed to it when compared to 1 tonne of shaped and welded steel plate.

So it is not the mass of material that is important, it is how much of it is needed to do the job at hand. As alluded to earlier the first job is to survive and learn. Thereafter the judicious paring back of the structure, or the substitution with more specialised materials may yield benefits.

Over time there should be no excessive material in the devices; however, an over-obsession with reducing material too early in the process will inevitably lead to failures in this largely unexplored and poorly defined load environment. The weight of the structure per kilowatt hour delivered may help compare some elements of early machines, but it should be used sparingly and in the full knowledge that it may be the entirely wrong parameter to measure.

12.4 Challenge four – environmental acceptability

Although there is an implicit understanding that renewables are good for the planet, there is a dangerous tendency to assume that everybody agrees that marine technology is the way.

Marine energy systems do change natural energy flows and therefore are not entirely benign. However, some change other aspects of the environment, often for the better. It is recognised there is potential for negative impacts on marine ecosystems and much public support will be required to bring the sector to a position where it can deliver all anticipated benefits [4].

Already the EU Ocean Energy Forum has identified a joint action for relevant planning and consenting authorities to de-risk environmental consenting. They aim to do this through developing guidance on planning, consenting, research, socio-economics and demonstration – to ensure that best practice and experience is shared and used to improve and streamline processes.

Structures in the sea become substrates for life. Colonisation of moorings and structures often needs to be controlled, but they also form significant colonies of life in otherwise comparatively empty water columns.

Fishing pressure may well be reduced due to the footprint of the structures and any safety zones established around sites. Trawling, a notoriously disruptive activity, is banned within the areas of marine renewables structures, largely due to the risk to cables. This allows the seabed to be left undisturbed by trawls and give a chance for biodiversity to return. However, the areas immediately around marine energy sites may see increased fishing activity due to the increased abundance of life within the no-take zone and the associated spill-out effects as happens around oil platforms in the Gulf of Mexico.

New fishing activities may become possible as a result of the structures themselves – either because of the introduction of the rock armouring/scour protection providing new habitat or else through aquaculture that may then become possibly attached to the generating structures themselves.

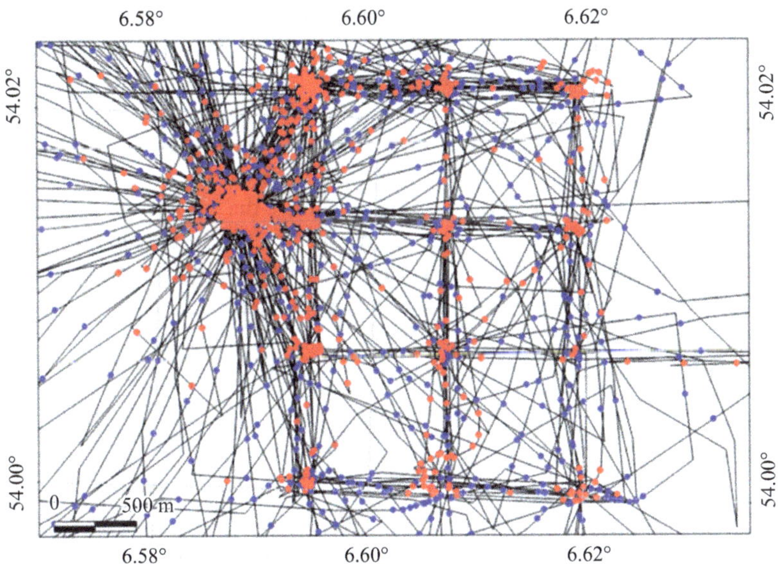

Figure 12.5 Seal feeding patterns around offshore wind turbines, D.J.F. Russell et al./Current Biology 2014

New nutrients may be introduced, particularly as a result of ocean thermal energy conversion (OTEC) plants which create synthetic upwellings from the deeper ocean into the warmer top layers.

And of course we should not just see this through human eyes. New marine energy structures have already proven to be of use and interest to other species, and this map (Figure 12.5) showing the modified feeding patterns of seals systematically visiting wind turbines is a perfect example of adaption to new opportunities [5].

A largely unexplored area is therefore what more could be done to utilise these new marine energy structures and actively build new fisheries. By choosing to texture structures it may be possible to selectively encourage particular species and so support otherwise threatened species.

12.5 Challenge five – social acceptability

The social acceptability of marine renewables is high. Studies routinely undertaken of public opinion show high levels of support for marine renewables. In the UK government's recent survey, supports for offshore wind were at their highest recorded levels since the tracker began (83 per cent). Support for other renewable energy developments remained high: 81 per cent said they supported wave and tidal energy, 76 per cent supported onshore wind, and 69 per cent support biomass [6].

It is interesting to note that these levels of support for schemes seem to be entirely organic. There does not appear to have been the sorts of campaign often run by government or business to convince the public they like something. The introduction of thousands of wind turbines into our coastal waters has gone almost unremarked, with only limited opposition, generally from more affluent areas where marine jobs are not significant.

Furthermore, in many coastal communities the opportunities to bring new sustainable employment is welcome, particularly when traditional industries are under pressure. There is widespread evidence of fishermen being able to diversify into well-paid renewables support activities. Given the opportunities that seem likely to open up for increased or improved fishing around the structures, there may well be a major opportunity for a coastal renaissance if handled well.

12.6 Challenge six – making it work commercially

Once a technology has reached TRL 8, or stage 4 in the WES model above, our next challenge is to show it can work commercially.

By this I mean developing a fully warranted technology which can be bought in an open market without (eventually) the need for public subsidy.

It is this stage – TRL 9 and beyond – where we expect to see multiple iterations of a full-scale working machine, allowing continual improvement, learning by doing, improved reliability and lowered cost.

This requires investors with nerve and deep pockets, and it is bridging this gap that remains a major challenge for the ocean energy sector.

This challenge has been described successfully by the Australian government, which has developed a commercial readiness index (CRI) as a natural extension to TRLs, represented here in Figures 12.6 and 12.7 [7].

It is fair to say that several of the technologies examined in the book have progressed as far as TRLs 8 and 9. However, few are managing to make the transition to Commercial Readiness Level (CRL) and, with the exception of offshore wind, none has made it to CRL 5.

For many technologies – their route along this commercialisation pathway will first depend on finding a specialist niche where their technology can outcompete others, or where it can meet a need that has not been previously met, as a first step towards serial production and cost reduction.

For ocean renewables, we can see where these niches may lie – remote communities or sea-based industries reliant on energy sources which may be unreliable, expensive or ecologically unacceptable.

These can offer a pathway where pure cost may not be the only driver for development and can help technologies improve reliability and drive down cost.

This journey can be lengthy. Indeed work done by the United Kingdom Energy Research Council in 2015 [8] showed that the average time for an energy product to come to market is 46 years.

In solar energy, e.g., Bell Labs piloted the first successful silicon PV cell (with 4 per cent efficiency) in 1954 based on a principle first reported in the 1870s, yet it was not until the 1970s that costs reduced sufficiently (from $100 to $20 per *watt*) to make solar power a commercial (rather than as purely space exploration)

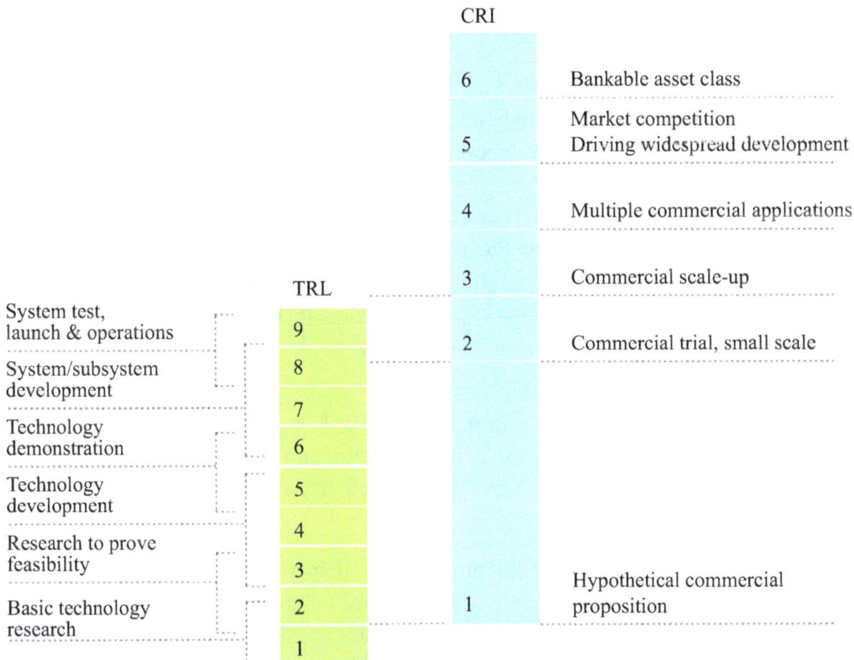

Figure 12.6 Stages towards a bankable asset class, Australian Renewable Energy Agency (ARENA)

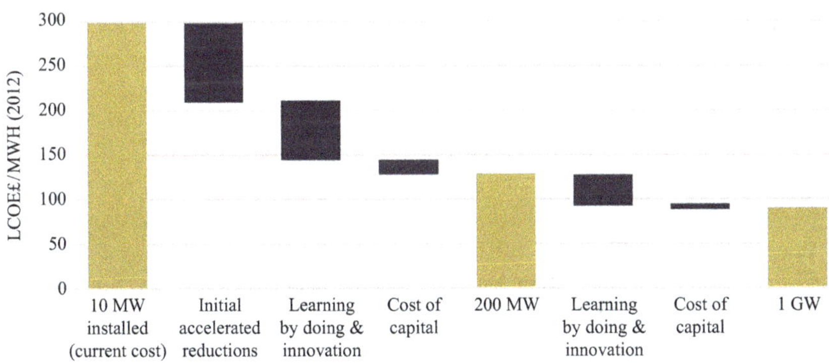

Figure 12.7 Tidal stream LCOE reduction, ORE Catapult

possibility. And it has taken until this decade – nearly 60 years on – to achieve cost parity with fossil fuel competitors. And it is expected to be cheaper than fossil fuels within the decade. And as we saw in Chapter 1, the commercialisation of onshore wind took a full 40 years to achieve. New ocean energy technologies will inevitably follow their own similar paths and effort is constantly being applied to understand and drive such savings.

12.7 Challenge seven – getting the price down

'Practice make perfect' is true, but in this context, practice makes things cheaper. The most effective way of getting the price down is repetition followed by more repetition and finally even more repetition.

Much has been written about learning rates, but it would benefit from being recounted here. In essence, there is empirical evidence that one gets better the more often one undertakes a task and the improvement is seen when one doubles the number of times it is undertaken. The improvement for each doubling appears to be approximately constant for each type of activity, or technology.

So, for example:

- Building the first machine will cost X
- The third (twice one + the original) will cost X–y per cent
- The seventh (twice three + the original) will cost (X–y per cent)–y per cent
- The fifteenth (twice seven + the original) will cost ((X–y per cent)–y per cent)–y per cent.

So, if a technology shows a learning rate of 10 per cent, the fifteenth machine will cost approximately 72 per cent of the first one. To get approximately 50 per cent of the cost of the first, one needs to build 63 machines and 25 per cent will happen at 8,000. If on the other hand the learning rate is 11 per cent, then the 25 per cent cost comes at around 4,000 machines.

So, the rate at which one learns has a huge impact on the cost of the latter machines. Therefore aggressively seeking to learn is important, but that only happens because you have *done* something. In other words, thinking about things helps, but doing it delivers.

And this is the conundrum we face at the moment. Most developers of the devices covered in earlier chapters have not achieved the amount of repetition needed to begin to drive down the cost. The exception is, of course, offshore wind where the sector has now deployed nearly 2,000 machines around the UK coast [9] and is racking up huge operational experience and seeing the price per megawatt fall dramatically as a result – and of course the entire offshore fleet around Europe now doubles those numbers.

The challenge for those other technologies that have yet to achieve the ubiquity of offshore wind is that they will struggle to show how they will get costs down until they can deploy in significant numbers. But nobody will commit to mass roll-out until the price drops. This is classic 'Catch-22':

> *You cannot get the price down until you deploy multiple machines, but*
>
> *You cannot get anybody to commit to build multiple machines until you make them cheaper.*

A recent study by the UK Offshore Renewable Energy Catapult [10] suggests that tidal stream has the potential to reach a levelised cost of energy (LCOE) of

£150/MWh by 100 MW installed, reducing to £90/MWh by 1 GW and £80/MWh by 2 GW – a pathway which will take until 2030 at least.

The report suggests that, with the proper support giving a route to market, significant cost reductions can be achieved in the near-term, followed by ongoing incremental reductions with growth in the industry.

Significant cost reductions are expected to be achieved through:

- initial accelerated reductions;
- learning by doing and innovation;
- reduced cost of capital.

The challenge for us all is how do we get from A to B – if A is a technology proven at TRL 8, and B is a commercially available product which can be bought at a reasonable price?

How do technologies take this decade-long journey to commercialisation, through learning by doing, and multiple installations, and get across the public/private funding gap known as 'the Valley of Death' (Figure 12.8) [11].

This idea – first posited in 2010 and illustrated above – describes a largely publicly funded R&D landscape and a largely privately financed commercial sector, with no clear bridge between.

I believe, as do many others, that a twin-track approach is required – technological development to reach TRL 9, and commercial support to aid scale-up and roll-out, with a clear pathway to encourage private investment.

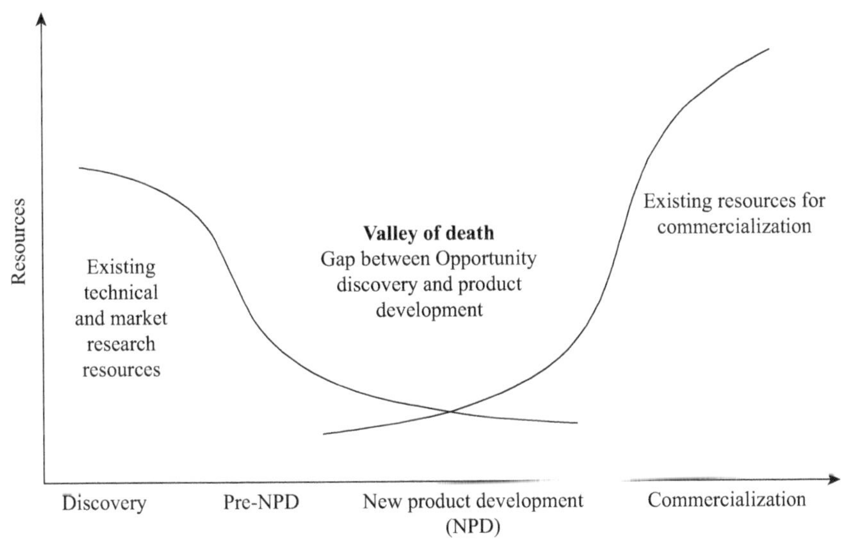

Figure 12.8 Valley of Death concept

12.8 Challenge eight – public support required

As a technology moves along this pathway, it is a paradox that the need for public and private funding temporarily *increases*, particularly as we near pre-commercial roll-out – as does the requirement for different types of support and involvement.

Figure 12.9 illustrates the relative share of public and private funding required through each phase of a technology's development [12].

As this shows – there is a critical point where a technology leaves the laboratory and meets the sea. By this stage, the first-stage prototype will be heavily reliant on public support (through R&D grants), yet at the same time must begin to attract private sector investment.

This challenge has been explicitly recognised in the wave sector, and Wave Energy Scotland has a clear remit: 'To secure, for Scotland, cost-competitive wave energy technology attractive to investors' [13].

At EMEC, I get to meet many ocean energy CEOs and have witnessed their personal transformations as they start off as jeans-wearing R&D engineers getting their hands dirty in the lab – and turn into suit-wearing sales professionals, delivering presentations to government officials, investors and financiers! It has always been thus.

Each one must cross 'the valley of death' and take their technology from prototypes and demonstration machines to pre-commercial devices and then industrial roll-out.

12.8.1 Financing arrays

Financing first arrays is one of the most challenging phases in ocean energy development – as a technology progresses towards TRL 9 and into the realms of

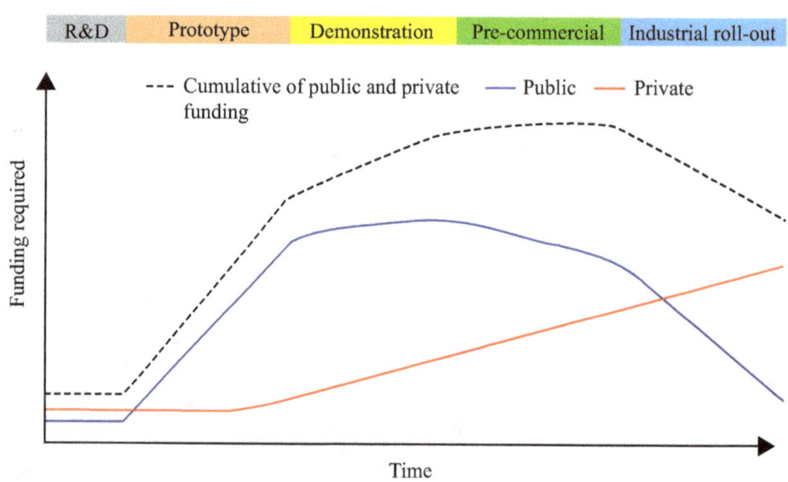

Figure 12.9 Indicative share of private and public sector funding required, per development phase, OEE

private finance. It is here that a technology must 'make its case' to investors, and secure the many millions required to enable a technology to progress.

As one of the pioneers in ocean energy, Scotland has tremendous experience in taking technologies from the lab and into the sea, and from there into the demonstration stage – particularly through the development and financing of the £51 million MeyGen array, owned by SIMEC Atlantis Energy. This 2016 project required a bespoke mix of project debt, grants and equity and involved a number of different supporters including the Scottish government's enterprise agency, its investment bank and the seabed owner the Crown Estate [14].

The complexity of this arrangement and the scale of funding required to make this project a reality underlines the necessity for the public sector to provide as many tools as possible to unlock the significant private investment needed.

Already there is growing recognition from the EU that a range of financial support mechanisms, including flexible capital, insurance guarantees and revenue support, are required to enable technologies to bridge 'the valley of death' – that long period between proving a technology works and developing a technology that is cost competitive.

The European Ocean Energy Forum's Strategic Roadmap [15] recommends the establishment of a €250m Investment Support Fund to provide flexible capital and to enable investment, alongside a €50–€70m Insurance and Guarantee Fund for demonstration and pre-commercial projects, covering risks that are not currently covered by insurance products or manufacturers' guarantees.

Given their importance to the sector, it is now anticipated these forms of support can be provided by EU research funding and other institutions such as the European Investment Bank.

12.8.2 Market pull

It is also agreed that 'market pull' (i.e. tariffs) is as important as 'technology push' (R&D grants) in developing a successful energy technology and it will be the role of policy makers at national and multinational levels to develop financial models to incentivise inward investment.

Investors will only invest if they can see a good prospect of a reasonable level of profit – via a tariff – to offset the cost and risks of developing first arrays.

The 'first wave' of renewables – solar power, hydro and onshore wind – were fortunate in securing supportive market tariffs for their power over many years, to the point where costs are now reaching parity with fossil fuel alternatives; whilst in many countries nuclear power continues to enjoy significant state support.

However recent rapid falls in the cost of solar power, onshore and offshore wind have left ocean energy in a difficult place – where new forms of generation are expected to compete against mature renewable technologies which in some cases have enjoyed decades of support.

Here, our ocean energy industries need to speak with a clear voice and outline the long-term benefits – including reduced emissions and industrial growth – that new ocean technologies will bring. Part of this narrative will be demonstrating, a clear route to cost reduction and commercialisation. Here I believe there is also an important role for strong lobbying by renewable trade associations and individual developers to make the case for ocean energy and for suitable support mechanisms which will support the future growth of the industry worldwide. This should include the development of suitable financial support models at all stage of development – including first devices, demonstration arrays and early commercial projects.

12.9 Challenge nine – market development

The final challenge is finding the customers to buy the technology.

There is a huge amount of effort being deployed in some parts of the marine renewables space, particularly around offshore wind. Unfortunately, the same cannot be said for the other sectors at present. The challenge is therefore to ramp up effort in these other sectors to seek to deliver the progress we have seen with offshore wind.

Around the world, we now have a growing network of test and demonstration areas offering grid-connected test sites with a range of conditions suited to small arrays (Figure 12.10), and in the years to 2020, there is no significant technical obstacle to the development and installation of the first small pre-commercial arrays. The obstructions here are only financial with a need for both private and public support.

Looking ahead, however, it is clear an increasingly commercial industry will require a range of development sites around the world – each with leasing and permitting processes suited to a range of potential nearshore and deepwater technologies, some at community scale and others offering grid-connected power.

Each of these sites will have a very specific set of prerequisites – and if we are to see commercial development sites flourish in the decade ahead a number of interconnected enablers will be required in each market with ocean energy potential.

These enablers, which will be specific to each country, include the following:

- a political vision in support of ocean energy;
- a plan with realistic targets;
- the identification of commercial development sites with good resource and potential to be developed on a commercial basis;
- a route to securing commercial leases of areas of sea/seabed;
- a route to securing all necessary planning permits and environmental consents;
- a clear route to securing grid connection, both for the site and for individual technology developers;
- supporting tariffs and market pull mechanisms.

Without the above, it will be hard for technologies to attract investors to develop and grow commercial energy farms.

However, it is clear to see that none of these are actually impossible to tackle. They are not matters of physics and chemistry, they are matters of human constructs of rules and many of them can be resolved once there is the will to do so.

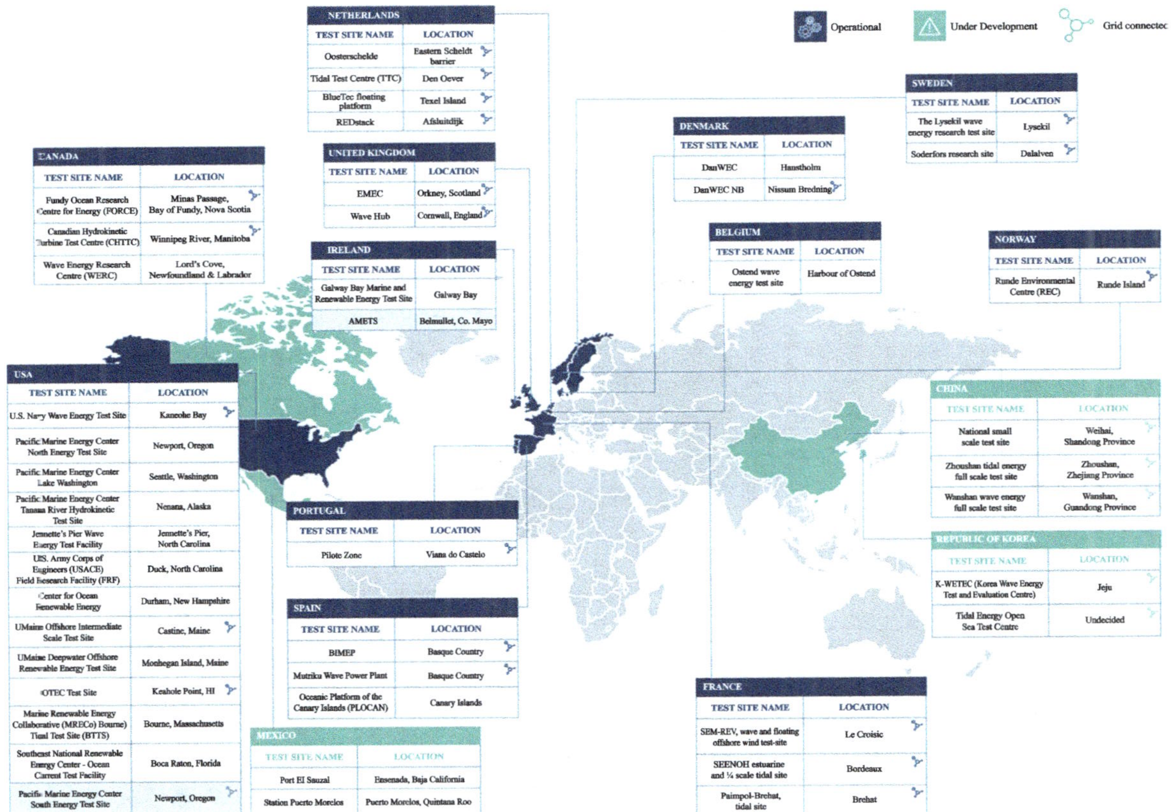

Figure 12.10 Global network of marine energy test centres, WavEC Offshore Renewables

On the plus side – in the same way the US/USSR 'space race' of the 1960s galvanised an extraordinary era of innovation and development – competitive tension between nation states can be utilised to foster support for particular technologies. As we saw in the example of onshore wind, Denmark was the first country to offer a route to market for early developers, and it is the country which puts these measures in place that will win the prize.

In this way governments in Britain, France and Canada, e.g., can be encouraged to support tidal technologies both through a desire to win first mover advantage (and the jobs and economic benefits this brings) as well as being seen not to 'lose' a new technology to competitor states. This is already being played out at sea, with supportive tariffs in the Bay of Fundy acting as a magnet for internationally mobile tidal energy firms.

12.10 Challenge ten – making it happen

So, as we stand here today – towards the end of the second decade of the twenty-first century – it is worthwhile considering where we have arrived, and how we got here.

Over millennia we have evolved from simple societies living lives based around the seasons to more advanced economies able to harness the power of fire, wind and water to drive innovation and growth.

By the nineteenth and twentieth centuries, rapid industrialisation and the discovery of vast reserves of oil, coal and gas led to a world that was living unsustainably – endangering the planet itself.

In the twenty-first century, we have seen a change – and as we slowly wean our economies off hydrocarbons, we can rely on new increasingly cheap forms of energy supplied by wind turbines and solar farms.

But I believe this is bringing with it a great moment of complacency and risk.

If – the argument goes – we are able to rely on cheap wind and solar (and not-so-cheap nuclear), then why should we (the public) support the development of different, costly renewables that don't (at the moment) work?

To my mind, this is what lies behind the rejection of feasible though costly tidal range schemes proposed in Wales – and the rowing back by governments of major support programmes for wave and tidal energy.

Why bother developing new forms of renewables when we already have renewable energy sources that work?

This – I am sure you will agree – is a facile argument. Today we are still living far beyond our means, ecologically speaking, and we need to do far, far more than we are doing at present. We need to decarbonise our transport systems and the way we heat our homes – alongside the more straightforward challenge of decarbonising our electricity supply.

For ocean energy, the coordinated actions we need to put in place – so that the technologies we know can work, can become mainstream forms of energy in the crucial decades ahead.

Fortunately for us, a large body of coordinated research has taken place this decade – all addressing specifically the challenges of the technological and financial advancement of the sector, led by the EU Ocean Energy Forum and TP Ocean – the European Technology and Innovation Platform for Ocean Energy.

Their key reports, the 2016 Ocean Energy Strategic Roadmap [16], 2016 Strategic Research Agenda [17] and 2017 report: An Integrated Framework of Ocean Energy Sector Challenges [18] set out clearly the work required to bring the sector forward, with concrete recommendations for policy makers and industry.

Whilst the focus of these studies in principally aimed at wave and tidal energy, and examines the world through a European lens, it makes sense to draw on these reports as a starting point for any examination of 'what next'?

What is lacking, in my view, is a sense of urgency, and a sentiment of shared endeavour, if we wish to overcome the threat of climate change.

As we have seen in the pages of this book, generating renewable energy from our oceans isn't just a technology challenge – it is a challenge for society. History has shown that challenges can be overcome if there is a will to do so, and a concerted effort to tackle them.

Many of the technologies we have looked at are already at TRLs 8 or 9. They can be made to work – and in time they will become cost effective – but we need to raise our eyes to a more distant horizon, 40 years hence, and look at where we want to be.

This requires a concerted pan-national approach bringing together the public sector, academia and private industry to create a shared endeavour to generate clean energy from our oceans.

But where can the money be found, or what model can we adopt, I hear you say?

Well, a glance at history shows that models can be found. The mission to place a man on the moon – which at its core was a race between the United States and Russia to prove whose economic model was better – cost the US government $23 billion and Russia around $10 billion. In the end, financial muscle and a coordinated 10-year programme won the race [19].

Similarly, we can look at nuclear power, which has enjoyed consistent, long-term government support since the 1940s.

Today, the ITER programme – a nuclear fusion megaproject which brings together 35 countries to build the International Thermonuclear Experimental Reactor in France – is able to rely on sustained multi-billion public support.

ITER was first mooted in 1987 and launched well over a decade ago as a multinational partnership between the European Union, the United States, China, Russia, Japan, India and South Korea. Its goal is to build a prototype fusion reactor to generate electricity in a process similar to the nuclear fusion that powers the sun, and by 2015 its budget had run to $US14 billion.

The facility is expected to finish its construction phase in 2025. The reactor will be commissioned that same year. It is then forecast to commence full deuterium–tritium fusion experiments in 2035.

So it is, in essence, a 50-year multinational research programme with an extraordinary budget to achieve the goal of low-carbon electricity. Critics say nuclear fusion is 'always 30 years away' – but I do not wish to pour water on this particular technology programme.

But I do ask – where would ocean energy be today if it had enjoyed a similar level of support, and how can we engender the sense of urgency needed to put a multi-decade, long-term research and commercialisation programme in place?

There needs to be, I fear, a new shock to the system. In the way the OPEC price shocks of the 70s spurred a decade-long flurry of research, and the Fukushima disaster changed national energy policy in Japan, our world maybe needs to recognise that change will not 'just happen'.

As long as climate disasters occur 'somewhere else' we may not see the concerted global effort I believe that ocean energy requires. But if, as I fear, we start to see droughts and water shortages in California, and increasingly extreme weather events across Europe, then that urgent wake-up call might just spark some action.

Making energy from oceans was science fiction, it is now an engineering fact. It is entirely achievable, and whilst it does bring some significant technical and financial challenges which necessitate a multi-party, coordinated approach it also brings enormous economic opportunities.

If we can convince our politicians, our industrialists and the public that ocean energy is a prize worth pursuing, then I have every confidence that in this century we will see new forms of reliable, abundant and clean energy from our seas.

References

[1] EMEC. http://www.emec.org.uk/services/

[2] https://www.independent.ie/business/irish/insolvent-energy-firms-in-liquidation-37158484.html

[3] https://www.telegraph.co.uk/business/2016/09/24/wind-energy-brings-off-shore-living-to-a-new-generation/

[4] https://www.nationalobserver.com/2018/01/17/news/research-needed-environmental-impact-tidal-energy-executive-director

[5] https://www.sciencenewsforstudents.org/article/wind-farms-restaurants-seals

[6] https://assets.publishing.service.gov.uk/government/uploads/system/uploads/attachment_data/file/702640/Wave_25_Summary_Report.pdf

[7] https://arena.gov.au/assets/2014/02/Commercial-Readiness-Index.pdf

[8] http://www.ukerc.ac.uk/

[9] RenewableUK. https://www.renewableuk.com/

[10] https://s3-eu-west-1.amazonaws.com/media.newore.catapult/app/uploads/2018/05/04120736/Tidal-Stream-and-Wave-Energy-Cost-Reduction-and-Ind-Benefit-FINAL-v03.02.pdf

[11] Traversing the Valley of Death: A practical guide for corporate innovation leaders, Markham, 2010. https://www.amazon.co.uk/Traversing-Valley-Death-practical-innovation/dp/0990985318

[12] EU Ocean Energy Strategic Roadmap. https://webgate.ec.europa.eu/maritimeforum/sites/maritimeforum/files/OceanEnergyForum_Roadmap_Online_Version_08Nov2016.pdf
[13] Tim Hurst, Managing Director WES, May 2016. https://www.all-energy.co.uk/RXUK/RXUK_All-Energy/2016/Presentations%202016/Wave%20and%20Tidal%20Day%201/Tim%20Hurst.pdf?v=635995925893067205
[14] https://simecatlantis.com/2014/09/18/51-million-meygen-financial-close-completed/
[15] https://webgate.ec.europa.eu/maritimeforum/sites/maritimeforum/files/Ocean-EnergyForum_Roadmap_Online_Version_08Nov2016.pdf
[16] https://webgate.ec.europa.eu/maritimeforum/sites/maritimeforum/files/Ocean-EnergyForum_Roadmap_Online_Version_08Nov2016.pdf
[17] https://www.oceanenergy-europe.eu/wp-content/uploads/2017/03/TPOcean-Strategic_Research_Agenda_Nov2016.pdf
[18] https://www.etipocean.eu/assets/Uploads/170221-Deliverable-2-1-Integrated-challenges-for-distribution4.pdf
[19] http://www.historyshotsinfoart.com/space/backstory.cfm

Index

www.ingramcontent.com/pod-product-compliance
Lightning Source LLC
LaVergne TN
LVHW011555200926
848484LV00001B/315